Lecture Notes in Computer Science 7817

Commenced Publication in 1973
Founding and Former Series Editors:
Gerhard Goos, Juris Hartmanis, and Jan van Leeuwen

Alexander Gelbukh (Ed.)

Computational Linguistics and Intelligent Text Processing

14th International Conference, CICLing 2013
Samos, Greece, March 24-30, 2013
Proceedings, Part II

 Springer

Volume Editor

Alexander Gelbukh
National Polytechnic Institute
Center for Computing Research
Av. Juan Dios Bátiz, Col. Nueva Industrial Vallejo
07738 Mexico D.F., Mexico

ISSN 0302-9743   e-ISSN 1611-3349
ISBN 978-3-642-37255-1   e-ISBN 978-3-642-37256-8
DOI 10.1007/978-3-642-37256-8
Springer Heidelberg Dordrecht London New York

Library of Congress Control Number: 2013933372

CR Subject Classification (1998): H.3, H.4, F.1, I.2, H.5, H.2.8, I.5

LNCS Sublibrary: SL 1 – Theoretical Computer Science and General Issues

Typesetting: Camera-ready by author, data conversion by Scientific Publishing Services, Chennai, India

Printed on acid-free paper

Springer is part of Springer Science+Business Media (www.springer.com)

Preface

CICLing 2013 was the 14th Annual Conference on Intelligent Text Processing and Computational Linguistics. The CICLing conferences provide a wide-scope forum for discussion of the art and craft of natural language processing research as well as the best practices in its applications.

This set of two books contains four invited papers and a selection of regular papers accepted for presentation at the conference. Since 2001, the proceedings of the CICLing conferences have been published in Springer's *Lecture Notes in Computer Science* series as volume numbers 2004, 2276, 2588, 2945, 3406, 3878, 4394, 4919, 5449, 6008, 6608, 6609, 7181, and 7182.

The set has been structured into 12 sections:

- General Techniques
- Lexical Resources
- Morphology and Tokenization
- Syntax and Named Entity Recognition
- Word Sense Disambiguation and Coreference Resolution
- Semantics and Discourse
- Sentiment, Polarity, Emotion, Subjectivity, and Opinion
- Machine Translation and Multilingualism
- Text Mining, Information Extraction, and Information Retrieval
- Text Summarization
- Stylometry and Text Simplification
- Applications

The 2013 event received a record high number of submissions in the 14-year history of the CICLing series. A total of 354 papers by 788 authors from 55 countries were submitted for evaluation by the International Program Committee; see Figure 1 and Tables 1 and 2. This two-volume set contains revised versions of 87 regular papers selected for presentation; thus the acceptance rate for this set was 24.6%.

The book features invited papers by

- Sophia Ananiadou, University of Manchester, UK
- Walter Daelemans, University of Antwerp, Belgium
- Roberto Navigli, Sapienza University of Rome, Italy
- Michael Thelwall, University of Wolverhampton, UK

who presented excellent keynote lectures at the conference. Publication of full-text invited papers in the proceedings is a distinctive feature of the CICLing conferences. Furthermore, in addition to presentation of their invited papers, the keynote speakers organized separate vivid informal events; this is also a distinctive feature of this conference series.

Table 1. Number of submissions and accepted papers by topic[1]

Accepted	Submitted	% accepted	Topic
18	75	24	Text mining
18	64	28	Semantics, pragmatics, discourse
17	80	21	Information extraction
17	67	25	Lexical resources
14	44	32	Other
14	35	40	Emotions, sentiment analysis, opinion mining
13	40	33	Practical applications
11	52	21	Information retrieval
11	51	22	Machine translation and multilingualism
8	30	27	Syntax and chunking
7	40	17	Underresourced languages
7	39	18	Clustering and categorization
6	23	26	Summarization
5	32	16	Morphology
5	24	21	Word sense disambiguation
5	19	26	Named entity recognition
4	20	20	Noisy text processing and cleaning
4	17	24	Social networks and microblogging
4	13	31	Natural language generation
3	11	27	Coreference resolution
3	9	33	Natural language interfaces
3	8	38	Question answering
2	23	9	Formalisms and knowledge representation
2	18	11	POS tagging
2	2	100	Computational humor
1	11	9	Speech processing
1	11	9	Computational terminology
1	8	12	Spelling and grammar checking
1	3	33	Textual entailment

[1] As indicated by the authors. A paper may belong to several topics.

With this event we continued with our policy of giving preference to papers with verifiable and reproducible results: in addition to the verbal description of their findings given in the paper, we encouraged the authors to provide a proof of their claims in electronic form. If the paper claimed experimental results, we asked the authors to make available to the community all the input data necessary to verify and reproduce these results; if it claimed to introduce an algorithm, we encouraged the authors to make the algorithm itself, in a programming language, available to the public. This additional electronic material will be permanently stored on the CICLing's server, www.CICLing.org, and will be available to the readers of the corresponding paper for download under a license that permits its free use for research purposes.

In the long run we expect that computational linguistics will have verifiability and clarity standards similar to those of mathematics: in mathematics, each

Table 2. Number of submitted and accepted papers by country or region

Country or region	Authors Subm.	Papers[2] Subm.	Accp.	Country or region	Authors Subm.	Papers[2] Subm.	Accp.
Algeria	4	4	–	Malaysia	7	1.67	1
Argentina	3	1	–	Malta	1	1	–
Australia	3	1	–	Mexico	14	6.25	3.25
Austria	1	1	–	Moldova	3	1	–
Belgium	3	1	1	Morocco	7	4	1
Brazil	13	6.83	2	Netherlands	8	4.50	1
Canada	11	4.53	1.2	New Zealand	5	1.67	–
China	57	21.72	3.55	Norway	6	2.92	0.92
Colombia	2	1	1	Pakistan	5	2	–
Croatia	5	2	2	Poland	8	3.75	0.75
Czech Rep.	10	5	2	Portugal	9	3	–
Egypt	22	11.67	1	Qatar	2	0.67	–
Finland	2	0.67	–	Romania	14	9.67	2
France	64	25.9	5.65	Russia	15	4.75	1
Georgia	1	1	0.5	Singapore	5	2.25	0.25
Germany	32	13.92	6.08	Slovakia	2	1	–
Greece	21	6.12	2.12	Spain	39	15.50	8.75
Hong Kong	9	2.53	0.2	Sweden	2	2	–
Hungary	12	6	–	Switzerland	8	3.83	1.33
India	98	49.2	5.6	Taiwan	1	1	–
Iran	14	11.33	–	Tunisia	24	11	2
Ireland	6	4.5	1.5	Turkey	11	6.25	3.25
Italy	22	11.37	4.5	Ukraine	2	1.25	0.50
Japan	48	20.5	5	UAE	1	0.33	–
Kazakhstan	10	3.75	–	UK	35	15.73	5.20
Korea, South	7	3	–	USA	54	18.98	8.90
Latvia	6	2	1	Viet Nam	8	3.50	–
Macao	6	2	–	*Total:*	788	354	87

[2] By the number of authors: e.g., a paper by two authors from the USA and one from UK is counted as 0.67 for the USA and 0.33 for UK.

claim is accompanied by a complete and verifiable proof (usually much longer than the claim itself); each theorem's complete and precise proof—and not just a vague description of its general idea—is made available to the reader. Electronic media allow computational linguists to provide material analogous to the proofs and formulas in mathematics in full length—which can amount to megabytes or gigabytes of data—separately from a 12-page description published in the book. More information can be found on www.CICLing.org/why_verify.htm.

To encourage providing algorithms and data along with the published papers, we selected a winner of our Verifiability, Reproducibility, and Working Description Award. The main factors in choosing the awarded submission were technical correctness and completeness, readability of the code and documentation, simplicity of installation and use, and exact correspondence to the claims of the

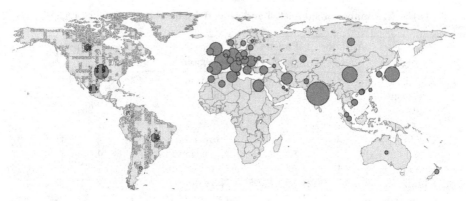

Fig. 1. Submissions by country or region. The area of a circle represents the number of submitted papers.

paper. Unnecessary sophistication of the user interface was discouraged; novelty and usefulness of the results were not evaluated—instead, they were evaluated for the paper itself and not for the data. This year's winning paper was published in a separate proceedings volume and is not included in this set.

The following papers received the Best Paper Awards, the Best Student Paper Award, as well as the Verifiability, Reproducibility, and Working Description Award, correspondingly (the best student paper was selected among papers of which the first author was a full-time student, excluding the papers that received a Best Paper Award):

1ˢᵗ Place: *Automatic Detection of Idiomatic Clauses,* by Anna Feldman and Jing Peng, USA;

2ⁿᵈ Place: *Topic-Oriented Words as Features for Named Entity Recognition,* by Ziqi Zhang, Trevor Cohn, and Fabio Ciravegna, UK;

3ʳᵈ Place: *Five Languages are Better than One: An Attempt to Bypass the Data Acquisition Bottleneck for WSD,* by Els Lefever, Veronique Hoste, and Martine De Cock, Belgium;

Student: *Domain Adaptation in Statistical Machine Translation Using Comparable Corpora: Case Study for English-Latvian IT Localisation,* by Mārcis Pinnis, Inguna Skadiņa, and Andrejs Vasiļjevs, Latvia;

Verifiability: *Linguistically-Driven Selection of Correct Arcs for Dependency Parsing,* by Felice Dell'Orletta, Giulia Venturi, and Simonetta Montemagni, Italy.

The authors of the awarded papers (except for the Verifiability Award) were given extended time for their presentations. In addition, the Best Presentation Award and the Best Poster Award winners were selected by a ballot among the attendees of the conference.

Besides its high scientific level, one of the success factors of CICLing conferences is their excellent cultural program. The attendees of the conference had a chance to visit unique historical places: the Greek island of Samos, the birthplace of Pythagoras (Pythagorean theorem!), Aristarchus (who first realized that the Earth rotates around the Sun and not vice versa), and Epicurus (one of the founders of the scientific method); the Greek island of Patmos, where John the Apostle received his visions of the Apocalypse; and the huge and magnificent archeological site of Ephesus in Turkey, where stood the Temple of Artemis, one of the Seven Wonders of the World (destroyed by Herostratus), and where the Virgin Mary is believed to have spent the last years of her life.

I would like to thank all those involved in the organization of this conference. In the first place these are the authors of the papers that constitute this book: it is the excellence of their research work that gives value to the book and sense to the work of all other people. I thank all those who served on the Program Committee, Software Reviewing Committee, Award Selection Committee, as well as additional reviewers, for their hard and very professional work. Special thanks go to Ted Pedersen, Adam Kilgarriff, Viktor Pekar, Ken Church, Horacio Rodriguez, Grigori Sidorov, and Thamar Solorio for their invaluable support in the reviewing process.

I would like to thank the conference staff, volunteers, and the members of the local organization committee headed by Dr. Efstathios Stamatatos. In particular, we are grateful to Dr. Ergina Kavallieratou for her great effort in planning the cultural program and Mrs. Manto Katsiani for her invaluable secretarial and logistics support. We are deeply grateful to the Department of Information and Communication Systems Engineering of the University of the Aegean for its generous support and sponsorship. Special thanks go to the Union of Vinicultural Cooperatives of Samos (EOSS), A. Giannoulis Ltd., and the Municipality of Samos for their kind sponsorship. We also acknowledge the support received from the project WIQ-EI (FP7-PEOPLE-2010-IRSES: Web Information Quality Evaluation Initiative).

The entire submission and reviewing process was supported for free by the EasyChair system (www.EasyChair.org). Last but not least, I deeply appreciate the Springer staff's patience and help in editing these volumes and getting them printed in record short time—it is always a great pleasure to work with Springer.

February 2013   Alexander Gelbukh

Organization

CICLing 2013 is hosted by the University of the Aegean and is organized by the CICLing 2013 Organizing Committee in conjunction with the Natural Language and Text Processing Laboratory of the CIC (Centro de Investigación en Computación) of the IPN (Instituto Politécnico Nacional), Mexico.

Organizing Chair

Efstathios Stamatatos

Organizing Committee

Efstathios Stamatatos (chair)
Ergina Kavallieratou
Manolis Maragoudakis

Program Chair

Alexander Gelbukh

Program Committee

Ajith Abraham
Marianna Apidianaki
Bogdan Babych
Ricardo Baeza-Yates
Kalika Bali
Sivaji Bandyopadhyay
Srinivas Bangalore
Leslie Barrett
Roberto Basili
Anja Belz
Pushpak Bhattacharyya
Igor Boguslavsky
António Branco
Nicoletta Calzolari
Nick Campbell
Michael Carl
Ken Church

Dan Cristea
Walter Daelemans
Anna Feldman
Alexander Gelbukh (chair)
Gregory Grefenstette
Eva Hajicova
Yasunari Harada
Koiti Hasida
Iris Hendrickx
Ales Horak
Veronique Hoste
Nancy Ide
Diana Inkpen
Hitoshi Isahara
Sylvain Kahane
Alma Kharrat
Adam Kilgarriff

Philipp Koehn
Valia Kordoni
Leila Kosseim
Mathieu Lafourcade
Krister Lindén
Elena Lloret
Bente Maegaard
Bernardo Magnini
Cerstin Mahlow
Sun Maosong
Katja Markert
Diana Mccarthy
Rada Mihalcea
Jean-Luc Minel
Ruslan Mitkov
Dunja Mladenic
Marie-Francine Moens
Masaki Murata
Preslav Nakov
Vivi Nastase
Costanza Navarretta
Roberto Navigli
Vincent Ng
Kjetil Nørvåg
Constantin Orasan
Ekaterina Ovchinnikova
Ted Pedersen
Viktor Pekar
Anselmo Peñas
Maria Pinango
Octavian Popescu

Irina Prodanof
James Pustejovsky
German Rigau
Fabio Rinaldi
Horacio Rodriguez
Paolo Rosso
Vasile Rus
Horacio Saggion
Franco Salvetti
Roser Sauri
Hinrich Schütze
Satoshi Sekine
Serge Sharoff
Grigori Sidorov
Kiril Simov
Vaclav Snasel
Thamar Solorio
Lucia Specia
Efstathios Stamatatos
Josef Steinberger
Ralf Steinberger
Vera Lúcia Strube de Lima
Mike Thelwall
George Tsatsaronis
Dan Tufis
Olga Uryupina
Karin Verspoor
Manuel Vilares Ferro
Aline Villavicencio
Piotr W. Fuglewicz
Annie Zaenen

Software Reviewing Committee

Ted Pedersen
Florian Holz
Miloš Jakubíček

Sergio Jiménez Vargas
Miikka Silfverberg
Ronald Winnemöller

Award Committee

Alexander Gelbukh
Eduard Hovy
Rada Mihalcea

Ted Pedersen
Yorick Wiks

Additional Referees

Rodrigo Agerri
Katsiaryna Aharodnik
Ahmed Ali
Tanveer Ali
Alexandre Allauzen
Maya Ando
Javier Artiles
Wilker Aziz
Vt Baisa
Alexandra Balahur
Somnath Banerjee
Liliana Barrio-Alvers
Adrián Blanco
Francis Bond
Dave Carter
Chen Chen
Jae-Woong Choe
Simon Clematide
Geert Coorman
Victor Darriba
Dipankar Das
Orphee De Clercq
Ariani Di Felippo
Maud Ehrmann
Daniel Eisinger
Ismail El Maarouf
Tilia Ellendorff
Milagros Fernández Gavilanes
Santiago Fernández Lanza
Daniel Fernández-González
Karën Fort
Koldo Gojenola
Gintare Grigonyte
Masato Hagiwara
Kazi Saidul Hasan
Eva Hasler
Stefan Höfler
Chris Hokamp
Adrian Iftene
Iustina Ilisei
Leonid Iomdin
Milos Jakubicek
Francisco Javier Guzman

Nattiya Kanhabua
Aharodnik Katya
Kurt Keena
Natalia Konstantinova
Vojtech Kovar
Kow Kuroda
Gorka Labaka
Shibamouli Lahiri
Egoitz Laparra
Els Lefever
Lucelene Lopes
Oier López de La Calle
John Lowe
Shamima Mithun
Tapabrata Mondal
Silvia Moraes
Mihai Alex Moruz
Koji Murakami
Sofia N. Galicia-Haro
Vasek Nemcik
Zuzana Neverilova
Anthony Nguyen
Inna Novalija
Neil O'Hare
John Osborne
Santanu Pal
Feng Pan
Thiago Pardo
Veronica Perez Rosas
Michael Piotrowski
Ionut Cristian Pistol
Soujanya Poria
Luz Rello
Noushin Rezapour Asheghi
Francisco Ribadas-Pena
Alexandra Roshchina
Tobias Roth
Jan Rupnik
Upendra Sapkota
Gerold Schneider
Djamé Seddah
Keiji Shinzato
João Silva

Website and Contact

The webpage of the CICLing conference series is www.CICLing.org. It contains information about past CICLing conferences and their satellite events, including published papers or their abstracts, photos, video recordings of keynote talks, as well as information about the forthcoming CICLing conferences and contact options.

Table of Contents – Part II

Sentiment, Polarity, Emotion, Subjectivity, and Opinion

Machine Translation and Multilingualism

Text Mining, Information Extraction, and Information Retrieval

Text Summarization

Stylometry and Text Simplification

Applications

Table of Contents – Part I

General Techniques

Lexical Resources

Morphology and Tokenization

Syntax and Named Entity Recognition

Word Sense Disambiguation and Coreference Resolution

Semantics and Discourse

Damping Sentiment Analysis in Online Communication: Discussions, Monologs and Dialogs

Mike Thelwall[1], Kevan Buckley[1], George Paltoglou[1], Marcin Skowron[2],
David Garcia[3], Stephane Gobron[4], Junghyun Ahn[5], Arvid Kappas[6],
Dennis Küster[6], and Janusz A. Holyst[7]

[1] Statistical Cybermetrics Research Group, University of Wolverhampton, Wolverhampton, UK
{m.thelwall,K.A.Buckley,g.paltoglou}@wlv.ac.uk
[2] Austrian Research Institute for Artificial Intelligence, Vienna, Austria
marcin.skowron@ofai.at
[3] Chair of Systems Design, ETH Zurich, Weinbergstrasse 56/58, 8092 Zurich, Switzerland
dgarcia@ethz.ch
[4] Information and Communication Systems Institute (ISIC), HE-Arc, HES-SO, Switzerland
stephane.gobron@gmail.com
[5] SCI IC RB Group, Ecole polytechnique fédérale de Lausanne EPFL, Switzerland
junghyun.ahn@epfl.ch
[6] School of Humanities and Social Sciences, Jacobs University Bremen, Bremen, Germany
{a.kappas,d.kuester}@jacobs-university.de
[7] Center of Excellence for Complex Systems Research, Faculty of Physics,
Warsaw University of Technology, Warsaw, Poland
jholyst@if.pw.edu.pl

Abstract. Sentiment analysis programs are now sometimes used to detect patterns of sentiment use over time in online communication and to help automated systems interact better with users. Nevertheless, it seems that no previous published study has assessed whether the position of individual texts within ongoing communication can be exploited to help detect their sentiments. This article assesses apparent sentiment anomalies in on-going communication – texts assigned significantly different sentiment strength to the average of previous texts – to see whether their classification can be improved. The results suggest that a damping procedure to reduce sudden large changes in sentiment can improve classification accuracy but that the optimal procedure will depend on the type of texts processed.

Keywords: Sentiment analysis, opinion mining, social web.

1 Introduction

The rapid development of sentiment analysis in the past decade has roots in the widespread availability of social web texts that are relevant to marketing needs. In particular, formal or informal product reviews online can now be mined with a wide range of sentiment analysis programs in multiple languages to give businesses information about what the public thinks about products and brands (Liu, 2012; Pang & Lee,

A. Gelbukh (Ed.): CICLing 2013, Part II, LNCS 7817, pp. 1–12, 2013.

2008). By harnessing real-time sources like Twitter, businesses can even be given daily updates about changes in average sentiment. More recently, however, sentiment analysis programs have been used to identify the sentiment expressed in texts, irrespective of whether any products are mentioned. One goal of this type of research has been to identify trends in sentiment over time in relation to a specific topic (Chmiel et al., 2011a; Garas, Garcia, Skowron, & Schweitzer, 2012) or more generally (Thelwall, Buckley, & Paltoglou, 2011) or in a particular genre (Dodds & Danforth, 2010; Kramer, 2010): both social sciences types of research. Another type of research detects users' sentiments in order to react to them in real time. As an example of the latter, dialog systems have been developed that react differently to users depending on the sentiment expressed (Skowron, 2010) and in one online environment, the facial expressions of an automatic chat partner in a three-dimensional virtual world respond to the sentiment expressed by the participants, as detected with a sentiment analysis program (Gobron et al., 2011; Skowron et al., 2011). In another computing application that is somewhat similar to this, the Yahoo! Answers system harnesses sentiment analysis to help identify people that receive positive feedback after submitting their answers so that these people can be identified and their answers given prominence in search results (Kucuktunc, Cambazoglu, Weber, & Ferhatosmanoglu, 2012). As a result of such applications, there is a need for sentiment analysis software that is optimised for general social web texts and that can take advantage of any regular patterns of sentiment expressions and reactions online in order to improve the accuracy of the predictions made.

Some research from psychology and from studies of online communication can shed light on how sentiment is best detected and measured in online environments. Psychologists have investigated emotions for over a century and today there is a field of emotion psychology (Cornelius, 1996; Fox, 2008). One important finding is that humans seem to process positive and negative sentiment separately and relatively independently. This means that although it is often practical and convenient to measure positive and negative sentiment together to give one combined overall result for each text, it is more natural to measure them separately and report two scores per text. Psychology research also confirms that emotions vary in strength (Cornelius, 1996; Fox, 2008) and so the natural way to measure emotion and hence sentiment is on a dual scale measuring the strength of positive and negative sentiment expressed. Emotion psychologists also recognize a range of different types of emotion (e.g., anger, hate) rather than just positivity and negativity but studies suggest that the fundamental divide is between positive and negative emotion with more fine-grained emotions being socially constructed to some extent (Fox, 2008). Thus it is reasonable from a psychology perspective to either focus on positive and negative sentiment or on more fine-grained sentiment, with the latter probably reflecting social conditioning more.

Research from non-psychologists has investigated emotion and sentiment online to see whether there are patterns in the use of sentiment in ongoing communications, with positive results. A common finding is that whilst different social web environment have different average levels of positive and negative sentiment (e.g., political discussions tend to be negative whereas comments between friends tend to be positive) (Thelwall, Buckley, & Paltoglou, 2012) above average levels of negativity associate with longer interactions: negativity seems to fuel longer discussions

(Chmiel et al., 2011ab; Thelwall, Sud, & Vis, 2012). Additionally, and perhaps unsurprisingly, some studies have found evidence of sentiment homophily between online friends: people tend to express similar levels of sentiment to that expressed by their friends, compared to the overall average (Bollen, Pepe, & Mao, 2011; Thelwall, 2010).

The above discussion suggests that the task of sentiment analysis in general social web texts may need to be tackled somewhat differently to that of product review sentiment analysis or opinion mining. Whilst there are programs, such as SentiStrength (discussed below), that are designed for social web texts it seems that all process each text separately and independently and none have attempted to improve sentiment detection by taking advantage of patterns of online communication, although some have successfully exploited discourse features (Somasundaran, Namata, Wiebe, & Getoor, 2009). This article assesses the potential for improving sentiment detection in this way. As an exploratory study, it uses four different types of social web context for evaluations (political forum discussions, non-political forum discussions, as well as dialogs and monologs in Twitter). It also assesses one simple method of exploiting the sentiment of previous texts when classifying the sentiment of new texts: damping. Defined precisely below, the damping method changes a sentiment prediction by bringing it closer to the average sentiment of the previous few texts if the prediction would otherwise be too different from this average. The experimental results suggest that the damping method works well in some contexts but not all and so should be used with care.

2   Sentiment Analysis

Previous sentiment analysis or opinion mining research has used many different methods in order to detect the sentiment of a text or the opinion expressed in a text towards a product or an aspect of a product. Lexical methods typically start with a pre-defined lexicon of terms with known typical sentiment polarity, such as Senti-WordNet (Baccianella, Esuli, & Sebastiani, 2010), sentiment terms from the General Inquirer lexicon (Choi & Cardie, 2008), LIWC (Pennebaker, Mehl, & Niederhoffer, 2003) as in (Thelwall, Buckley, Paltoglou, Cai, & Kappas, 2010), or a human-created list of sentiment terms (Taboada, Brooke, Tofiloski, Voll, & Stede, 2011). These lists are then matched with terms in texts to be classified and then a set of rules applied to classify the texts. Classifications are typically either binary (positive or negative), or trinary (positive, negative or neutral/objective) although some also detect sentiment strength in addition to polarity.

A non-lexical approach is to use machine learning methods to decide which words are the most relevant for sentiment based upon a set of linguistic or non-linguistic rules and a large set of pre-classified texts for training. An advantage of not using a pre-defined lexicon, which is particularly relevant when developing a sentiment classifier for reviews of a particular type of product, is that non-sentiment terms may be identified that carry implied sentiment by expressing a judgment, such as "heavy" in the phrase "the phone was very heavy". The limitation of needing a corpus of human-coded texts to train a non-lexical classifier can be avoided in some cases by exploiting free online product review sites in which reviewers score products in addition to giving text reviews. In the absence of these, other unsupervised methods (Turney, 2002) and domain transfer methods (Glorot, Bordes, & Bengio, 2011; Ponomareva & Thelwall, 2012) have also been developed. Two disadvantages of the non-lexical approach for social science research purposes, however, are that they can introduce systematic anomalies through exploiting non-sentiment words (Thelwall et al., 2010) and that they seem to be less transparent than lexical methods, which can often give a clear explanation as to why a sentence has been classified in a certain way, by reference to the predefined list of sentiment terms (e.g., "this sentence was classified as positive because it contains the word 'happy', which is in the lexicon of positive terms"). Sentiment analysis methods can exploit linguistic structure to make choices about the types of words to analyze, such as just the adjectives (Wiebe, Wilson, Bruce, Bell, & Martin, 2004).

Although most sentiment analysis programs seem to classify entire texts as positive, negative or neutral, aspect-based sentiment analysis classifies texts differently based upon the aspects of a product discussed. For instance, an aspect-based classifier might detect that "cheap" is negative in the context of a phone design but positive in the context of the phone's price. Other programs are more fine-grained in a different sense: classifying multiple emotions, such as anger, sadness, hate, joy and happiness (Neviarouskaya, Prendinger, & Ishizuka, 2010) and/or sentiment strength (Wilson, Wiebe, & Hwa, 2006).

Some sentiment analysis programs have attempted to use the position of a text in order to help classify sentiment, but only for the larger texts containing classified smaller texts. In movie reviews, sentences near the end typically carry more weight than earlier sentences and hence movie review classifiers that work by detecting the sentiment of individual sentences and then aggregating the results to predict the sentiment of the overall review can improve their performance by giving higher weights to later texts (Pang, Lee, & Vaithyanathan, 2002). Discourse structure has been successfully used in one case to classify contributions in work-based meetings as positive, negative or neutral, producing a substantial increase in accuracy in comparison to baseline approaches (Somasundaran et al., 2009). This promising approach has not been tried for social web texts, however, and may work best in formal discussions. Another investigation uses discourse structure to help separate discussion participants into different camps but not to help classify the sentiment of their texts (Agrawal, Rajagopalan, Srikant, & Xu, 2003). Despite these examples, no sentiment analysis seem to exploit the occurrence of many texts in communication chains, such as monologs, dialogs or multi-participant discussions, in order to predict their sentiment more accurately.

3   Sentiment Strength Detection with SentiStrength

The damping method described below was tested by being applied to SentiStrength (Thelwall & Buckley, in press; Thelwall et al., 2010; Thelwall et al., 2012). This sentiment analysis program was chosen because it is designed to detect the strength of positive and negative sentiment in short informal text and has been tested on a range of different social web text types: Tweets, MySpace comments, RunnersWorld forum posts, BBC discussion forum posts, Digg posts, and comments on YouTube videos. SentiStrength assigns a score of 1, 2, 3, 4, or 5 for the strength of positive sentiment and -1, -2, -3, -4 or -5 for the strength of negative sentiment, with each text receiving one score for each. For instance, the text "I hate Tony but like Satnam" might get a score of (-4, 3), indicating strong negative sentiment and moderate positive sentiment.

SentiStrength's dual positive/negative scoring scheme is unusual for sentiment strength detection and stems from the psychology input to the design of the software because psychologists accept that humans process positive and negative sentiment in parallel rather than in a combined way (Norman et al., 2011); hence positive and negative sentiment do not necessarily cancel each other out. As mentioned above, for a psychological analysis of sentiment, and hence for a social science analysis of sentiment, it is reasonable to detect positive and negative sentiment separately. SentiStrength has been used to analyze social web texts to detect patterns of communication but no previous study has attempted to improve its performance by taking advantage of sentiment patterns in on-going communications.

SentiStrength works primarily through a lexicon of terms with positive and negative weights assigned to them. In the above example, "hate" is in the lexicon with strength -4 and "like" has strength +3. Each text is given a score equal to the largest positive and negative value of the sentiment words contained in it, subject to some additional rules. These rules include methods for dealing with negation (e.g., don't), booster words (e.g., very), emoticons, and informal expressions of sentiment (e.g., "I'm haaaaaapy!!!").

3.1   Sentiment Damping

The adjustment method is based upon the assumption that a text in a series that has a significantly different sentiment level than the previous texts, according to a classifier, may be an anomaly in the sense of having been misclassified and may have a real sentiment that is closer to the average. This is operationalized by two rules:

- If the classified positive sentiment of text A differs by at least 1.5 from the average positive sentiment of the previous 3 posts, then adjust the positive sentiment prediction of text A by 1 point to bring it closer to the positive average of the previous 3 terms.
- If the classified negative sentiment of text A differs by at least 1.5 from the average negative sentiment of the previous 3 posts, then adjust the negative sentiment prediction of text A by 1 point to bring it closer to the negative average of the previous 3 terms.

For example, if four consecutive texts are classified as 1, 2, 1, 4 for positive sentiment then rule 1 would be triggered since 4 is more than 1.5 greater than the average of 1, 2, and 1, and hence the prediction of 4 would be adjusted by 1 towards the average. Hence the adjusted predictions would be 1, 2, 1, 3. Figure 1 is another example from the Twitter dialogs data set.

Tweet (first 3 from Stacey, last from Claire)	Neg. score
@Claire she bores me too! Haha x	-2
@Claire text me wen your on your way x x x	-1
@Claire u watch BB tonight? I tried one of them bars..reem! x x x	-1
@Stacey lush in they ... do u watch American horror story ... Cbb was awsum tonight bunch of bitches !!	-4

Fig. 1. A dialog between two tweeters with SentiStrength negative classifications that would trigger damping for the final contribution. The term *horror* triggered a strong negative score in the final contribution but human coders judged that this was not strongly negative, presumably because it was part of a TV series name. This type of anomaly would be corrected by the damping method (names changed and contributions slightly changed to anonymize participants).

4   Data Sets

Multiple data sets were created to reflect different kinds of web-based informal communication: discussions, dialogs and monologs.

4.1   BBC World News Discussions (BWNpf)

This data set consists of contributions to the BBC World News online discussion forum. This was chosen as an example of a political forum discussion in which multiple participants can contribute. Contributions were selected for coding if the adjustment rule would trigger a positive or negative change in them. In addition, a random set of non-adjusted texts was also selected for coding. A text was not chosen if any of the previous 3 contributions to the discussion had been chosen. This was to avoid taking too many contributions from the same part of the discussion.

4.2   RunnersWorld (RWtf)

This data set consists of contributions to the RunnersWorld online marathon running discussion forum. This was chosen as an example of a non-political topical discussion forum in which multiple participants can contribute. Although the forum focuses on a single topic, this is probably true for most online discussion forums and so it represents a popular type of online discussion despite its specialist nature. Contributions were selected in the same way as for the BWNpf data set.

4.3   Twitter Monologs (Tm)

This data set consists of tweets in English from randomly selected Twitter users tweeting in English and geolocated in the US. This data set was obtained by monitoring the Twitter API with a blank US geolocation search during early 2012. Each "monolog" in the dataset consists of all tweets from the random user, and at least 10 tweets per user. This represents tweeting in the sense of broadcasting comments rather than necessarily interacting with other tweeters, although some comments may also be interactions. Tweets were selected for coding as for BWNpf.

4.4 Twitter Dialogs (Td)

This data set is similar to Tm but represents a set of dialogs between pairs of users. For each user in the Td data set, a random target (i.e., a Tweeter, indicated using the @ convention) of one of their tweets was selected and all of this user's tweets were downloaded. If the target user also targeted the original user then their tweets were combined and arranged in chronological order to form a Twitter "dialog" in this data set, discarding all tweets not directed at the other dialog partner. For instance, if the two contributors were User1 and User2, then tweets from User1 were discarded unless they contained @User2 and tweets from User2 were discarded unless they contained @User1. Contributions were randomly selected from these dialogs for coding subject to the restriction that a contribution must be either preceded to followed by a contribution from the other dialog participant (so that they would not be part of a mini-monolog rather than a genuine dialog).

4.5 Preliminary Analysis of Data Sets

Table 1 reports some basic statistics from SentiStrength (without damping) applied to the four data sets. The table reports the average of all statistics calculated separately for each thread/monolog/dialog in each sample. The results show differences between the data sets in all statistics. For example, the RunnersWorld forum threads have the highest average positive sentiment strength and the BBC World News forum has the highest average negative sentiment strength, probably reflecting their discussion topics. The negative correlations between positive and negative scores for the first two data sets in comparison to positive correlations between positive and negative scores last two probably reflects the length limit on tweets: a slight tendency for tweets to contain either positive or negative sentiment but not both. In contrast, for the first two forums, if a person expresses negative sentiment then they are also likely to express positive sentiment and vice versa. This would be consistent with some texts being factual or objective and others being subjective.

Of most interest here are the lag 1 autocorrelations: these are correlations between the sentiment scores and the sentiment scores offset by one. High correlations (close to 1) would suggest that the sentiment of a post tends to be similar to the sentiment of the previous post, supporting the damping method for sentiment analysis. Although all the autocorrelations are significantly non-zero they seem to be small enough to be irrelevant in practice. This suggests that within these data sets, texts with similar sentiment levels have only a small tendency to cluster together.

Table 1. Statistics and autocorrelations for the threads/monologs/dialogs with at least 30 contributions. All correlations and autocorrelations are significantly different from 0 at p=0.001

Data set	Sample size*	Mean positive	Mean negative	Positive-negative correlation	Lag 1 positive autocorr.	Lag 1 negative autocorr.
BWNpf	4580	1.918	-2.414	-.2378	.0331	.0529
RWtf	4958	2.200	-1.666	-.1867	.0924	.0634
Tm	675	1.691	-1.364	.0328	.0558	.0529
Td	329	1.778	-1.367	.0349	.0299	.0389

* Sample size is number of threads for BWNpf and RWtf, the number of dialogs for Tm and the number of monologs for Td.

4.6   Inter-coder Consistency

The texts selected as described above for each data set were given to two experienced coders who were not associated with the project and who were not told the purpose of the project. The coders were given the texts to code, along with the previous texts in the dialog/monolog/thread in order to reveal the context of each text for more accurate coding. The coders were asked to score each text with the standard SentiStrength scheme of two whole numbers: [no positive sentiment] 1 – 2 – 3 – 4 – 5 [very strong positive sentiment] and [no negative sentiment] -1 – -2 – -3 – -4 – -5 [very strong negative sentiment]. The coders were each given a standard codebook to describe and motivate the task and were requested to code for a maximum of one hour per day, to minimise the risk of mistakes through fatigue.

Krippendorff's inter-coder weighted alpha (Krippendorff, 2004) was used to calculate the extent of agreement between the coders, using the difference between the categories assigned as the weights. The results showed that the level of inter-coder agreement was good but not excellent, probably because sentiment is a subjective phenomenon. It is therefore reasonable to use the values of the coders to assess the sentiment analysis results. The values of the second coder were chosen because this person coded more texts.

Table 2. Krippendorff inter-coder weighted alpha values for the similarity between codes from the two coders

Data set	Positive sentiment α	Negative sentiment α
BWNpf (n=466)	0.655	0.559
RWtf (n=379)	0.572	0.659
Tm (n=445)	0.695	0.744
Td (n=508)	0.689	0.738

5   Experimental Results

Table 3 reports a comparison of the results for damped SentiStrength with undamped SentiStrength for the random selection of human coded texts that were damped by

SentiStrength (i.e., only the changed values). The table reports damping increases in
sentiment strength separately from damping decreases in sentiment strength. For each
type of damping, the result is either a more accurate or a less accurate prediction and
Table 3 reports the proportion of each. The results are mixed: an overall improvement in
9 of the 16 cases examined (although three are marginal: 51%, 51% and 54%) and no
clear pattern about which of the four types of damping are always effective. Nevertheless, there are six cases in which the improvement is substantial – 65% to 75% – and
this suggests that if damping is applied selectively by choosing which of the four types
to use for a given data set then this should improve sentiment classification accuracy.

Table 3. Percentage of sentiment classification improvements when damping increases
sentiment scores and when damping decreases sentiment scores. Figures above 50% indicate an
overall increase in classification accuracy.

Data set	Positive sentiment increase improvement	Positive sentiment decrease improvement	Negative sentiment increase improvement	Negative sentiment decrease improvement
BWNpf	38% (n=74)	73% (n=127)	75% (n=165)	51% (n=166)
RWtf	71% (n=175)	43% (n=153)	54% (n=139)	65% (n=280)
Tm	71% (n=97)	33% (n=319)	51% (n=55)	41% (n=300)
Td	69% (n=81)	33% (n=304)	47% (n=43)	44% (n=331)

6   Conclusions

The results clearly show that damping can improve sentiment strength detection for
social web texts, although some forms of damping have no effect on particular types
of text or make the results worse. Hence, when optimising sentiment analysis for a
new dataset, experiments should be run to decide which of the four types of damping
to include and which to exclude (i.e., damping sentiment increases, damping sentiment decreases, for both positive and negative sentiment). A limitation of this approach is that the performance improvement caused by damping is likely to be minor
because only a minority of predictions will be damped, depending on the corpus used.
Moreover, a practical limitation is that human-coded texts will be needed to identify
the types of damping to use. This human coding is resource-intensive because it must
be conducted specifically for the damping, with a dataset of texts potentially subject
to damping changes, and hence would not be a random set of texts that could be used
for other evaluations.

For future work, it would be useful to conduct a larger scale and more systematic
evaluation of different types of texts in order to produce recommendations for the
contexts in which the different types of damping should be used. This would save

future researchers the time needed to test each new data set to select which damping methods to use. It would also be useful to compare this approach to the use of discourse markers (Somasundaran et al., 2009) and attempt to combine both to improve on the performance of each one.

Acknowledgement. This work was supported by a European Union grant by the 7th Framework Programme, Theme 3: Science of complex systems for socially intelligent ICT. It is part of the CyberEmotions project (contract 231323).

References

1. Agrawal, R., Rajagopalan, S., Srikant, R., Xu, Y.: Mining newsgroups using networks arising from social behavior. In: Proceedings of WWW, pp. 529–535 (2003)
2. Baccianella, S., Esuli, A., Sebastiani, F.: SentiWordNet 3.0: An enhanced lexical resource for sentiment analysis and opinion mining. In: Proceedings of the Seventh Conference on International Language Resources and Evaluation (2010), http://www.lrec-conf.org/proceedings/lrec2010/pdf/769_Paper.pdf (retrieved May 25, 2010)
3. Bollen, J., Pepe, A., Mao, H.: Modeling public mood and emotion: Twitter sentiment and socioeconomic phenomena. In: ICWSM 2011, Barcelona, Spain (2011), http://arxiv.org/abs/0911.1583 (retrieved June 2, 2011)
4. Chmiel, A., Sienkiewicz, J., Thelwall, M., Paltoglou, G., Buckley, K., Kappas, A., Hołyst, J.A.: Collective emotions online and their influence on community life. PLoS ONE 6(7), e22207 (2011a)
5. Chmiel, A., Sienkiewicz, J., Paltoglou, G., Buckley, K., Thelwall, M., Holyst, J.A.: Negative emotions boost user activity at BBC forum. Physica A 390(16), 2936–2944 (2011b)
6. Choi, Y., Cardie, C.: Learning with compositional semantics as structural inference for subsentential sentiment analysis. In: Proceedings of the Conference on Empirical Methods in Natural Language Processing, pp. 793–801 (2008)
7. Cornelius, R.R.: The science of emotion. Prentice Hall, Upper Saddle River (1996)
8. Dodds, P.S., Danforth, C.M.: Measuring the happiness of large-scale written expression: Songs, blogs, and presidents. Journal of Happiness Studies 11(4), 441–456 (2010)
9. Fox, E.: Emotion science. Palgrave Macmillan, Basingstoke (2008)
10. Garas, A., Garcia, D., Skowron, M., Schweitzer, F.: Emotional persistence in online chatting communities. Scientific Reports 2, article 402 (2012), doi:10.1038/srep00402
11. Glorot, X., Bordes, A., Bengio, Y.: Domain adaptation for large-scale sentiment classification: A deep learning approach. In: Proceedings of the 28th International Conference on Machine Learning (ICML 2011) (2011)
12. Gobron, S., Ahn, A., Silvestre, Q., Thalmann, D., Rank, S., Skowron, M., Thelwall, M.: An interdisciplinary VR-architecture for 3D chatting with non-verbal communication. In: Proceedings of the Joint Virtual Reality Conference of EuroVR (EGVE 2011), Nottingham, UK, pp. 87–94 (2011)
13. Kramer, A.D.I.: An unobtrusive behavioral model of "gross national happiness". In: Proceedings of CHI 2010, pp. 287–290. ACM Press, New York (2010)
14. Krippendorff, K.: Content analysis: An introduction to its methodology. Sage, Thousand Oaks (2004)

15. Kucuktunc, O., Cambazoglu, B.B., Weber, I., Ferhatosmanoglu, H.: A large-scale senti-
    ment analysis for Yahoo! Answers. Paper Presented at the Web Search and Data Mining
    (WSDM 2012), Seattle, Washington, pp. 633–642 (2012)
16. Liu, B.: Sentiment analysis and opinion mining. Morgan and Claypool, New York (2012)
17. Neviarouskaya, A., Prendinger, H., Ishizuka, M.: Recognition of fine-grained emotions
    from text: An approach based on the compositionality principle. In: Nishida, T., Jain, L.,
    Faucher, C. (eds.) Modelling Machine Emotions for Realizing Intelligence: Foundations
    and Applications, pp. 179–207 (2010)
18. Norman, G.J., Norris, C., Gollan, J., Ito, T., Hawkley, L., Larsen, J., Berntson, G.G.: Cur-
    rent emotion research in psychophysiology: The neurobiology of evaluative bivalence.
    Emotion Review 3, 3349–3359 (2011), doi:10.1177/1754073911402403
19. Pang, B., Lee, L.: Opinion mining and sentiment analysis. Foundations and Trends in In-
    formation Retrieval 1(1-2), 1–135 (2008)
20. Pang, B., Lee, L., Vaithyanathan, S.: Thumbs up? Sentiment classification using machine
    learning techniques. In: Proceedings of the Conference on Empirical Methods in Natural
    Language Processing, pp. 79–86. ACL, Morristown (2002)
21. Pennebaker, J., Mehl, M., Niederhoffer, K.: Psychological aspects of natural language use:
    Our words, our selves. Annual Review of Psychology 54, 547–577 (2003)
22. Ponomareva, N., Thelwall, M.: Do neighbours help? an exploration of graph-based algo-
    rithms for cross-domain sentiment classification. In: The 2012 Conference on Empirical
    Methods on Natural Language Processing and Computational Natural Language Learning
    (EMNLP-CoNLL 2012) (2012)
23. Skowron, M.: Affect listeners: Acquisition of affective states by means of conversational
    systems. In: Esposito, A., Campbell, N., Vogel, C., Hussain, A., Nijholt, A. (eds.) Second
    COST 2102. LNCS, vol. 5967, pp. 169–181. Springer, Heidelberg (2010)
24. Skowron, M., Pirker, H., Rank, S., Paltoglou, G., Ahn, J., Gobron, S.: No peanuts! Affec-
    tive cues for the virtual bartender. In: Murray, R.C., McCarthy, P.M. (eds.) Proceedings of
    the Florida Artificial Intelligence Research Society Conference (FLAIRS-24), pp. 117–
    122. AAAI Press, Menlo Park (2011)
25. Somasundaran, S., Namata, G., Wiebe, J., Getoor, L.: Supervised and unsupervised me-
    thods in employing discourse relations for improving opinion polarity classification. In:
    Empirical Methods in Natural Language Processing (EMNLP 2009), pp. 170–179 (2009)
26. Taboada, M., Brooke, J., Tofiloski, M., Voll, K., Stede, M.: Lexicon-based methods for
    sentiment analysis. Computational Linguistics 37(2), 267–307 (2011)
27. Thelwall, M., Buckley, K.: Topic-based sentiment analysis for the social web: The role of
    mood and issue-related words. Journal of the American Society for Information Science
    and Technology (in press)
28. Thelwall, M.: Emotion homophily in social network site messages. First Monday 10(4)
    (2010),
    http://firstmonday.org/htbin/cgiwrap/bin/ojs/index.php/fm/ar
    ticle/view/2897/2483 (retrieved March 6, 2011)
29. Thelwall, M., Buckley, K., Paltoglou, G.: Sentiment in twitter events. Journal of the Amer-
    ican Society for Information Science and Technology 62(2), 406–418 (2011)
30. Thelwall, M., Buckley, K., Paltoglou, G.: Sentiment strength detection for the social web.
    Journal of the American Society for Information Science and Technology 63(1), 163–173
    (2012)
31. Thelwall, M., Buckley, K., Paltoglou, G., Cai, D., Kappas, A.: Sentiment strength detec-
    tion in short informal text. Journal of the American Society for Information Science and
    Technology 61(12), 2544–2558 (2010)

32. Thelwall, M., Sud, P., Vis, F.: Commenting on YouTube videos: From Guatemalan rock to el big bang. Journal of the American Society for Information Science and Technology 63(3), 616–629 (2012)
33. Turney, P.D.: Thumbs up or thumbs down? Semantic orientation applied to unsupervised classification of reviews. In: Proceedings of the 40th Annual Meeting of the Association for Computational Linguistics (ACL), Philadelphia, PA, July 6-12, pp. 417–424 (2002)
34. Wiebe, J., Wilson, T., Bruce, R., Bell, M., Martin, M.: Learning subjective language. Computational Linguistics 30(3), 277–308 (2004)
35. Wilson, T., Wiebe, J., Hwa, R.: Recognizing strong and weak opinion clauses. Computational Intelligence 22(2), 73–99 (2006)

Optimal Feature Selection for Sentiment Analysis

Basant Agarwal and Namita Mittal

Malaviya National Institute of Technology, Jaipur, India
thebasant@gmail.com, nmittal@mnit.ac.in

Abstract. Sentiment Analysis (SA) research has increased tremendously in recent times. Sentiment analysis deals with the methods that automatically process the text contents and extract the opinion of the users. In this paper, *unigram* and *bi-grams* are extracted from the text, and composite features are created using them. Part of Speech (POS) based features adjectives and adverbs are also extracted. Information Gain (IG) and Minimum Redundancy Maximum Relevancy (mRMR) feature selection methods are used to extract prominent features. Further, effect of various feature sets for sentiment classification is investigated using machine learning methods. Effects of different categories of features are investigated on four standard datasets i.e. Movie review, product (book, DVD and electronics) review dataset. Experimental results show that composite features created from prominent features of *unigram* and *bi-gram* perform better than other features for sentiment classification. mRMR is better feature selection method as compared to IG for sentiment classification. Boolean Multinomial Naïve Bayes (BMNB) algorithm performs better than Support Vector Machine (SVM) classifier for sentiment analysis in terms of accuracy and execution time.

Keywords: Sentiment Analysis, feature selection methods, machine learning, Information Gain, Minimum Redundancy Maximum Relevancy (mRMR), composite features.

1 Introduction

Sentiment Analysis (SA) is a task that finds the opinion (e.g. positive or negative) from the text documents like product reviews /movie reviews [1], [9]. As user generated data is increasing day by day on the web, it is needed to analyze those contents to know the opinion of the users, and hence it increases the demand of sentiment analysis research. People express their opinion about movies and products etc. on the web blogs, social networking websites, content sharing sites and discussion forums etc. These reviews are beneficial for users and companies. Users can know about various features of products that can help in taking decision of purchasing items. Companies can improve their products and services based on the reviews. Sentiment analysis is very important for e-Commerce companies to know the online trends about the products and services. Example of sentiment analysis includes identifying movie popularity from online reviews; which model of a camera is liked by most of the users and which music is liked the most by people etc.

A. Gelbukh (Ed.): CICLing 2013, Part II, LNCS 7817, pp. 13–24, 2013.

Sentiment classification is to assign a document into categories (positive, negative and neutral) by its subjective information. The challenge in movie review polarity classification is that the generally real facts are also mixed with actual review data. It is difficult to extract opinion from reviews when there is a discussion of the plot of the movie, discussion of the good qualities of actors of the movie but in the end over-all movie is disliked. One of the biggest challenges of this task is to handle negated opinion. Product review domain considerably differs from movie review dataset. In product reviews, reviewer generally writes both positives and negative opinion, because some features of the product are liked and some are disliked. It is diffi-cult to classify that review into positive and negative class. Also, some feature specific comments are written in the review, for example like battery life of the laptop is less, but overall performance is good. To identify overall sentiment of these type of reviews are difficult. Generally, product review dataset contains more comparative sentences than movie review dataset, which is difficult to classify [6].

Machine learning methods have been extensively used for sentiment classification [1], [2], [9]. The Bag of Words representation is commonly used for sentiment classi-fication, resulting very high dimensionality of the feature space. Machine learning algorithm can handle this high-dimensional feature space by using feature selection methods which eliminate the noisy and irrelevant features [17].

In proposed approach, *unigram* and *bi-grams* feature set are extracted from text, and various composite feature sets are created. Effect of various feature sets are inves-tigated for sentiment classification using Boolean Multinomial Naïve Bayes (BMNB) [18] and Support Vector Machine (SVM) [11] classifiers. Information Gain (IG) and Minimum Redundancy Maximum Relevancy (mRMR) feature selection techniques are used to extract prominent features.

Contributions of this paper are as follows.

1. Different composite feature set are created using *unigram* and *bi*-gram that per-form better than other features.

2. Used mRMR feature selection method for sentiment analysis, and compared its performance with the IG.

3. Compared the performance of BMNB and SVM for sentiment analysis, and found that BMNB classifier performs better than state of art SVM classifier.

4. Proposed method is evaluated on four standard datasets on varied domain re-views.

The paper is organized as follows: A brief discussion of the earlier research work is given in Section 2. Feature selection methods used for reducing the feature vector size are discussed in Section 3. Section 4 describes the machine learning algorithm used in the experiments. Dataset, Experimental setup and results are discussed in Section 5. Finally, Section 6 describes conclusions.

2   Related Work

A lot of work has been done for feature selection for sentiment classification [9],[10], [13], [16], [17] using machine learning methods [1], [2], [5], [13]. Pang and Lee [2] used *unigrams*, *bi*-grams and adjectives for creating feature vector. Authors used different machine learning algorithms like NB, SVM, and Maximum-Entropy (ME) for sentiment analysis of movie review dataset. Further, they investigated that presence or absence of a term in the feature vector gives better classification results than using term frequency, and concluded that SVM performs best amongst classifiers. Sentiment classification using machine learning methods face problem of dealing high dimension of the feature vector [1], [13]. Many researchers worked on reducing feature vector size with different feature selection methods. The performance comparison of standard machine learning techniques with different feature selection methods have been discussed [1], [5], [9], [13]. Pang and Lee [4] used minimum cut method for sentiment polarity detection. Authors eliminated the objective sentences from the documents. In [3], Categorical Probability Proportion Difference (CPPD) feature selection method is proposed, which is capable of selecting the features which are relevant and capable of discriminating the class.

O' keefe et al. [15] compared three feature selection methods and feature weighting scheme for sentiment classification. Wang et al. [14] proposed a new Fisher's discriminant ratio based feature selection method for text sentiment classification. Abbasi et al. [17] found that information gain or genetic algorithm improves the accuracy of sentiment classification. They also proposed Entropy Weighted Genetic Algorithm (EWGA) by combining the two, which produces high accuracy. S. Tan [13], discussed four feature selection methods Mutual Information (MI), IG, Chi square (CHI), and Document Frequency (DF) for sentiment classification on Chinese documents, using five machine learning algorithms i.e. K- nearest neighbour, Centroid classifier, Winnow classifier, NB and SVM. Authors observed that IG performs best among all the feature selection methods and SVM gives best results among machine learning algorithms.

Verma et al. [6] used semantic score for initial pruning of semantically less important terms, further by using information gain feature selection technique important features are extracted, for better classification accuracy. Part-of-speech (POS) information is commonly used in sentiment analysis and opinion mining [5], [9]. There are several comparisons of efficiency of adjectives, adverbs, verbs and other POS [1], [9], [20]. Turney [7] proposed a sentiment classification method using phrases based on POS patterns, mostly including adjective and adverbs.

3   Feature Selection Method

Feature selection methods select important features by eliminating irrelevant features. Reduced feature vector comprising relevant features improves the computation speed and increases the accuracy of machine learning methods [10], [17].

3.1   Minimum Redundancy Maximum Relevance (mRMR)

The Minimum Redundancy Maximum Relevance (mRMR) feature selection method [12] is used to identify the discriminant features of a class. mRMR method selects features those have high dependency to class (maximum relevancy) and minimum dependency among features (minimum redundancy). Sometimes relevant features with maximum relevancy with the class may have redundancy among features. When two features have redundancy then if one feature is eliminated, there is not much difference in class discrimination [12].

Mutual information is used for calculating the correlation/dependency between features and class attribute, and among features. mRMR feature selection technique selects features which have high mutual information (maximum relevant) with the class attribute and eliminate features which have high mutual information (highly correlated) among themselves (minimum redundant).

3.2   Information Gain (IG)

Information gain (IG) is one of the important feature selection techniques for sentiment classification. IG is used to select important features with respect to class attribute. It is measured by the reduction in the uncertainty in identifying the class attribute when the value of the feature is known. The top ranked (important) features are selected for reducing the feature vector size in turn better classification results. Information gain of a term can be calculated by using equation 1 [11].

$$
\begin{aligned}
IG(t) = &-\sum_{J=1}^{K} P(C_J) log(P(C_J)) + \\
&P(w)\sum_{J=1}^{K} P(C_J \mid w) log(P(C_J \mid w)) + P(\overline{w})\sum_{J=1}^{K} P(C_J \mid \overline{w}) log\, P(C_J \mid \overline{w})
\end{aligned}
\tag{1}
$$

Here, $P(C_J)$ is the fraction of number of documents that belongs to class C_j out of total documents and $P(w)$ is fraction of documents in which term w occurs. $P(C_j \mid w)$ is computed as fraction of documents from class C_j that have term w.

4   Machine Learning Algorithms

4.1   Multinomial Naïve Bayes

Naive Bayes [11] is frequently used for text classification problems. It is computationally very efficient and easy to use. The Naïve Bayes assumption is that features are conditionally independent of one another, given the class [11]. A Multinomial Naive Bayes classifier [18] with Term Frequency is a probability based learning method, which constructs a model by using term frequency of a feature/word to represent documents.

In Boolean Multinomial Naïve Bayes (BMNB) [18], TF of a word in a document is counted as 1 if that term is present else it is counted as zero.

4.2   Support Vector Machine (SVM)

SVM is a supervised learning method [9], [11]. SVM finds a hyperplane that divides the training documents in such a way that both the class data points are maximum separable. SVM has shown to be superior in comparison to other machine learning algorithms, in case of limited but sufficient training samples. SVM has been widely used for text classification [11], [18] and sentiment analysis [1], [2], [9].

5   Dataset, Experimental Setup and Results

5.1   Dataset Used

To evaluate the prominent features, feature selection method and best machine learning algorithm, one of the most popular publically available movie review dataset [4] is used. This standard dataset, known as Cornell Movie Review Dataset is consisting of 2000 reviews that contain 1000 positive and 1000 negative reviews collected from Internet Movie Database. To make experiment scientifically more stable, we used product review dataset consisting amazon products reviews provided by Blitzer et al. [22]. Reviews are available for different domains. We used product reviews of books, DVD and electronics for experiments. Each domain has 1000 positive and 1000 negative labelled reviews. An average number of words per document are larger in Movie review dataset as compared to product review dataset.

5.2   Features Extraction and Selection in Proposed Approach

In proposed approach, each review is pre-processed in such a way that machine learning algorithm is applied. Negation word (no, not, never, didn't, don't, can't) reverses the polarity of the sentence that is important to handle for sentiment classification. It is done by concatenating first word after the negation word that should not be a stop word. For example, "this is not a good movie", polarity of word "good" is reversed by "not", and it becomes "notgood" after negation handling [6]. Boolean weighting scheme is used for representing text document.

Features are categorized on the basis of the way we have extracted them from the text. The categories are (i) words occurring in the document i.e. *unigrams*, *bi*-gram. (ii) POS based words, i.e. adjectives, adverbs.

(i) In the first category, firstly negation handling is performed. Then document is tokenized, and stop words are removed. Each word is stemmed according to Porter's algorithm [8]. In the pre-processing phase, Document Frequency (DF) is used for initial pruning of unimportant features by eliminating features occurring in less number of documents. Firstly feature set using *unigram* (*F1* feature set) and *bi*-gram (*F2* feature set) features are generated. *Bi-gram* based features (*F2*) are capable of handling negation words in the context of the text [2] that is why there is no need of negation handling explicitly in this case.

Further, prominent feature sets and composite feature sets are created from *unigram* and *bi-gram* features. Prominent features are extracted from *unigrams* with IG and mRMR, we call it as *PIGF1* (*Prominent **IG** Features* 1-gram) and *PmRMRF1* (*Prominent **mRMR** Features* 1-gram) respectively. Similarly, optimal features are extracted from *bi-grams* with IG and mRMR, those are *PIGF2* (*Prominent **IG** Features* 2-gram) and *PmRMRF2* (*Prominent **mRMR** Features* 2-gram) respectively. Further, by combining *unigrams* and *bi-grams*, **Com**posite **F**eature set (*ComF*) is created. Then, by combining prominent *unigram* and *bi*-gram **IG** features (*PIGF1* and *PIGF2*), **P**rominent **Com**posite IG features (*ComPIG*) are created. Similarly, by using Prominent *unigram* and *bi-gram* mRMR features (*PmRMRF1, PmRMRF2*), Prominent **Com**posite **mRMR** features *ComPmRMR* feature set is created.

(ii) In the second category, Stanford POS tagging software[1] is used for tagging each term according to Part of Speech. Stop word removal and stemming is not performed in this method for extracting features, as the same word can occur with different POS. For example, die as Noun is different than die as Verb. Adjective and Adverbs are extracted because these are considered as important features for sentiment classification [1, 5, 9]. Feature sets namely *P1* and *P2* are generated using adjectives and adverbs respectively. Further, **Com**posite **F**eature set (*ComP*) is also created by combining POS features (*P1 and P2*).

5.3   Evaluation Metrics

Precision, Recall, Accuracy and F- measure are used for evaluating performance of sentiment classification [11]. Precision for a class C is the fraction of total number of documents that are correctly classified and total number of documents that classified to the class C (sum of True Positives (TP) and False Positives (FP)). Recall is the fraction of total number of correctly classified documents to the total number of documents that belongs to class C (sum of True Positives and False Negative (FN)). F –measure is the combination of both precision and recall, is given by

$$F - Measure = 2*(precision * recall)/(precision + recall) \qquad (2)$$

F-measure is used to report the performance of classifiers for the sentiment classification.

5.4   Results and Discussions

Different feature vector generated after pre-processing are further used for the classification. Among different machine learning algorithms Support Vector Machine (SVM) classifiers are the mostly used for sentiment classification [2], [5], [6], [10], [13], [17]. In our Experiments, BMNB and SVM are used for classifying review documents into positive or negative sentiment polarity, since BMNB can perform better than SVM in case some appropriate feature selection method is used. Evaluation of classification is done by 10 fold cross validation [21]. Linear SVM and Naïve Bayes Multinomial are used for all the experiments with default setting in WEKA [19].

---

[1] http://nlp.stanford.edu/software/

Determination of Prominent Feature and Classifiers

The performance of different feature sets are compared with respect to F-measure values using BMNB and SVM classifiers. F-measure values for all the features with BMNB and SVM classifiers for four datasets are shown in Table 1.

For unigram features (*F1*), BMNB is performing better than SVM for all the dataset except movie review dataset. This is because SVM performs better with large feature vector as number of unique terms in movie review dataset is larger over product review datasets (refer Table 2). When we consider *PIGF1* and *PmRMRF1* features, performance of BMNB increased significantly compare to their unigram features because BMNB is very sensitive with the noisy features. If noisy and irrelevant features are removed from the feature vector, BMNB can perform better. Also, performance of SVM is increased compared to its performance with unigram features (refer Table 1). Performance is increased due to IG and mRMR methods removed noisy and irrelevant features from the feature vector which deteriorate the performance of a classifier.

It can be observed from Table 1 that *bi-gram* feature set individually doesn't give better performance as compared to unigram features. However, when prominent *bi-grams* are extracted in *PIGF2* and *PmRMRF2* with IG and mRMR, F-measure values are increased due to the fact that feature selection methods (IG and mRMR) reduce the noisy and irrelevant features.

Further, when composite feature vector *ComF* (combining *unigram* and *bi-gram*) is considered, performance of both the classifier (SVM and BMNB) improves but at the cost of execution overhead as given in Table 1. As Feature vector size of *ComF* features is large, so it is required to filter the irrelevant and noisy features for better classification results. That is done by creating feature vector by combining only prominent features of both unigram and bigrams denoted as *ComPIG, ComPmRMR*. *ComPIG* and *ComPmRMR* features produce significantly good results with small feature vector size. Performance (in terms of F-measure) of *ComPmRMR* presents greater than *ComPIG*. F-measure for BMNB classifier is 82.7% with *unigram (F1)* features, while with the same classifier *ComPmRMR* gives 91.1% (+10.15%) with movie review dataset. Similarly, for other datasets, *ComPmRMR* outperforms other feature selection methods.

mRMR feature selection method performs better than IG as IG selects relevant features based on reduction in uncertainty in identifying the class after knowing the value of the feature. It does not eliminate redundant features. However, mRMR discards redundant features which are highly correlated among features, and retain relevant features having minimum correlation. It is intuitive that when *unigram* and *bi-gram* features are combined, redundancy remains there. So, in case of composite features more information is included but at the cost of redundancy, which is removed with the use of mRMR feature selection method. Since, IG only considers relevancy of the feature with the class, it only includes important features of both *unigram* and *bi-gram* but not considering the effect of redundancy. In case of mRMR method, it includes prominent features of both unigram and bi-gram, with eliminating the redundant features.

Table 1. F-measure (%) for different features sets and feature selection methods

	Movie		Book		DVD		Electronics	
	BMNB	SVM	BMNB	SVM	BMNB	SVM	BMNB	SVM
F1	82.7	84.2	80.9	76.2	78.9	77.3	80.8	76.5
PIGF1	89.2	85.8	89.3	84.2	89.1	84.5	86.4	84.6
PmRMRF1	90.2	87.1	90.1	84.1	90.1	85.3	87.2	84.9
F2	79.2	78.8	68.6	66.8	67.1	68.0	72.6	70.4
PIGF2	81.1	80.4	80.4	75.4	74.8	77.1	79.2	74.9
PmRMRF2	80.1	81.4	81.1	76.0	76.1	75.5	80.2	76.0
ComF	87.0	86.7	82.6	79.5	79.9	79.3	85.2	80.8
ComPIG	90.6	89.2	92.1	87.1	90.4	87.3	91.3	88.1
ComPmRMR	91.1	90.2	92.5	88.3	91.5	88.0	91.8	89.0
P1	80.8	81.1	79.4	77.9	74.0	74.6	78.6	77.5
P2	70.4	68.2	72.5	71.2	68.0	67.9	68.2	66.4
ComP	82.1	82.4	81.4	80.9	77.8	79.0	79.0	81.2

When only adjectives are considered to generate feature vector, it is observed that performance is degraded as compared to *unigrams* features. Adverbs individually are performing worse as compared to adjectives and *unigram* features. Combining Adjectives and adverbs gives performance near to base *unigram* features (refer Table1). Composite features (*ComP*) perform better as compared to the features considered independently with respect to F- measure value.

Both mRMR and IG perform considerably better with optimal features for classifying instances compare to results reported in previous literature. We observed during experiments that mRMR and IG selects approximately 65-70% features in common for all the dataset considered. However, remaining 30-35% features in IG features set were those features, which were correlated with other features. mRMR feature selection method was able to remove those redundant features to included more relevant features which IG method was unable to do. mRMR discards unwanted noisy features and retains only relevant feature with minimum correlation among features. That is why mRMR feature selection method performed better as compared to IG.

Dependency among attributes inevitably decrease the power of NB classifier [11]. mRMR selects the prominent features out of complete feature set those are not correlated among features. It is observed from the experiments that performance of BMNB increased significantly after removing the irrelevant and noisy features. This is due to the fact that prominent features are less likely to be depended among themselves. BMNB after mRMR feature selection method performs best because mRMR feature selection technique is capable of removing the correlation among the features. In addition, BMNB is significantly faster than SVM.

Effect of Feature Vector Size on Classification Performance
For deciding, in what ratio prominent features should be selected from unigrams and bigrams? We empirically experimented with different combination of prominent features vector sizes. It is observed that unigrams are more important than bigram that is also resembles with the results of Table1. So, we decided to include *unigram* and *bi-gram* in 60:40 percent ratio. For example, to create *ComPIG* feature vector size of 1000, top 600 features are selected from *PIGF1* and top 400 features are selected from *PIGF2*.

Table 2. Feature vector size for all the features for different datasets

S.No	Features	Movie Review	Book	DVD	Electronics
1	*F1*	9045	5391	5955	4270
2	*PIGF1 and PmRMRF1*	600	480	720	480
3	*F2*	6050	6484	8888	5513
4	*PIGF2 and PmRMRF2*	400	320	480	320
5	*ComF*	15095	11875	14843	9783
6	*ComPIG and ComPmRMR*	1000	800	1200	800
7	*P1*	1330	1120	1280	980
8	*P2*	377	350	400	310
9	*ComP*	1707	1470	1680	1290

Effect of feature vector size is also experimented with feature selection technique on performance of classifier. Understanding the limitation of space, we report the performance of IG and mRMR for composite features i.e. *ComPmRMR* and *ComPIG* for BMNB classifier since composite features performed best among all the features and BMNB to be better than SVM. Feature vector size for all the features is shown in Table 2. Effect of different feature vector size with IG and mRMR on the performance of BMNB classifiers on different dataset is shown in Figure 1-2.

How many features should be selected for classification is taken based on these observations? For taking this decision, it is observed from Figure 1-2 that if feature size is not reduced much, F-measure value is varying in a narrow range, and that is approximately 10-15% of total features. Therefore, with empirically experimenting, we selected very less number of features for creating feature vector. Feature vector sizes used for our experiments are shown in Table 2.

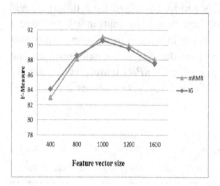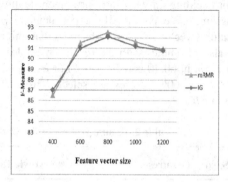

Fig. 1. Effect of feature size for *ComPmRMR* feature with BMNB classifier on Movie Review and book dataset respectively

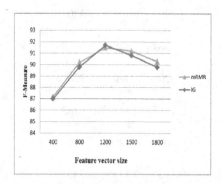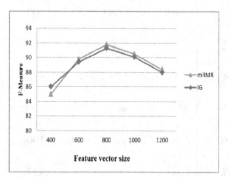

Fig. 2. Effect of feature size for *ComPmRMR* feature with BMNB classifier on DVD and electronics dataset respectively

6   Conclusion

In this paper, different features like unigrams, bigrams, adjectives, adverbs were extracted and composite features were created. Effect of various categories of features was investigated on four different standard dataset of different domains. Composite feature of prominent features of *unigram* and *bi-gram* gives better performance as compared to unigrams, bigrams, adjectives, adverbs individually with respect to F-measure. IG and mRMR feature selection methods are used for extracting predominant features. Comparative performance of IG and mRMR is investigated for sentiment classification, and it is observed that mRMR performs better than IG. It is due to the fact that mRMR feature selection method is capable of selecting relevant features as well as it can eliminate redundant features unlike IG which can only compute importance of the feature. SVM and BMNB classifiers are used for sentiment classification. Performance of BMNB is better as compared to SVM in terms of performance, and significantly better than SVM in terms of execution time. The advantage of using *unigrams* and *bi-grams* over other POS based features are that they are easy to extract, while POS based features require tagger to extract the features, and POS tagging is very slow process. BMNB perfomed best with prominent mRMR composite features (*ComPmRMR*) in terms of execution time and accuracy for sentiment classification. We wish to compare the performance of these features on more datasets of different domain, and also study the affect of proposed method on non-english documents.

References

1. Pang, B., Lee, L.: Opinion mining and sentiment analysis. Foundations and Trends in Information Retrieval 2(1-2), 1–135 (2008)
2. Pang, B., Lee, L., Vaithyanathan, S.: Thumbs up? Sentiment classification using machine learning techniques. In: Proceedings of the Conference on Empirical Methods in Natural Language Processing (EMNLP), pp. 79–86 (2002)
3. Agarwal, B., Mittal, N.: Categorical Probability Proportion Difference (CPPD): A Feature Selection Method for Sentiment Classification. In: Proceedings of the 2nd Workshop on Sentiment Analysis where AI Meets Psychology (SAAIP 2012), COLING 2012, Mumbai, pp. 17–26 (2012)
4. Pang, B., Lee, L.: A sentimental education: sentiment analysis using subjectivity summarization based on minimum cuts. In: Proceedings of the Association for Computational Linguistics (ACL), pp. 271–278 (2004)
5. Mullen, T., Collier, N.: Sentiment analysis using support vector machines with diverse information sources. In: Proceedings of the Conference on Empirical Methods in Natural Language Processing (EMNLP), pp. 412–418 (2004)
6. Verma, S., Bhattacharyya, P.: Incorporating semantic knowledge for sentiment analysis. In: Proceedings of ICON 2009, Hyderabad, India (2009)
7. Turney, P.: Thumbs up or Thumbs Down? Semantic Orientation Applied to Unsupervised Classification of Reviews. In: ACL 2002, pp. 417–424 (2002)
8. Porter, M.F.: An Algorithm for Suffix Stripping. Program 14(3), 130–137 (1980)
9. Liu, B.: Sentiment Analysis and Subjectivity. In: Indurkhya, N., Damerau, F.J. (eds.) Handbook of Natural Language Processing, 2nd edn., pp. 627–666. Chapman & Hall (2010)
10. Abbasi, A., France, S., Zhang, Z., Chen, H.: Selecting Attributes for Sentiment Classification Using Feature Relation Networks. IEEE Transactions on Knowledge and Data Engineering 23, 447–462 (2011)

11. Witten, I.H., Frank, E.: Data mining: Practical Machine Learning Tools and techniques. Morgan Kaufmann, San Francisco (2005)
12. Peng, H., Long, F., Ding, C.: Feature selection based on mutual information: criteria of max-dependency, max-relevance, and min-redundancy. IEEE Trans. Pattern Analysis and Machine Intelligence 27(8), 1226–1238 (2005)
13. Tan, S., Zhang, J.: An empirical study of sentiment analysis for chinese documents. Expert Systems with Applications 34, 2622–2629 (2008)
14. Wang, S., Li, D., Wei, Y., Li, H.: A feature selection method based on fisher's discriminant ratio for text sentiment classification. In: Liu, W., Luo, X., Wang, F.L., Lei, J. (eds.) WISM 2009. LNCS, vol. 5854, pp. 88–97. Springer, Heidelberg (2009)
15. O'Keefe, T., Koprinska, I.: Feature Selection and Weighting Methods in Sentiment Analysis. In: Proceedings of the 14th Australasian Document Computing Symposium (2009)
16. Nicholls, C., Song, F.: Comparison of feature selection methods for sentiment analysis. In: Farzindar, A., Kešelj, V. (eds.) Canadian AI 2010. LNCS, vol. 6085, pp. 286–289. Springer, Heidelberg (2010)
17. Abbasi, A., Chen, H.C., Salem, A.: Sentiment analysis in multiple languages: Feature selection for opinion classification in web forums. ACM Transactions on Information Systems 26(3), Article no:12 (2008)
18. Manning, C.D., Raghvan, P., Schutze, H.: Introduction to information retrieval. Cambridge University Press, Cambridge (2008)
19. WEKA. Open Source Machine Learning Software Weka, http://www.cs.waikato.ac.nz/ml/weka/
20. Benamara, F., Cesarano, C., Picariello, A., Reforgiato, D., Subrahmanian, V.S.: Sentiment analysis: Adjectives and adverbs are better than adjectives alone. In: Proceedings of the International Conference on Weblogs and Social Media (ICWSM) (2007)
21. Kohavi, R.: A study of cross-validation and bootstrap for accuracy estimation and model selection. In: IJCAI 1995 Proceedings of the 14th International Joint Conference on Artificial Intelligence, vol. 2, pp. 1137–1143 (1995)
22. Blitzer, J., Dredze, M., Pereira, F.: Biographies, Bollywood, boom-boxes and blenders: Domain adaptation for sentiment classification. In: Proceedings of the Association for Computational Linguistics (ACL), pp. 440–447 (2007)

Measuring the Effect of Discourse Structure on Sentiment Analysis

Baptiste Chardon[1,2], Farah Benamara[1], Yannick Mathieu[3],
Vladimir Popescu[1], and Nicholas Asher[1]

[1] IRIT-CNRS, Toulouse University
[2] Synapse Développement, Toulouse
[3] LLF-CNRS, Paris 7 University
{chardon,benamara,popescu,asher}@irit.fr,
yannick.mathieu@linguist.jussieu.fr

Abstract. The aim of this paper is twofold: measuring the effect of discourse structure when assessing the overall opinion of a document and analyzing to what extent these effects depend on the corpus genre. Using Segmented Discourse Representation Theory as our formal framework, we propose several strategies to compute the overall rating. Our results show that discourse-based strategies lead to better scores in terms of accuracy and Pearson's correlation than state-of-the-art approaches.

1 Introduction

Discourse structure can be a good indicator of the subjectivity and / or the polarity orientation of a sentence. It can also be used to recognize implicit opinions and to enhance the recognition of the overall stance of texts. For instance, sentences related by a *Contrast*, *Parallel* or a *Continuation* relation often share the same subjective orientation, as in *Mary liked the movie. His husband too*, where the *Parallel* relation allows us to detect the implicit opinions conveyed by the second sentence. Polarity is reversed in case of *Contrast* and usually preserved in case of *Parallel* and *Continuation*. *Result* on the other hand doesn't have a strong effect on subjectivity and polarity is not preserved. For instance, in *Your life is miserable. You don't have a girlfriend. So, go see this movie*, the prior positive polarity of the recommendation follows negative opinions. Hence, *Result* can help to determine the contextual polarity of opinionated sentences. Finally, in case of *Elaboration*, subjectivity is not preserved, in contrast to polarity (It is difficult to say *The movie was excellent. The actors were bad*).

We aim in this paper to empirically measuring the effect of discourse structure on assessing the overall opinion of a document and by analyzing to what extent these effects depend on the corpus genre. To our knowledge, this is the first research effort that empirically validates the importance of discourse for sentiment analysis. Our analysis relies on manually annotated discourse information following the Segmented Discourse Representation Theory (SDRT) [1]. This is a first and a necessary step before moving to real scenarios that rely on automatic annotations

A. Gelbukh (Ed.): CICLing 2013, Part II, LNCS 7817, pp. 25–37, 2013.
© Springer-Verlag Berlin Heidelberg 2013

(we recall that as far as we know the only existing powerful discourse parser based on SDRT theory is the one that has been developed for a dialogue corpus (Verbmobil corpus [2]). This first step allowed us to show the real added value of discourse in computing both the overall polarity and the overall rating.

2   Related Works

Although rhetorical relations seem to be very useful in sentiment analysis, most extant research efforts on both document-level and sentence-level sentiment classification do not use discourse information. Among the few research reports on discourse-based opinion analysis, let us cite the following. [3] proposed a shallow semantic representation of subjective discourse segments using a feature structure and five types of SDRT-like rhetorical relations. [4] as well as [5] have used an RST discourse parser in order to calculate semantic orientation at the document level by weighting the nuclei more heavily. [6] proposed the notion of opinion frames as a representation of documents at the discourse level in order to improve sentence-based polarity classification and to recognize the overall stance. Two sets of 'home-made' relations were used: relations between targets and relations between opinion expressions. [7] used the semantic sequential representations to recognize RST-based discourse relations for eliminating intra-sentence polarity ambiguities. [8] propose a context-based approach to sentiment analysis and show that discursive features improve subjectivity classification. [9] discuss the application of the Linguistic Discourse Model theory to sentiment analysis in movie reviews. Finally, [10] examine how two types of RST-like rhetorical relations (*conditional* and *concessive*) contribute to the expression of appraisal in movie and book reviews.

We aim here to go further by answering the following questions: (1) *What does the discourse structure tell us about opinion?* (2) *What is the impact of discourse structure when assessing the overall opinion of a document?* (3) *Does our analysis depend on the corpus genre?*. The first question is addressed in section 3 while the last two ones in section 4.

3   Discourse Structure and Opinion

Our data comes from two corpora: movie reviews (*MR*) taken from *AlloCiné.fr* and news reactions (*NR*) taken from the politics, economy and international section of *Lemonde.fr* newspaper. In order to guarantee that the discourse structure is informative enough, we only selected movies and articles that are associated to more than 10 reviews / reactions. We also filtered out documents containing less than three sentences. In addition, we balanced the number of positive and negative reviews according to their corresponding general evaluation when available (in *NR* users were not asked to give a general evaluation). This selection yielded a total of 180 documents for *MR* and 131 documents for *NR*.

3.1   Annotation Scheme

Our basic annotation level is the Elementary Discourse Units (EDU). We chose to automatically identify EDUs and then to manually correct the segmentation if necessary. We relied on an already existing discourse segmenter [8] that yields an average F-measure of 86.45%. We have a two-level annotation scheme: *at the segment level* and *at the document level*. Annotators used the GLOZZ platform (www.glozz.org) which provides a discourse graph as part of its graphical user interface.

EDU Annotation Level. For each EDU, annotators were asked to specify its subjectivity orientation as well as polarity and strength; Subjectivity can be either one of the following: **SE** – EDUs contain explicitly lexicalized *subjective and evaluative* expressions, as in *very bad movie*; **SI** – EDUs do not contain any explicit subjective cues but opinions are inferred from the context, as in *The movie should win the Oscar*; **O** – EDUs do not contain any lexicalized subjective term, neither do an implied opinion. **SN** – subjective, but non-evaluative EDUs that are used to introduce opinions, as in the segment a in *[I suppose]$_a$ [that the employment policy will be a disaster]$_b$*; and finally **SEandSI** which are segments that contain both explicit and implicit evaluations on the same topic or on different topics, as in *[Fantastic pub !]$_a$ [The pretty waitresses will not hesitate to drink with you]$_b$*. Polarity can be of four different values: **+, –, both** which indicates a mixed polarity as in *this stupid President made a wonderful talk*, and **no polarity** which indicates that the segment does not convey any sentiment. Finally, strength has to be stated on a four-level scale going from 0 to 3 where 0 is the score associated to O segments, 1, 2 and 3 respectively indicates a weak, a medium and a strong strength.

Document Annotation Level. First, annotators have to give the overall opinion orientation of the document (the initial star ratings in MR corpus were removed) by using a six-level scale, going from -3 to -1 for negative opinion documents and from $+1$ to $+3$ for positive ones. Then, they have to build the discourse structure of the document by respecting the structural principles of SDRT, such as the right frontier principle and structural constraints involving complex discourse units (CDUs) (which are build from EDUs in recursive fashion). It's important to recall that SDRT allows for the creation of full discourse graphs (and not trees as in the RST [11]) which allow to capture complex discourse phenomena, such as long-distance attachments and long-distance discourse pop-ups, as well as crossed dependencies.

During the elaboration of our manual, we faced a dilemma: should we annotate opinion texts using a small set of discourse relations, as already done by [3], [6] and [10] or should we use a larger set of discourse relations? Given our goals, we chose the second solution. We used 17 oriented and mostly backward-looking relations grouped into coordinating relations that link arguments of equal importance (*Contrast, Continuation, Conditional, Narration, Alternative, Goal, Result, Parallel, Flashback*) and subordinating relations that link an important argument to a less important one (*Elaboration*, *E* − *Elab*, *Correction*, *Frame*, *Explanation*, *Background*, *Commentary*, *Attribution*). To deal with the situation where the annotators are not able to decide which relation is more appropriate to link two constituents, we added a relation labeled *Unknown*.

3.2 Results of the Annotation Campaign

Each document of our corpus was doubly annotated by three undergraduate linguistic students who were provided with a complete and revised annotation manual as well as an annotation guide explaining the inner workings of GLOZZ. Annotators were first trained on 12 movie reviews and then they were asked to annotate separately 168 documents from MR. Then, they were trained on 10 news reactions. Afterwards, they continued to annotate separately 121 documents from NR. The training phase for MR was longer than for NR since annotators had to learn about the annotation guide and the annotation tool.

Results at the EDU Level. Table 1 gives a quantitative overview (in percents) of the annotations provided by our three annotators. We get a total number of 3478 annotated segments for MR and 2150 for NR.

Table 1. Quantitative overview of the annotated data (in percents)

	SE	SN	SI	O	SEandSI	+	−	both	no polarity
MR	50	2	29	14	5	45.48	33.78	4	16.74
NR	22	6	49	2	12	17.40	55	4	23.60

The Cohen's Kappa on segment type averaged over the three annotators was 0.69 for MR and 0.44 for NR. For segment polarity we get 0.74 for MR and 0.49 for NR. Since the "both" and the *SEandSI* category are very rare in our data, they have been counted with "+" (resp. *SE*). For MR, we get very good results for both *SE* (0.79) and the polarity (positive (0.78) and negative (0.77)) of the segment. *SN* class's kappa is also very good (0.73). However, the agreements for the *SI* and *O* classes are moderate (resp. 0.62 and 0.61) because annotators often fail to decide whether a segment is purely objective and thus if it conveys only facts or if a segment holds an implicit opinion. This can also explain the lower kappa measure we get for "no polarity" (0.66). Nonetheless, these figures are well in the range of state-of-the-art research reports in distinguishing between explicit and implicit opinions (see [12]).

For NR, our results are moderate for the *SE* and *SN* classes (0.55 for each class) and fair for the *SI* and *O* classes (resp. 0.33 and 0.34). We have the same observations for the agreements on segment polarities where we obtain moderate kappas on all the three classes (0.49). This shows that the newspaper reactions corpus was a bit more difficult to annotate because the main topic is more difficult to determine (even by the annotators) − it can be one of the subjects of the article, the article itself, its author(s), a previous comment or even a different topic, related to various degrees to the subject of the article. Hence, implicit opinions, which are more frequent, can be of a different nature: ironic statements, suggestions, hopes and personal stances, especially for comments to political articles.

We finally compute the inter-annotator agreements on the overall document rating. After collapsing the ratings -1 to -3 and +1 to +3 into respectively positive and negative ratings, we get a kappa of 0.73 for MR and 0.58 for NR for both classes when averaged over our three annotators. We have also observed, that the agreement on extreme points of our six-level scale (namely -3 and +3) are relatively good (for example, we get respectively 0.8 and 0.72 for MR) whereas the kappa on the other points is fair. We get the same observation when computing agreements on segment's strength.

Results at the Discourse Level. Our goal here is to show the importance of discourse for opinion analysis and not to build a discourse bank that examine how well SDRT predicts the intuition of subjects, regardless of their knowledge of discourse theories. Therefore, computing inter-annotator agreements is out of the scope of this paper (for a detailed description of non-expert annotations using SDRT, see [13]). The analysis of the frequency of discourse relations per corpus genre shows that *Continuation* and *Commentary* are the most frequent relations (resp. 18% and 30% for MR and 23% and 24% for NR). However, *Explanation, Elaboration, $E - Elab$* (entity elaboration), *Comment, Contrast, Result* and *Goal* also have non-negligible frequencies going from 3% to 15% for each corpus genre. These results are essentially stable from one corpus to the other. Also, *Conditional, Alternative* and *Attribution* are more frequent in NR than in MR, which is consistent with a logically more structured discourse structure for news reactions than for movie reviews.

We have also analysed the ratio of complex segments to the total number of rhetorical relation arguments in our annotations. We have observed that, for both corpus genres, rhetorical relation instances between EDUs only are a minority and that CDUs are yet more numerous in NR – 56%, than for MR – 53%. This underscores the importance of CDUs for our task. We have finally analysed the impact of rhetorical relations on both subjectivity and polarity of their arguments only in case of relations linking two EDUs. Table 2 gives statistics (in percent) as a / b. a stands for on the stability (St) (that is (SE, SE), (SI, SI), (SE, SI) and (SI, SE)) and the variation (Var) of the subjectivity class (i.e. for the (O, other) and the (other, O) couples, where "other" spans the set of subjectivity classes, other than O). b stands for the polarities class but only between subjective (SN, SE, SI) EDUs only : the ($+$, $+$) and ($-$, $-$) couples for stability and the ($+$, $-$) and ($+$, $-$) couples for polarity change. We observe that our predictions (as stated in the introduction) are by and large confirmed.

3.3 The Gold Standard

The gold standard used for our experiments was made after discussion between the three annotators. This process was supervised by two experts in discourse

Table 2. Impact of rhetorical relations on both subjectivity and polarity

	MR		NR	
	St	**Var**	**St**	**Var**
Continuation	81 / 97	19 / 3	79 / 90	21 / 10
Commentary	61 / 82	39 / 18	75 / 96	25 / 4
Elaboration	50 / 100	50 / 0	82 / 100	18 / 0
Contrast	76 / 15	24 / 85	76 / 59	24 / 41
Result	81 / 100	19 / 0	47 / 100	53 / 0
Attribution	14 / 50	86/ 50	18 / 100	82 / 0
Parallel	100 / 100	0 / 0	73 / 100	27 / 0
Explanation	76 / 80	24 / 20	78 / 83	22 / 17
Frame	39 / 100	61 / 0	47 / 86	53 / 14

analysis and opinion mining. At the EDU level, the main difficulty was to achieve a consensus on implicit and objective segments, especially for NR. At the discourse level, annotators often produce equivalent discourse structures (two of our annotators used to systematically group constituents in CDUs while the others often produced flat structures). While building the gold standard, annotators used CDUs as often as possible. Finally, annotators have to agree on the overall document score. The graph in Figure 1 illustrates an annotation from the gold standard. Segments 1 to 12 are EDUs while segments 13 to 16 are CDUs.

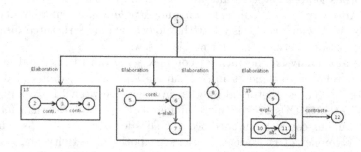

Fig. 1. Two examples of produced discourse annotation

In order to measure the effects of topic information (also called target) to compute the overall opinion, we have asked the annotators to specify, within each EDU, text spans that correspond to the topic. Topic can be of three types: the **main** topic of the document, such as *the movie*, a **partof** topic in case of features related to the main topic, such as *the actors*, and finally an **other** topic that has no mereological relation with the main topic. Once all the topics have been identified, the next step is to link them to the subjective segments of the document. For example, in *[I saw (Grey's Anatomy)_t_1 yesterday]_1. [It was boring]_2 [and (the actors)_t_2 were bad]_3*, we get topic(2, t_1 : main) and

$topic(3, t_2 : partof)$. This annotation was made by consensus due to the difficulty of the task, especially for NR. For MR, the gold standard contains 151
documents, 1905 EDUs (SE: 53.85%, SI: 26.20%), 1766 discourse relations and
1386 topics (main: 26.26%, partof: 62.62%, other: 11.11%). For NR, we have 112
documents, 835 EDUs (SE: 20.24%, SI: 51.25%), 924 relations and 586 topics
(main: 5.63%, partof: 59.55%, other: 34.81%). The distribution of the overall
rating is: 37% positive opinion and 63% negative opinion for MR, versus 33%
positive opinion and 67% negative opinion for NR.

4   Computing the Overall Opinion

For each document D, we aim at computing the overall opinion $score\_D$ of D
such as $score_D \in [-3, +3]$. We consider D as an oriented graph $(\aleph, \Re)$ such
that: $\aleph = E \cup C$ is the set of EDUs and CDUs of D and $\Re$ is the set of rhetorical
relations that link elements from $\aleph$. $\forall edu \in E$, $edu =< T, S, Val >$ where
$T = topic(edu)$ denotes the topic of edu and $T \in \{main, partof, other\}$, $S =$
$subj(edu)$ is the subjectivity orientation of edu and $S \in \{SE, SI, O, SN\}$
(SE and SI segments are considered to be of the SE type) and $Val = score(edu)$
is the opinion score of edu stated on the same discrete interval as $score_D$. Each
$cdu \in C$ has the same properties as an edu i.e. $cdu =< T_{cdu}, S_{cdu}, Val_{cdu} >$,
however, T_{cdu}, S_{cdu} and Val_{cdu} (which is in this case a set of scores) are not
given by the annotations but are the result of a reduction process of the cdu to
an edu (see Section 4.3).

We propose three strategies to compute $score_D$: (1) **Bag-of-segments (BOS)**
that does not take into account the discourse structure. The overall rating is computed using a numerical function that takes the set E as argument and outputs
the value $score_D$. (2) **Partial discourse** which takes the discourse graph as
input and then prunes it in order to select a subset $\aleph' \subseteq \aleph$ of nodes that are
relevant for computing $score_D$. This score is then computed by applying a numerical function only to $\aleph'$. (3) **Full discourse** which is based on the full use
of discourse structure where a rule-based approach guided by the semantics of
rhetorical relations aggregates the opinion scores of all the elements in $\aleph$ in a
bottom-up fashion.

4.1   Bag-of-Segments

Here we consider $D = E = \{edu_1, \ldots, edu_i\}$. In order to evaluate the impact of
segments' subjectivity and topic on our task, we propose to filter out some elements of D by applying a subjectivity filter and / or a topic filter. We have three
subjectivity filters: $\emptyset$ that keeps all the segments (i.e the filter is not activated), **se**
and **si** that respectively keep SE and SI segments. We also have four topic filters:
$\emptyset$ where the filter is not activated, **m** and **p** that respectively keep segments that
contain main and part-of topics, and finally **mp** that keeps segments that contain main or part-of topics. Each filter can be applied alone or in sequence with
other filters. For example, if we apply se and then m, we get the subset $D' \subseteq D$

such that $D' = \{edu_i \in D \; ; topic(edu_i) = main \; and \; subj(edu_i) = SE\}$. Filtering can drastically reduce the number of segments in D' ($D' = \emptyset$ or $\forall edu \in D'$ $subj(edu) = O \; or \; subj(edu) = SN$). Hence, some filters are relaxed if necessary.

Computing $score_D$ in the BOS strategy consists in applying a numerical function to all elements of D or to a subset of D obtained after filtering. Let D' be a subset of D. We have seven functions on it: (1) $A(D')$ and (2) $M(D')$, which respectively compute the average and the median of the scores associated to each EDU edu in D'. Unlike the average, the median is more suitable in case of skewed distributions. (3) $MSc(D')$ computes the maximum positive scores Max_Pos and the maximum negative scores Max_Neg of elements of D' and then returns $Max(Max\_Pos, Max\_Neg)$. In case of equality, we choose the scores with positive polarity for MR and with negative polarity for NR which correspond to the general polarity orientation of each corpus genre (see section 3). (4) $MSc\_A(D')$ computes $Sc(D')$ when the elements of D' have the same polarity orientation and $A(D')$ otherwise. (5) $Fr(D')$ returns the most frequent opinion score found in D'. In case of equality, it chooses the score that is the closest to the second most frequent score in D'. (6) $Frt(D')$ and (7) $Lst(D')$ returns the score of the first and the last element edu of D' such that $subj(edu) = SE$ or $subj(edu) = SI$. We consider here that the order of elements in D' follows the reading order of the document.

4.2   Partial Discourse (PD)

This strategy takes the discourse graph D as input and proceeds by pruning it in order to select the most important nodes for computing the overall rating. We consider two main types of pruning: (a) one based on the distinction between *subordinating* and *coordinating* relations and (b) another one based on *top-level constituents*. (a) can be done either by a **Sub1** pruning that selects from $\aleph$ only EDUs (or CDUs) that are the first argument of a subordinating relation or by a **Sub2** pruning where the selected segments are the first argument of a subordinating relation and at the same time do not appear as the second argument of a subordinating relation. The aim here is to deal with a 'cascade' of subordinations. On the other hand, (b) aims at deleting from $\aleph$ nodes that are right arguments of subordinating relations or nodes that are left arguments of already pruned constituents. Pruning in (b) can be done either by using a **Top1** strategy that preserves all the constituents of the CDUs or by using a **Top2** strategy that reduces CDUs by recursively applying **Top1** to all the elements of the CDU. The resulting set of segments $\aleph' \subseteq \aleph$, obtained after using one of the previous four pruning strategies, can be filtered by using either a subjectivity and / or a topic filter (see Section 4.1).

As in *BOS*, some filters can be relaxed if necessary. It is important to notice that our pruning / filtering process guarantees the connectivity of the graph since the non-selected nodes are not physically removed. Instead, their subjectivity type is set to O. $score_D$ is then computed by applying to all the elements of $\aleph'$ one of the seven numerical functions lastly presented.

4.3   Full Discourse (FD)

The third strategy as well has the discourse graph D as input. FD does not prune the graph and does not use any filter but it recursively determines the topic, the subjectivity and the score of each node in a bottom-up fashion. This process is guided by a set of rules that are associated to each rhetorical relation $r(a, b) \in \Re$. A rule merges the opinion information of a (i.e $< topic(a),\ subj(a),\ score(a) >$) and b (i.e $< topic(b),\ subj(b),\ score(b) >$) and computes a triple $< T_{ab}, S_{ab}, Val_{ab} >$ depending on the semantics of r. Since the rules are recursively applied to all nodes of the graph, they thus have to deal with CDUs. For instance, in case we have $r(a, b)$ where $a \in C$ and / or $b \in C$, we first need to reduce the complex segment a and / or b by computing its corresponding triple $< T_{cdu}, S_{cdu}, Val_{cdu} >$ using the rules associated to each relation which links the segments belonging to a (resp. b). Let $cdu \in C$ and let $\aleph_{cdu} = E_{cdu} \cup C_{cdu}$ be the set of nodes of the segment cdu and let $\Re_{cdu}$ be the set of relations that link elements of $\aleph_{cdu}$. The reduction process is done in a depth-first traversal of the sub-graph of cdu according to the functions $reduce(cdu)$ and $merge(cdu)$ defined below:

```
reduce(cdu){                             merge(cdu){
While (C_cdu ≠ ∅)                        Let e' ∈ ℵ_cdu
   ∀cdu' ∈ C_cdu reduce(cdu')            ∀r(e, e') ∈ ℜ_cdu and r is subordinating {
Let e ∈ ℵ_cdu a left-most node              merge(e')
return(merge(e)) }                          e = ApplyRule(r,e,e') }
                                         If (∃r(e, e') ∈ ℜ_cdu and r is coordinating {
                                            e= ApplyRule(r,e,e')}
                                         return(e) }
```

Once each CDU in C is reduced, we consider the resulting graph as a unique CDU that needs to be reduced again following the same process. The result of the FD strategy is a triple $< T_D, S_D, Val_D >$ containing the overall topic, subjectivity and score of D. Finally, $score_D$ is inferred from Val_D which is a set of scores obtained after reductions. If $|Val_D| = \emptyset$, $score_D$ is not computed, because in this case, the document does not contain relevant opinion instances (e.g. opinions on a topic of the *other* category). Otherwise, if $|Val_D| = 1$, then $score_D = Val_D$, else $score_D = \Gamma(Val_D)$ such that Γ is one of our seven functions.

Drawing on the already established effect on both subjectivity and polarity of the rhetorical relations used in the annotation campaign, we have designed 17 rules (which correspond to $ApplyRule(r, e, e')$ in the $merge$ function above). We show below the rule associated to $Contrast(e, e')$. Until now, $\forall e, e' \in E$, if $subj(e) = SE$ (resp. SI) and $subj(e') = SI$ (resp. SE) then $subj(e) = subj(e') = su$.

In addition to the very strong effect of this relation on opinion, we have also observed that this effect may depend on the syntactic order of its arguments. For instance, the overall opinion on the movie is more negative in *The idea is original, but there are some meaningless sequences* than in *There are some meaningless sequences but the idea is original*. Hence, the positivity / negativity of $Contrast(e, e')$ is determined by e'. Then, $ApplyRule(Contrast,\ e,\ e') = < T, S, Val >$ where:

– if $topic(e) = topic(e')$ then $T = topic(e)$, if $topic(e) = main$ or $topic(e') = main$ then
$T = main$ (as in *The idea is original, but the movie was bad*),
– $S = su$ if $subj(e) = su$ or $subj(e') = su$, $S = O$ otherwise.
– If $topic(e) = topic(e') = main$ or $topic(e) = topic(e') = partof$, then, if $(score(e) > score(e'))$ then $Val = score(e')$, otherwise $Val = Int^-(score(e))$. Finally, if $topic(e) = main$ then $Val = Int^- score(e)$, if $topic(e) = partof$ then $Val = Int^- score(e')$.

5   Evaluation

We have used these three strategies for each document D of our gold standard. For BOS and FD, we have first applied a subjectivity filter ($\emptyset$, *se* and *si*). Then for each subjectivity filter, we have applied a topic filter ($\emptyset$, m, p and mp) (the order of application of our two filters does not matter). Consequently, we get 12 configurations corresponding to 12 subsets $D' \subseteq D$ for the BOS and to 12 subsets $\aleph' \subseteq \aleph$ for the PD. If one of these sets is empty, or if it only contains objective segments, we proceed by relaxing some filters (see Section 4.1). For each subset, we have applied one of the seven functions described in Section 4.1. We have thus computed 84 scores per strategy. For the FD strategy, the result set Val_D can be reduced by the same set of functions, thus yielding 7 different computed scores.

We have assessed the reliability of our three strategies by comparing their results (namely $score_D$) against the score given in the gold standard. We have also compared our results against a baseline which consists in applying BOS with the subjectivity filter *se* followed by the topic filter $\emptyset$. This baseline is similar to state-of-the-art approaches in rating-inference problems [14] that aggregate the strengths of the opinion words in a review with respect to a given polarity and then assign an overall rating to the review to reflect the dominant polarity. We used two evaluation metrics: accuracy and Pearson's correlation. Accuracy corresponds to the total number of correctly classified documents divided by the total number of documents while Pearson's correlation (r) reflects the degree of linear relationship between the set of scores computed by our strategies and the set given by the gold standard. The closer r is to +1 (or to -1), the better the correlation.

We have performed two experiments: (1) *an overall polarity rating* where we consider that the overall ratings -3 to -1 represent the -1 score (i.e. negative documents) and the ratings +1 to +3 correspond to the +1 score (positive documents); and (2) *a an overall multi-scale rating* where the ratings are considered to be in the continuous interval $[-3, +3]$. Among the 84 experiments made for BOS and PD and among the 7 experiments made for FD, Tables 3 and 4 give the configuration that leads to the best results for polarity and multi-scale ratings, respectively. For a strategy s, the notation (a, b, c) indicates that the given accuracy (resp. correlation) is computed when applying to s the subjectivity filter a followed by the topic filter b and by using the function c. The results below are statistically significance since we get a p-value < 0.01 for reviews corpus and < 0.05 for news reactions.

Table 3. Overall polarity ratings in both corpus genres

	MR		NR	
	Accuracy	**Pearson**	**Accuracy**	**Pearson**
Baseline	0.89 (A)	0.81 (A)	0.88 (MSc)	0.52 (MSc)
BOS	0.92	0.87	0.94	0.77
	(∅, ∅, Fr)	(∅, ∅, A)	(∅, ∅, MSc)	(∅, ∅, MSc)
Sub1	0.91	0.84	0.92	0.74
	(∅, ∅, M)	(∅, ∅, M)	(∅, ∅, A)	(∅, ∅, A)
Sub2	0.91	0.84	0.92	0.74
	(∅, ∅, M)	(∅, ∅, M)	(∅, ∅, A)	(∅, ∅, A)
Top1	0.96	0.94	0.92	0.77
	(∅, ∅, Fr)	(∅, ∅, M)	(∅, ∅, A)	(∅, ∅, MSc)
Top2	0.90	0.80	0.96	0.82
	(∅, ∅, A)	(∅, ∅, A)	(∅, ∅, MSc)	(∅, ∅, MSc)
FD	0.90 (MSc)	0.86 (MSc_A)	0.94 (Fr)	0.82 (MSc_A)

We observe that for assessing the overall polarity, the baseline results for MR in term of accuracy (when applying the *average*) are as good as those obtained by other strategies, whereas for NR, the results are worse. In terms of Pearson's correlation, we observe that the results are quite good (the baseline beats the *Top 2* strategy when applying the *average*), whereas for NR the correlations are not good compared to other strategies. PD strategy beats the *BOS*. For instance, for MR, *Top1* outperforms *BOS* by 4% for accuracy and 8% for correlation while for NR, *Top2* is the best with more than 2% for accuracy and 5% for correlation. The *FD* strategy is less efficient in MR than in NR when comparing its results to *BOS*. This difference shows that *FD* is very sensitive to the complexity of the discourse structure. The more elaborate the discourse is, (as in NR) the better the results yielded by the rule-based approach are. In addition, for both *BOS* and *PD*, the best combination of filters consists in keeping all segments' types (the *K_all* strategy) and then keeping all the types of topics (*K_all*) (similar results were obtained when applying the other topic filters i.e. *K_M, K_P* and *K_MP*). This entails that both explicit and implicit opinions are important for computing the overall polarity, whereas using topic information does not seem to be very useful. For instance, in MR, we get, for *BOS*, an accuracy of 0.84 when applying *K_SI* with the *MaxSc* function and hence − 4 % compared to *K_SE* while for NR we get 0.93 when using the *MaxSc* function and hence + 5% over applying *K_SE*. The same holds for the Pearson's correlation. This brings us to the conclusion that the importance of implicit opinions varies, depending on the corpus genre: for movie reviews, more direct and sometimes terse, explicit opinions are better correlated to the global opinion scores, whereas for news reactions, implicit opinions are more important when negative opinions are concerned. This could indicate a tendency to conceal negative opinions as apparently objective statements, which can be related to social conventions (politeness, in particular).

For overall multi-scale ratings, the baselines results are not good compared to the other strategies. In addition, we observe that discourse-based strategies yield better results for both corpus genres. For MR, *FD* gives a significant improvement

Table 4. Overall multi-scale ratings in both corpus genres

	MR		NR	
	Accuracy	**Pearson**	**Accuracy**	**Pearson**
Baseline	0.63 (Fr)	0.84 (A)	0.60 (MSc)	0.66 (MSc)
BOS	0.63 (se, ∅, Fr)	0.91 (∅, ∅, M)	0.70 (si, ∅, MSc)	0.82 (∅, ∅, MSc)
Sub1	0.63 (∅, p, MSc)	0.90 (∅, ∅, M)	0.69 (si, ∅, MSc)	0.78 (∅, p, A)
Sub2	0.63 (∅, p, Fr)	0.90 (∅, ∅, M)	0.69 (si, ∅, MSc)	0.77 (∅, p, A)
Top1	0.63 (se, ∅, M)	<u>0.94</u> (∅, ∅, M)	0.70 (∅, ∅, MSc)	0.78 (si, ∅, MSc)
Top2	0.65 (si, ∅, MSc)	0.84 (∅, ∅, M)	0.68 (∅, ∅, MSc)	0.80 (si, ∅, MSc)
FD	<u>0.75</u>(MSc)	0.91 (Avg)	<u>0.73</u> (A)	<u>0.84</u> (A)

of 12% over the baseline in terms of accuracy while NR gets an improvement of 13%. In terms of Pearson's correlations, the best results are obtained when applying *Top1* to MR and *FD* to NR. Concerning the filters, we observe that we get different configurations than in overall polarity. Indeed, in terms of accuracy, the best results in MR are given by the K_SE followed by the topic filter K_all or the configuration K_all for subjectivity and K_P for topic for all the strategies (except for the *Top2*). Similar observations hold for NR, where we have in addition the subjectivity filter K_SI. Unlike for polarity overall ratings, the weight of implicit opinions seems to be less important for MR and more important for NR. On the other hand, taking into account *partof* topics has a stronger effect on multi-scale ratings, especially for MR. This might be because opinions focused on *partof* topics are more often used to express intensity nuances.

The discourse-based strategies (PD and FD) fail to capture the overall score in four main cases. The first one, concerns situations where the writer expresses implicit opinions towards other topics or when he is in a position of observer or recessed relative to the discussion. Second, sometimes, opinions in a document do not reflect the writer's point of view but the feelings of other persons. Hence, identifying the holder can yield an improvement. Third, ironic and sarcasm documents, where most subjective segments in a document are implicit. Finally, other cases of errors come from documents that are neither positive nor negative towards the main or a *partof* topic (about 4% of MR).

6   Conclusion

In this paper, we proposed the first research effort that empirically validates the importance of discourse for sentiment analysis. Based on a manual annotation campaign conducted on two corpus genres (movie reviews and news reactions), we have first shown that discourse has a strong effect on both polarity and subjectivity analysis. Then, we have proposed three strategies to compute document overall rating, namely bag of segments, partial discourse and full discourse.

Our results show that discourse-based strategies lead to better scores in terms of accuracy and Pearson's correlation on both corpus genres. Our results are more salient for overall scale rating than for polarity rating. In addition, this added value is more important for newspaper reactions than for movie reviews. The next step is to validate our results on automatically parsed data. We attempt to do this by adapting [15]'s parser to opinion texts.

Acknowledgement. This work was supported by a DGA-RAPID project under grant number 0102906143.

References

1. Asher, N., Lascarides, A.: Logics of Conversation. Cambridge University Press (2003)
2. Baldridge, J., Lascarides, A.: Annotating discourse structures for robust semantic interpretation. In: IWCS (2005)
3. Asher, N., Benamara, F., Mathieu, Y.Y.: Distilling opinion in discourse: A preliminary study. In: CoLing, pp. 7–10 (2008)
4. Taboada, M., Voll, K., Brooke, J.: Extracting sentiment as a function of discourse structure and topicality. School of Computing Science Technical Report 2008-20 (2008)
5. Heerschop, B., Goossen, F., Hogenboom, A., Frasincar, F., Kaymak, U., de Jong, F.: Polarity analysis of texts using discourse structure. In: Proceedings of the 20th ACM International Conference on Information and Knowledge Management, pp. 1061–1070 (2011)
6. Somasundaran, S.: Discourse-level relations for Opinion Analysis. PhD Thesis, University of Pittsburgh (2010)
7. Zhou, L., Li, B., Gao, W., Wei, Z., Wong, K.F.: Unsupervised discovery of discourse relations for eliminating intra-sentence polarity ambiguities. In: Proceedings of EMNLP, pp. 162–171 (2011)
8. Benamara, F., Chardon, B., Mathieu, Y., Popescu, V.: Towards context-based subjectivity analysis. In: Proceedings of the IJCNLP, pp. 1180–1188 (2011)
9. Polanyi, L., van den Berg, M.: Discourse structure and sentiment. In: Data Mining Workshops (ICDMW), pp. 97–102 (2011)
10. Trnavac, R., Taboada, M.: The contribution of nonveridical rhetorical relations to evaluation in discourse. Language Sciences 34 (3), 301–318 (2012)
11. Carlson, L., Marcu, D., Okurowski, M.E.: Building a discourse-tagged corpus in the framework of rhetorical structure theory. In: van Kuppevelt, J., Smith, R. (eds.) Current Directions in Discourse and Dialogue, pp. 85–112. Kluwer Academic Publishers (2003)
12. Toprak, C., Jakob, N., Gurevych, I.: Sentence and expression level annotation of opinions in user-generated discourse. In: ACL, Morristown, NJ, USA, pp. 575–584 (2010)
13. Afantenos, S., et al.: An empirical resource for discovering cognitive principles of discourse organisation: the annodis corpus. In: LREC (2012)
14. Leung, C.W.K., Chan, S.C.F., Chung, F.L., Ngai, G.: A probabilistic rating inference framework for mining user preferences from reviews. World Wide Web 14(2), 187–215 (2011)
15. Muller, P., Afantenos, S., Denis, P., Asher, N.: Constrained decoding for text-level discourse parsing. In: Proceedings of COLING (2012)

Lost in Translation: Viability of Machine Translation for Cross Language Sentiment Analysis

Balamurali A.R.[1,2], Mitesh M. Khapra[3], and Pushpak Bhattacharyya[1]

[1] Indian Institute of Technology Bombay, India
[2] IITB-Monash Research Academy, India
[3] IBM Research, India
{balamurali,pb}@cse.iitb.ac.in, mikhapra@in.ibm.com

Abstract. Recently there has been a lot of interest in Cross Language Sentiment Analysis (CLSA) using Machine Translation (MT) to facilitate Sentiment Analysis in resource deprived languages. The idea is to use the annotated resources of one language (say, L_1) for performing Sentiment Analysis in another language (say, L_2) which does not have annotated resources. The success of such a scheme crucially depends on the availability of a MT system between L_1 and L_2. We argue that such a strategy ignores the fact that a Machine Translation system is much more demanding in terms of resources than a Sentiment Analysis engine. Moreover, these approaches fail to take into account the divergence in the expression of sentiments across languages. We provide strong experimental evidence to prove that even the best of such systems do not outperform a system trained using only a few polarity annotated documents in the target language. Having a very large number of documents in L_1 also does not help because most Machine Learning approaches converge (or reach a plateau) after a certain training size (as demonstrated by our results). Based on our study, we take the stand that languages which have a genuine need for a Sentiment Analysis engine should focus on collecting a few polarity annotated documents in their language instead of relying on CLSA.

1 Introduction

In these times of multilingual information processing, there is a keen interest in bringing NLP capability to resource deprived languages by leveraging the resources of a rich language. This is true in the case of Sentiment Analysis (SA) also, where, polarity annotated documents in one language are used for building a SA engine for another language through the instrument of Machine Translation [1]. This task is known as Cross Language Sentiment Analysis (CLSA) wherein the following steps are commonly observed:

1. The polarity marked documents of a resource rich language L_1 are translated to L_2
2. An SA machine M is trained on these translated documents
3. M is then applied to a test document D of language L_2 to detect its polarity

Another alternative is to (i) train a SA machine M for the resource rich language L_1 (ii) given a document D in L_2, first translate it to L_1 and (iii) apply M to this translated D

A. Gelbukh (Ed.): CICLing 2013, Part II, LNCS 7817, pp. 38–49, 2013.

to detect its polarity. However, the first alternative is better because it does not involve any translation at test time and hence has lesser test-time complexity and cost (it just has a fixed training time cost).

We claim with quantitative analysis that MT based CLSA at document level is fundamentally not a sound idea. One will instead do better by investing in creating *direct* resources for sentiment analysis. More explicitly, we say that *"if you want to do sentiment analysis in your language and have a limited amount of money, spend the money in creating polarity marked documents for your language, instead of using MT and then doing CLSA"*.

Our focus is on document level SA wherein documents are classified into polarity classes (positive and negative) [2]. It is obvious that a case for developing sentiment analysis engine exists for a given language, if many polar documents (e.g., product or movie reviews) are available in electronic form in that language. Given such documents, the effort in annotating them with correct polarity is very little, especially compared to the effort in building an MT system needed for CLSA. For example, it is possible for a single lexicographer to annotate 500 reviews with correct polarity using minimal effort[1]. Our experiments suggest that 500 polarity annotated reviews are sufficient for building a good SA engine for a language (see section 5). Any additional document produces very marginal gain- *the proverbial case of saturation* (see Figure 12 which shows that this happens for three different languages).

Given that the effort involved in collecting polarity annotated documents is quite small, the next question is of performance. We define this performance of a SA engine in terms of its sentiment classification accuracy. Our experiments involving 4 languages suggest that the performance of a SA engine trained using *in-language* polarity annotated documents is better than that obtained using CLSA (see section 5). This is not contrary to intuition, and the reasons are not far to seek:

1. Training a sentiment analysis engine on the own-language corpus ensures that divergences due to cultural differences between two languages are minimal.
2. MT systems are not very accurate and as a result there is always noise in the polarity annotated documents translated from the source language.

We substantiate our arguments by extensive evaluation of well-established CLSA techniques (described in section 3) for four languages, *viz., English, French, German and Russian*.

The remainder of this paper is organized as follows. In section 2, we discuss related work on CLSA. In section 3 we present the CLSA approaches employed in our work. Section 4 describes the experimental setup and datasets used for evaluation. In section 5 we present the results, followed by discussions in section 6. Section 7 concludes the paper.

---

[1] For instance, the authors of this paper were able to annotate 50 reviews with their correct polarity in 1 hour. It would thus take 10 hours to annotate 500 documents with their respective polarity labels. Compare this effort with the effort required to collect or generate parallel corpora for creating an SMT system, which is much larger.

2   Related Work

To reduce the need of developing annotated resources for SA in multiple languages, cross-lingual approaches [3–6] have been proposed. To use the model trained on L_1 on the test data from L_2, a Machine Translation (MT) system or a bilingual dictionary is used for transfer between the two languages.

In [6], a cross-lingual approach based on Structured Correspondence Learning (SCL) was proposed, which aims at eliminating the noise introduced due to faulty translations by finding a common low dimensional representation shared by the two languages. In [7], lexicon based and supervised approaches for cross language sentiment classification are compared. Their results show that lexicon based approaches perform better. In [3] and [4], cross-lingual methods which exploit existing tools and resources in English to perform subjectivity analysis in Romanian are proposed.

The state of the art in CLSA is an approach based on co-training. For example, in [5] labeled English data and unlabeled Chinese data was used to perform sentiment classification in Chinese. Here, the English features and the Chinese features are considered as two different views of the same document (one view is formed by English features and the other view is formed by Chinese features extracted after translating the document). Two classifiers are trained using these two views, and each classifier is then applied to the unlabeled Chinese data. The instances which get tagged with high confidence by both the classifiers are then added to the initial training data. Note that the approach requires two MT systems ($L_1 \rightarrow L_2$ and $L_2 \rightarrow L_1$).

Most, if not all, of the above methods advocate that even a low quality translation engine is adequate for performing CLSA. Our experiments involving 4 languages and 24 combinations of source-target pairs suggest that this argument is not correct. Further, we believe that it is hard to capture sentiment in a language using documents in another language, because of the disparate ways in which sentiments are expressed across languages, a result of cultural diversity amongst different languages. A good example, which we found in our data is that English users use the word '*suck*' frequently to express negative opinion (as in 'This X sucks' where X could refer to a movie, actor, director, *etc*). However, the translation of 'This X sucks' (which contains the French word *suce/sucer/succion*) was never seen in the French corpus. This suggests that French speakers do not use the equivalent of 'This X sucks' to express negative sentiment. Hence, training an English SA by translating training documents from French would most likely not work on an English documents if the word '*sucks*' is the only negative sentiment bearing word in the document.

3   CLSA Techniques We Use

Depending on the available tools and resources, (*viz.*, annotated corpus in L_1, MT between L_1 and L_2, bilingual dictionary, unannotated corpus in L_2, we discuss four established methods [3–5] of performing document level CLSA.

1. Resource rich L_1 helps resource disadvantaged L_2 using MT (*MT-X*): Build a Sentiment Analysis system for L_2 by leveraging the annotated resources of L_1 and a Machine Translation (MT) system from L_1 to L_2. The approach is outlined in Algorithm 1:

Algorithm 1. *MT-X*

$LD_1 :=$ Polarity annotated data from L_1

$LD'_2 := translateUsingMT_{L_1}^{L_2}(LD_1)$

$\phi :=$ model trained using LD'_2

$test(\phi, testDocument_{L_2})$

MT-X stands for "a resource rich language $\underline{X}$ assists a target language using $\underline{MT}$".

2. Resource rich language helps a resource disadvantaged language using a bilingual dictionary (*BD-X*): Here, the aim is same as above, but instead of using a MT system, a bilingual dictionary (BD)[2] is used for translating polarity annotated documents from L_1 to L_2. This method thus caters to situations where a MT system is not available for a language pair. Every word in an L_1 document is replaced by its translation in L_2 as found in a bilingual dictionary. The approach is outlined in Algorithm 2:

Algorithm 2. *BD-X*

$LD_1 :=$ Polarity annotated data from L_1

$LD'_2 := translateUsingBiDict_{L_1}^{L_2}(LD_1)$

$\phi :=$ model trained using LD'_2

$test(\phi, testDocument_{L_2})$

BD-X stands for "a resource rich language $\underline{X}$ assists a target language using a bilingual dictionary $(\underline{BD})$".

3. Multiple resource rich languages assist a resource deprived language using MT (*MMT-X*):

Here, instead of using the labeled data available in one language, we use the labeled data available in multiple resource rich languages to help a resource deprived language. *MMT-XYZ* stands for "$\underline{M}$ultiple resource rich languages $\underline{X}$, $\underline{Y}$ and $\underline{Z}$ assist a target language using Machine Translation ($\underline{MT}$)".

4. Co-Training (*CoTr-X*): Here, a co-training based approach is used which harnesses the unlabeled data in L_2. The steps involved in this algorithm are as follows:

---

[2] BD is created by taking all the unique words present in the resource disadvantaged language and translating them at word-level to resource rich language using Microsoft's online translation services (http://www.bing.com/translator).

Algorithm 3. *MMT-XYZ*

LD'_2 := empty

n := number of assisting languages ($n > 1$)

for $i = 1 \rightarrow n$ **do**

    LD_i := Polarity annotated data from L_i

    $LD'_2 := LD'_2 + translateUsingMT^{L_2}_{L_i}(LD_i)$

end for

ϕ := model trained using LD'_2

$test(\phi, testDocument_{L_2})$

Training

- **Step 1**: Translate annotated data (LD_1) from L_1 to L_2 (LD'_2) using an MT system.
- **Step 2**: Translate unannotated data (UD_2) from L_2 to L_1 (UD'_1) using an MT system.
- **Step 3**: Train models θ_1 and θ_2 using LD_1 and LD'_2 respectively.
- **Step 4**: Use θ_1 and θ_2 to label the reviews in UD'_1 and UD_2 respectively.
- **Step 5**: Find p positive and n negative reviews from UD'_1 which were labeled with the highest confidence by θ_1. Add these to LD_1 and add their translations to LD'_2.
- **Step 5**: Find p positive and n negative reviews from UD_2 which were labeled with the highest confidence by θ_2. Add these to LD'_2 and add their translations to LD_1.
- **Step 6**: Repeat Steps 1 to 5 for i iterations.

Testing

- **Step 7**: Test data from from L_2 using θ_2.

The basic idea here is to treat LD_1 and LD'_2 as two different views of the same data. The unlabeled instances which are classified with a high confidence by a classifier trained on one view can then help to improve the classifier trained on the other view. Note that *CoTr-X* stands for "a resource rich language $\underline{X}$ assists a target language using *Co-Tr*aining." Two MT systems ($L_1 \rightarrow L_2$ and $L_2 \rightarrow L_1$) are needed for this approach thus making it heavily dependent on MT systems.

4   Experimental Setup

We performed an extensive evaluation using four languages, *viz., English, French, German and Russian*. We downloaded movie reviews for English, French and German from IMDB[3]. The reviews for these languages were downloaded separately and randomly. Reviews with rating greater than 7 (out of 10) were labeled as positive. and those with the rating of less than 3 were labeled as negative. We ignored reviews having ratings between 3-7 as we found them to be ambiguous. For Russian, since we did not find enough movie review data, we focused on book reviews [8], a domain closely related to movie reviews[4].

---

[3] http://www.imdb.com, http://www.imdb.fr, http://www.imdb.de

[4] This gave us chance to study cross domain CLSA.

We collected 3000 positive and 3000 negative reviews for English, French and German and 500 positive and 500 negative reviews for Russian. The data in each language was translated to all of the other 3 languages using the Bing[5] translation service. We did not use Google translate because the APIs are no longer freely available. Even though we collected upto 3000 positive and 3000 negative reviews, we found that in almost all cases the performance showed saturation after 400 documents.

We report CLSA results by increasing the training documents in the *source* language L_1 from 50 to 400 in steps of 50. The number of test documents in each language were 200 (*i.e.*, 100 positive and 100 negative reviews). Further, to ensure that our results are not biased to a particular training set and test set we created 10 different sets of 400 positive and negative reviews in each language as well as 10 different sets of 100 positive and negative reviews in each language. Training set 1 in L_1 was then used to perform CLSA on test set 1 in L_2. We repeated this procedure with all the 10 sets and reported the average accuracy obtained over the 10 sets (similar to 10 fold cross validation albeit in a cross language setting).

We used SVM as the classifier because it is known to give the best results for sentiment classification [2]. Specifically, we used C-SVM (linear kernel with parameters optimized over training set using 5 fold cross validation) available as a part of the Lib-SVM[6] package. The feature set comprises of unigrams extracted from the seed labeled data. We also experimented with bigram features but did not find much difference in the performance. Further, using higher n-grams features would be unfair to the CLSA systems because most existing MT systems do not produce translations having a good syntactic structure. Hence, we stick to unigram features in this work.

5   Results

The results of our experiments are presented in Figures 1 to 12. Figure 1 compares the performance of *MT-X, BD-X* and *MMT-XYZ* using different source languages and English as the target language. Figures 2, 3 and 4 present the same comparison with French, German and Russian as the target language. Next, we also wanted to see if one or more resource rich languages can help in improving the performance of another resource rich language (as opposed to assisting a resource poor language). To test this we used k polarity annotated documents from the target language and added k polarity annotated documents each translated from one or more source languages. These results are presented in Figures 5 to 8. For ease of understanding and representation, we report the overall accuracy over both positive and negative test documents. In all the graphs, we use the following language codes for representing languages: **En**→*English*, **Fr**→*French*, **Ge**→*German* and **Ru**→*Russian*. Along the X-axis, we represent the number of documents used for training and along the Y-axis we represent the accuracy. To help the reader in interpreting the graphs we explain the different curves in Figure 1 and Figure 5 with English as the target language. The curves in the other graphs can be interpreted similarly.

---

[5] http://www.microsofttranslator.com/

[6] http://www.csie.ntu.edu.tw/~cjlin/libsvm

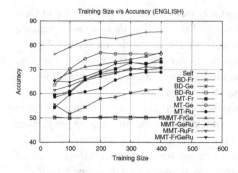

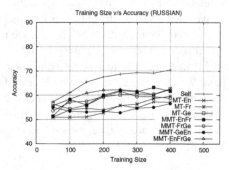

Fig. 1. Comparing the performance of different algorithms with English as the target language

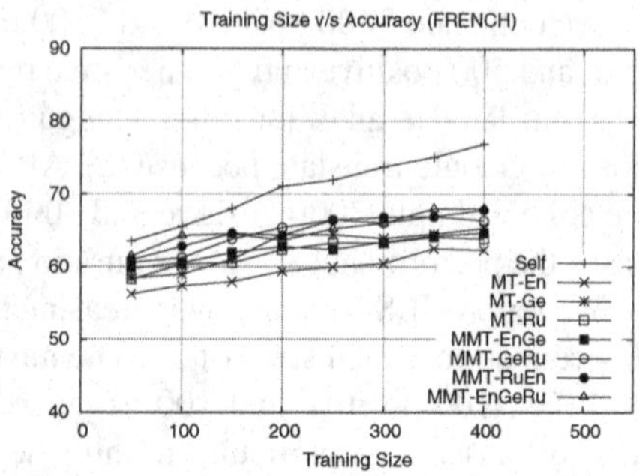

Fig. 2. Comparing the performance of different algorithms with French as the target language

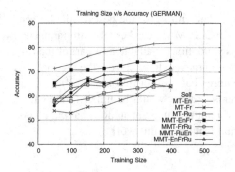

Fig. 3. Comparing the performance of different algorithms with German as the target language

Fig. 4. Comparing the performance of different algorithms with Russian as the target language

– *Self* : The accuracy obtained by training a sentiment analysis engine using polarity annotated documents in the target language itself.
– *MT-Fr* : The accuracy obtained by training a sentiment analysis engine using the polarity annotated documents translated from French (Fr) to English using MT.
– *BD-Fr* : The accuracy obtained by training a sentiment analysis engine using the polarity annotated documents translated from French (Fr) to English using a bilingual dictionary.
– *MMT-FrGe* : The accuracy obtained by training a sentiment analysis engine using the polarity annotated documents translated from French (Fr) and German (Ge) to English using MT.
– *MMT-FrGeRu* : The accuracy obtained by training a sentiment analysis engine using the polarity annotated documents translated from French (Fr), German (Ge) and Russian (Ru) to English using MT.
– *Self + MMT-FrGeRu* : This curve in Figure 5-8 plots the accuracy obtained by training a sentiment analysis engine using the polarity annotated documents in English

plus the polarity annotated documents translated from French (Fr), German (Ge) and Russian (Ru) to English using a MT system.

- *CoTr-Fr* : This curve in Figure 9 plots the accuracy obtained by training a sentiment analysis engine using the Co-Training approach which uses the polarity annotated documents in French plus the unannotated documents in English.

6 Discussions

In this section, we discuss some important observations made from our evaluation.

1. *In-language* sentiment analysis clearly outperforms *cross language* sentiment analysis: We first compare the performance of *MT-X* and *BD-X* with *Self*. In all the graphs (see Figures 1 to 4), the curve of *MT-X* and *BD-X* is much below the curve of *Self*. Specifically, if we compare the performance obtained by using 400 (positive and negative) in-language documents (*i.e., Self*) with that obtained using 400 (positive and negative) cross-language documents, the performance of *Self* is better than *MT-X* by 8-10%. The same difference between *Self* and *BD-X* is much higher. The poor results for *BD-X* suggest that a strategy that simply uses word based translations and ignores the syntactic and semantic structure performs poorly. Thus, the argument that even a very low quality translation engine which ignores syntactic and semantic structure suffices for cross language sentiment analysis does not seem to hold true.

Next, we wanted to see if using data from multiple assisting languages as opposed to a single assisting language can help. The intuition was that taking training examples from multiple languages would increase the diversity in the collection and perhaps be a better strategy for cross language sentiment analysis. However, the results here are not consistent. In some cases, using cross-language data from multiple assisting languages, performs better than taking data from a single assisting language while in other cases it does not. For example, in Figure 1 taking a total of 400 documents from French, German and Russian (*MMT-FrGeRu*) performs better than individually using 400 documents from French or Russian(*MT-Fr, MT-Ru*). On the other hand, *MT-Ru* performs better than *MMT-FrGeRu*. However, for all the target languages, the results are in agreement with the stand taken in this paper, *i.e.*, the performance of cross language sentiment analysis using single/multiple assisting language/languages is lower when compared to in-language sentiment analysis.

2. Does having unannotated data in the target language help?
We wanted to check the importance of unannotated data in the target language. Over all Co-Training seems to be the best CLSA technique, but, in general, it still does not outperform in-language sentiment analysis(Figure 9-11). Specifically, at small training sizes (50, 100), Co-Training does better than in language sentiment analysis but as the training size increases in-language Sentiment Analysis performs better than CLSA. These results contradict previously made claims that CLSA using Co-Training clearly outperforms in-language SA. Further, it should be noted that Co-Training requires (1) two MT systems and (2) untagged corpus in L_2. As mentioned earlier, if untagged documents are already available in L_2 then the effort involved in annotating them is much

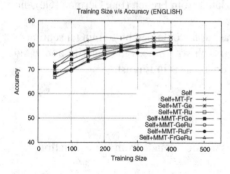

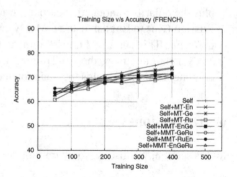

Fig. 5. Comparing the performance of different algorithms with English as the target language when self training data in English is also available

Fig. 6. Comparing the performance of different algorithms with Russian as the target language when self training data in Russian is also available

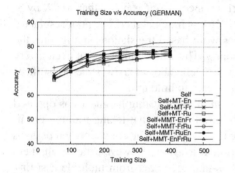

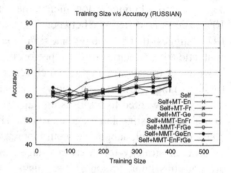

Fig. 7. Comparing the performance of different algorithms with German as the target language when self training data in German is also available

Fig. 8. Comparing the performance of different algorithms with Russian as the target language when self training data in Russian is also available

less than the effort involved in building two MT systems.

3. Additional data from other languages does not improve the performance of in-language sentiment analysis: Figures 5 to 8 suggest that in the presence of annotated data in the target language, adding additional data from other languages harms the performance. For all the target languages, the performance of *Self* is always better than *Self + MT-X* or *Self + MMT-XYZ*. There could be two possible reasons why the additional

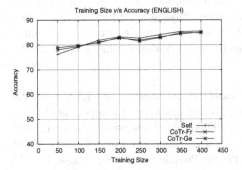

Fig. 9. Comparing the performance of *CoTr-X* and *Self* with English as the target language

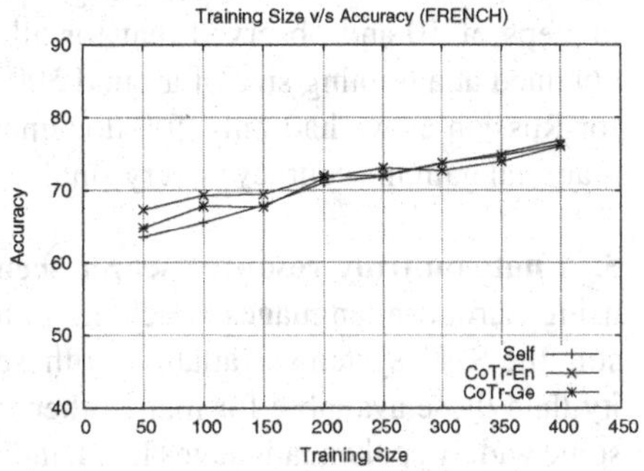

Fig. 10. Comparing the performance of *CoTr-X* and *Self* with French as the target language

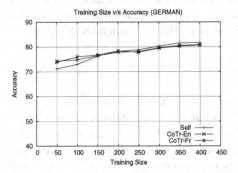

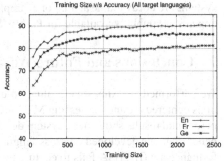

Fig. 11. Comparing the performance of *CoTr-X* and *Self* with German as the target language

Fig. 12. Effect of training size on accuracy (beyond 500 documents there is very little improvement in the accuracy

training data from other languages harms the performance. Firstly, the translations obtained using the MT system maybe erroneous and thereby add noise to the training process. One reason of for this is the incorrect spellings present in the reviews which can affect the translation quality but probably may not affect the self training because the same incorrect spellings may be present in the test set. Secondly, there might be cultural differences in the manner in which sentiment is expressed in different languages. For example, in some languages double negation is a common phenomenon. A unigram feature based Cross Language Sentiment Analysis trained in a language where such a phenomenon is rare may harm the classification accuracy. These differences again make the training data noisy leading to poor learning and consequently poor performance.

4. How much in-language data does one really need?

The answer to this question is important for making an informed choice regarding the number of documents needed to get a reasonably good accuracy in a language. Specifically, we are interested in the number of documents beyond which the marginal gain in accuracy is negligible. To do so, we plotted the accuracies obtained using increasing amounts of data in the target language. We varied the training data size from 50 to 2500 in steps of 50 and observed that for all the three languages the knee of the curve is obtained at a training size of around 500 documents (we could not run this experiment for Russian as we had only 400 documents in Russian). Beyond this training size the marginal gain in accuracy is very small.

5. A note on truly resource scarce scenarios: Our experiments on CLSA were done using European languages which are politically and commercially important. As a result, the SMT systems available for these languages are of comparatively higher quality than those available for many other widely used languages. For example, consider some widely spoken languages like Hindi, Pashto, Punjabi, Sundanese, Hausa, Marathi, Gujarati, *etc*, which have a native speaker base of more than 25 million people[7]. Good quality translation engines are not available for these languages. The results obtained for European languages which have good MT systems suggests that such CLSA systems have very less hope in truly resource deprived scenarios. Further for many languages MT systems are not available at all. For example, to the best of our knowledge, *no* translation engines are publicly available for Pashto, Sundanese, Hausa, Marathi and Punjabi.

7   Conclusions and Future Work

We performed an exhaustive evaluation using four languages and different configurations centered around harnessing MT for Cross Language Sentiment Analysis. Our experimental results show that a system developed using *in-language* data performs much better than one developed on cross-language data. Two main reasons for the better performance are (i) CLSA fails to capture the cultural divergence between languages with respect to expression of sentiments and (ii) MT systems are not very accurate and hence introduce noise in the training data. Further, our study falsifies the claim that a crude translation using bilingual dictionary suffices to perform SA in the target language. We also observed that in the presence of training data in a language, adding additional data from other languages actually harms the performance. We would like to emphasize that our experiments were performed on languages which are commercially dominant and hence have much better MT systems than a host of other languages. The poor performance of CLSA in the presence of such better quality MT systems gives rise to the following question: *if there is a genuine interest in developing sentiment analysis engines for these languages then isn't it wiser to invest in collecting polarity annotated documents than to rely on a MT system which is much more complex and hard to obtain?*

References

1. Brown, P.F., Pietra, V.J.D., Pietra, S.A.D., Mercer, R.L.: The mathematics of statistical machine translation: parameter estimation. Comput. Linguist. 19, 263–311 (1993)
2. Pang, B., Lee, L.: Thumbs up? sentiment classification using machine learning techniques. In: Proceedings of EMNLP 2002, pp. 79–86 (2002)

---

[7] http://en.wikipedia.org/wiki/List_of_languages_by_number_of_native_speakers

3. Mihalcea, R., Banea, C., Wiebe, J.: Learning multilingual subjective language via cross-lingual projections. In: ACL (2007)
4. Banea, C., Mihalcea, R., Wiebe, J., Hassan, S.: Multilingual subjectivity analysis using machine translation. In: EMNLP, pp. 127–135 (2008)
5. Wan, X.: Co-training for cross-lingual sentiment classification. In: ACL/AFNLP, pp. 235–243 (2009)
6. Wei, B., Pal, C.: Cross lingual adaptation: an experiment on sentiment classifications. In: Proceedings of the ACL 2010 Conference Short Papers, ACLShort 2010, pp. 258–262. Association for Computational Linguistics, Stroudsburg (2010)
7. Brooke, J., Tofiloski, M., Taboada, M.: Cross-linguistic sentiment analysis: From english to spanish. In: International Conference on Recent Advances in NLP (2009)
8. Zagibalov, T.: Unsupervised and knowledge-poor approaches to sentiment analysis. PhD thesis, University of Sussex (2010)

An Enhanced Semantic Tree Kernel
for Sentiment Polarity Classification

Luis A. Trindade, Hui Wang, William Blackburn, and Niall Rooney

University of Ulster, School of Computing and Mathematics,
Faculty of Computing and Engineering
Trindade-L@email.ulster.ac.uk,
{H.wang,Wt.blackburn,Nf.rooney}@ulster.ac.uk

Abstract. Sentiment analysis has gained a lot of attention in recent years, mainly due to the many practical applications it supports and a growing demand for
such applications. This growing demand is supported by an increasing amount
and availability of opinionated online information, mainly due to the proliferation and popularity of social media. The majority of work in sentiment analysis
considers the polarity of word terms rather than the polarity of specific senses
of the word in context. However there has been an increased effort in distinguishing between different senses of a word as well as their different opinion-
related properties. Syntactic parse trees are a widely used natural language
processing construct that has been effectively employed for text classification
tasks. This paper proposes a novel methodology for extending syntactic parse
trees, based on word sense disambiguation and context specific opinion-related
features. We evaluate the methodology on three publicly available corpuses, by
employing the sub-set tree kernel as a similarity function in a support vector
machine. We also evaluate the effectiveness of several publicly available sense
specific sentiment lexicons. Experimental results show that all our extended
parse tree representations surpass the baseline performance for every measure
and across all corpuses, and compared well to other state-of-the-art techniques.

Keywords: Information Retrieval, Social Media, Sentiment Analysis, Opinion
Mining, Polarity Classification, Kernel Methods, Word Sense Disambiguation.

1   Introduction

Text consists of either facts or opinions. Facts are objective descriptions of entities,
events and their properties; opinions are subjective expressions of people's sentiments, appraisals or feelings toward entities, events and their properties [11]. Determining the opinion contained within a piece of text is the aim of *sentiment analysis*
(or *opinion mining*), which is assisted by techniques drawn from *natural language
processing* (NLP), *information retrieval* (IR) and *computational linguistics* (CL).

Sentiment analysis has gained a lot of attention in recent years. This is mainly due
to the many practical applications it supports. Examples include: helping companies
and organizations find customer opinions of commercial products or services; tracking opinions in online forums, blogs and social networks; and helping individuals
decide on which product to buy or which movie to watch.

A. Gelbukh (Ed.): CICLing 2013, Part II, LNCS 7817, pp. 50–62, 2013.
© Springer-Verlag Berlin Heidelberg 2013

This growing demand for automated sentiment analysis is supported by an increasing amount and availability of opinionated information online, mainly due to the proliferation of social media websites [11], [18]. Some of the most common tasks in sentiment analysis include: *subjectivity classification* [16]; *polarity classification* [16]; *polarity intensity classification* [17]; *feature/aspect-based sentiment analysis* [10]. These tasks can also be performed in combination, for example, one can start by classifying expressions as being either objective or subjective in nature; expressions classified as subjective can then be further classified as neutral or polar; and finally polar expressions can be classified as either positive, negative or both. Moreover polarity classification can be performed at various levels, for example: *word-level*, *phrase-level*, *sentence-level* and *document-level*. Note that classifying the sentiment of documents is a very different task from recognizing the *contextual polarity* of words and phrases, for instance, when working at the sentence level (or sub-sentence level) there is very little contextual information.

Polarity classification is commonly considered a binary text classification task, amounting to the classification of the polarity of a given piece of text as either positive or negative. *Support vector machine* (SVM) is a popular kernel method for text classification tasks [22]. Kernel methods are based on the use of a kernel function, which allows the mapping of data from the original data space into a higher dimensional feature space. The comparison of data can be done by computing the inner product in the high dimensional feature space, albeit implicitly through the so-called *kernel trick*. The choice of kernel function depends on the application and since this mapping (from data space to high dimensional feature space) is very general, kernel methods can be applied to complex structured objects such as sequences, images, graphs and textual documents [23]. This makes them well suited for structured NLP [25] and they have been applied to various tasks such as Question Answering, Summarization and Recognizing Textual Entailment. This paper focuses on *tree kernels* (TK) and explores their use for sentence (and phrase) level sentiment classification tasks. TK measure the similarity between two parse trees by aggregating the frequency of their matching sub-structures (for example in terms of subset trees or subtrees). A common approach is to consider the syntactic or dependency parse trees of two pieces of text. Advantages in the use of kernel approaches to natural language based classification, include the avoidance of complex feature engineering.

Despite recent efforts [2], [4], [5], [6], [12], [21] the majority of work in sentiment analysis still considers the polarity of word terms rather than the polarity of specific senses of the word. It is clear that different senses of a word can have different opinion-related properties, for example, the verb "kill" can mean a source of pain (e.g. these new shoes are killing me) but it can also mean overwhelm with hilarity, pleasure, or admiration (e.g. "*the comedian was so funny, he was killing me*"). This paper explores a range of features based on *word sense disambiguation* (WSD) and *sentiment lexicons* with sense specific opinion-related properties. We make use of those features to augment the syntactic parse trees used by the TKs and make them more efficient for sentiment polarity classification tasks. The features we consider are the WordNet [13] senses (defined as a concatenation of the word's lemma, its reduced part of speech (POS) tag and its sense number, see section 3.3) and their contextual polarity (processed for negation). We evaluate our extended parse tree representations on a binary text classification task, the determination of sentence level polarity for various corpuses. Our methodology surpasses the baseline performance for every measure and across all corpuses. To the best of our knowledge no previous study has considered the extensions to parse trees in the way that we do.

The rest of this paper is structured as follows. Section 2 gives a brief introduction to tree kernels and the trees and substructures they make use of. Section 3 describes the methodology as well as the text classification task and experimental setting considered for evaluation. Section 4 reports the experimental results. Section 5 concludes this paper with a discussion of the results and possible future work.

2   Tree Kernels

The main underlying idea of tree kernels is to compute the number of common substructures (fragments) between two trees, for example parse trees. These are usually constructed according to either the constituency parse tree or a dependency parse tree or graph. For the purposes of this paper we consider constituent syntactic parse trees. In constituent syntactic parse trees each non leaf node and its children are associated with a grammar production rule, where the symbol on the left-hand side corresponds to the parent node and the symbols on right-hand side are associated with its children (e.g. NP => DT JJ NN). These trees make the distinction between terminal and non-terminal nodes. The interior nodes are labelled by non-terminal categories of the grammar, while the leaf nodes are labelled by terminal categories. For example, Figure 1 illustrates the syntactic parse tree of an example sentence "This is not a bad movie ".

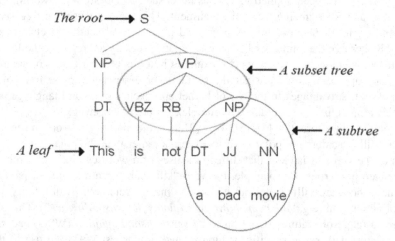

Fig. 1. Syntactic parse tree of an example sentence ("This is not a bad movie")

2.1   Substructures

This paper considers two types of parse tree substructures, the subtrees (STs) and the subset trees (SSTs). A ST is defined as any node of a tree along with all its descendants. For example, the ST rooted in the NP node, which is circled in Figure 1. A SST is a more general structure where the leaves can be associated with non-terminal symbols. The SSTs satisfy the constraint that they follow the same grammatical rules set which generated the original tree. For example, [VP [VBZ RB NP]] is a SST of the tree in Figure 1 which has three non-terminal symbols, VBZ, RB and NP, as leaves.

Given a syntactic tree we can use the set of all its STs or SSTs as a feature representation. For instance, in the example sentence ("This is not a bad movie") there are ten STs but there are hundreds of SSTs. This substantial difference in the number of substructures between the two tree-based representations, indicates a difference in the level of information these substructures convey.

2.2   The Tree Kernel Function

The main idea of tree kernels is to compute the number of the common substructures between two trees T_1 and T_2 without explicitly considering the whole fragment space. For this purpose, Moschitti [15], slightly modified the kernel function proposed by Collins & Duffy [8] by introducing a parameter σ which enables the evaluation of the subtree kernel (STK) or the subset tree kernel (SSTK). Given the set of fragments $F = \{f_1, f_2, \ldots, f_{|F|}\}$, the indicator function $x_i(n)$ is equal 1 if the target f_i is rooted at node n and 0 otherwise. Let the tree kernel function TK be defined as:

$$TK(T_1, T_2) = \sum_{n_1 \in N_{T_1}} \sum_{n_2 \in N_{T_2}} \Delta(n_1, n_2) \tag{1}$$

where N_{T_1} and N_{T_2} are the sets of the T_1's and T_2's nodes, respectively and $\Delta(n_1, n_2) = \sum_{i=1}^{|F|} x_i(n_1) x_i(n_2)$. This latter is equal to the number of common fragments rooted in the n_1 and n_2 nodes. Δ can be computed as follows:

1. If the productions at n_1 and n_2 are different then $\Delta(n_1, n_2) = 0$;
2. If the productions at n_1 and n_2 are the same, and n_1 and n_2 have only leaf children (meaning they are pre-terminals symbols) then $\Delta(n_1, n_2) = 1$;
3. If the productions at n_1 and n_2 are the same, and n_1 and n_2 are not pre-terminals then:

$$\Delta(n_1, n_2) = \prod_{j=1}^{nc(n_1)} (\sigma + \Delta(c_{n_1}^j, c_{n_2}^j)) \tag{2}$$

where $\sigma \in \{0,1\}$, $nc(n_1)$ is the number of the children of n_1 and c_n^j is the j-th child of the node n. Note that, since the productions are the same, $nc(n_1) = nc(n_2)$.

When σ is equal to 0, $\Delta(n_1, n_2)$ is equal to 1 only if $\forall j \, \Delta(c_{n_1}^j, c_{n_2}^j) = 1$, meaning that all the productions associated with the children are identical. From the recursive application of this property, it follows that the subtrees in n_1 and n_2 are identical. Thus, equation 1 evaluates the STK when $\sigma = 0$. When σ is equal to 1, $\Delta(n_1, n_2)$ evaluates the number of SSTs common to n_1 and n_2 as proved in Collins and Duffy [8].

The computational complexity of $TK(T_1, T_2)$ is $O(|N_{T1}| \times |N_{T2}|)$. Although this basic implementation has quadratic complexity, this scenario is quite unlikely for the syntactic trees of natural language sentences, as Collins & Duffy [8] noted. In practice it is possible to design algorithms that run in linear time on average [15]. Moschitti has implemented these algorithms, effectively encoding the STK and SSTK in a popular SVM (SVM-light), and made them freely available online (http://disi.unitn.it/moschitti/Tree-Kernel.htm).

3   Methodology

This paper presents a novel methodology for enriching (syntactic) parse trees with WSD and sense specific opinion-related properties, in order to improve the effectiveness of TKs for polarity classification tasks. It explores a range of features, used separately or in combination, to extend the leaf nodes (words) of syntactic parse trees with the corresponding WordNet senses and/or their contextual polarity (processed for negation). Note that this makes the extended features the new leaf nodes in the parse tree. For a visual interpretation see Figure 2 below.

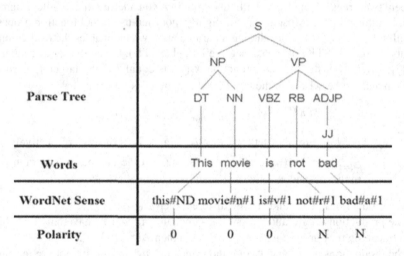

Fig. 2. Example of a parse tree extended with WordNet senses and polarity

This allows the kernel to not only match the surface words (at the leaf node) but also senses of the words as well as the polarity of the word senses. The idea is that having a set of features that is tailored for the task will increase the overall performance.

3.1   Sentence/Phrase Level Sentiment Polarity Corpuses

To evaluate our approach we compare our extended trees with the plain syntactic parse trees on a binary text classification task, the determination of whether a sentence/phrase expresses a positive or negative sentiment. We conduct a series of

10-fold cross-validation tests on three publicly available corpuses from different do-mains.,namely:

- *Movie Reviews corpus* (sentence polarity dataset v1.0) [17] – This corpus contains 5331 positive and 5331 negative processed sentences/snippets taken from several movie reviews.
- *SemEval-2007 Affective Task corpus* [24] – This corpus contains 1000 positive and 1000 negative news headlines, extracted from news web sites (such as Google news and CNN) and/or newspapers.
- *Mixed Product Reviews* [26] – This corpus contains 923 positive and 1320 negative sentences. These sentences are extracted from 294 product reviews from various online sources, manually annotated with sentence level sentiment.

3.2   Word Sense Disambiguation

We start by obtaining the syntactic parse trees for each sentence/phrase in the corpuses using the Stanford CoreNLP package (nlp.stanford.edu/software/corenlp.shtml). We then perform WSD with a WordNet-based method (WordNet::SenseRelate::AllWords [20]) in order to obtain the WordNet sense corresponding to the words in the corpuses. We choose the same combination of parameters that achieved the best result reported in [20], using the Lesk measure [19] as the similarity function, which tends to result in much higher recall, (since it is able to measure the similarity between words with any POS); and a window size of 15 (the number of words, to be taken into consideration when performing the WSD). In order to increase the compatibility of the sentences in the corpuses with WordNet::SenseRelate::AllWords, we replace contracted expressions with their full version (e.g. "won't" replaced with "will not").

3.3   Sentiment Lexicons

Despite recent efforts, most work still makes use of the words' prior polarity in order to classify the polarity of sentences or documents. Often overlooking the fact that the polarity of a word depends on the context in which it is expressed [28]. In order to address this issue this paper makes use of several WordNet-based sentiment lexicons that take into account the polarity of particular senses of the words. The lexicons in question are Micro-WNOp [7], Q-WordNet [1] and SentiWordNet [3], [9].

   In SentiWordNet and Micro-WNOp each WordNet synset is associated polarity scores (ranging from 0 to 1) that describe how positive and negative the senses are. This paper instead assigns each WordNet sense a value based on an aggregated score (A-score = P-score – N-score) similar to the approach taken by Agerri et al. [1]. Namely assigning a:

- **P** to positive senses (A-score > 0) – e.g. true#a#2 which has a P-score of 1 and a N-score of 0;
- **N** to negative senses (A-score < 0) – e.g. cynical#a#1 which has a P-score of 0 and a N-score of 1; and
- **O** to objective and neutral senses (A-score = 0) – e.g. real#a#7 which has a P-score of 0 and a N-score of 0.

We also consider an alternative representation by assigning a **B** for senses that can have both polarities (A-score = 0, P-score ≠ 0, N-score ≠ 0, and P-score = N-score) – e.g. literal#a#1 which has a P-score of 0.25 and a N-score of 0.25. This alternative representation seems to have little to no effect in preliminary experiments, as such it is not considered for the final experiments.

We analyse the effectiveness and coverage of the polarities obtained from the different sentiment lexicons, by themselves and in combination as depicted in Table 1.

Table 1. Sentiment Lexicons Considered

Lexicon ID	Lexicon	Senses
L1	Micro-WNOp (MWN)	2800
L2	Q-WordNet (QWN)	15511
L3	SentiWordNet (SWN)	49447
CL1	Micro-WNOp + Q-WordNet (MWN + QWN)	18062
CL2	Micro-WNOp + SentiWordNet (MWN + SWN)	51001
CL3	Q-WordNet + SentiWordNet (QWN+SWN)	60738
CL4	Micro-WNOp + Q-WordNet + SentiWordNet (MWN+QWN+SWN)	62194

The polarity lexicons are in the format Lemma#ReducedPart-of-SpeechTag#SenseNumber Polarity {P or N or O (or B)}. Note that the combined lexicon QWN+SWN (CL3), for example, does not have the same meaning as SWN+QWN. QWN+SWN is generated by using the polarities in Q-WordNet as a starting point and then adding to it the polarities extracted from SentiWordNet for words that are present in SentiWordNet and not in Q-WordNet. This means that there are other possible combinations that are not featured in this table, since they proved to be less efficient. The most efficient combinations are those that give priority to the most fine-grained and smallest lexicons especially when considering SWN, for example QWN (15511) + SWN (49447) results in 60738 total unique WordNet sense polarities. This might be due to the fact that SWN was not manually annotated and some senses are misclassified, so by giving priority to the senses in MWN and QWN we reduce this negative influence.

To examine the quality and coverage of the polarities obtained from the different sentiment lexicons, prior to the final experiments, we consider a simple measure based on Turney's [27]. The total percentage of sentences in the corpuses that are positive and whose sum of polarities (of the individual WordNet senses of terms in the sentence) is more than 0, in combination with those that are negative and whose sum of polarities is less than 0 relative to the total number of examples. The lexicon that scores best using this measure is CL4 (MWN+QWN+SWN) which also offers the most coverage of the data, as broken down in Table 2.

Table 2. Polarity lexicon quality and coverage in term of the percentage of correctly classified positive (Pos), negative (Neg), overall neutral (Neu) and total examples (Tot)

Lexicon	Movie Reviews				SemEval				Mixed Reviews			
	Pos	Neg	Neu	Tot	Pos	Neg	Neu	Tot	Pos	Neg	Neu	Tot
L1	19.95	0.30	40.48	10.11	5.98	0.00	45.77	2.82	20.15	0.08	33.44	8.34
L2	48.09	4.78	26.47	26.4	23.93	3.23	41.45	12.98	48.00	3.18	25.32	21.62
L3	61.53	15.37	16.91	38.42	36.32	5.89	35.61	20.22	64.90	10.23	16.54	32.72
CL1	50.83	5.34	24.78	28.05	26.5	3.80	41.45	14.49	49.19	3.79	24.83	22.47
CL2	61.80	15.39	16.77	38.56	36.54	6.27	35.81	20.52	65.11	10.15	16.63	32.77
CL3	**65.69**	16.71	14.14	**41.16**	42.52	12.17	**29.48**	26.46	**67.71**	9.47	15.87	33.44
CL4	65.44	**16.82**	**14.11**	41.09	**42.52**	**12.36**	30.18	**26.56**	67.61	**9.70**	15.74	**33.53**

3.4 Negation Processing

It should be clear from the breakdown presented in Table 2 that even with CL4 a greater percentage of the positive examples (42-67%) are correctly classified, as opposed to a very small percentage of negative examples (9-16%). In an effort to address this issue and balance these measures, we make use of the dependencies generated by the Stanford CoreNLP, in order to process each sentence for negation, namely the dependency modifier "neg", which allows us to easily determine the presence of several simple types of negation. We found that the average number of negations per sentence greatly varies with the domain of the corpus. While the Movie Reviews and Mixed Reviews corpuses have around 1 negation every 5 sentences, the SemEval News corpus has only 1 negation every 50 sentences.

We tested different negation schemas in preliminary tests and found that the most efficient schema is when we emphasize the negation. When the negated word is positive (e.g. good) or neutral, the resulting polarity for the negating word (e.g. not) and negated word will both be negative; and when the negated word is negative, the resulting polarity for the negating word (e.g. not) and negated word (e.g. bad) will be positive. This is illustrated in the following examples:

Table 3. Feature breakdown of two example sentences, higlighting negation

Features	Sentence				
Word	This	movie	is	not	good
Word Sense	this#ND	movie#n#1	is#v#1	not#r#1	good#a#1
Polarity	O	O	O	N	P
Polarity with Negation	O	O	O	N	N
Word	This	movie	is	not	bad
Word Sense	this#ND	movie#n#1	is#v#1	not#r#1	bad#a#1
Polarity	O	O	O	N	N
Polarity with Negation	O	O	O	P	P

Processing negation offers significant improvement when the lexicon considered has a low coverage for the data, but gradually decreases in influence as the lexicon considered grows in size. This is illustrated in Table 4.

Table 4. Lexicon polarity quality and coverage with and without negation processing, in terms of the percentage of correctly classified examples

Lexicon	Movie Reviews		SemEval News		Mixed Reviews	
	Plain Polarities	With Negation	Plain Polarities	With Negation	Plain Polarities	With Negation
L1	10.11	16.06	8.34	20.91	2.80	4.30
L2	26.40	28.71	21.62	28.85	12.90	13.70
L3	38.42	39.12	32.72	35.00	20.10	20.70
CL1	28.05	29.94	22.47	29.78	14.40	15.20
CL2	38.56	39.22	32.77	34.95	20.40	21.00
CL3	41.16	**41.54**	33.44	35.13	26.30	26.60
CL4	41.09	41.43	33.52	**35.27**	26.40	**26.70**

Again the lexicon that scores best across most corpuses is CL4 (MWN+QWN+SWN), which also offers the most coverage of the data and thus is the lexicon chosen for the actual parse tree extension experiments.

3.5   Support Vector Machine

The SVM implementation chosen to run the classification tasks is SVMlight-TK 1.2 [14]. This SVM package contains the implementations of the STK and SSTK as part of it. Since we are mostly interested in comparing the performance of our extended parse trees against the plain parse trees, we leave the parameters in both the SVM and the kernels as default.

4   Experimental Evaluation

We evaluate the impact of the proposed methodology, for extending syntactic parse trees with WSD and polarity features, for polarity classification tasks. We start by evaluating the performance of the different sentiment lexicons. We also evaluate the impact of the features in separate and combination as well as the impact of negation processing. Finally we compare the performance of TKs for sentiment polarity classification compared to the other kernel based approaches. We use 10-fold cross-validation classification accuracy (%) as a measure of performance throughout our experimental evaluations. Note that early experiments revealed that the SSTK is much more accurate than the STK (by about 10%) so we decided to use only the SSTK in our final experiments. This is not surprising since the SSTK is a specialized kernel which is more appropriate to explore constituent syntactic parse trees [14].

The sentiment lexicon evaluation confirmed our initial analysis of the quality and coverage of the lexicons we consider. However, this is true only when the polarity is used in combination with the word senses.

Table 5. Sentiment lexicon evaluation - parse trees extended with WSD and polarity with and without negation processing

Lexicon	Movie Reviews		SemEval News		Mixed Reviews	
	WSD+Pol	WSD+Pol-N	WSD+Pol	WSD+Pol-N	WSD+Pol	WSD+Pol-N
L1	74.18	74.23	65.00	64.80	72.40	72.45
L2	74.18	74.24	65.00	64.90	72.40	72.36
L3	74.18	74.28	65.00	64.80	72.40	72.62
CL1	74.18	74.27	65.00	64.90	72.40	72.36
CL2	74.18	74.28	64.20	64.80	72.40	72.62
CL3	74.18	74.26	65.00	65.00	72.13	72.62
CL4	**74.19**	**74.29**	**65.00**	**65.00**	**72.40**	**72.63**

Table 6. Evaluation of our parse tree extensions

Features	Movie Reviews	SemEval News	Mixed Reviews
Tree Kernel Baseline	71.70	62.60	71.29
WSD	73.27	63.90	71.24
Polarity	73.35	64.30	72.13
Polarity with Neg	73.44	64.10	72.00
WSD + Polarity	74.19	65.00	72.40
WSD + Pol with Neg	**74.29**	**65.00**	**72.63**

Table 7. Comparison of our approach and other popular kernels for polarity classification tasks

Methodology	Movie Reviews	SemEval	Mixed Reviews
Linear / Bag of Words	50.47	54.10	59.20
TK Syntactic Parse tree	71.70	62.60	71.29
TK Extended Parse tree (WSD + Pol with Neg)	74.29	65.00	72.63
Sequence Kernel / bigrams	76.21	67.60	74.45

As we can see our parse tree extensions provide an improvement over the baseline (the syntactic parse tree with no augmentation) results across all corpuses. The results also seem to indicate that the WordNet senses and polarities are complementary features, since the improvement provided by extending the parse trees with both WordNet senses and polarities, is always larger than when these features are used to extend the parse trees separately. Furthermore negation seems to offer some benefits in most cases, especially when combined with the WSD features. Note that early experiments with the STK still show the same (or higher) improvement but the results were much lower in general. This can be attributed to the different substructures that each kernel considers.

5 Discussion

Document level and sentence level polarity classification are two very different tasks. When working at the sentence level (and sub-sentence) there is very little contextual information, leading in most cases to lower results. Furthermore the majority of the work in sentiment analysis considers the polarity of word terms rather than the polarity of specific senses of the word. It should be clear that different senses of a word can have different opinion-related properties. This paper addressed the issues of word sense and contextual polarity by making use of a novel combination of features drawn from external knowledge sources.

We evaluated three sentiment lexicons and four combinations of these. We found that the combined lexicon CL4 comprising Micro-WNop, Q-WordNet and Senti-WordNet, achieves the best performance. Prior to the final experiments, we used a simple measure to analyse the quality and coverage of the polarities obtained from the different sentiment lexicons. We noticed that a great percentage of the positive examples are correctly classified (42-67%), as opposed to a very small percentage of negative examples (9-16%). We addressed this issue and managed to balance these measures, by processing each sentence for negation with the use of the dependencies generated by the Stanford CoreNLP. We also tested different negation schemas in preliminary tests and found that the most efficient schema is when we emphasize the negation. As such when the negated word is positive (e.g. good) or neutral, the resulting polarity for the negating word (e.g. not) and negated word will both be negative; and when the negated word is negative, the resulting polarity for the negating word (e.g. not) and negated word (e.g. bad) will be positive.

Note that despite WSD being reportedly only about 50-70% accurate [5], [20], [21] the experimental evaluation shows that our parse tree extensions provide an improvement over the baseline results (for all measures) across all corpuses. The improvement provided by extending the parse trees with both WordNet senses and polarities is always larger than when these features are used to extend the parse trees separately, suggesting that the features we selected are complementary. This confirms that WSD offers improvements for polarity classification tasks, however since the WSD is an intermediate task, disambiguation errors can affect the quality of the corresponding sense specific opinion-related properties and thus the classification quality. Furthermore the results indicate that our local negation processing offers some benefits, especially when combined with the WSD features. Particularly in the Movie Reviews and Mixed Reviews corpuses where there was a significant improvement in performance. This appears to relate with the number of negations in the corpuses, while the Movie Reviews and Mixed Reviews corpuses have around 1 negation every 5 sentences; the SemEval News corpus has only 1 negation every 50 sentences.

Finally, our methodology has the added benefit of working with most TKs, so advances in TKs that make use of syntactic parse trees, might be further enhanced by our extended parse trees.

Possible work for the future includes: developing different extension representations; enhancing dependency trees; developing our own unique tree representations, rather than extending parse trees; including more features (e.g. Named Entities); applying the methodology for multi-class polarity classification tasks; and adapting the methodology to document-level polarity classification.

References

1. Agerri, R., Garc, A.: Q-WordNet: Extracting polarity from WordNet senses. In: Seventh Conference on International Language Resources and Evaluation Malta (2009)
2. Akkaya, C., Wiebe, J., Conrad, A., Mihalcea, R.: Improving the impact of subjectivity word sense disambiguation on contextual opinion analysis. In: Proceedings of the Fifteenth Conference on Computational Natural Language Learning, pp. 87–96. Association for Computational Linguistics (2011)
3. Baccianella, S., Esuli, A., Sebastiani, F.: SentiWordNet 3.0: An Enhanced Lexical Resource for Sentiment Analysis and Opinion Mining. In: Proceedings of the Seventh conference on International Language Resources and Evaluation LREC 2010 European Language Resources Association (ELRA), pp. 2200–2204 (2008)
4. Balamurali, A., Joshi, A., Bhattacharyya, P.: Robust Sense-Based Sentiment Classification. In: ACL HLT 2011, p. 132 (2011)
5. Balamurali, A.R., Joshi, A., Bhattacharyya, P.: Harnessing WordNet senses for supervised sentiment classification. In: Conference on Empirical Methods in Natural Language Processing, EMNLP 2011, July 27-31, pp. 1081–1091. Association for Computational Linguistics (ACL), Edinburgh (2011)
6. Carrillo de Albornoz, J., Plaza, L., Gervás, P.: A hybrid approach to emotional sentence polarity and intensity classification. In: Proceedings of the Fourteenth Conference on Computational Natural Language Learning, pp. 153–161. Association for Computational Linguistics (2010)
7. Cerini, S., Compagnoni, V., Demontis, A., Formentelli, M., Gandini, G.: Micro-WNOp: A gold standard for the evaluation of automatically compiled lexical resources for opinion mining. In: Language Resources and Linguistic Theory: Typology, Second Language Acquisition, English linguistics, Franco Angeli Editore, Milano, IT (2007)
8. Collins, M., Duffy, N.: New ranking algorithms for parsing and tagging: Kernels over discrete structures, and the voted perceptron. In: Proceedings of the 40th Annual Meeting on Association for Computational Linguistics, pp. 263–270. Association for Computational Linguistics (2002)
9. Esuli, A., Sebastiani, F.: Sentiwordnet: A publicly available lexical resource for opinion mining. In: Proceedings of LREC, pp. 417–422. Citeseer (2006)
10. Hu, M., Liu, B.: Mining and summarizing customer reviews. In: Proceedings of the tenth ACM SIGKDD International Conference on Knowledge Discovery and Data Mining, pp. 168–177. ACM (2004)
11. Liu, B.: Sentiment analysis and subjectivity. In: Handbook of Natural Language Processing (2010)
12. Martın-Wanton, T., Balahur-Dobrescu, A., Montoyo-Guijarro, A., Pons-Porrata, A.: Word sense disambiguation in opinion mining: Pros and cons. In: Special Issue: Natural Language Processing and its Applications, pp. 119–130 (2010)
13. Miller, G.A.: WordNet: a lexical database for English. Communications of the ACM 38(11), 39–41 (1995)
14. Moschitti, A.: Efficient convolution kernels for dependency and constituent syntactic trees. In: Fürnkranz, J., Scheffer, T., Spiliopoulou, M. (eds.) ECML 2006. LNCS (LNAI), vol. 4212, pp. 318–329. Springer, Heidelberg (2006)
15. Moschitti, A.: Making tree kernels practical for natural language learning. In: Proceedings of EACL, pp. 113–120 (2006)

16. Pang, B., Lee, L.: A sentimental education: Sentiment analysis using subjectivity summarization based on minimum cuts. In: Proceedings of the 42nd Annual Meeting on Association for Computational Linguistics, pp. 271–278. Association for Computational Linguistics (2004)

17. Pang, B., Lee, L.: Seeing stars: Exploiting class relationships for sentiment categorization with respect to rating scales. In: Proceedings of the 43rd Annual Meeting on Association for Computational Linguistics, pp. 115–124. Association for Computational Linguistics (2005)

18. Pang, B., Lee, L.: Opinion mining and sentiment analysis. Foundations and Trends in Information Retrieval 2(1-2), 1–135 (2008)

19. Patwardhan, S., Banerjee, S., Pedersen, T.: Using measures of semantic relatedness for word sense disambiguation. In: Gelbukh, A. (ed.) CICLing 2003. LNCS, vol. 2588, pp. 241–257. Springer, Heidelberg (2003)

20. Pedersen, T., Kolhatkar, V.: WordNet: SenseRelate: AllWords: a broad coverage word sense tagger that maximizes semantic relatedness. In: Proceedings of Human Language Technologies: The 2009 Annual Conference of the North American Chapter of the Association for Computational Linguistics, Companion Volume: Demonstration Session, pp. 17–20. Association for Computational Linguistics (2009)

21. Rentoumi, V., Giannakopoulos, G., Karkaletsis, V., Vouros, G.A.: Sentiment analysis of figurative language using a word sense disambiguation approach. In: Proc. of the International Conference RANLP, pp. 370–375 (2009)

22. Sebastiani, F.: Machine learning in automated text categorization. ACM Computing Surveys (CSUR) 34(1), 1–47 (2002)

23. Shawe-Taylor, J., Cristianini, N.: Kernel methods for pattern analysis. Cambridge University Press (2004)

24. Strapparava, C., Mihalcea, R.: Semeval-2007 task 14: Affective text. In: Proceedings of SemEval, vol. 7 (2007)

25. Suzuki, J., Hirao, T., Sasaki, Y., Maeda, E.: Hierarchical directed acyclic graph kernel: Methods for structured natural language data. In: Proceedings of the 41st Annual Meeting on Association for Computational Linguistics, vol. 1, pp. 32–39. Association for Computational Linguistics (2003)

26. Täckström, O., McDonald, R.: Discovering fine-grained sentiment with latent variable structured prediction models. In: Clough, P., Foley, C., Gurrin, C., Jones, G.J.F., Kraaij, W., Lee, H., Mudoch, V. (eds.) ECIR 2011. LNCS, vol. 6611, pp. 368–374. Springer, Heidelberg (2011)

27. Turney, P.D.: Thumbs up or thumbs down?: semantic orientation applied to unsupervised classification of reviews. In: Proceedings of the 40th Annual Meeting on Association for Computational Linguistics, pp. 417–424. Association for Computational Linguistics (2002)

28. Wiegand, M., Klakow, D.: Convolution kernels for opinion holder extraction. In: Human Language Technologies: The 2010 Annual Conference of the North American Chapter of the Association for Computational Linguistics, pp. 795–803. Association for Computational Linguistics (2010)

Combining Supervised and Unsupervised Polarity Classification for non-English Reviews

José M. Perea-Ortega, Eugenio Martínez-Cámara,
María-Teresa Martín-Valdivia, and L. Alfonso Ureña-López

SINAI Research Group, Computer Science Department, University of Jaén
Escuela Politécnica Superior, Campus Las Lagunillas s/n, 23071, Jaén, Spain
{jmperea,emcamara,maite,laurena}@ujaen.es

Abstract. Two main approaches are used in order to detect the sentiment polarity from reviews. The supervised methods apply machine learning algorithms when training data are provided and the unsupervised methods are usually applied when linguistic resources are available and training data are not provided. Each one of them has its own advantages and disadvantages and for this reason we propose the use of meta-classifiers that combine both of them in order to classify the polarity of reviews. Firstly, the non-English corpus is translated to English with the aim of taking advantage of English linguistic resources. Then, it is generated two machine learning models over the two corpora (original and translated), and an unsupervised technique is only applied to the translated version. Finally, the three models are combined with a voting algorithm. Several experiments have been carried out using Spanish and Arabic corpora showing that the proposed combination approach achieves better results than those obtained by using the methods separately.

1   Introduction

Opinion Mining (OM), also known as Sentiment Analysis (SA) is a challenging task that combines data mining and Natural Language Processing (NLP) techniques in order to computationally treat subjectivity in textual documents [1]. This new area of research is becoming more and more important mainly due to the growth of social media where users continually generate contents on the web in the form of comments, opinions, emotions, etc. There are several issues related to OM like subjectivity detection, opinion extraction, irony detection and so on. However, perhaps the most widely-studied task is sentiment polarity classification. This task aims to determine which is the overall sentiment-orientation (positive or negative) of the opinions contained within a given document. The document contains subjective information such as product reviews or opinionated posts in blogs.

Although different approaches have been applied to polarity classification, the mainstream basically consists of two major methodologies. On the one hand, the Machine Learning (ML) approach (also known as the supervised approach) is based on using a collection of data to train the classifiers [2]. On the other hand,

A. Gelbukh (Ed.): CICLing 2013, Part II, LNCS 7817, pp. 63–74, 2013.

the approach based on Semantic Orientation (SO) does not need prior training, but takes into account the positive or negative orientation of words [3]. This method, also known as the unsupervised approach, makes use of lexical resources like lists of opinionated words, lexicons, dictionaries, etc. Both methodologies have their advantages and drawbacks. For example, the ML approach depends on the availability of labeled data sets (training data), which in many cases are impossible or difficult to achieve, partially due to the novelty of the task. On the other hand, the SO strategy requires a large amount of linguistic resources which generally depend on the language, and often this approach obtains lower recall because it depends on the presence of the words comprising the lexicon in the document in order to determine the orientation of opinion. In order to overcome the weaknesses of both approaches, we have performed several experiments, combining ML and SO through different strategies.

Most of the studies on polarity classification only deal with English documents, perhaps due to the lack of resources in other languages. However, people increasingly comment on their experiences, opinions, and points of views not only in English but in many other languages. Consequently, the management and study of subjectivity and SA in languages other than English is a growing need. The work presented herein is mainly motivated by the need to develop polarity detection systems in languages other than English.

According to Mihalcea, Banea and Wiebe [4], there are two main approaches in the context of multilingual SA. The first one is a Lexicon-based approach, where a target-language subjectivity classifier is generated by translating an existing lexicon into another idiom. The second one is a Corpus-based approach, where a subjectivity-annotated corpus for the target language is built through projection, training a statistical classifier on the resulting corpus. In this paper we follow this second approach and we generate an English parallel corpus by applying machine translation to the original corpus.

The aim of this study is to evaluate an approach based on the combination of supervised and unsupervised methods to improve the results obtained using these methods separately. Specifically, this study has been carried out on two different corpora of reviews in Arabic and Spanish. The main idea is to translate the original corpus into English, generating a parallel corpus. Thus, we could apply the supervised approach to the original corpus and the unsupervised one to the translated version of the original corpus, since it is more feasible to find linguistic resources for this language. languages that have few lexical resources for tackling the polarity classification problem.

The rest of the paper is organized as follows: the next section presents work related to polarity detection dealing with languages other than English and multilingual opinion mining. Section 3 presents the approach proposed in this work. Section 4 describes the different resources used in our experiments including the MC and MCE corpora and SentiWordNet. The different experiments carried out and the results obtained are expounded in Section 5. In Section 6 the obtained results are analyzed. Finally, the main conclusions and ideas for further work are expounded in Section 7.

2   Background

Most of the research papers on SA that we can find in the literature have been applied to English exclusively, although works on other languages are growing increasingly. There are some interesting papers that have studied the problem of polarity classification using non-English collections such as German, French, Chinese, Arabic or Spanish. Below, we summarize some of the most interesting related works.

Kim and Hovy [5] compared opinion expressions between an aligned corpus of emails in German and English. They developed two models: for the first one they translated German emails into English and then applied opinion-bearing words. For the second one they translated English opinion-bearing words into German and then analyzed the German emails using the German opinion-bearing words. The results showed that the first model worked slightly better than the second one. Following this work, Denecke [6] worked on German comments collected from Amazon. These reviews were translated into English using standard machine translation software. Then the translated reviews were classified as positive or negative, using three different classifiers: LingPipe, SentiWordNet with classification rule, and SentiWordNet with machine learning.

Tan and Zhang [7] were among the first researchers to study opinion mining in Chinese. They carried out a widely experimental revision using lots of different models. Zhang et al. [8] applied Chinese SA on two datasets. In the first one, euthanasia reviews were collected from different web sites, while the second dataset was about six product categories collected from Amazon (Chinese reviews). They proposed a rule-based approach including two phases: firstly, by determining each sentence's sentiment based on word dependency, and secondly, by aggregating sentences in order to predict the document sentiment. Wan [9] studied the sentiment polarity identification of Chinese product reviews using a semantic orientation. He made use of bilingual knowledge including both Chinese resources and English resources. The corpus was composed of 886 Chinese documents that were translated into English by using Google Translate and Yahoo Babel Fish. In addition, the approach used ensemble methods to combine the individual results over Chinese and English datasets. The results for the combination methods improved the performance of individual results.

Ghorbel and Jacot [10] used a corpus with movie reviews in French. They applied a supervised classification combined with SentiWordNet in order to determinate the polarity of the reviews. French is also managed in Balahur and Turchi [11], along with Spanish and German. Different machine translation systems and meta-classifiers were tested in order to demonstrate that multilingual SA using these techniques is comparable to the English performance.

In Rushdi-Saleh et al. [12] a corpus of movies reviews in Arabic annotated with polarity was presented and several experiments using machine learning techniques were performed. Subsequently, they generated the parallel EVOCA corpus (English version of OCA) by translating the OCA corpus automatically into English. The results showed that, although the results obtained with EVOCA were worse than those obtained with OCA, they are comparable to other English experiments, since the loss of precision due to the translation process is very slight, as can be seen in Rushdi-Saleh et al. [13].

Regarding opinion mining focused on Spanish, there are also some remarkable studies. For example, Banea et al. [14] proposed several approaches to cross lingual subjectivity analysis by directly applying the translations of opinion corpus in English to training an opinion classifier in Romanian and Spanish. This study showed that automatic translation is a viable alternative for the construction of resources and tools for subjectivity analysis in a new target language. Brooke et al. [15] presented several experiments dealing with Spanish and English resources. They concluded that although the ML techniques can provide a good baseline performance, it is necessary to integrate language-specific knowledge and resources in order to achieve an improvement. Finally, Cruz et al. [16] generated the MuchoCine corpus by recollecting manually Spanish movie reviews from the MuchoCine website. This corpus was generated in order to develop a sentiment polarity classifier based on semantic orientation. On the other hand, Martínez-Cámara, Martín-Valdivia and Ureña-López [17] applied the supervised approach to the MuchoCine corpus using different ML algorithms, obtaining better results than those obtained by applying the unsupervised approach proposed by Cruz et al.

One of the drawbacks for the investigation in SA over non-English texts is the lack of linguistic resources. In Steinberger et al. [18] is presented a novelty method to develop multilingual and comparable sentiment dictionaries, which consists of using two high-level gold-standard sentiment dictionaries for two languages (English and Spanish) and then translated them automatically into third languages. The third languages dictionaries are formed by the overlap of the translations, i.e. via triangulation. The obtained dictionaries are manually filtered and expanded.

3   Combination of Supervised and Unsupervised Methods

The aim of the approach proposed in this study is to improve the polarity classification of the reviews provided by a corpus whose documents are in a language other than English. The main proposal is to translate the original corpus into English and work with parallel corpora, generating several learning models by using both corpora. Furthermore, since we have a corpus translated into English, we can make use of semantic resources for opinion mining tasks such as Senti-WordNet[1] in order to apply a non-supervised approach to that corpus. In this way, the models (supervised and unsupervised) generated using the parallel corpora can be combined in a meta-classifier that could apply different algorithms to establish the final polarity classification. Figure 1 illustrates this approach.

One of the advantages of our architecture is its modularity, allowing the use of different supervised algorithms for both corpora (original and translated) and even in the meta-classifier, for combining previous generated models. As can be seen in Figure 1, we apply a processing to the corpora, which usually consists

---

[1] http://sentiwordnet.isti.cnr.it

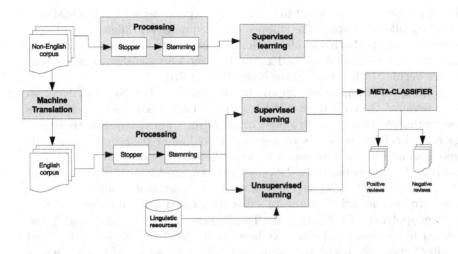

Fig. 1. Overview of the approach proposed

of a stemming process for extracting the root of each word after removing the words without semantic meaning (stopwords).

Once the corpora were processed, we generated the learning models that were used later in the meta-classifier. The supervised approach was applied to both corpora using different learning algorithms such as SVM or NB. However, the unsupervised approach was applied solely to the translated corpus because the linguistic resources, such as SentiWordNet or WordNet-Affect[2], are available in English only. Finally, the meta-classifier process combined several features from the supervised and unsupervised models previously generated, allowing to apply different combination algorithms.

The approach proposed in this paper is especially suitable when we work with non-English corpora because using the translated version of the original corpus we could apply unsupervised approaches on it, since there are very few linguistic and semantic resources for non-English corpora. In this way, we could improve the results obtained by using the supervised methods and to gain some independence from the domain.

4   Experiment Framework

In order to verify the performance of the proposed approach, we decided to apply it on two non-English corpora, specifically on the MuchoCine corpus in Spanish and the OCA corpus in Arabic. In this section we explain the main tools used in carrying out the experiments presented in this study. Then, we describe both corpora employed for the experiments.

---

[2] http://wndomains.fbk.eu/wnaffect.html

For the processing carried out to the parallel corpora we used the RapidMiner[3] tool, which allows to apply the stopper and stemming for different languages. The supervised approach was also performed using this tool, since it allows to apply the cross-validation method using different learning algorithms such as Support Vector Machines (SVM) or Naïve Bayes (NB).

Regarding the unsupervised approach, we used SentiWordNet 3.0 [19] as semantic resource. SentiWordNet (SWN) is a lexical resource for SA which assigns three sentiment scores to each *synset* of WordNet[4]: positivity, objectivity and negativity. Each of the scores ranges from 0 to 1, and their sum is equals 1. A good example is the word *beautiful*, which belongs to two *synsets* (00217728, 01800764). For the *synset* 00217728, the SWN score of *beautiful* is $(0.75, 0.25, 0)$ and for the synset *01800764* is $(0.625, 0.375, 0)$. We used nouns, adjectives, verbs and adverbs as linguistic features. In a first step, the translated corpus was processed by applying a POS tagger like TreeTagger[5]. The aim of this process was to obtain all the nouns, adjectives, verbs and adverbs of each review. The second step after tagging the translated corpus was to generate a total of 15 sub-corpora by making a combination of the four possibilities (nouns, adjectives, verbs and adverbs) in order to analyze the impact of each type of word. Finally, we calculated the SWN score for each review as the polarity score of the document. This score was obtained following the method proposed by Denecke [6] based on the calculation of a triplet of positivity, negativity and objectivity scores.

Below, we explain the main features of the both parallel corpora used for the experiments carried out in this study.

4.1   The OCA-EVOCA Corpus

The Arabic corpus called OCA (Opinion Corpus for Arabic) was generated by Rushdi-Saleh et al. [12] to be freely used for the research community related to OM[6]. It is composed of 500 film reviews that were extracted from different blogs in Arabic found on the Internet. 250 reviews were labeled as positive and the other 250 as negative. In Rushdi-Saleh et al. [12] can be found more details about the process of generation of OCA and its evaluation carried out by applying the cross-validation method.

The same authors conducted the machine translation of OCA into English, generating the parallel corpus called EVOCA (English Version of OCA), also available for research purposes[7]. This translation was carried out using the PROMT[8] tool. In Rushdi-Saleh et al. [13] can be found the evaluation performed on the EVOCA corpus also using the cross-validation method.

---

[3] http://rapid-i.com

[4] WordNet is a large lexical database of English. Nouns, verbs, adjectives and adverbs are grouped into sets of cognitive synonyms (*synsets*), each expressing a distinct concept. It is available in http://wordnet.princeton.edu

[5] http://www.ims.uni-stuttgart.de/projekte/corplex/TreeTagger

[6] http://sinai.ujaen.es/wiki/index.php/OCA_Corpus_(English_version)

[7] http://sinai.ujaen.es/wiki/index.php/EVOCA_Corpus_(English_version)

[8] http://translation2.paralink.com

4.2   The MC-MCE Corpus

The MuchoCine corpus (MC) was presented in Cruz et al. [16] and it is freely available for the research community. It is composed of 3,878 movie reviews collected from the MuchoCine website[9]. The reviews are written by web users instead of professional film critics. This increases the difficulty of the task because the sentences found in the documents may not always be grammatically correct, or they may include spelling mistakes or informal expressions. The corpus contains about 2 million words and an average of 546 words per review.

The opinions are rated on a scale from 1 to 5. One point means that the movie is very bad and 5 means very good. Films with a rating of 3 can be considered as *"neutral"*, which means that the user considers the film is neither bad nor good. In our experiments we have discarded the neutral examples because the polarity classification task is binary, i.e. we have to classify the reviews as positive or negative only. Therefore, the opinions with ratings of 1 or 2 were considered as negative and those with ratings of 4 or 5 were considered as positive.

The MuchoCine English corpus (MCE) is the English version of MC. We generated MCE by applying a machine translation process using the Microsoft Translator[10] tool, formerly known as Bing Translator. Specifically we used the Java API provided for that tool. The MCE corpus is also freely available[11].

5   Experiments and Results

In this section we describe the experiments carried out and the results obtained after applying the proposed approach to the OCA-EVOCA and MC-MCE corpora. In the first subsection, the best individual results obtained for each parallel corpus are shown. Then, in the second subsection, we show the results obtained using the proposed approach.

5.1   Individual Results

According to the evaluation carried out by Rushdi-Saleh et al. [13] using supervised approaches over OCA and EVOCA, the configuration that reported the best results for the OCA corpus used SVM and TF·IDF as learning algorithm and weighting scheme, respectively, and did not apply the stemming process. The score obtained for the F1 measure was 0.9073. However, for the EVOCA corpus, the best F1 score (0.8840) was obtained by applying the stemming process and also using SVM and TF·IDF. For the unsupervised method, we carried out several experiments, as explained at the beginning of Section 4, and the configuration that reported the best F1 score used nouns and adjectives solely, obtaining a F1 score of 0.6698, which is lower than that obtained using the supervised approach, as expected.

---

[9] http://www.muchocine.net

[10] http://www.bing.com/translator

[11] http://sinai.ujaen.es/wiki/index.php/MCE_Corpus_(English_version)

Regarding the evaluation of the MC corpus, Martínez-Cámara et al. [20] followed a similar procedure based on the cross-validation method for the supervised approach. The best configuration for the MC corpus used SVM, TF·IDF, stopper and did not apply the stemming process. The best F1 score was 0.8767. For the translated version of MC (MCE), we considered the same configuration as the best one, achieving 0.8698 of F1 score. Regarding the semantic orientation approach, we carried out the same experiments as for the EVOCA corpus and the configuration that reported the best F1 score used adjectives and verbs solely, achieving a F1 value of 0.6879.

Table 1 summarizes the best individual results obtained for both corpora, showing the score obtained for the typical measures in classification tasks, such as *precision* (P), *recall* (R) and F1.

Table 1. Best results obtained for both parallel corpora individually

Corpora	Approach	Setting	P	R	F1
OCA	supervised	SVM, TF·IDF and no stemming	0.8699	0.9480	**0.9073**
EVOCA	supervised	SVM, TF·IDF and stemming	0.9007	0.8680	0.8840
	unsupervised	nouns + adjectives	0.5535	0.8480	0.6698
MC	supervised	SVM, TF·IDF and no stemming	0.8771	0.8763	**0.8767**
MCE	supervised	SVM, TF·IDF and no stemming	0.8704	0.8693	0.8698
	unsupervised	adjectives + verbs	0.5669	0.8744	0.6879

5.2   Results Obtained Using the Proposed Approach

After carrying out the individual experiments we propose the following method: if we use several classifiers for the same data then we will obtain several models that have learned different patterns from that data. In this manner it is very likely that the correct combination of the models achieves better results than those obtained by each classifier individually. Therefore we adapted the idea of the ensemble classifiers, but working with parallel corpora instead of the same corpus.

Taking into account the best results obtained individually over the OCA-EVOCA and MC-MCE corpora, we decided to combine them in order to improve the performance achieved separately. Specifically we tried *voting* as one of the most widely used combination algorithms in order to carry out the meta-classifier process that combines the three models generated from each corpora. The proposed algorithm makes use of the well-known voting system called majority rule [21]. Then we proposed two possible combinations for both corpora:

- Combination of the three models generated: the supervised approach applied to the original corpus (OCA-SVM and MC-SVM), the supervised approach applied to the translated corpus (EVOCA-SVM and MCE-SVM), and the unsupervised approach applied to the translated corpus (EVOCA-SWN and MCE-SWN).

- Combination of the supervised models: OCA-SVM + EVOCA-SVM for the OCA-EVOCA corpora, and MC-SVM + MCE-SVM for the MC-MCE corpora.

Due to the fact that the number of voters in the first combination is odd, the application of the voting system always returns a single-winner. However, in the second combination (supervised models from original and translated corpora) it is possible to obtain a draw because the predicted class for the OCA-SVM/MC-SVM voters may be different from that obtained by the EVOCA-SVM/MCE-SVM voters, respectively. In order to solve this problem we have considered two possible heuristics:

- Assign a final positive prediction only if both voters return a positive prediction (otherwise negative prediction), or
- Assign a final positive prediction if at least one of the voters returns a positive prediction (negative prediction only when both voters return a negative prediction)

Taking into account these possible combinations and heuristics, Table 2 shows the results obtained by applying the proposed approach to the OCA-EVOCA and MC-MCE corpora.

Table 2. Results obtained by applying the proposed approach

Corpora	Combination	Heuristic	P	R	F1
OCA-EVOCA	OCA-SVM + EVOCA-SVM + EVOCA-SWN	-	0.8566	0.9800	**0.9142**
	OCA-SVM + EVOCA-SVM	pos. if both voters	0.8984	0.9200	0.9091
		pos. if one voter	0.8483	0.9840	0.9111
MC-MCE	MC-SVM + MCE-SVM + MCE-SWN	-	0.8160	0.9608	0.8825
	MC-SVM + MCE-SVM	pos. if both voters	0.8551	0.8893	0.8719
		pos. if one voter	0.8003	0.9843	**0.8828**

6   Analysis of the Results

In this section we analyze the results obtained for both individual and combined experiments. Regarding the individual experiments is noteworthy the good behavior of the supervised approach versus the unsupervised one, as expected. Taking into account the translated versions of the corpora evaluated, the difference obtained for the supervised approach was around +32% and +26% regarding the unsupervised one for the EVOCA and MCE corpora, respectively. If we compare the supervised approach between the original corpus and its translation, the results obtained for the original corpus improve slightly those obtained for the translated version. For the OCA-EVOCA corpora, this improvement was around +3%, while for the MC-MCE corpora was around +0.8%. This behavior is also expected due to the noise that almost all automatic translation tools introduce during the process, although, specifically for the corpora evaluated, it is important to note the good performance of this translation process.

If we compare the results obtained by using the proposed combination approach with those obtained by using the supervised and unsupervised approaches separately, we can observe the improvement achieved by using the proposed approach. As can be seen in Table 3, for the OCA-EVOCA corpora we obtained an improvement of +0.76% regarding the supervised approach applied to the OCA corpus (OCA-SVM). On the other hand, for the MC-MCE corpora we obtained an improvement of +0.70% regarding the supervised approach applied to MC corpus (MC-SVM). This means that the proposed approach can be considered an interesting strategy for applying in polarity classification tasks when we work with parallel corpora.

Table 3. Comparison between the best results obtained by applying the proposed combination approach and those obtained by using the supervised and unsupervised approaches separately

Corpora	Approach	P	R	F1
OCA-EVOCA	OCA-SVM	0.8699	0.9480	0.9073
	EVOCA-SVM	0.9007	0.8680	0.8840
	OCA-SVM + EVOCA-SVM + EVOCA-SWN (combined)	0.8566	0.9800	**0.9142**
MC-MCE	MC-SVM	0.8771	0.8763	0.8767
	MCE-SVM	0.8704	0.8693	0.8698
	MC-SVM + MCE-SVM (combined)	0.8003	0.9843	**0.8828**

7   Conclusions and Further Work

In this paper we have presented a study about polarity classification over corpora written in different languages of English. In the proposed approach, firstly we translated the original corpus into English in order to generate its parallel corpus. Then, several experiments were carried out in order to build supervised and unsupervised models using these corpora. SentiWordNet was used as linguistic resource for the unsupervised experiments. Finally, the individual models were combined by applying a voting algorithm based on the majority rule. Although the results obtained with individual models were very promising, we have shown that the combination approach improved the performances achieved individually. In addition, this improvement was achieved in two parallel corpora so the robustness of the proposed method was evaluated in different frameworks.

For further work, we would like to test the performance using linguistic resources other than SentiWordNet, like for example WordNet-Affect or General

Inquirer. Moreover, it could be interesting to generate several lists of affective words for languages other than English. Thus, we could apply a semantic orientation approach directly to the original corpus and obtain a new model to consider in the meta-classifier architecture.

Acknowledgments. This study has been partially supported by a grant from the Fondo Europeo de Desarrollo Regional (FEDER), TEXT-COOL 2.0 project (TIN2009-13391-C04-02), ATTOS project (TIN2012-38536-C03-0) from the Spanish Government. Also, this study is partially funded by the European Commission under the Seventh (FP7 - 2007-2013) Framework Programme for Research and Technological Development through the FIRST project (FP7-287607). This publication reflects the views only of the authors, and the Commission cannot be held responsible for any use which may be made of the information contained therein.

References

1. Pang, B., Lee, L.: Opinion mining and sentiment analysis. Found.Trends Inf.Retr. 2, 1–135 (2008)
2. Pang, B., Lee, L., Vaithyanathan, S.: Thumbs up?: Sentiment classification using machine learning techniques. In: Proceedings of the ACL 2002 Conference on Empirical Methods in Natural Language Processing, EMNLP 2002, vol. 10, pp. 79–86. Association for Computational Linguistics, Stroudsburg (2002)
3. Turney, P.D.: Thumbs up or thumbs down?: semantic orientation applied to unsupervised classification of reviews. In: Proceedings of the 40th Annual Meeting on Association for Computational Linguistics, ACL 2002, pp. 417–424. Association for Computational Linguistics, Stroudsburg (2002)
4. Mihalcea, R., Banea, C., Wiebe, J.: Learning multilingual subjective language via cross-lingual projections. In: Proceedings of the 45th Annual Meeting of the Association of Computational Linguistics, pp. 976–983. Association for Computational Linguistics, Prague (2007)
5. Kim, S.M., Hovy, E.: Identifying and analyzing judgment opinions. In: Proceedings of the Main Conference on Human Language Technology Conference of the North American Chapter of the Association of Computational Linguistics, HLT-NAACL 2006, pp. 200–207. Association for Computational Linguistics, Stroudsburg (2006)
6. Denecke, K.: Using sentiwordnet for multilingual sentiment analysis. In: ICDE Workshops, pp. 507–512 (2008)
7. Tan, S., Zhang, J.: An empirical study of sentiment analysis for chinese documents. Expert Systems with Applications 34, 2622–2629 (2008)
8. Zhang, C., Zeng, D., Li, J., Wang, F.Y., Zuo, W.: Sentiment analysis of chinese documents: From sentence to document level. Journal of the American Society for Information Science and Technology 60, 2474–2487 (2009)
9. Wan, X.: Co-training for cross-lingual sentiment classification. In: Proceedings of the Joint Conference of the 47th Annual Meeting of the ACL and the 4th International Joint Conference on Natural Language Processing of the AFNLP, ACL 2009, vol. 1, pp. 235–243. Association for Computational Linguistics, Stroudsburg (2009)
10. Ghorbel, D.J.H.: Sentiment analysis of french movie reviews. In: Proceedings of the 4th International Workshop on Distributed Agent-based Retrieval Tools (DART 2010) (2010)

11. Balahur, A., Turchi, M.: Multilingual sentiment analysis using machine translation? In: Proceedings of the 3rd Workshop in Computational Approaches to Subjectivity and Sentiment Analysis, WASSA 2012, pp. 52–60. Association for Computational Linguistics, Stroudsburg (2012)

12. Rushdi-Saleh, M., Martín-Valdivia, M.T., Ureña López, L.A., Perea-Ortega, J.M.: OCA: Opinion corpus for Arabic. Journal of the American Society for Information Science and Technology 62, 2045–2054 (2011)

13. Rushdi-Saleh, M., Martín-Valdivia, M.T., Ureña-López, L.A., Perea-Ortega, J.M.: Bilingual Experiments with an Arabic-English Corpus for Opinion Mining. In: Angelova, G., Bontcheva, K., Mitkov, R., Nicolov, N. (eds.) RANLP 2011 Organising Committee, pp. 740–745 (2011)

14. Banea, C., Mihalcea, R., Wiebe, J., Hassan, S.: Multilingual subjectivity analysis using machine translation. In: Proceedings of the Conference on Empirical Methods in Natural Language Processing, EMNLP 2008, pp. 127–135. Association for Computational Linguistics, Stroudsburg (2008)

15. Brooke, J., Tofiloski, M., Taboada, M.: Cross-linguistic sentiment analysis: From english to spanish. In: International Conference RANLP, pp. 50–54 (2009)

16. Cruz, F.L., Troyano, J.A., Enriquez, F., Ortega, J.: Clasificación de documentos basada en la opinión: experimentos con un corpus de críticas de cine en español. Procesamiento del Lenguaje Natural 41, 73–80 (2008)

17. Martínez-Cámara, E., Martín-Valdivia, M.T., Ureña-López, L.A.: Opinion classification techniques applied to a spanish corpus. In: Muñoz, R., Montoyo, A., Métais, E. (eds.) NLDB 2011. LNCS, vol. 6716, pp. 169–176. Springer, Heidelberg (2011)

18. Steinberger, J., Ebrahim, M., Ehrmann, M., Hurriyetoglu, A., Kabadjov, M., Lenkova, P., Steinberger, R., Tanev, H., Vázquez, S., Zavarella, V.: Creating sentiment dictionaries via triangulation. Decision Support Systems 53, 689–694 (2012)

19. Baccianella, S., Esuli, A., Sebastiani, F.: Sentiwordnet 3.0: An enhanced lexical resource for sentiment analysis and opinion mining. In: Chair, N.C.C., Choukri, K., Maegaard, B., Mariani, J., Odijk, J., Piperidis, S., Rosner, M., Tapias, D. (eds.) Proceedings of the Seventh International Conference on Language Resources and Evaluation (LREC 2010), European Language Resources Association (ELRA), Valletta (2010)

20. Martínez-Cámara, E., Martín-Valdivia, M.T., Perea-Ortega, J.M., Ureña-López, L.A.: Opinion classification techniques applied to a Spanish corpus. Procesamiento del Lenguaje Natural 47 (2011)

21. Johnson, P.E.: Voting systems. a textbook-style overview of voting methods and their mathematical properties. Technical report, University of Kansas (2005)

Word Polarity Detection Using a Multilingual Approach

Cüneyd Murad Özsert and Arzucan Özgür

Department of Computer Engineering, Boğaziçi University,
Bebek, 34342 İstanbul, Turkey
`muradozsert@gmail.com, arzucan.ozgur@boun.edu.tr`

Abstract. Determining polarity of words is an important task in sentiment analysis with applications in several areas such as text categorization and review analysis. In this paper, we propose a multilingual approach for word polarity detection. We construct a word relatedness graph by using the relations in WordNet of a given language. We extend the graph by connecting the WordNets of different languages with the help of the Inter-Lingual-Index based on English WordNet. We develop a semi-automated procedure to produce a set of positive and negative seed words for foreign languages by using a set of English seed words. To identify the polarity of unlabeled words, we propose a method based on random walk model with commute time metric as proximity measure. We evaluate our multilingual approach for English and Turkish and show that it leads to improvement in performance for both languages.

Keywords: Semantic orientation, word polarity, sentiment analysis, random walk model, commute time, hitting time, WordNet.

1   Introduction

Identifying the semantic orientation or polarity of words is one of the most important topics in sentiment analysis. Many applications such as analyzing product/movie reviews (Morinaga et al., 2002; Turney, 2002; Popescu and Etzioni, 2005), and determining the attitudes of participants in online discussions (Hassan et al., 2010) are based on the polarities of the individual words.

Most previous studies on word polarity detection have been carried on for English and make use of language-specific resources such as WordNet (Miller, 1995) and General Inquirer (Stone et al., 1966). Wordnet, is a large lexical database for English, consisting of synsets (i.e. set of synonyms) each belonging to a distinct meaning. General Inquirer is an English lexicon, where words have been tagged with semantic categories such as positive and negative. In polarity detection studies WordNet has mainly been used to construct word relatedness graphs by connecting semantically related words and General Inquirer has been used to obtain labeled seed words for supervised settings and for evaluation purposes (Takamura et al., 2005; Hassan and Radev, 2010). Many languages do not have semantically tagged lexicons such as General Inquirer. Even though some of these languages have WordNets, they are in general not as comprehensive as the English WordNet. Most foreign WordNets such

A. Gelbukh (Ed.): CICLing 2013, Part II, LNCS 7817, pp. 75–82, 2013.

as EuroWordNet (Vossen, 1998) and BalkaNet (Tufiş et al., 2004) are structured in the same way as English WordNet (Miller, 1995) and are linked to each other with an Inter-Lingual-Index based on English WordNet.

In this work, we take advantage of the compatibility in WordNets and develop a multilingual approach for detecting polarities of English as well as foreign words. We construct a word-relatedness graph by not only connecting semantically related words in one WordNet but by also linking words from WordNets of different languages. We also propose a semi-automated method to generate labeled seed words for other languages by using the list of English seed words and the Inter-Lingual-Index. Then, we define a random walk over the word-relatedness graph from any given word to the set of positive and negative seed words. We use commute time as a proximity measure and classify a given word as positive if it is closer to the set of positive seed words compared to the negative seed words, and classify it as negative otherwise. We evaluate our approach for English and Turkish. Turkish WordNet (Bilgin et al., 2004) is completed within the BalkaNet project (Tufiş et al., 2004). It is constructed as being fully compatible with EuroWordNet, which in turn is compatible with English Word-Net. We first show that our commute time model achieves performance comparable to the state-of-the-art in the literature. Then, we demonstrate that creating a multilingual word relatedness graph by connecting the WordNets of English and Turkish boosted the performance of word polarity detection for both languages. To our knowledge, we report the first results for Turkish word polarity detection and achieve an accuracy of 95%.

2   Related Work

Word polarity detection has been studied by several researchers in the past few years. Most of these studies have been evaluated for English words and are based on language resources available for English. For example, Turney and Littman (2003) propose an unsupervised algorithm, where they define seven positive and seven negative paradigm seed words. They use the English web corpus to query any given word with the paradigm words by using the near operator in a search engine. If the word tends to co-occur with positive paradigm words, it is classified as positive, and it is classified as negative otherwise. Takamura et al. (2005) propose a method, which regards semantic orientation as spin of electrons. They consider each word as an electron and its polarity as a spin value. They construct a word relatedness graph by using gloss definitions, thesaurus, and co-occurrence statistic for English. Words are classified as positive or negative according to their spin values. Hassan and Radev (2010) introduce a semi-supervised method where random walk model is used to find the polarities of English words. They construct a word relatedness graph by using the relations in English WordNet and use mean hitting time for polarity estimation.

Hassan et al. (2011) propose an algorithm to find semantic orientation of foreign words and evaluate their approach for Arabic and Hindi with a set of 300 manually labeled seed words for each language. They use random walk model with hitting time for polarity detection. They construct a multilingual network by connecting English and foreign words by using a Foreign-English dictionary. For every foreign word, they look up its possible meanings in the dictionary and connect this foreign word to its possible meanings. Instead, we develop a new approach to establish Foreign-English connections. We propose to use Inter-Lingual-Index for multilingual connections. With the help of this index, WordNets are easily and effectively connected to each other by linking the words in one WordNet to their similar meanings in the other WordNets. We use Turkish as a foreign language and generate a list of 2812 semi-automatically labeled seed words. We propose using commute time as a proximity measure with random walk model for word polarity detection. We show that besides improving the performance for Turkish, our approach also improves the performance for English.

3 Approach

3.1 Monolingual Graph Construction

We construct an undirected weighted graph $G = (V, E)$ comprising a set V of vertices and a set E of edges. Vertices correspond to word and part-of-speech pairs in Word-Net. Two words are connected with if they have one or more of the *synonym, hypernym, also see, similar to and derivation* relations in WordNet. Weight of an edge between two words is directly proportional to the number of WordNet relations between them.

3.2 Multilingual Graph Construction

Foreign WordNets are in general not as comprehensive as the English WordNet. However, most WordNets such as EuroWordNet (Vossen, 1998) and BalkaNet (Tufiş et al., 2004) are designed to be compatible with English WordNet. This compatibility provides a simple and effective way to integrate such WordNets to the powerful English WordNet. We extend our word relatedness graph by connecting the words in English WordNet with similar words in foreign WordNet by using the Inter-Lingual-Index. With the help of this index, it is possible to reach from a synset in any Word-Net to the synsets of the same meaning in the other WordNets.

3.3 Random Walk with Commute Time

Consider a random walk (Lovazs, 1996) on graph G. If we are on vertex i, the probability of moving to the neighbor vertex j in the next step is directly proportional to the weight of the edge between i and j. Thus, the transition probability p_{ij} of moving from vertex i to vertex j is as follows:

$$p_{ij} = \frac{w_{ij}}{\sum_k w_{ik}}$$

Here, W_{ij} is the weight of the edge between vertices i and j, and k denotes all the neighbors of vertex i. *Hitting time* and *commute time* are two proximity measures originating from random walks. *Hitting time* between vertex i and vertex j, denoted by h_{ij}, is the expected number of steps in a random walk before vertex j is visited for the first time starting from vertex i (Sarkar , 2010). It can be calculated recursively as follows:

$$h_{ij} = \begin{cases} 0, & i = j \\ 1 + \sum_k p_{ik} h_{kj}, & i \neq j \end{cases}$$

where k denotes all neighbors of vertex i. Hitting time has been used to find word polarity by Hassan and Radev (2010), who have shown that it achieves the state of art performance in the literature. A drawback of hitting time is that it is not symmetric. It is possible to end up with situations where vertex i is close to vertex j (h_{ij} is small), but vertex j is far away from vertex i (h_{ji} is big). We propose using the commute time proximity measure, which is a symmetric extension of hitting time.

Commute time between vertex i and vertex j, denoted by c_{ij}, is the expected number of steps in a random walk to reach vertex j for the first time starting from vertex i and return to vertex i again. It can be calculated by using hitting time:

$$c_{ij} = h_{ij} + h_{ji}$$

Hitting and commute time are sensitive to long paths far away from the starting node (Sarkar, 2010). In general, similar words tend to be close to each other on a word relatedness graph. Therefore, we use *T-truncated hitting and commute* time, which only consider paths shorter than T.

To find the polarity of a given word, we start a random walk from that word and compute the commute time to the set of positive (P) and negative (N) seed words. Let $c_{i|P}$ be the average of truncated commute times from i to each seed in P and $c_{i|N}$ be the average of truncated commute times from i to each seed in N. If $c_{i|P}$ is less than $c_{i|N}$ word i is classified as positive, otherwise it is classified as negative. When the graph and the size of the seed list is large calculation of $c_{i|P}$ and $c_{i|N}$ is time consuming. We use a sampling approach to estimate $c_{i|P}$ and $c_{i|N}$ similar to previous works (Hassan and Radev, 2010; Sarkar, 2010).

We start M independent random walks with maximum length of T. Hitting one of the labeled seed words and returning to the starting word is the stopping condition. The length of a random walk in which the stopping condition is not met is estimated as T. Let's assume that m of M random walks met the stopping condition and the length of each random walk is $\langle t_1, t_2, \ldots, t_m \rangle$. S denotes set of positive and negative seed words. Then truncated commute time is estimated as:

$$c_{i|S}^* = \frac{\sum_{i=1}^m t_i}{M} + (1 - \frac{m}{M})T$$

The summary of our approach to find polarity of a given word is shown in Algorithm 1.

- For any given word i
- Start M random walks with length T on G.
- Calculate $c_{i|P}^*$ as estimated commute time to set of positive seeds.
- Start M random walks with length T on G.
- Calculate $c_{i|N}^*$ as estimated commute time to set of negative seeds.
- If $c_{i|P}^* > c_{i|N}^*$ classify word i as negative.
- Else classify word i as positive

Algorithm 1. Polarity detection using random walk model with estimated commute time

4 Experiments

We apply our approach to detect polarities of English and Turkish words. We use the WordNets of each language to construct monolingual word-relatedness graphs. A multilingual graph is obtained by connecting these graphs with the Inter-Lingual-Index. We use General Inquirer as a source for English seed words. Like in previous works (Hassan and Radev, 2010; Turney and Litman, 2003), we ignore some ambiguous words and end up with 2085 negative and 1730 positive words. Like most foreign languages, Turkish does not have a resource such as General Inquirer to obtain seed words. Algorithm 2 summarizes the semi-automated method that we propose to produce foreign seed words using the Inter-Lingual-Index. By using this algorithm, we generate 1398 positive and 1414 negative seed words for Turkish.

We use random walk model over the monolingual graphs and the English-Turkish multilingual graph to identify the polarities of words. We propose using commute

- For each word i in positive English seed words.
- Find all synsets in English WordNet that contain i.
- For each synset, find similar synset j in Foreign WordNet by using Inter-Lingual-Index.
- Select each word in synset j as a possible seed word.
- Repeat the same procedure for negative seeds.
- Process the generated foreign seed lists manually to remove the ambiguous words.

Algorithm 2. Foreign Seed Generation Algorithm

time as a proximity measure and compare it with hitting time that was shown to out-perform the previous approaches for English word polarity detection by Hassan and Radev, 2010. We use 10 fold cross validation in our experiments and report the accuracies of polarity detection for the English and Turkish seed words both when the monolingual and the multilingual graphs are used.

Our experimental results are summarized in Figure 1. The proposed commute time algorithm performs similarly to the hitting time method. The accuracy for English when the monolingual graph is used is 89.7%, which is comparable to 91.1% achieved by hitting time[1]. The accuracy for Turkish when the monolingual graph is used is 86.6%, which is slightly better than 84.5% achieved by hitting time. Turkish WordNet is not as rich as English WordNet. Therefore, the accuracies for Turkish are lower than the ones for English when we use the monolingual graphs.

Figure 1 shows that the multilingual approach leads to improvements for both languages. The improvement for Turkish is more significant since we take advantage of the dense English graph. Accuracy for Turkish is improved from 86.6% to 95% with the commute time method, and it is improved from 84.5% to 95.5% with the hitting time method. Accuracy for English is improved from 89.7% to 92.3% with the commute time method, and from 91.1% to 92.8% with the hitting time method. These results demonstrate that the richness of the English WordNet is a valuable resource for Turkish word polarity detection. Interestingly, Turkish WordNet is also able to boost the performance for English word polarity detection.

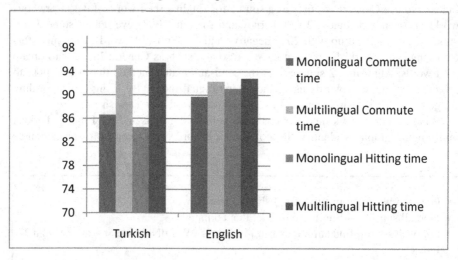

Fig. 1. Accuracies of the monolingual and multilingual approaches using commute time and hitting time methods for Turkish and English

---

[1]   The accuracy for English when hitting time is used is reported as 93.1% in (Hassan and Radev, 2010). The difference might be due to a different version of WordNet or the seed list.

5   Conclusions

We addressed the problem of identifying the polarities of English and foreign words. Most previous studies on polarity detection focus on English and depend on language specific resources such as WordNet. Many foreign languages have WordNets. However, they are not as comprehensive as the English WordNet. In this study, we develop an approach that utilizes the compatibility of English and foreign WordNets to build a multilingual word relatedness graph. We propose using random walk model with commute time proximity measure over this graph to predict word polarities. We evaluate our approach for English and Turkish. We show that the random walk model with commute time achieves similar performance to the state of art method for English in the literature. Our multilingual approach based on connecting the English and Turkish word relatedness graphs led to significant improvement in performance for both languages.

Acknowledgements. We would like to thank Amjad Abu-Jbara and Ahmed Hassan from the University of Michigan for their assistance in providing the implementation details of their algorithm.

References

1. Morinaga, S., Yamanishi, K., Tateishi, K., Fukushima, T.: Mining Product Reputations on the Web. In: Proceedings of the 8th ACM SIGKDD International Conference on Knowledge Discovery and Data Mining, pp. 341–349 (2002)
2. Turney, P.D.: Thumbs Up or Thumbs Down? Semantic Orientation Applied to Unsupervised Classification of Reviews. In: Proceedings of the 40th Annual Meeting on Association for Computational Linguistics, pp. 417–424 (2002)
3. Popescu, A., Etzioni, O.: Extracting Product Features and Opinions from Reviews. In: Proceedings of the Conference on Human Language Technology and Empirical Methods in Natural Language Processing Association for Computational Linguistics, pp. 339–346 (2005)
4. Hassan, A., Qazvinian, V., Radev, D.: What's with the Attitude? Identifying Sentences with Attitude in Online Discussions. In: Proceedings of the 2010 Conference on Empirical Methods in Natural Language Processing, pp. 1245–1255 (2010)
5. Miller, G.A.: WordNet: A Lexical Database for English. Communications of the ACM 38(11), 39–41 (1995)
6. Stone, P., Dunphy, D., Smith, M., Ogilvie, D.: The General Inquirer: A Computer Approach to Content Analysis. The MIT Press, Cambridge (1966)
7. Takamura, H., Inui, T., Okumura, M.: Extracting Semantic Orientations of Words Using Spin Model. In: Proceedings of the 43rd Annual Meeting on Association for Computational Linguistics, pp. 133–140 (2005)
8. Hassan, A., Radev, D.: Identifying Text Polarity Using Random Walks. In: Proceedings of the 48th Annual Meeting of the Association for Computational Linguistics, pp. 395–403 (2010)

9. Vossen, P.: Eurowordnet: A Multilingual Database with Lexical Semantic Networks. Kluwer Academic Publishers, Norwell (1998)
10. Tufiş, D., Cristea, D., Stamou, S.: Balkanet: Aims, Methods, Results and Perspectives. A General Overview. *Romanian* Journal on Science and Technology of Information 7, 9–43 (2004)
11. Bilgin, O., Çetinoglu, Ö., Oflazer, K.: Building a Wordnet for Turkish. Romanian Journal on Information Science and Technology 7, 163–172 (2004)
12. Turney, P.D., Littman, M.L.: Measuring Praise and Criticism: Inference of Semantic Orientation from Association. ACM Transactions on Information Systems 21(4), 315–346 (2003)
13. Hassan, A., Abu-Jbara, A., Jha, R., Radev, D.: Identifying the Semantic Orientation of Foreign Words. In: Proceedings of the 49th Annual Meeting of the Association for Computational Linguistics, vol. 2, pp. 592–597 (2011)
14. Lovasz, L.: Random Walks on Graphs: A Survey. Bolyai Society Mathematical Studies 2, 353–398 (1996)
15. Sarkar, P.: Tractable Algorithms for Proximity Search on Large Graphs. Ph.D. Thesis, Carnegie Mellon University (2010)

Mining Automatic Speech Transcripts
for the Retrieval of Problematic Calls

Frederik Cailliau and Ariane Cavet

Sinequa, 12 rue d'Athènes, F-75009 Paris
{cailliau,cavet}@sinequa.com

Abstract. In order to assure and to improve the quality of service, call center operators need to automatically identify the problematic calls in the mass of information flowing through the call center. Our method to select and rank those critical conversations uses linguistic text mining to detect sentiment markers on French automatic speech transcripts. The markers' weight and orientation are used to calculate the semantic orientation of the speech turns. The course of a conversation can then be graphically represented with positive and negative curves. We have established and evaluated on a manually annotated corpus three heuristics for the automatic selection of problematic conversations. Two proved to be very useful and complementary for the retrieval of conversations having segments with anger and tension. Their precision is high enough for use in real world systems and the ranking evaluated by mean precision follows the usual relevance behavior of a search engine.

Keywords: Sentiment analysis, conversational speech, call center transcripts, customer satisfaction.

1 Introduction

Call centers are often the primary communication interface between large companies and their customers. Every center may employ hundreds of agents who are continuously communicating with the clients.

A small part of the calls in this information stream concerns unsatisfied customers blaming the company for some trouble it presumably has caused. These calls are important for the company for several reasons. Firstly, they may reveal recurrent customer problems due to general dysfunctions in the company's operational procedures. Secondly, the use of real world examples is a must for training call center agents. Completely manual sampling of these calls is however unsatisfying, not to say impossible, because of the high number of calls.

Spontaneous speech as recorded in call centers gives automatic speech recognition (ASR) systems a hard time. It is characterized by typical discourse markers and disfluencies like repetitions, restarts, filler words, filler sounds, etc. The records' quality may also be low, e.g. because of background noise or cell phone use. This leads to performance drops of 10 to 20 % in terms of Word Error Rate (WER) compared to automatic broadcast transcription. In the French evaluation campaigns ESTER 1 and

A. Gelbukh (Ed.): CICLing 2013, Part II, LNCS 7817, pp. 83–95, 2013.

2, the best system showed a WER of respectively 11.9 % [9] and 12.1 % [10] on broadcast transcripts. When adapted to call center conversations, the same system reduced an initial WER of 51 % to 21% with an 18.9 real-time factor [13]. Tests done in the French Infom@gic project on a 10 hour corpus of calls similar to ours have shown a WER of 27 % for the call center agents and 33 % for the clients. Further results on French broadcasts vary from 31% to 41% WER depending on whether the speech is totally spontaneous or a little prepared [1].

Text mining, mostly based on pattern matching, is very sensitive to transcription errors. Independently from each other, the authors of [2] and [7] have studied the impact of ASR on the detection of linguistic patterns on call center speech. They took the manual transcripts as reference and studied the degradation on the automatic transcripts. According to these studies, approximately one pattern out of four is not detected and one detected pattern out of five is incorrect.

For our research, we had the opportunity to work on a corpus of 1 000 hours of transcribed speech collected in a call center of a French energy supplier, corresponding to 8 556 conversations. The maximum duration of a conversation is half an hour, due to technical cut-off. Recording has been done on one channel, introducing speech overlap between speakers and therewith unintelligible speech and ditto transcript.

This article presents a text analytics approach using sentiment analysis on automatic speech transcripts in order to select and score problematic calls in the everyday life of a call center.

2   Related Work on Call Center Speech Analysis

The motivation of our work is surprisingly close to the research described in [14] and [27] to improve AT&T's spoken dialogue system by identifying task failure dialogues and dialogues with low user satisfaction on operational installations. Since only a small fraction of the calls can be listened to, the problematic ones need to be identified. Whereas they apply statistics on features in the system's logs of the human-machine dialogues, our study object is the conversation itself.

The recent development of ASR on call-center speech has opened up new research perspectives for applying classical IR or text-mining techniques on the automatic speech transcripts. Alternative to this approach, keyword spotting is also able of recognizing patterns in speech. This can be done on the audio (acoustic) or by matching the automatic phonetic transcripts [12]. The different approaches are compared in [20] in an experimental setup for keyword spotting in informal continuous speech.

Call-center speech analysis on automatic speech transcripts has a recent history. The following overview shows that it has treated heterogeneous subjects, with corpora having very different characteristics (recording set-up, domain, number of calls, etc.). There is no reference corpus publically available for text mining on call-center transcripts. Research on call-center speech is mostly done in an industrial environment. Unlike corpora like the Fischer corpus [6] created on a voluntary basis on general subjects, the public distribution of call-center speech and its transcripts is hindered by the presence of private or confidential information concerning the customers and the company.

In [23], the calls to a university's IT help desk are classified into 98 different call types. The corpus totals 4 359 conversation sides (dual-channel recording) for 283 hours of audio. The authors of [15] identify the issues raised by the callers by assigning a significance level to the fragments of the calls. They worked on 2 276 calls from the IBM internal customer support service. Dealing with incoming calls covering different domains (e.g. mobile phones, car rental), the authors of [19] automatically build domain-specific models for topic identification. In [22], important segments of the conversations are identified with automatic selection of the features. Viewpoints are then extracted using dictionaries prepared by experts. The corpus consists of nearly thousand conversations from a car rental service center. The authors of [16] identify procedure steps by clustering the transcripts firstly by topic and then separately the agent and client transcripts.

In [28], automatic quality monitoring is performed on IBM's call centers. Human monitors listen to a random sample of the calls then evaluate the quality of each call by answering a set of 31 questions. Two thirds of these questions can reliably be answered by simple pattern matching methods. They also score the calls by estimating a bad outcome with maximum entropy. The features used are textual patterns as well as generic ASR features like the number of hesitations and the duration of silences. The test set consists of 195 manually annotated calls. Precision is about 60% on the first percentages of the presumed bad calls. It drops to about 50% for the bottom 10%, and to 40% for the bottom 20% of calls. The authors claim to triple the efficiency of human monitors by preselecting the bad calls.

The authors of [17] explicitly position themselves as applying interaction mining on call-center analytics. They automatically annotate the argumentative structure of the call, identify controversial topics and calculate a score of cooperativeness for all speakers. The corpus is made of 213 manually transcribed conversations of a help desk call center in the banking domain.

The French Infom@gic-Callsurf research project [11] resulted in a transcription, search and information discovery system for French call-center speech, including theme identification of the calls' segments and a complete audio-enabled interface with faceted search [3].

Left apart some precursor activities, sentiment analysis has been booming with the advent of web 2.0 thanks to publicly available reviews and user feedback on movies, products and services. An overview of the research on sentiment analysis can be found in [18] and [24]. Some methods use machine learning to create classification models or to select sentiment triggers, others use hand-made lexicons. The lexicon-based approach, has been proven to be robust on different domains and unseen data, even when handmade [21].

Of course, sentiment analysis can also be performed on the audio. Until very recently, corpora were too small to demonstrate reliable emotion recognition, as mentioned the overview made in [26] on emotional speech analysis. Since, ASR for spontaneous speech has been improving, opening the way for emotion mining on call-center speech, as presented in [5], [8] and [25].

3   Linguistics-Based Selection of Problematic Conversations

We have established and evaluated three heuristics to select problematic conversations. All of them exploit the detection of linguistic patterns.

3.1   Lexicon and Associated Grammars

The sentiment patterns of our lexicons and grammars have been collected on manually selected extracts of about 300 of the 8 556 conversations. The calls have firstly been selected by using a seed list of sentiment keywords chosen by introspection. This list has been progressively augmented with new seeds from the retrieved calls and has eventually been transformed into the sentiment lexicon and its detection grammar. The result is a lexicon containing more than 1 000 sentiment words and expressions which are typically found in call-center speech. The lexicon does not contain any domain-dependent vocabulary, in our case expressions related to energy.

Each pattern belongs to one of the following five classes, whose design is inspired by the evaluative modalities for discourse analysis as defined in [4]:

- Acceptance – Refusal (a)
- Agreement – Disagreement (b)
- Favorable – Unfavorable Appreciation (c)
- Opinion (d)
- Surprise (e)

When matched, the patterns are normalized into a subclass of these five classes. All subclasses are illustrated in Table 1.

Table 1. Entity classes and subclasses

a	Acceptance	**pourquoi pas** oui *why not yes*
	Refusal	je **refuse** de payer la somme qu'on me demande *I refuse to pay the amount demanded*
b	Total agreement	**tout à fait** le relevé compteur date du mois de novembre *exactly the meter reading dates from November*
	Approximate agreement	je le **conçois** j'ai compris la situation *I hear you I've understood the situation*
	Amending	ce serait **plutôt** pour son appartement qu'il faudrait vérifier *you'd rather check for his apartment*
	Disagreement	je suis **pas d'accord** *I don't agree*
c	Favorable appreciation	Ça c'est **sympa** *That's nice*
	Unfavorable appreciation	je trouve ça **inadmissible** *I think it's unacceptable*

Table 1. (*Continued*)

	Conviction	je vous dis **franchement** c'est trop pour moi *I tell you **straight out** this is too much for me*
	Strong certainty	vous avez **sûrement** un fournisseur pour le gaz *you **certainly** have a gas supplier*
d	Medium certainty	c'est une estimation **je suppose** *it's an estimate **I assume***
	Low certainty	je vous **garantis pas** que ce soit ça *I don't **guarantee** that it is right*
	Doubt	oui mais je **m'interroge** sur les chiffres *yes but **I wonder** about the figures*
	Positive surprise	vous allez avoir une **bonne surprise** *you will have a **good surprise***
e	Neutral surprise	ce qui **m'intrigue** c'est que la banque vous facture des frais *the thing which **intrigues** me is that the bank charges you fees*
	Negative surprise	C'est **bizarre** j'ai pas eu le courrier *it's **odd** that I've not received the mail*

Each subclass has a positive or negative orientation, exception made for the *opinion* subclasses and the *neutral surprise* subclass whose orientation is calculated on the context as explained in 3.2.

The weights are predefined: a weight of 1 is given to subclasses of weak modality (*opinion* subclasses, *approximate agreement* and *amending*), and a weight of 2 to subclasses of strong modality (all the others). When a pattern is considered as emotive, which we define as having a high emotional intensity, its weight is doubled. For example, the weight of "génial" ("*brilliant*", *emotive favorable appreciation*) is double the weight of "intéressant" ("*interesting*", *favorable appreciation*).

Further details on these grammars can be found in [2].

3.2 Calculus of the Sentiment by Speech Turn

We compute a positive score and a negative score for each speech turn. The negative score is the sum of the weights of the negative patterns found in the turn, and the positive score is the sum of the weights of the positive patterns. The orientation for the patterns without orientation depends on the highest frequency of positive or negative patterns in the speech turn. The calculus is illustrated in the following two tables .

Table 2. Example for sentiment calculus

Automatic speech transcript	Translation
c'est **vrai que** non **sinon** c'est **facile** archives tout ça va vraiment y a **aucun problème**	*it is **true that** <no> apart from that it is **easy** <archives> all of this <goes> really there is **no problem***

Table 3. Patterns and their weights in a speech turn

Pattern	Translation	Orientation	Weight
vrai que	true that	-	1
sinon	apart from that	negative	1
facile	easy	positive	2
aucun problème	no problem	positive	2

The negative score of this speech turn is 1 and the positive score is 5. The weight of the pattern "vrai que" has been added to the positive score because the positive polarity is more frequent than the negative in this speech turn.

If a pattern belongs to multiple classes with the same orientation, its weight will be counted only once in the calculus of the speech turns' sentiment weight. When the pattern belongs to classes with a different orientation, the corresponding weight is summed with the speech turn's most frequent polarity.

3.3   Smoothing

The error rate of the automatic transcription, estimated at 30% WER, directly impacts the pattern detection. Its precision falls with 17% and its recall with 28% as evaluated in [3] on the same corpus. The following phenomena occur.

A sentiment expression can be uttered but not transcribed as such leading to silence in the pattern recognition (a). When a non-sentiment expression is uttered and wrongly transcribed into a sentiment pattern, we get noise (b). Less frequently, a sentiment expression may be transcribed into another sentiment pattern, without any guarantee on weight or orientation of the output (c): it may be correct or false. Table 4 shows an example of each of these phenomena. Extracted patterns are in bold.

Table 4. Examples of transcription errors

	Manual transcript	Automatic speech transcript
a	qui **ne me plaît absolument pas** *which **does not please me at all***	y connaît absolument pas *does not know it at all*
b	pour faire mon virement *to make my transfer payment*	pour faire **mentir** *to make **lie***
c	c'est **pas malin** *it is **not clever***	c'est **pas mal** *that's **not bad***

In order to soften the impact of these errors, we take the average scores on a sliding window of five speech turns. We assume that emotions do not appear isolated, and if they do, then they are most probably the result of a speech recognition error.

Fig. 1 is the graphical representation of the evolution of the smoothed positive and negative scores in a conversation. The conversation begins very badly, with a high peak of negative sentiment which drops after a while and shows a happy end.

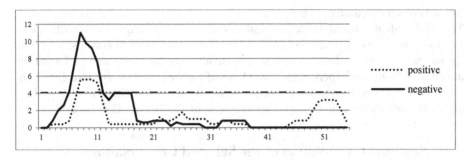

Fig. 1. Polarity curves on a conversation, by speaker turns

3.4   Heuristics for Selecting Problematic Conversations

We've established and tested the three heuristics to select problematic conversations. For this, we only exploit the negative scores.

One or More Peaks (OoMP)
Our first method is based on the curves described above. It returns all the conversations in which the negative curve crosses at least one time an empirically defined threshold. The aim is to find conversations in which a very high number of negative entities occur at the same time. Our experiences showed that 4 seemed to be the best value for this threshold. The conversation illustrated in Fig. 1 would typically be selected by this method. We rank the conversations retrieved by this method by the value of the conversation's highest peak.

Relatively Frequent Medium-to-High Values (MtH)
Our second method is also based on the negative and positive curves. It returns the conversations in which the negative curve is above the medium threshold of 2.5 for at least 8% of the conversation's speech turns. Its aim is to find the conversations in which a certain amount of negative entities are uttered throughout the entire dialogue. Fig. 2 shows an example of a conversation selected by this method. We rank the conversations retrieved by this method by the percentage of turns that surpass the 2.5 threshold.

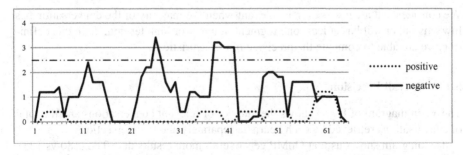

Fig. 2. Conversation selected by MtH, polarity curves, by speaker turns

Term Frequency-emotive (TFe)
The third method is based on the classical measure of Term Frequency. We adapt it to focus on the emotive entities, in order to detect strong negative sentiments: we count the number of emotive detected patterns and divide it by the total number of words in the conversation. Our experiences showed that if the frequency of emotive entities is over 7% in a conversation, this conversation has a high probability of being problematic. We rank the conversations retrieved by this method by the TFe score.

4   Sentiment Annotation of the Selected Conversations

The three above described heuristics selected 264 unique conversations on the total working corpus of 8 556 conversations.

Four computational linguists have listened to these conversations without having access to the automatic transcripts, and annotated them by signaling the presence of anger, high tension, low tension or off-topic segments in the conversation. They also mentioned whether these segments exist once, multiple times or cover the majority of the conversation.

From the start, an annotation guide clearly defined the limits between the different sentiments. *Anger* is annotated when a speaker loses his cool, is aggressive, upset or very annoyed. If at least one speaker is in an awkward position, if there is some animosity or annoyance, then the conversation is annotated with *high tension. Low tension* is annotated when at least one of the speakers is slightly on the defensive or if the exchange is tricky. When a segment is not about the energy supplier, other energy suppliers nor on the topic of energy, then it is considered *off topic.*

Since sentiment perception is subjective, we took some precautions to keep the annotation homogenous. After an initial briefing, the four annotators annotated eight conversations and discussed their annotations in order to apply the same graduation. All other conversations have been annotated by two arbitrarily chosen annotators, who compared and unified their results. For a very small minority of conversations, the opinion of a third annotator was necessary to decide on a persisting disagreement.

5   Evaluation Results

We consider that a conversation is relevant when the majority of the conversation has low tension, or if it has at least one segment of anger or high tension. With this definition, we are able to compute the precision of our heuristics.

5.1   Overall Precision

The Venn diagram of Fig. 3 shows the number of relevant conversations and the total of conversations retrieved by each heuristic separately and by intersection.

The diagram shows us that OoMP gets overall more results than TFe (180 vs 125), but for a lower precision (65% vs 75%).

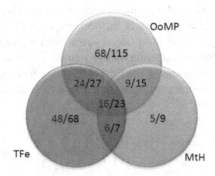

Fig. 3. Relevant conversations / retrieved conversations

MtH seems not very interesting in comparison to the two other heuristics: it doesn't catch a lot of conversations (54), and its global relevancy of 66% of drops to 56% on the few (9) conversations it catches separately.

When tuning the heuristics, we noticed that some off-topic conversations had been retrieved. These are private conversations between agents whose progress and vocabulary do not correspond to the usual call-center conversations. They can contain tension or anger, but or not interesting in the scenario of bettering the customer service. As described below, their elimination improves the mean average precision on the top results. Fig. 4 shows the number of off-topic conversations retrieved by heuristic.

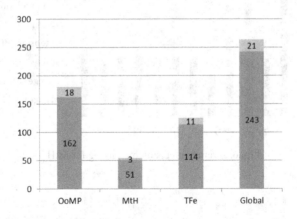

Fig. 4. Part of off-topic conversations (upper)

5.2   Precision by Ranking

For each heuristic we evaluate the ranking of the conversations by calculating the average precision with a 5 documents interval. This means that, exception made for the first document, the average is taken on the first five, ten, fifteen, etc. conversations. This measure seems especially relevant to us, since the results will be integrated in a search engine.

The following figures show the average precision for our three heuristics. Relevant conversations, as defined before, have at least one segment of anger and are designated with "anger", or they have at least one segment with high tension or are low tension on the majority of the conversation, designated with "tension". We have included the part of off-topic conversations in the figures to show their influence on the results. Their parts are indicated with a very light color above each class.

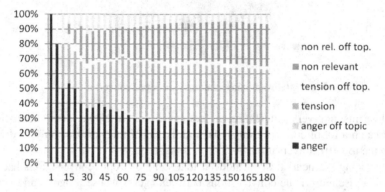

Fig. 5. Average precision by ranking for OoMP

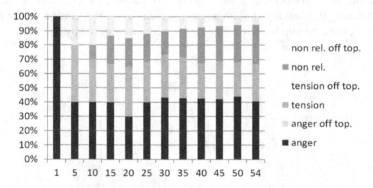

Fig. 6. Average precision by ranking for MtH

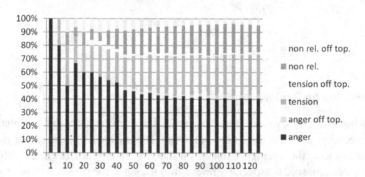

Fig. 7. Average precision by ranking for TFe

These results are satisfying. The precision score on the top results for OoMP are at 80% on the top 15, then fall rapidly to 70% on the top 25, but remains close to this limit on the top 140. TFe has better scores with 80% of precision on the top 30 and a score that stays over 70% for all 125 retrieved conversations. We also see that the TFe selects globally 10% more "anger" conversations than OoMP does. Both methods are clearly complementary, since they don't select the same conversations but are sufficiently relevant to be useful. As for any search engine, users expect to find relevant results in the top of the list. OoMP and TFe are heuristics on linguistic objects that can successfully be used to re-rank the results of a search engine. As we can see, the off-topic conversations rapidly degrade the results and the system may gain performance by identifying them automatically.

These results outperform the results presented in [28] as discussed in the related work section. Their work on automatic quality monitoring is closest to our objectives. We must however bear in mind that we treat other languages and that our results heavily depend on the quality of our lexicon.

6   Summary and Future Work

There is a real need for automating the selection of problematic calls to improve the quality of call center services. We have shown that simple heuristics exploiting linguistic sentiment modeling of call center speech is useful for integration in a real world system, despite the high error rate of the automatic transcription. Our evaluation shows that our system follows the expected behavior of a search engine.

Although this work has proven to be useful on industrial proof-of-concept test cases in the domain of postal services and do-it-yourself businesses, it lacks a thorough evaluation on an independently annotated corpus. Our evaluation should gain in perspective if such a corpus should become available.

Future work includes the identification of off-topic conversations, since this can highly improve the average precision of the top results. This is not necessarily a straightforward classification task, since some conversations have large off-topic segments whereas the rest of the conversation is relevant. Therefore, better results may be achieved by identifying the off-topic parts of conversations and not including those segments in the sentiment calculus.

Acknowledgements. This work has been accomplished within VoxFactory, a French FUI 6 research project labeled by the French business cluster Cap Digital.

References

1. Bazillon, T., Jousse, V., Béchet, F., Estève, Y., Linarès, G., Luzzati, D.: La parole spontanée : transcription et traitement. TAL 49(3), 47–76 (2008)
2. Cailliau, F., Cavet, A.: Analyse des sentiments et transcription automatique: modélisation du déroulement de conversations téléphoniques. TAL 51(3), 131–154 (2010)

3. Cailliau, F., Giraudel, A.: Enhanced Search and Navigation on Conversational Speech. In: Proc. of SSCS 2008, SIGIR 2008 Workshop, Singapour, pp. 66–70 (2008)
4. Charaudeau, P.: Grammaire du sens et de l'expression. Hachette Education, Paris (1992)
5. Chastagnol, C., Devillers, L.: Analysis of Anger across several agent-customer interactions in French call centers. In: ICASSP 2011, pp. 4960–4963 (2011)
6. Cieri, C., Miller, D., Walker, K.: The Fisher Corpus: a Resource for the Next Generations of Speech-to-Text. In: Proc. of LREC 2004, Lisbon, pp. 69–71 (2004)
7. Danesi, C., Clavel, C.: Impact of spontaneous speech features on business concept detection: a study of call-centre data. In: Proc. of SSCS 2010, pp. 11–14. ACM, New York (2010)
8. Devillers, L., Vaudable, C., Chastagnol, C.: Real-life emotion-related states detection in call centers: a cross-corpora study. In: INTERSPEECH 2010, pp. 2350–2353 (2010)
9. Galliano, S., Geoffrois, E., Mostefa, D., Choukri, K., Bonastre, J.-F., Gravier, G.: The ESTER Phase II Evaluation Campaign for the Rich Transcription of French Broadcast News. In: Proc. of Interspeech 2005, Lisbonne, pp. 1149–1152 (2005)
10. Galliano, S., Gravier, G., Chaubard, L.: The ESTER 2 evaluation campaign for the rich transcription of French radio broadcasts. In: Proc. of INTERSPEECH, Brighton, pp. 2583–2586 (2009)
11. Garnier-Rizet, M., Adda, G., Cailliau, F., Guillemin-Lanne, S., Waast, C.: CallSurf - Automatic transcription, indexing and structuration of call center conversational speech for knowledge extraction and query by content. In: Proc. of LREC 2008, Marrakech (2008)
12. Gavalda, M., Schlueter, J.: The truth is out there: Using advanced speech analytics to learn why customers call help-line desks and how effectively they are being served by the call center agent. In: Neustein, A. (ed.) Advances in Speech Recognition: Mobile Environments, Call Centers and Clinics, pp. 221–243. Springer (2010)
13. Gauvain, J.-L., Adda, G., Lamel, L., Lefèvre, F., Schwenk, H.: Transcription de la parole conversationnelle. TAL 45(3), 35–47 (2004)
14. Hastie, H.W., Prasad, R., Walker, M.: What's the trouble: automatically identifying problematic dialogues in DARPA communicator dialogue systems. In: Proc. of ACL 2002, pp. 384–391. ACL, Stroudsburg (2002)
15. Mishne, G., Carmel, D., Hoory, R., Roytman, A., Soffer, A.: Automatic analysis of call-center conversations. In: Proc. of CIKM 2005, pp. 453–459. ACM, New York (2005)
16. Padmanabhan, D., Kummamuru, K.: Mining conversational text for procedures with applications in contact centers. Int. J. Doc. Anal. Recognit. 10(3), 227–238 (2007)
17. Pallotta, V., Delmonte, R., Vrieling, L., Walker, D.: Interaction Mining: the new frontier of Call Center Analytics. In: Proc. of DART 2011, Palermo (2011)
18. Pang, B., Lee, L.: Opinion Mining and Sentiment Analysis. Found. Trends Inf. Retr. 2(1-2), 1–135 (2008)
19. Roy, S., Subramaniam, L.V.: Automatic generation of domain models for call centers from noisy transcriptions. In: Proc. ACL 2006, pp. 737–744. ACL, Stroudsburg (2006)
20. Szöke, I., Burget, L., Černocký, J., Fapšo, M., Karafiát, M., Matějka, P., Schwarz, P.: Comparison of Keyword Spotting Approaches for Informal Continuous Speech. In: Proc. of INTERSPEECH 2005, Lisbon, pp. 633–636 (2005)
21. Taboada, M., Brooke, J., Tofiloski, M., Voll, K., Stede, M.: Lexicon-Based Methods for Sentiment Analysis. Computational Linguistics 37(2), 267–307 (2011)
22. Takeuchi, H., Subramaniam, L.V., Nasukawa, T., Roy, S.: Automatic identification of important segments and expressions for mining of bussiness-oriented conversations at contact centers. In: Proc. of EMNLP-CoNLL, pp. 458–467 (2007)

23. Tang, M., Pellom, B., Hacioglu, K.: Call-type Classification and Unsupervised Training for the Call Center Domain. In: IEEE-ASRU 2003. St. Thomas, US Virgin Islands (2003)
24. Tang, H., Tan, S., Cheng, X.: A survey on sentiment detection of reviews. Expert Systems with Applications 36(7), 10760–10773 (2009)
25. Vaudable, C., Rollet, N., Devillers, L.: Annotation of Affective Interaction in Real-life Dialogs Collected in a Call-center. In: Proc. 3rd Intern. Workshop on Emotion, LREC 2010, Valletta, Malta (2010)
26. Ververidis, D., Kotropoulos, C.: Emotional Speech Recognition: Resources, Features, and Methods. Speech Communication 48, 1162–1181 (2006)
27. Walker, M.A., Langkilde-Geary, I., Wright Hastie, H., Wright, J., Gorin, A.: Automatically Training a Problematic Dialogue Predictor for a Spoken Dialogue System. Journal of Artificial Intelligence Research 16, 293–319 (2002)
28. Zweig, G., Siohan, O., Saon, G., Ramabhadran, B., Povey, D., Mangu, L., Kingsbury, B.: Automated quality monitoring for call centers using speech and NLP technologies. In: Proc. of NAACL-Demonstrations 2006, pp. 292–295. ACL, Stroudsburg (2006)

Cross-Lingual Projections vs. Corpora Extracted Subjectivity Lexicons for Less-Resourced Languages

Xabier Saralegi, Iñaki San Vicente, and Irati Ugarteburu

Elhuyar Foundation,
Osinalde Industrialdea 3,
20160 Usurbil, Spain
{x.saralegi,i.sanvicente,i.ugarteburu}@elhuyar.com

Abstract. Subjectivity tagging is a prior step for sentiment annotation. Both machine learning based approaches and linguistic knowledge based ones profit from using subjectivity lexicons. However, most of these kinds of resources are often available only for English or other major languages. This work analyses two strategies for building subjectivity lexicons in an automatic way: by projecting existing subjectivity lexicons from English to a new language, and building subjectivity lexicons from corpora. We evaluate which of the strategies performs best for the task of building a subjectivity lexicon for a less-resourced language (Basque). The lexicons are evaluated in an extrinsic manner by classifying subjective and objective text units belonging to various domains, at document- or sentence-level. A manual intrinsic evaluation is also provided which consists of evaluating the correctness of the words included in the created lexicons.

Keywords: Sentiment Analysis, Subjectivity Detection, Less Resourced Languages.

1 Introduction

Opinion mining or sentiment analysis are tasks involving subjectivity detection and polarity estimation. Both tasks are necessary in many sentiment analysis applications, including sentiment aggregation and summarization or product comparisons. Researchers into sentiment analysis have pointed out the frequent benefit of a two-stage approach, in which subjective instances are distinguished from objective ones, after which the subjective instances are further classified according to polarity ([1,2,3]). Pang and Lee [2] obtain an improvement from 82.8% to 86.4% for polarity classification by applying a subjectivity classifier in advance. So, developing a method for subjectivity detection seems an adequate first step for building an Opinion mining system for a certain language.

When dealing with subjectivity, some authors proposed rule-based methods [4] which use subjectivity lexicons. Other authors propose supervised methods based on machine learning techniques [1]. In both cases, subjectivity lexicons

A. Gelbukh (Ed.): CICLing 2013, Part II, LNCS 7817, pp. 96–108, 2013.

are an important knowledge resource. So it is clear that subjectivity lexicons are a key resource for tackling this task. Nowadays, there are widely used lexicons, such as OpinionFinder [5], Sentiwordnet [6] and General Inquirer [7], but, as is the case with many NLP resources, those lexicons are geared towards major languages. This means that new subjectivity lexicons must be developed when dealing with many other languages.

As manual building is very costly and often uneconomic for most languages, especially less-resourced languages, machine building methods offer a viable alternative. In that sense, several methods [8,9,10,11,12] have been proposed for building subjectivity lexicons. The methods rely on two main strategies: building the lexicon from corpora or trying to project existing subjectivity resources to a new language. The first approach often produces domain specific results, and so, its performance in out-of-domain environments is expected to be poorer. Projecting a lexicon to another language would produce a resource that would *a priori* be more consistent in all environments. However, as the projection involves a translation process, the errors ocurring at that step could reduce the quality of the final lexicon as shown by Mihalcea et al. [10].

In our research we compared these two cost-effective strategies for building a subjectivity lexicon for a less-resourced language. We assumed that for languages of this type the availability of parallel corpora and MT systems is very limited, and that was why we avoided using such resources. Our contribution lies in a robust cross-domain evaluation of the two strategies. This experiment was carried out using Basque. First, we compared the correctness of the resulting lexicons at word level. Then, the lexicons were applied in a task to classify subjectivity and objectivity text units belonging to different domains: newspapers, blogs, reviews, tweets and subtitles.

The paper is organized as follows. The next chapter offers a brief review of the literature related to this research, and discusses the specific contributions of this work. The third section presents the resources we used for building the subjectivity lexicons, the experiments we designed and the methodology we followed. In the fourth chapter, we describe the different evaluations we carried out and the results obtained. Finally, some conclusions are drawn and we indicate some future research directions.

2   State of the Art

Wilson et al. [13] define a subjective expression as any word or phrase used to express an opinion, emotion, evaluation, stance, speculation, etc. A general covering term for such states is private state. Quirk et al. [14] define a private state as a state that is not open to objective observation or verification: "a person may be observed to assert that God exists, but not to believe that God exists". Belief is in this sense 'private'. So, subjectivity tagging or detection consists of distinguishing text units (words, phrases sentences...) used to present opinions and other forms of subjectivity from text units used to objectively present factual information. Detection is part of a more complex task which

Wilson [15] called subjectivity analysis, which consists of determining when a private state is being expressed and identifying the attributes of that private state. Identifying attributes such as the target of the opinion, the polarity of the subjective unit or its intensity, is outside the range of this work.

2.1 Subjectivity Detection Methods

Methods for subjectivity detection can be divided into two main approaches. Rule-based methods which rely on subjectivity lexicons, and supervised methods based on classifiers trained from annotated corpora.

Wiebe et al. [16] use manually annotated sentences for training Naive Bayes classifiers. Pang and Lee [2] successfully apply Naive Bayes and SVMs for classifying sentences in movie reviews. Wang and Fu [17] present a sentiment density-based naive Bayesian classifier for Chinese subjectivity classification. Das and Bandyopadhyay [18] propose a Conditional Random Field (CRF)-based subjectivity detection approach tested on English and Bengali corpora belonging to multiple domains.

Lexicon-based systems are also proposed in the literature. Turney [8] computed the average semantic orientation of product reviews based on the orientation of phrases containing adjectives and adverbs. The classifier proposed by Riloff and Wiebe [4] uses lists of lexical items that are good subjectivity clues. It classifies a sentence as subjective if it contains two or more of the strongly subjective clues. Das and Bandyopadhyay [19] proposed a classifier which uses sentiment lexicons, theme clusters and POS tag labels.

A third alternative would be to combine both approaches. Yu and Hatzivassiloglou [1] obtain 97% precision and recall using a Bayesian classifier that uses lexical information. This proves that subjectivity lexicons are indeed important resources.

According to Yu and Kübler [20], opinion detection strategies designed for one data domain generally do not perform well in another domain, due to the variation of the lexicons across domains and different registers. They evaluated the subjectivity classification in news articles, semi-structured movie reviews and blog posts using Semi-Supervised Learning (SSL) methods, and obtained results that vary from domain to domain. Jijkoun and de Rijke [21] propose a method to automatically generate subjectivity clues for a specific topic by extending a general purpose subjectivity lexicon.

2.2 Methods for Subjectivity Lexicon Building

Text corpora are useful for obtaining subjectivity and polarity information associated with words and phrases. Riloff et al. [22] adopt a bootstrapping strategy based on patterns to extend a seed set of 20 terms classified as strongly subjective. Baroni and Vegnaduzzo [23] apply the PMI (Pointwise Mutual Information) method to determine term subjectivity. Subjectivity level is measured according to the association degree with respect to a seed set of 35 adjectives marked as subjective.

When tackling the problem of the lack of annotated corpora, many authors propose using MT techniques. Mihalcea and others [10] annotate an English corpus using OpinionFinder [5] and use cross-lingual projection across parallel corpora to obtain a Romanian corpus annotated for subjectivity. Following the same idea, Banea et al. [11] use machine translation to obtain the required parallel corpora. In this case they apply the method for Romanian and Spanish. Wan [12] also proposed the generation of Chinese reviews from English texts by Machine Translation.

Another approach to building a subjective word list in a language is the translation of an existing source language lexicon by using a bilingual dictionary. Mihalcea et al. [10] used a direct translation process to obtain a subjectivity lexicon in Romanian. Their experiments concluded that the Romanian subjectivity clues derived through translation are less reliable than the original set of English clues, due to ambiguity errors in the translation process. Das and Bandyopadhyay [18] proposed improving the translation of ambiguous words by using a stemming cluster technique followed by SentiWordNet validation. Jijkoun and Hofmann [24] apply a PageRank-like algorithm to expand the set of words obtained through machine translation.

Banea et al. [25] compare different methods of subjectivity classification for Romanian. Among subjectivity lexicon building methods, there are bootstrapping a lexicon by using corpus-based word similarity, and translating an existing lexicon. They conclude that the corpus-based bootstrapping approach provides better lexicons than projection.

In this work we wanted to analyse strategies for developing a subjectivity lexicon for a Less-Resourced Language. We assumed that such languages can only avail themselves of monolingual corpora and bilingual lexicons. So parallel corpora, MT system-based approaches and approaches based on large subjectivity annotated corpora are not contemplated. We focused on a corpus-based approach and projection onto the target language.

3   Experiments

Projection-based lexicon building requires a subjectivity lexicon L_S_s in a source language s and a bilingual dictionary $D_{s \to t}$ from s to the target language t. In our experiments we took the English subjectivity lexicon (L_S_{en}) introduced in [5] as a starting point. L_S_{en} contains 6,831 words (4,743 strong subjective and 2,188 weak subjective). According to the authors, those subjective words were collected from manually developed resources and also from corpora. Strong subjective clues have subjective meanings with high probability, and weak subjective clues have a lower probability of having subjective meanings. As for the bilingual dictionary, a bilingual English-Basque dictionary $D_{en \to eu}$ which includes 53,435 pairs and 17,146 headwords was used.

Corpora-based lexicon extraction requires subjective and objective corpora. Subjective and objective corpora can be built by using simple heuristics. News from newspapers or Wikipedia articles can be taken as objective documents.

Opinion articles from newspapers can be taken as subjective articles. Those heuristics are not trouble free, but then again, they allow us to create low-cost annotated corpora. Using news as an objective corpus can be a rough heuristic because, according to Wiebe et al. [26], many sentences (44%) included in news are subjective. On the other hand, as Wikipedia belongs to a different domain from that of newspaper opinion articles, some divergent words can be incorrectly identified as subjective if we compare a Wikipedia corpus with a subjective corpus comprising opinion articles, due to the fact that they are a feature in the journalism domain but not in Wikipedia texts.

We built a subjective corpus TC_S_{eu} by taking 10,661 opinion articles from the Basque newspaper Berria[1]. Two objective corpora were built: one by collecting 50,054 news items from the same newspaper TCN_O_{eu}, and the other by gathering all the articles (143,740) from the Basque Wikipedia TCW_O_{eu}. A subset of TCN_O_{eu} containing the same number of articles as TC_S_{eu} was also prepared for parameter tuning purposes which we will name $TCN\_O'_{eu}$.

3.1   Cross-Lingual Projection of the Subjectivity Lexicon

We translated the English subjectivity lexicon L_S_{en} by means of a bilingual dictionary $D_{en \to eu}$ to create a Basque subjectivity lexicon L_P_{eu}. Ambiguities are resolved by taking the first translation[2]. Using this method we obtained translations for 36.67% of the subjective English words: L_P_{eu} includes 1,402 strong and 1,169 weak subjective words. The number of translations obtained was low, especially for strong subjective words. Most of these words are inflected (e.g., "terrified", "winners", ...) forms or derived words where prefixes or suffixes have been added (e.g., "inexact", "afloat", ...).

According to Mihalcea et al. [10] translation ambiguity is another problem that distorts the projection process. In their experiments Romanian subjectivity clues derived through translation were less reliable than the original set of English clues. In order to measure to what extent that problem would affect our projection, we randomly selected 100 English words and their corresponding translations. Most of the translations (93%) were correct and subjective according to a manual annotation involving two annotators (97% inter-tagger agreement, Cohen's k=0.83). So we can say that the translation selection process is not critical. We annotated as correct translations those corresponding to the subjective sense of the English source word. Unlike Mihalcea et al. [10], we did not analyse whether the translated word had less subjective connotation than the source word.

3.2   Corpus-Based Lexicon Building

Our approach was based on inferring subjective words from a corpus which includes subjective and objective documents. So, we identified as subjective words

---

[1] http://berria.info

[2] The bilingual dictionary has its translations sorted according to their frequency of use, so the first translation method should provide us with the most common translations of the source words.

those whose relevance in subjective documents is significantly higher than in objective documents. We adopted a corpus-based strategy, because it is affordable and easily applicable to less-resourced languages. We extracted Basque subjectivity lexicons in accordance with various relevance measures and objective corpora. TC_S_{eu} was used as the subjective corpus, and TCW_O_{eu} (Wikipedia) or TCN_O_{eu} (News) as objective corpora. For each word w in the subjective corpus we measured its degree of relevance with respect to the subjective corpus as compared with the objective corpus. That way we obtained the most salient words in a certain corpus, the subjective corpus in this case. We took that degree of relevance as the subjectivity degree $bal(w)$. That degree was calculated by the Log Likelihood ratio (LLR) or by the percentage difference ($\%DIFF$). Maks and Vossen [27] compared LLR and $\%DIFF$ for that purpose, and obtained better results by using $\%DIFF$.

In order to evaluate the adequacy of the measurements (LLR or $\%DIFF$) and the various corpus combinations (Wikipedia or News for the objective part), we analysed how subjective and objective words are distributed through the rankings corresponding to the different combinations (LLR_News, $DIFF\_News$, $DIFF\_Wiki$ and LLR_Wiki). For that aim, two references were prepared. The first one includes only subjective words, while the second one includes both objective and subjective words. The first reference was built automatically by taking the strong subjective words of L_P_{eu}. For the second reference three annotators manually tagged subjective and objective words in a sample of 500 words selected randomly from the intersection of all candidate dictionaries ($DIFF\_Wiki$, $DIFF\_News$, LLR_Wiki and LLR_News). The overall inter-agreement between the annotators was 81.6% (Fleiss' k=0.63). Simple majority was used for resolving disagreements (27% of the words evaluated).

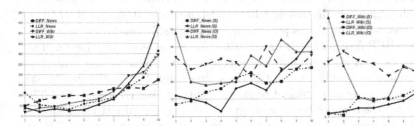

Fig. 1. Distribution of subjective words with various measure and corpus combinations

Fig. 2. Distribution of subjective and objective words using TCN_O_{eu} as objective corpus

Fig. 3. Distribution of subjective and objective words using TCW_O_{eu} as objective corpus

According to the results shown in Figures 1, 2 and 3 Wikipedia seems to be a more adequate objective corpus. It provides a higher concentration of subjective words in the first positions of the rankings[3] (i.e. last intervals) than News when

---

[3] In Figures 1, 2, 3 and 4, higher intervals contain words scoring higher in the rankings.

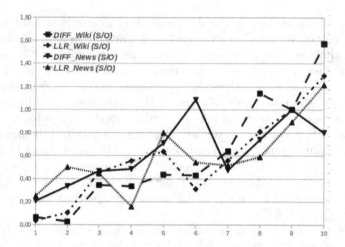

Fig. 4. Subjective/objective ratio with respect to ranking intervals

using both measurements and for both references. In addition, the concentration of objective words in the first positions is slightly lower when using TCW_O_{eu}, compared with using TCN_O_{eu} as the objective reference corpus.

Regarding the measurements, LLR provides better distributions of subjective words than $\%DIFF$ for both reference corpora. The highest concentration of the subjective words is in the first positions of the rankings. However $\%DIFF$ seems to be more efficient for removing objective ones from first ranking positions. Figure 4 plots the distribution of subjective/objective word rates across different ranking intervals. The best ratio distribution is achieved by the $\%DIFF$ measurement when used in combination with TCW_O_{eu}.

In terms of size, corpora-based lexicons are bigger than the projection-based one. For high confidence thresholds, $LLR > 3.84$, $p\text{-value} < 0.05$; and $\%DIFF > 100$ [27], corpora-based lexicons provide 9,761; 6,532; 8,346 and 6,748 words for $DIFF\_Wiki$, $DIFF\_News$, LLR_Wiki and LLR_News, respectively. These will be the dictionaries used in the evaluation presented in the next section. The sizes of these dictionaries are close to that of the source English lexicon L_S_{en} (6,831 words). However, after projecting it to Basque, this number goes down to 2,571. So it seems that the corpora-based strategy provides bigger subjectivity lexicons. Then again, we have to take into account that corpus-based lexicons include several objective words (See Figure 1.). In addition, corpus-based lexicons are biased towards the domain of journalism.

4   Evaluation

4.1   Classifier

In this work, we adopted a simple lexicon-based classifier similar to the one proposed in [28]. We propose the following ratio for measuring the subjectivity of a text unit tu:

$$subrat(tu) = \sum_{w \in tu} bal(w)/|tu| \tag{1}$$

where $bal(w)$ is 1 if w is included in the subjectivity lexicon[4].

Those units that reach a threshold are classified as subjective. Otherwise, the units are taken as objective. Thresholds are tuned by maximising accuracy when classifying the training data at document level. Even if most of the evaluation data collections are tagged at sentence level, the lack of a sentence level annotated training corpus led us to choose this parameter optimisation method. In order to tune the threshold with respect to a balanced accuracy for subjective and objective classification, tuning is done with respect to a balanced training corpus comprising TC_S_{eu} and $TCN\_O'_{eu}$, which we will call $Train\_D$.

4.2   Annotation Scheme

We evaluated the subjectivity lexicons obtained by the different methods in an extrinsic manner by applying them within the framework of a classification task. That way we measured the adequacy of each lexicon in a real task. The gold-standard used for measuring the performance comprises subjective and objective text units that belong to different domains. As we mentioned in section 2.1, the performance of subjectivity classification systems is very sensitive to the application domain. In order to analyse that aspect, we prepared the following test collections:

– Journalism documents ($Jour\_D$) and sentences ($Jour\_S$): texts collected from the Basque newspaper Gara[5].
– Blog sentences ($Blog\_S$): texts collected from Basque blogs included in the website of Berria.
– Twitter sentences ($Tweet\_S$): tweets collected from the aggregator of Basque tweets Umap[6]. Only tweets written in standard Basque are accepted.
– Sentences of music reviews (Rev_S): reviews collected from the Gaztezulo[7] review site.
– Sentences of subtitles (Sub_S): subtitles of different films are collected from the azpitituluak.com site.

In the case of documents, no manual annotation was done. Following the method explained in section 3, we regarded all opinion articles as subjective, and all news articles as objective. The sentences were manually annotated. Our annotation scheme is simple compared to that used in MPQA [5] which represents private states and attributions. In contrast, our annotation is limited to tagging a sentence as subjective if it contains one or more private state expression; otherwise, the sentence is objective. A private state covers opinions, beliefs, thoughts, feelings, emotions, goals, evaluations, and judgements.

---

[4] We experimented using weights based on the strength of subjectivity but no improvement was achieved, and so, these results are not reported.
[5] http://www.gara.net
[6] http://umap.eu/
[7] http://www.gaztezulo.com/

Table 1. Statistics and class distribution of the reference collections

Source	Unit	Domain	# units	# sub+	# sub	# obj	# obj+
Train_D	document	Journalism	21,320	10,660		10,660	
Jour_D	document	Journalism	9,338	4,669		4,669	
Jour_S	sentence	Journalism	192	60	46	35	51
Blog_S	sentence	Blog	206	94	50	20	42
Tweet_S	sentence	Twitter	200	69	40	21	70
Rev_S	sentence	Music Reviews	138	54	36	24	24
Sub_S	sentence	Subtitles	200	98	31	20	51

We classified sentences according to four categories, depending on aspects such as the number of private state expressions, their intensity, etc.: completely subjective (sub+); subjective but containing some objective element (sub); mostly objective but containing some subjective element (obj); and completely objective (obj+). In order to obtain a robust annotation, three references per annotation were done by three different annotators. Disagreement cases were solved in two different ways. Firstly, annotators discussed all sentences including three different annotations or two equal annotations and a third that was to a distance of more than one category, until consensus was achieved. For dealing with the rest of the disagreement cases, majority voting was used. Table 1 shows the statistics for the test collections and the results of our annotation work.

4.3   Results

By means of our average ratio classifier, we classified the text units in the seven collections presented in the previous section. As mentioned in section 4.1, the units in the test collections were classified according to the subjectivity threshold tuned over the documents in *Train_D*. The optimum subjectivity threshold is computed for each lexicon we evaluated (L_P_{eu}, *DIFF_News*, *LLR_News*, *DIFF_Wiki* and *LLR_Wiki*).

Table 2 and 3 present overall accuracy results and F-score results of the subjective units achieved by the different lexicons in the various test collections.

Table 2. Accuracy results for subjectivity and objectivity classification

	L_P_{eu}	*DIFF_Wiki*	*DIFF_News*	*LLR_Wiki*	*LLR_News*
Train_D	0.63	0.66	**0.90**	0.64	0.87
Jour_D	0.74	0.76	0.80	0.74	**0.87**
Jour_S	0.63	0.59	0.57	0.58	**0.64**
Blog_S	0.65	**0.73**	0.66	**0.73**	0.72
Tweet_S	**0.68**	0.58	0.62	0.59	0.60
Rev_S	**0.70**	**0.70**	0.67	0.67	0.67
Sub_S	0.67	**0.71**	0.70	0.67	0.67

Table 3. F-score results for subjectivity classification

	L_P_{eu}	$DIFF\_Wiki$	$DIFF\_News$	LLR_Wiki	LLR_News
$Train\_D$	0.65	0.68	**0.90**	0.68	0.87
$Jour\_D$	0.75	0.77	0.82	0.75	**0.86**
$Jour\_S$	0.73	0.71	0.58	0.72	**0.74**
$Blog\_S$	0.76	0.82	0.77	**0.83**	0.83
$Tweet\_S$	**0.73**	0.69	0.70	0.70	0.71
Rev_S	0.79	0.77	0.78	0.75	**0.80**
Sub_S	0.78	**0.81**	0.79	0.78	0.79

In this evaluation, only a binary classification was performed, text units belonging to **obj** and **obj+** classes were grouped into a single category, and the same was done for **sub** and **sub+**. Firstly, according to those results, corpus-based lexicons compiled using TCN_O_{eu} (News) as objective reference (columns 3 and 5) are very effective for document classification. The projected lexicon L_P_{eu} performs significantly worse. Those results were expected, since the corpora-based lexicons have the domain advantage. However, L_P_{eu}'s performance is comparable to corpus-based lexicons' on non-journalistic domains. Moreover, it is better than the corpus-based lexicons in the Twitter domain, both in terms of accuracy and F-score of subjective units. Taking all the results into account, we can see that despite the better performance of corpus-based lexicons in most the domains, the performance of the projected lexicon is more stable across domains than the performance of corpus-based lexicons.

With regard to the corpus used as objective reference (columns 2 and 4 versus columns 3 and 5), the use of the wikipedia corpus TCW_O_{eu} improves the results of the News corpus only in non-journalistic domains and in terms of accuracy. Furthermore, Table 3 shows that if we only take into account the classification of subjective text units, TCN_O_{eu} performs better in all cases except for the subtitle domain collection.

Differences between LLR and $\%DIFF$ vary across the domains. In terms of accuracy, $\%DIFF$ provides better performance when dealing with tweets, reviews, and subtitles. On the contrary, in terms of F-score of subjective units, $\%DIFF$ is only better over subtitles.

We used 4 categories to annotate the references with different degrees of subjectivity. It is interesting how the performance of subjectivity detection changes depending on the required subjectivity degree. In some scenarios only the detection of highly subjective expressions is demanded. In order to adapt the system to those scenarios, we optimised the subjectivity threshold by maximising the $F_{0.5}$-score against training data. Table 4 shows precision and recall results for subjectivity detection if we only accept the ones that belong to the class **sub+** as subjective sentences. According to those results, with the new optimisation of the threshold, the system's performance for classifying **sub+** is similar to that of the initial system.

Table 4. Precision, recall and F-score results for detecting clearly subjective sentences

	L_P_{eu} sub+			LLR_News sub+		
	P	R	F	P	R	F
$Jour\_S$	0.61	0.90	0.73	0.65	0.84	0.73
$Blog\_S$	0.73	0.80	0.76	0.74	0.96	0.83
$Tweet\_S$	0.67	0.82	0.73	0.64	0.83	0.72
Rev_S	0.73	0.86	0.79	0.65	0.99	0.79
Sub_S	0.69	0.88	0.78	0.68	0.99	0.80

5   Conclusions and Future Work

This paper has presented the comparison between two techniques to automatically build subjectivity lexicons. Both techniques only rely on easily obtainable resources, and are adequate for less-resourced languages.

Our results show that subjectivity lexicons extracted from corpora provide a higher performance than the projected lexicon over most of the domains. Accuracies obtained with this method range from 87%, in case of the document classification, to 60-67%, in case of sentences. Projection provides a slight better performance only when dealing with non-journalistic domains. So, it could be an alternative for those domains. If we are interested in identifying only very subjective sentences, both methods offer a good performance (0.72-0.83 in terms of F-score), in particular, the corpora extracted subjectivity lexicons. Hence, the resources obtained with our methods could be applied in social-media analysis tasks where precision is the priority.

Regarding to ongoing and future work, as we have already mentioned, the methods we have researched in this paper are applicable to less-resourced languages because they only require widely available resources. At the moment, we are analysing the effect the characteristics (size, domain,...) of the resources used have on the quality of the final subjectivity lexicon. In the future, we plan to evaluate the Bootstrapping method proposed by Banea et al. [11], which also relies on corpora.

Acknowledgements. This work has been partially funded by the Industry Department of the Basque Government under grants IE12-333 (Ber2tek project) and SA-2012/00180 (BOM2 project).

References

1. Yu, H., Hatzivassiloglou, V.: Towards answering opinion questions: separating facts from opinions and identifying the polarity of opinion sentences. In: Proceedings of the 2003 Conference on Empirical Methods in Natural Language Processing, EMNLP 2003, pp. 129–136 (2003)

2. Pang, B., Lee, L.: A sentimental education: sentiment analysis using subjectivity summarization based on minimum cuts. In: Proceedings of the 42nd Annual Meeting of the Association for Computational Linguistics, ACL 2004 (2004)

3. Wilson, T., Wiebe, J., Hoffmann, P.: Recognizing contextual polarity in phrase-level sentiment analysis. In: Proceedings of HLT/EMNLP 2005, pp. 347–354 (2005)

4. Riloff, E., Wiebe, J.: Learning extraction patterns for subjective expressions. In: Proceedings of the 2003 Conference on Empirical Methods in Natural Language Processing, pp. 105–112 (2003)

5. Wiebe, J., Wilson, T., Cardie, C.: Annotating expressions of opinions and emotions in language. Language Resources and Evaluation 39(2), 165–210 (2005)

6. Esuli, A., Sebastiani, F.: SENTIWORDNET: a publicly available lexical resource for opinion mining. In: Proceedings of LREC 2006, pp. 417–422 (2006)

7. Stone, P., Dunphy, D., Smith, M.: The general inquirer: A computer approach to content analysis (1966)

8. Turney, P.D.: Thumbs up or thumbs down?: semantic orientation applied to unsupervised classification of reviews. In: Proceedings of the 40th Annual Meeting on Association for Computational Linguistics, ACL 2002, Philadelphia, Pennsylvania, p. 417 (2002)

9. Kaji, N., Kitsuregawa, M.: Building lexicon for sentiment analysis from massive collection of HTML documents. In: Proceedings of the 2007 Joint Conference on Empirical Methods in Natural Language Processing and Computational Natural Language Learning, EMNLP-CoNLL 2007, Prague, Czech Republic, pp. 1075–1083 (2007)

10. Mihalcea, R., Banea, C., Wiebe, J.: Learning multilingual subjective language via cross-lingual projections. In: Proceedings of the 45th Annual Meeting of the Association for Computational Linguistics, vol. 45, p. 976 (2007)

11. Banea, C., Mihalcea, R., Wiebe, J., Hassan, S.: Multilingual subjectivity analysis using machine translation. In: Proceedings of the Conference on Empirical Methods in Natural Language Processing, EMNLP 2008, pp. 127–135 (2008)

12. Wan, X.: Using bilingual knowledge and ensemble techniques for unsupervised chinese sentiment analysis. In: Proceedings of the Conference on Empirical Methods in Natural Language Processing, EMNLP 2008, pp. 553–561 (2008)

13. Wilson, T., Hoffmann, P., Somasundaran, S., Kessler, J., Wiebe, J., Choi, Y., Cardie, C., Riloff, E., Patwardhan, S.: OpinionFinder. In: Proceedings of HLT/EMNLP on Interactive Demonstrations, Vancouver, British Columbia, Canada, pp. 34–35 (2005)

14. Quirk, R., Greenbaum, S., Leech, G., Svartvik, J.: A comprehensive grammar of the English language. Pearson Education India (1985)

15. Wilson, T.A.: Fine-grained Subjectivity and Sentiment Analysis: Recognizing the Intensity, Polarity, and Attitudes of Private States. ProQuest (2008)

16. Wiebe, J.M., Bruce, R.F., O'Hara, T.P.: Development and use of a gold-standard data set for subjectivity classifications. In: Proceedings of the 37th Annual Meeting of the Association for Computational Linguistics on Computational Linguistics, ACL 1999, pp. 246–253 (1999)

17. Wang, X., Fu, G.H.: Chinese subjectivity detection using a sentiment density-based naive bayesian classifier. In: 2010 International Conference on Machine Learning and Cybernetics (ICMLC), vol. 6, pp. 3299–3304 (2010)

18. Das, A., Bandyopadhyay, S.: Subjectivity detection in english and bengali: A CRF-based approach. In: Proceeding of ICON (2009)

19. Das, A., Bandyopadhyay, S.: Theme detection an exploration of opinion subjectivity. In: 3rd International Conference on Affective Computing and Intelligent Interaction and Workshops, ACII 2009, pp. 1–6 (September 2009)
20. Yu, N., Kübler, S.: Filling the gap: Semi-supervised learning for opinion detection across domains. In: Proceedings of the Fifteenth Conference on Computational Natural Language Learning, pp. 200–209 (2011)
21. Jijkoun, V., de Rijke, M.: Bootstrapping subjectivity detection. In: Proceedings of the 34th International ACM SIGIR Conference on Research and Development in Information Retrieval, SIGIR 2011, New York, NY, USA, pp. 1125–1126 (2011)
22. Riloff, E., Wiebe, J., Wilson, T.: Learning subjective nouns using extraction pattern bootstrapping. In: Proceedings of the Seventh Conference on Natural Language Learning at HLT-NAACL 2003, Edmonton, Canada, pp. 25–32 (2003)
23. Baroni, M., Vegnaduzzo, S.: Identifying subjective adjectives through web-based mutual information. In: Proceedings of the 7th Konferenz zur Verarbeitung Natürlicher Sprache, KONVENS 2004, pp. 613–619 (2004)
24. Jijkoun, V., Hofmann, K.: Generating a non-english subjectivity lexicon: relations that matter. In: Proceedings of the 12th Conference of the European Chapter of the Association for Computational Linguistics, EACL 2009, pp. 398–405 (2009)
25. Banea, C., Mihalcea, R., Wiebe, J.: Multilingual sentiment and subjectivity analysis. In: Multilingual Natural Language Processing (2011)
26. Wiebe, J., Wilson, T., Bell, M.: Identifying collocations for recognizing opinions. In: Proceedings of the ACL 2001 Workshop on Collocation: Computational Extraction, Analysis and Exploitation, pp. 24–31 (2001)
27. Maks, I., Vossen, P.: Building a fine-grained subjectivity lexicon from a web corpus. In: Proceedings of the Eight International Conference on Language Resources and Evaluation (LREC 2012), Istanbul, Turkey (May 2012)
28. Wang, D., Liu, Y.: A cross-corpus study of unsupervised subjectivity identification based on calibrated em. In: Proceedings of the 2nd Workshop on Computational Approaches to Subjectivity and Sentiment Analysis, WASSA 2011, pp. 161–167 (2011)

Predicting Subjectivity Orientation
of Online Forum Threads

Prakhar Biyani[1], Cornelia Caragea[2], and Prasenjit Mitra[1]

[1] The Pennsylvania State University, US
[2] University of North Texas, US
{pxb5080,pmitra}@ist.psu.edu, ccaragea@unt.edu

Abstract. Online forums contain huge amounts of valuable information in the form of discussions between forum users. The topics of discussions can be subjective seeking opinions of other users on some issue or non-subjective seeking factual answer to specific questions. Internet users search these forums for different types of information such as opinions, evaluations, speculations, facts, etc. Hence, knowing subjectivity orientation of forum threads would improve information search in online forums. In this paper, we study methods to analyze subjectivity of online forum threads. We build binary classifiers on textual features extracted from thread content to classify threads as subjective or non-subjective. We demonstrate the effectiveness of our methods on two popular online forums.

1 Introduction

Online forums contain huge amounts of discussions between Internet users on various domain-specific problems such as Mac OS products, cameras, operating systems, music, traveling, health, as well as daily life experiences. Such information is difficult to find in other online sources (e.g., product manuals, Wikipedia, etc), and hence, these forums are increasingly becoming popular among Internet users. Topics of discussion in online forum threads can be *subjective* or *non-subjective*. Subjective topics seek personal opinions or viewpoints, whereas non-subjective topics seek factual information.

Different users have different needs. Some search the web for subjective information like discussions on a certain topic to educate themselves about multiple points of view related to the topic, people's emotions, etc. Others pose queries that are objective and have short factual answers. Specifically, a user may want to learn what other people think about some problem, e.g., *"which is the best camera for beginners?"* or they may want un-opinionated information such as facts or verifiable information, e.g., *"what do the numbers on camera lenses mean?"*. We call the former question as *subjective* and the latter as *non-subjective*.

Subjective information needs are more likely to be satisfied by forum threads discussing subjective topics and non-subjective information needs are more likely to be satisfied by forum threads discussing non-subjective topics. Let us consider this example. A user has two information needs related to Canon 7D camera that

A. Gelbukh (Ed.): CICLing 2013, Part II, LNCS 7817, pp. 109–120, 2013.

he conveys to some camera forum's search engine by issuing the following queries: 1. "How is the resolution of canon 7D?", and 2. "What is the resolution of canon 7D?". Both queries are about the resolution of canon 7D (and may look similar at first sight) but the user's intent is different across the two queries. In the first query, the user seeks opinions of different camera users on the resolution of the Canon 7D camera, i.e., how different users feel about the resolution, what are their experiences (good, bad, excellent, etc.) with Canon 7D as far as its resolution is concerned; hence, the query is subjective. In the second query, the user does not seek opinions but an answer to a specific question, which in this case, is the value of the resolution and therefore the query is non-subjective. Hence, prior knowledge of the subjectivity of threads would help in satisfying users' information needs more effectively by taking into account the user's intent in addition to the keywords in the query. In order to answer such queries effectively, forum search engines need to identify subjective threads in online forums and differentiate them from threads providing non-subjective information. Threads can be filtered by matching their subjectivity orientation with that of the query or they can be ranked by combining scores of lexical relevance and subjectivity match with the query.

Here, we address the first part of this vision; we show how to identify the subjectivity of threads in an online forum with high accuracy using simple word features. Recent works on online forum thread retrieval have taken into account the distinctive properties of online threads such as conversational structure [1], and hyperlinking patterns and non-textual metadata [2] to improve their retrieval. Previous works on subjectivity analysis in social media have mainly focused on online review sites for opinion mining and sentiment analysis [3,4,5] and on improving question-answering in community QA [6,7,8,9]. In contrast, our focus is on analyzing subjectivity in online forums using content based features.

We propose a simple and effective classification method using textual features obtained from online forum threads to identify subjective threads of discussion. We model the task as a binary classification of threads in one of the two classes: subjective and non-subjective. We say a thread is subjective if its topic of discussion is subjective and non-subjective if its topic is non-subjective. We used combinations of words and their parts-of-speech tags as features. The features were generated from the text in: (i) the title of a thread, (ii) the title and initial post of a thread and (iii) the entire thread. We performed experiments on two popular online forums (Dpreview and Trip Advisor–New York forums). We used ensemble techniques to improve learning of classifiers on unbalanced datasets and also explored the effects of feature selection to improve the performance of our classifiers. Our experiments show that our classifiers using textual features produce highly accurate results with respect to F1-measure.

Our contributions are as follows. We show that simple features generated from n-grams and parts-of-speech tags work *effectively* for identifying subjective and non-subjective discussion threads in online forums. We believe that online forum search engines can improve their ranking functions by taking into account the subjectivity match between users' queries and threads.

2   Related Work

Subjectivity analysis has received a lot of attention in the recent literature. For example, subjectivity analysis of sentences has been widely researched in the field of Sentiment Analysis [3,10,4,5]. An integral part of sentiment analysis is to separate opinionated (generally subjective) sentences from un-opinionated (non-subjective) sentences [10] by classifying sentences as subjective or non-subjective and then sentiments in the opinionated sentences are classified as positive or negative. Finally, a summary of sentiments is generated [4]. Previous works in this field have mainly focused on online product reviews sites where the aim is to summarize product reviews given by the users [3,5]. In contrast, our work aims at predicting subjectivity orientation of forum threads for use in improving retrieval. In sentiment analysis, only subjective sentences are of interest because sentiments are generally expressed in subjective languages whereas in our case, a user's query governs the interest, i.e., threads having similar subjectivity orientation (subjective or non-subjective) as that of a user's query are of interest.

Other recent works have used subjectivity analysis to improve question-answering in social media [6,7,8,9,11] and multi-document summarization [12,13]. For example, Stoyanov et al., [8] identify opinions and facts in questions and answers to make multi-perspective question-answering more effective. They showed that answers to opinion questions have different properties than answers to factual questions, e.g., opinion answers were approximately twice as long as fact answers. They used these differences to filter factual answers for opinion questions thereby improving answer retrieval for opinion questions. Somasundaran et al., [11] recognized two types of attitudes in opinion sentences: sentiment and arguing and used it to improve answering of attitude questions by matching the attitude type of the questions and answers in multi-perspective QA. Li et al. [6] used classification to identify subjectivity orientation of questions in community QA. Gurevych et al. [7] used an unsupervised lexicon based approach to classify questions as subjective or factoid (non-subjective). They manually extracted patterns of words that are indicative of subjectivity from annotated questions and scored test questions based on the number of patterns present in them. These works analyzed the subjectivity of questions and answers that are usually given by *single authors in community sites*. In contrast, we analyze the subjectivity of *online forum threads that contain replies from multiple authors*.

In our previous work [14], we performed thread level subjectivity classification using thread-specific non-lexical features. In contrast, in this work, we use ensembles of classifiers built on balanced samples using lexical features.

Next, we state our problem and describe various features used in the subjectivity classification task.

3   Problem Statement and Approach

An online forum thread starts with a topic of discussion posted by the (thread) starter in the title and initial post of the thread. The topic can either be subjective or non-subjective. Following the definitions of subjective and objective sentences given by Bruce et. al.[15], we say that a thread's topic is *subjective* if the thread starter seeks private states of minds of other people such as opinions, evaluations, speculations, etc. and *non-subjective* if the thread starter seeks factual and/or verifiable information. We call a thread subjective if its topic of discussion is subjective and non-subjective if it discusses a non-subjective topic. We assume that subjective threads have discussions, mainly, in subjective languages whereas non-subjective threads discuss, mainly, in factual languages. We note that there may be cases where this assumption does not hold good, however, analysis of such exceptional cases is not the focus of this paper and is left for future work.

Problem Statement: Given an online forum thread T, classify it into one of the two classes: subjective (denoted by $+1$) or non-subjective (denoted by -1).

In this work, we assume that a thread discusses a single topic which is specified by the thread starter in the title and the initial post. Analyzing subjectivity of threads with multiple topics is a separate research problem that is out of scope of this work.

3.1 Feature Generation

Intuitively, in online forums, threads discussing subjective topics would contain more subjective sentences compared to threads discussing non-subjective topics. This difference usually results in different vocabulary and grammatical structures of these two types of sentences [16]. To capture this intuition, we used words, parts-of-speech tags and their combinations as the features for classification. These features have been shown to perform well in other subjectivity analysis tasks [17,18,19]. We used the *Lingua-en-tagger* package from CPAN[1] for part-of-speech tagging. The following features were extracted for a sentence in different structural elements (title, initial post, reply posts) of a thread:

- **Bag of Words (BoW):** all words of a sentence.
- **Unigrams + POS tags (BoW+POS):** all words of a sentence and their parts-of-speech tags.
- **Unigrams + bigrams (BoW+Bi):** all words and sequences of 2 consecutive words in a sentence.
- **Unigrams + bigrams + POS tags (BoW+Bi+POS):** all words, their parts-of-speech tags and sequences of 2 consecutive words in a sentence.

Table 1 describes feature generation on a sentence containing three words W_i, W_{i+1} and W_{i+2} and POS_i, POS_{i+1} and POS_{i+2} are the parts-of-speech tags for the words W_i, W_{i+1} and W_{i+2}, respectively. For feature representation we used term frequency (as we empirically found it to be more effective than *tf-idf* and *binary*) as the weighting scheme and used minimum document frequency for a term as 3 (we experimented with minimum document frequency 3, 5 and 10 and 3 gave the best results).

---

[1] http://search.cpan.org/dist/Lingua-EN-Tagger/Tagger.pm

Table 1. Feature generation for sentence W_i W_{i+1} W_{i+2}

Feature type	Generated feature
BoW	W_i, W_{i+1}, W_{i+2}
BoW+POS	$W_i, POS_i, W_{i+1}, POS_{i+1}, W_{i+2}, POS_{i+2}$
BoW+Bi	$W_i, W_{i+1}, W_{i+1}, W_iW_{i+1}, W_{i+1}W_{i+2}$
BoW+Bi+POS	$W_i, POS_i, W_{i+1}, POS_{i+1}, W_{i+2}, POS_{i+2}, W_iW_{i+1}, W_iPOS_{i+1},$
	$POS_iW_{i+1}, W_{i+1}W_{i+2}, W_{i+1}POS_{i+2}, POS_{i+1}W_{i+2}$

3.2 Model Training

We used a Naive Bayes classifier [20] for classification as it performs well on word features. We experimented with Support Vector Machines and Logistic Classifiers with *tf*, *tf-idf*, and *binary* as the feature encoding schemes, and found that the Naive Bayes classifier gave the best results. The Naive Bayes classifier outputs the following two probabilities for a test thread T: $P(+1|T)$, i.e., the probability of thread T belonging to the subjective class and $P(-1|T)$, i.e., the probability of thread T belonging to the non-subjective class, where $P(+1|T) + P(-1|T) = 1$.

Our datasets are highly unbalanced (as described in Section 4) with a majority of the threads belonging to the subjective class. In this setting, even a classifier labeling all the instances as subjective would give reasonably high overall accuracy while performing poorly on the minority class (the non-subjective class). To address this problem, one way is to create a balanced dataset by undersampling from the majority class an equal number of instances to the minority class size and then train a classifier on that dataset. Such a classifier is highly dependent on the small sample.

To address this problem, we used an *ensemble* of classifiers approach [21]. We created multiple balanced samples by taking all the threads of the minority class and sampling (multiple times) an equal number of threads from the majority class. We trained a classifier on each balanced sample. However, our test sets retain the "natural" distribution of the data, which is unbalanced. On the test set, we combined the predictions of all the classifiers for each instance. More precisely, we created n balanced datasets $D_1, \cdots, D_n$ and trained n classifiers $C_1, \cdots, C_n$ such that C_i is trained on D_i. For a test instance T, the final prediction of the ensemble is computed by averaging the prediction of all the classifiers. That is: $P_{ens}(+1|T) = \frac{1}{n}\sum_{i=1}^n P_{C_i}(+1|T)$, where $P_{C_i}(+1|T)$ is the probability estimate given by classifier C_i of thread T belonging to the subjective class. $P_{ens}(-1|T) = 1 - P_{ens}(+1|T)$. For classification, we used a threshold of 0.5 on the ensemble's prediction.

4 Datasets

To evaluate our approach, we used threads from the two popular online forums: Digital Photography Review (denoted by **dpreview**) and Trip Advisor–New

Table 2. Sample queries used for data collection from dpreview forum

Subjective Queries	Non-subjective Queries
nikon DSLR vs. sony DSLR	what is flash focal length
which camera should I buy for all round photography?	what does a wide angle lens do
carl zeiss better than canon	what is exposure compensation

York (denoted by **trip-advisor**), described below. The choice for these forums is that we wanted to evaluate our models across the two popular genres of on-line forums, namely, technical and non-technical online forums, **dpreview** is a technical forum whereas **trip-advisor** is a non-technical forum.

1. **dpreview** is an online forum with discussions related to digital cameras and digital photography[2]. We manually framed 39 queries, mix of subjective and non-subjective, on topics related to digital cameras (see Table 2 for several examples) and ran them on the Google search engine. We limited the search space of Google to the website *http://forums.dpreview.com/forums*, ensuring the results are discussion threads from the dpreview forum only. For each query, the top 200 returned threads were crawled and processed to identify structural elements (such as title, posts, authors, etc). Note that, in some cases, less than 200 threads were retrieved by the search engine.

2. **trip-advisor** is an online forum having travel related discussions mainly for New York city[3]. We used a publicly available dataset[4] [2] that had 83072 threads from which we randomly selected 700 threads for our experiments. The processing of threads for identifying thread elements (i.e., title, posts, authors, etc) is the same as for dpreview.

Data Annotation. Threads in our datasets were annotated by two human annotators. The annotators were asked to annotate a thread as subjective if its topic of discussion is subjective and non-subjective if the topic of discussion is non-subjective. The annotators were provided with a set of instructions for annotations. The set contained definitions of subjective and non-subjective topics with examples and guidelines for doing annotations[5].

The annotations for each dataset were conducted in three stages. First, the annotators were asked to annotate a sample of 20 threads (for which we already had annotations) from the dataset using the instruction set. Second, separate discussions were held between the first author and each annotator. Each annotator was asked to provide arguments (for the annotations) and, in case of inconsistencies, they were educated through discussions to attain a common understanding of subjectivity. Third, they were given the full dataset for annotation.

---

[2] http://forums.dpreview.com/forums/
[3] http://www.tripadvisor.com/ShowForum-g60763-i5-New_York_City_New_York.html
[4] http://www.cse.psu.edu/ sub194/datasets/ForumData.tar.gz
[5] blindreview.com

Table 3. Distribution of threads in the two classes

	Dpreview Trip–Advisor New York	
No. of subjective threads	3320	412
No. of non-subjective threads	536	197

The overall percentage agreement between the annotators was 90% on the **dpreview** dataset and 87% on **trip-advisor** dataset. For our experiments, *we used only the data on which the annotators agreed.* Table 3 shows the number of threads in the two classes. There are much more subjective than non-subjective threads in the two forums, which confirms that online forum users tend to discuss subjective topics. This observation is consistent with previous works on subjectivity analysis of other online social media such as community question answering sites. For example, Li *et al.* [6] found that 66% of the questions asked in Yahoo! Answers were subjective.

5   Experiments and Results

In this section, we describe our experimental setting and present the results.

5.1   Experimental Setting

We used k-fold cross validation to evaluate our classification models. k-fold cross validation is a popular method for performance evaluation of classifiers when the data do not have dependencies. Since the method randomly partitions the data into training and test set, if there are dependent data points in the training and test, the prediction of the classifier will be biased. In our case, there were dependencies in the **dpreview** dataset. Threads corresponding to a query discussed similar topics and, hence, would contain similar words and would have similar subjectivity orientations. Their presence in both training and test sets would make the sets dependent. In such a setting, a classifier's performance may be overestimated because of the dependence bias. To address this problem, we used *leave-one-out* cross validation at the query level. Threads corresponding to a query were held-out and the classifier was trained on the remaining threads. Testing was done on the held-out set. This holding out was done for each query and the average of the classifiers' performance over all queries was computed. For the **trip-advisor** dataset, since there were not any inbuilt dependencies, we used k-*fold* cross validation with $k = 5$. We used the Weka data mining toolkit [22] with default settings to conduct our experiments.

As described in Section 3, we conducted experiments with four kinds of features: (i) bag of words (BoW), (ii) unigrams and POS tags (BoW+POS), (iii) unigrams and bigrams (BoW+Bi), (iv) unigrams, bigrams and POS tags (Bow+Bi+POS) extracted from the textual content of different structural elements (title (t), initial post (I), reply posts (R)) of the threads. First, we trained a basic model where we used only the text of the titles (denoted by t) for classification; that is our baseline. Then, we incorporated the text of initial posts (denoted by t+I) and finally, we used the textual content of the entire thread (denoted by t+I+R) for classification. For each dataset, we performed experiments using: (i) a single classifier trained on a balanced sample, (ii) a single classifier trained on the entire unbalanced dataset, and (iii) an ensemble of n classifiers, with each classifier in the ensemble being trained on a balanced sample of the data. For the ensemble, we empirically determined the value of n, that is, we conducted experiments with different values of n and used the value corresponding to the best results, $n = 20$ for **dpreview** and $n = 7$ for **trip-advisor**. Also, we investigated the effect of feature selection on the classification performance. We ranked the features using Information Gain [23] to get the most informative ones with respect to the class variable. We trained classifiers for various numbers of selected features, starting from 100 and ending at 2000, in steps of 100.

5.2   Results

Table 4 is divided into two halves. The upper half shows the results for **dpreview** and the lower half shows the results for **trip-advisor**. We used macro averaged F1-measure to report the classification performance of our models.

Effect of Different Features: For **dpreview**, the combination of unigrams, bigrams and part-of-speech tags (BoW+Bi+POS) extracted from title and the initial post gave the best F1-measure (0.884), using an ensemble of classifiers, whereas for **trip-advisor**, the same combination of unigrams, bigrams and part-of-speech tags (BoW+Bi+POS) this time extracted from title, the initial post, and the reply posts gave the best F1-measure (0.745), using again an ensemble of classifiers. However, for **trip-advisor**, the improvement in performance by incorporating parts-of-speech tags over BoW+Bi is not statistically significant.

Effect of Different Structural Units: In Table 4, we see that incorporating text from the first post (t+I) improves the classification performance over the baseline (t) for the two datasets. This observation suggests that initial posts along with titles convey more information than titles alone about the subjectivity orientation of online threads, which is intuitive as titles contain only a few keywords about the topic whereas initial posts contain full details about the topic. Incorporation of text from the reply posts has different effects for the two datasets. For **dpreview**, the classification performance remains almost the same as compared to t+I setting. However, for **trip-advisor**, there is a high improvement in performance. In principle, this observation says that for the dpreview forum the subjectivity orientation of threads is mainly determined by their titles and initial posts combined, and the reply posts do not convey any significant additional information about the subjectivity orientation. For the trip-advisor forum, the subjectivity orientation of threads is determined by the entire thread including its reply posts. We conjecture the reason of this difference to be the *more* informal nature of trip-advisor than dpreview as the former is a non technical forum and the latter is a technical forum. In trip-advisor threads, there is

Table 4. Classification performance (F1-measure) of different features extracted from different structural components of the forum threads. t, I and R are title, initial post and set of all reply posts of a thread, respectively. BoW, BoW+POS, BoW+Bi and BoW+Bi+POS are the different kinds of features that we used (explained in Table 1). [Sin] and [Ens] denote experiments with single balanced sample and with ensembling (i.e., using multiple balanced samples) respectively.

Dpreview dataset (leave-one-out cross validation)				
	BoW	BoW+POS	BoW+Bi	BoW+Bi+POS
t[Sin]	0.791	0.802	0.787	0.793
t+I[Sin]	**0.862**	**0.865**	**0.871**	**0.877**
t+I+R[Sin]	0.859	0.859	0.876	0.875
t [Ens]	0.807	0.811	0.807	0.801
t+I [Ens]	**0.865**	**0.865**	**0.877**	**0.884**
t+I+R [Ens]	0.867	0.863	0.876	0.878

Trip Advisor–New York dataset (5-fold cross validation)				
	BoW	BoW+POS	BoW+Bi	BoW+Bi+POS
t[Sin]	0.557	0.572	0.561	0.552
t+I[Sin]	0.606	0.618	0.642	0.666
t+I+R[Sin]	**0.701**	**0.702**	**0.729**	**0.738**
t [Ens]	0.565	0.564	0.568	0.566
t+I [Ens]	0.633	0.641	0.674	0.691
t+I+R [Ens]	**0.723**	**0.717**	**0.74**	**0.745**

generally more topic drift, i.e., there are discussions that are not related to the topic specified by the titles and initial posts of the threads. Hence, the subjectivity orientation is no longer, mainly, determined by titles and initial posts of the threads. We plan to investigate this difference in more detail as part of future research on subjectivity analysis of online forums.

To verify that these differences (in results) are not due to the difference in sizes of the two datasets, we conducted additional experiments with the **dpreview** dataset. We experimented with a small fraction of dpreview, i.e., 0.35, obtained by under-sampling [24] from the entire dataset. Specifically, we first under-sampled from the minority class of dpreview a small subset that approximately matched the size of the minority class in **trip-advisor**; we then under-sampled from the majority class of **dpreview** to obtained a balanced subset (same number of instances from both classes). Hence, on **dpreview**, we trained classifiers on approximately the same sized balanced samples as in **trip-advisor**, where the size of balanced sample is 394 (197 subjective and 197 non-subjective). The under-sampling was performed only on the training set (the test set remained unbalanced). Table 5 provides results for this experiment.

Effect of ensembling: For both datasets, using an ensemble of classifiers, with each classifier trained on a balanced sample, improves the performance of a single classifier trained on a balanced sample. However, the improvement is generally small, especially for **dpreview** (see Table 4). This implies that the classifiers learn almost the same patterns from the different random samples of the majority class.

Table 5. The performance of classifiers (in terms of F1-measure) trained on smaller balanced samples of the **dpreview** dataset. The number of threads in the balanced sample is 376 (188 subjective and 188 non-subjective). As can be seen, performance of t+I is similar to that of t+I+R.

	BoW	BoW+POS	BoW+Bi	BoW+Bi+POS
t	0.772	0.777	0.764	0.764
t+I	0.863	0.863	0.869	0.87
t+I+R	0.876	0.878	0.859	0.857

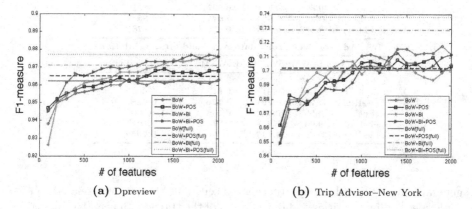

(a) Dpreview   (b) Trip Advisor–New York

Fig. 1. Classification performance of top 2000 features for the two datasets for settings t+I (for dpreview) and t+I+R (for Trip Advisor–New York). Straight lines represent performance corresponding to all the features for a particular kind of representation (Table 4).

Effect of Feature Selection: Figures 1(a) and 1(b) show the performance of *single* classifiers (not ensembling) as a function of the number of features, ranging from 100 to 2000 in steps of 100, for **dpreview** and **trip-advisor**, respectively. Due to space constraints, we only report the results for the two best performing experimental settings for the two datasets: t+I for **dpreview** and t+I+R for **trip-advisor**. We used all the feature representations described in Table 1. For **dpreview**, the performance of the BoW+Bi+POS-based classifier using all the features ($\approx 100,000$ features) is matched by that of the BoW+Bi+POS-based classifier using only the top 1700 selected features (F1-measure = 0.877). On the other hand, for **trip-advisor**, the BoW-based classifier using feature selection (with the number of features ranging between 100 and 2000) achieves the highest performance (F1-measure =0.718) using 1900 features, which is worse than that of BoW+Bi+POS-based classifier using all the features (F1-measure =0.738). However, in every case (for the two datasets) the number of features corresponding to the best performance is much smaller compared to the total number of features.

Table 6. True positive rates (for minority class) of classifiers trained on unbalanced and balanced data for the two datasets for BoW features

	Dpreview		Trip Advisor–New York	
	Unbalanced	Balanced	Unbalanced	Balanced
t	0.53	0.752	0.305	0.635
t+I	0.56	0.73	0.467	0.66
t+I+R	0.558	0.618	0.426	0.545

Unbalanced Dataset vs. Balanced Dataset: Table 6 compares true positive rates (for the minority class) of single classifiers trained on balanced and unbalanced (entire) data for the two datasets. As expected, classifiers built on unbalanced data performed worse on the minority class when compared to those trained on balanced datasets. We show the results only for BoW features for the three experimental settings, (t), (t+I), (t+I+R), but the same behavior was observed for other types of features.

6   Conclusion and Future Work

In this paper, we presented a supervised machine-learning approach to classifying online forum threads as subjective or non-subjective. Our methods showed that features generated from n-grams and parts-of-speech tags of the textual content of forum threads give promising results. In the future, we plan to use the subjectivity analysis to improve search in online forums.

Acknowledgement. This material is based upon work supported by the National Science Foundation under Grant No. 0845487.

References

1. Seo, J., Croft, W., Smith, D.: Online community search using thread structure. In: CIKM, pp. 1907–1910. ACM (2009)
2. Bhatia, S., Mitra, P.: Adopting inference networks for online thread retrieval. In: Proceedings of the 24th AAAI, pp. 1300–1305 (2010)
3. Hu, M., Liu, B.: Mining and summarizing customer reviews. In: SIGKDD, pp. 168–177. ACM (2004)
4. Liu, B.: Sentiment analysis and subjectivity. In: Handbook of Natural Language Processing (2010) ISBN: 978–1420085921
5. Ly, D., Sugiyama, K., Lin, Z., Kan, M.: Product review summarization from a deeper perspective. In: JCDL, pp. 311–314 (2011)
6. Li, B., Liu, Y., Ram, A., Garcia, E., Agichtein, E.: Exploring question subjectivity prediction in community qa. In: SIGIR, pp. 735–736. ACM (2008)

7. Gurevych, I., Bernhard, D., Ignatova, K., Toprak, C.: Educational question answering based on social media content. In: Proceedings of the 14th International Conference on Artificial Intelligence in Education. Building Learning Systems that Care: From Knowledge Representation to Affective Modelling, pp. 133–140 (2009)

8. Stoyanov, V., Cardie, C., Wiebe, J.: Multi-perspective question answering using the opqa corpus. In: EMNLP, pp. 923–930. Association for Computational Linguistics (2005)

9. Yu, H., Hatzivassiloglou, V.: Towards answering opinion questions: Separating facts from opinions and identifying the polarity of opinion sentences. In: EMNLP, pp. 129–136. Association for Computational Linguistics (2003)

10. Pang, B., Lee, L.: Opinion mining and sentiment analysis. Foundations and Trends in Information Retrieval 2, 1–135 (2008)

11. Somasundaran, S., Wilson, T., Wiebe, J., Stoyanov, V.: Qa with attitude: Exploiting opinion type analysis for improving question answering in on-line discussions and the news. In: ICWSM (2007)

12. Seki, Y., Eguchi, K., Kando, N., Aono, M.: Multi-document summarization with subjectivity analysis at duc 2005. In: DUC. Citeseer (2005)

13. Carenini, G., Ng, R., Pauls, A.: Multi-document summarization of evaluative text. In: EACL, pp. 305–312 (2006)

14. Biyani, P., Bhatia, S., Caragea, C., Mitra, P.: Thread specific features are helpful for identifying subjectivity orientation of online forum threads. In: COLING, pp. 295–310. ACL, Mumbai (2012)

15. Bruce, R., Wiebe, J.: Recognizing subjectivity: a case study in manual tagging. Natural Language Engineering 5, 187–205 (1999)

16. Riloff, E., Wiebe, J.: Learning extraction patterns for subjective expressions. In: EMNLP, pp. 105–112. ACL (2003)

17. Li, B., Liu, Y., Agichtein, E.: Cocqa: co-training over questions and answers with an application to predicting question subjectivity orientation. In: EMNLP, pp. 937–946. ACL, Stroudsburg (2008)

18. Yu, H., Hatzivassiloglou, V.: Towards answering opinion questions: separating facts from opinions and identifying the polarity of opinion sentences. In: EMNLP, pp. 129–136. ACL, Stroudsburg (2003)

19. Aikawa, N., Sakai, T., Yamana, H.: Community qa question classification: Is the asker looking for subjective answers or not? IPSJ Online Transactions 4, 160–168 (2011)

20. Mitchell, T.: Machine learning. McGraw-Hill, Burr Ridge (1997)

21. Dietterich, T.G.: Ensemble methods in machine learning. In: Kittler, J., Roli, F. (eds.) MCS 2000. LNCS, vol. 1857, pp. 1–15. Springer, Heidelberg (2000)

22. Hall, M., Frank, E., Holmes, G., Pfahringer, B., Reutemann, P., Witten, I.: The weka data mining software: an update. ACM SIGKDD Explorations Newsletter 11, 10–18 (2009)

23. Yang, Y., Pedersen, J.: Feature selection in statistical learning of text categorization. In: Proc. ICML, pp. 412–420 (1997)

24. Kubat, M., Matwin, S.: Addressing the curse of imbalanced training sets: One-sided selection. In: ICML, pp. 179–186. Morgan Kaufmann (1997)

Distant Supervision for Emotion Classification with Discrete Binary Values

Jared Suttles and Nancy Ide

Department of Computer Science, Vassar College
Poughkeepsie, New York 12604 USA
{jasuttles,ide}@cs.vassar.edu

Abstract. In this paper, we present an experiment to identify emotions in tweets. Unlike previous studies, which typically use the six basic emotion classes defined by Ekman, we classify emotions according to a set of eight basic *bipolar* emotions defined by Plutchik (Plutchik's "wheel of emotions"). This allows us to treat the inherently multi-class problem of emotion classification as a binary problem for four opposing emotion pairs. Our approach applies *distant supervision*, which has been shown to be an effective way to overcome the need for a large set of manually labeled data to produce accurate classifiers. We build on previous work by treating not only emoticons and hashtags but also emoji, which are increasingly used in social media, as an alternative for explicit, manual labels. Since these labels may be noisy, we first perform an experiment to investigate the correspondence among particular labels of different types assumed to be indicative of the same emotion. We then test and compare the accuracy of independent binary classifiers for each of Plutchik's four binary emotion pairs trained with different combinations of label types. Our best performing classifiers produce results between 75-91%, depending on the emotion pair; these classifiers can be combined to emulate a single multi-label classifier for Plutchik's eight emotions that achieves accuracies superior to those reported in previous multi-way classification studies.

1 Introduction

The development of web- and mobile-based media devoted to persistent social interaction among users ("social networks") has provided a massive, continuous stream of data reflecting the public's opinions about and reactions to phenomena from political and world events to movies and consumer products. Over the past ten years, there has been no shortage of studies attempting to mine this data to inform decisions about product design, brand identity, corporate strategy, government policies, etc., as well as improve social-psychological correlational studies and predictive models of human behavior. Recently, many analyses have focused on the microblogging service Twitter, which provides a continuous stream of user-generated content in the form of short texts under 140 characters in length. Much of this work involves *sentiment* analysis, in which user attitudes toward a particular topic or product are classified as positive, negative, or

A. Gelbukh (Ed.): CICLing 2013, Part II, LNCS 7817, pp. 121–136, 2013.

neutral (e.g., [12,20]). Other studies have tackled the broader problem of detecting *emotions* in tweets, often for the purpose of modeling collective emotional trends [3,4,9,23].

In this paper, we present an experiment to identify emotions in tweets. Unlike previous studies, which typically use the six basic emotion classes defined by Ekman [10,11], we classify emotions according to a set of eight basic *bipolar* emotions defined by Plutchik (Plutchik's "wheel of emotions" [21]). This allows us to treat the inherently multi-class problem of emotion classification as a binary problem for four opposing emotion pairs. Our approach applies *distant supervision* (see e.g. [16]), which has been shown to be an effective way to overcome the need for a large set of manually labeled data to produce accurate classifiers (e.g., [12,23]). We build on previous work by treating not only emoticons and hashtags but also emoji, which are increasingly used in social media, as an alternative for explicit, manual labels. Since these labels may be noisy, we first perform an experiment to investigate the correspondence among particular labels of different types assumed to be indicative of the same emotion. We then test and compare the accuracy of independent binary classifiers for each of Plutchik's four binary emotion pairs trained with different combinations of label types. Our best performing classifiers produce results between 75-91%, depending on the emotion pair; these classifiers can be combined to emulate a single multi-label classifier for Plutchik's eight emotions that achieves accuracies superior to those reported in previous multi-way classification studies.

2   Previous Work

Several studies have focused on the task of identifying *emotions* in different text types, including stories [2,17,24], spoken data [6,7,14], blogs [15,19], and microblogs (tweets) [18,23,26]. Earlier studies relied on datasets that were manually annotated for emotion and were typically *keyword-based*, identifying the presence of an emotion based on the appearance of pre-determined lexical markers. It is well-recognized that this approach has drawbacks: determining the contents of the emotional lexicon is subjective, and there is no guarantee that the lexicon is comprehensive; furthermore, the selected words may be ambiguous. These problems are compounded when performing sentence-level analyses where very little context is available, which is clearly a factor in studies involving context-poor Twitter messages.

To address this and the problem of generating large annotated datasets for training, several studies have attempted to exploit the widespread use of emoticons and other indicators of emotional content in tweets by treating them as noisy labels in order to automatically obtain very large training sets (see e.g., [12,18, 23]). This strategy of *distant supervision* [16] has been used to achieve accuracy scores as high as 80-83% for distinguishing positive and negative sentiment [12]. Studies using distant supervision commonly rely on a set of Western-style emoticons (e.g., ":-)", ": (", etc.) and Eastern-style emoticons (e.g., "(^_^)", "(>_<)", etc.) as emotional labels [20, 25, 28]. The means by which these labels are associated with specific emotions varies from study to study–the most common strategy is to manually classify emoticons such as those available from on-line emoticon lists (e.g., Wikipedia *List of Emoticons*[1], Yahoo messenger classification[2]) as indicative of a specific emotion. The most commonly-used scheme for emotion classification is Ekman's [10,11], which identifies six primary emotions based on facial expressions.

Recently, there has been work exploring the use of Twitter *hashtags* to collect datasets indicative of emotional states for distant supervision. Hashtags, consisting of a tag or word prepended with "#" are typically used to indicate the tweet's topic in order to facilitate search and increase visibility. However, the practice of using hashtags has extended to other kinds of labeling, in particular, noting attitudes such as #sarcasm and #irony as well as emotional states (#angry, #happy, etc.). Previous studies collected tweets with specific hashtags to create datasets of sarcastic tweets [13]; recently, this approach has been applied to hashtags signaling the presence of particular emotions [18, 23, 29]. Again, the means by which hashtags are associated with particular emotions varies, but most studies use the names of Ekman's six basic emotions as relevant hashtags [5,18], sometimes together with a few closely related terms [23]. However, the number of messages containing this small set of words as hashtags is typically very small, as noted in [23]. To increase the number of relevant terms, others have relied on pre-compiled lists of emotion words from psychological literature [29]. Our strategy, described in Section 3.2, differs from previous studies by using hashtags extracted from a large database of current tweets that have been manually labeled for emotional content.

3   Methodology

3.1   Emotional Binaries

Our work relies on a set of eight basic *bipolar* emotions as defined in Plutchik's psychoevolutionary theory of emotion [21] rather than the six basic emotion classes defined by Ekman [10] or previously-used minor variants [2,26]. Ekman's basic emotions include ANGER, DISGUST, FEAR, HAPPINESS, SADNESS, and SUR-PRISE; Plutchik's theory defines eight primary emotions, consisting of a superset of Ekman's and with two additions: TRUST and ANTICIPATION. These eight emotions are organized into four bipolar sets: JOY vs. SADNESS, ANGER vs. FEAR, TRUST vs. DISGUST, and SURPRISE vs. ANTICIPATION. Plutchik's "wheel of emotions" (see Figure 1) represents the relations among emotions as a color wheel; like colors, emotions can vary in intensity (proximity to the center indicates intensity) and mix to create additional emotions (*primary dyads*, appearing in the white spaces between primary emotions). Most relevant to our work is Plutchik's definition of *emotional opposites*, represented in the spatial oppositions in the wheel, which are considered to be mutually exclusive.[3]

---

[1] http://en.wikipedia.org/wiki/List_of_emoticons

[2] http://messenger.yahoo.com/features/emoticons/

[3] A recent study [26] adapted Ekman's classification to define an emotional ontology and a set of emotional oppositions very similar to those in Plutchik's Wheel.

We adopted Plutchik's model over Ekman's for several reasons. First, it includes LOVE, an emotion very frequently expressed on Twitter. In Plutchik's scheme, LOVE is defined as a *primary dyad*, i.e., a combination of the two primary emotions JOY and TRUST; Ekman's set of six emotions, grounded in physiological rather than psychological research, omits LOVE and is in general more focused on negative emotions. The main advantage of using Plutchik's theory for our work is that it allows us to exploit his notion of *emotional polar opposites* to treat emotion detection as a binary rather than multi-way classification problem. Whereas previous studies used multiple category classification for emotion detection (see, e.g., [6,7,27]) or simulated binary classification by distinguishing one emotion class from all others (e.g. ANGER v. NOT ANGER, [18,23]), the classifiers used in this study make binary decisions concerning which of each pair of opposing emotions is most probable, thereby likening the problem to that of distinguishing two opposing classes (e.g., positive vs. negative sentiment) rather than presence or absence of a class among several others. This simplification enables development of four independent binary classifiers, one for each binary emotion pair, that can be combined to emulate a single multi-label classifier for Plutchik's eight primary emotions.

3.2   Emotion Lexicon

Our lexicon comprises a combination of emotional labels including hashtags, traditional emoticons, and emoji. It is assumed that the use of any of these symbols reflects the emotion of the author of the tweet, even when the emotional state of another individual is the topic. Support for this assumption is provided by studies on internet-based social interactions and the representation of emotions (e.g., [8]), which show that emoticons are used to increase the intensity of emotions already conveyed by the lexical content. It has also been suggested that "emotional punctuations" (e.g., noting laughter) in spoken transcriptions are similar to written emoticons, with both acting as punctuation for the surrounding language [22].

Our lexicon of 69 emoticons was derived from Wikipedia[4]. The emotion class assignments were based on those used in previous studies [1,12,20,23]. Our lexicon also includes *emoji*[5], which originally developed in Japan but have come into widespread use since their inclusion in Unicode Standard 6.0 and ISO/IEC 10646 (Universal Character Set) and subsequent support in newer operating systems and mobile phones. Despite their increasing prevalence, emoji have not been used in previous work[6] In the absence of existing categorizations, we labeled 70 emoji (consisting of facial expressions and a few additional symbols such as hearts, kissing lips, etc.) with the eight Plutchik primary emotion categories.

Our initial approach to determining the hashtags to be included in the lexicon used the eight primary emotion names defined by Plutchik (ANGER, DISGUST,

---

[4] http://en.wikipedia.org/wiki/List_of_emoticons

[5] http://en.wikipedia.org/wiki/Emoji

[6] Of the 38.9 million emotional tweets in our dataset, 7% include emoji from our lexicon and 7.8% contain emoticons from our lexicon.

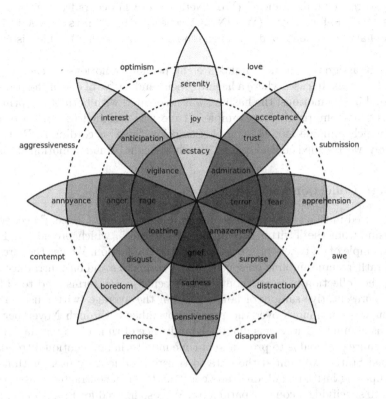

Fig. 1. Plutchik's Wheel of Emotions (Image from Wikimedia Commons)

FEAR, JOY, SADNESS, SURPRISE, TRUST, ANTICIPATION) as seed words, and added the WordNet 3.1 synsets and hyponyms of each name in order to create a set of terms for each emotion. This resulted in a large set of over 60 terms for each emotion. We later abandoned this method because the WordNet terms–and to a lesser extent the emotion names themselves–occurred infrequently in the data. Rather, users tend to use shorter, more colloquial hashtags instead of the words in WordNet synsets; for example, users prefer the tag #ew to a longer and more formal term #disgusted. We therefore turned to the data itself to determine a set of hashtags that reflect actual user behavior. Using a list of the most frequent hashtags in our training set, we identified those that are likely to be emotional labels (e.g., #happytweet, #ugh, #yuck, #fml). This method of determining a set of relevant hashtags maximized our ability to collect a large number of labeled tweets, since the hashtags were guaranteed to appear frequently in our dataset. It also provides a more representative sampling of typical tweets in terms of word use and content and avoids selecting for unusual tweets containing infrequent hashtags. We also filtered out ambiguous tags such as #sad, which, in addition to occurring in its sense of "experiencing or showing sorrow or unhappiness" (WordNet3.1 sense 1) occurs frequently in the sense of "bad; unfortunate" (WordNet3.1 sense 3–e.g., "`Christina Aguilera used to have the best body.. then she got fat. #sad`"), which is closer to DISGUST.

We next assigned one of Plutchik's eight primary emotions to each of the selected hashtags. In cases where a hashtag seems indicative of one of the primary dyad (combined) emotions, the hashtag was associated with both of the primary emotions that comprise it–for example, `#love` was assigned to both JOY and TRUST, which combine into the complex emotion "love" according to Plutchik. Ultimately, we assigned 56 hashtags to Plutchik's eight primary emotion classes.[7]

3.3   Data Collection and Preparation

The data used in this study consists of microblog messages ("tweets") collected in real-time from the Twitter Streaming API service[8], which provides a 1-2% random sample of all tweets produced during the connection. We use the streaming API rather than sampling on specific query terms to avoid bias introduced by limiting the collection to tweets containing specific search terms, and to obtain a more representative sample of language from the average twitter user. Data collection was continuous over the period November 9 through November 30, 2012, thus eliminating any bias due to the influence of time of day or day of the week. Because our goal is to provide real-time monitoring of emotional trends in the Untied States, we limited the data to tweets produced by users within the US by imposing latitude and longitude constraints on the extracted messages in addition to specifying a country parameter. We also filtered for English language messages through the language parameter. The resulting dataset consists of 38.9 million tweets.

We extracted a dataset D_k consisting of 5.9 million tweets from the 38.9 million tweet dataset containing any of the emotional tokens in our lexicon and labeled each with the corresponding emotion. Tweets with multiple emotional tokens were assigned a label for each of the associated emotion classes. We included tweets with labels appearing both within (i.e., as a part of the message, as in "`I am so #angry about that!`") and at the end of the tweet; it has been suggested that in-line labels are less reliable indicators for sarcasm [13], but examination of our data does not support this observation for emotions. Tweets containing one or more emotional tokens from both classes of an opposing binary pair were discarded, since the emotional content was considered to be undecidable based on Plutchik's assumption of exclusivity of opposite emotions. Table 1 shows the distribution of labels for each emotion in the initial dataset.

The data were tokenized and normalized as follows: following [1,12,20], we replaced usernames (names prepended with "@") with the token USERNAME and web addresses (e.g. http://t.co/zDO9b7xD) with the token URL, and replaced

---

[7] The complete emotional lexicon used in this study is available at
http://www.emotitweets.com

[8] See http://dev.twitter.com/docs/streaming-api

Table 1. Distribution of emotional labels in D_k

Label Type	Joy	Sadness	Anticipation	Surprise
Hashtag	54,172	29,325	24,008	35,871
Emoticon	1,692,711	352,527	128,287	68,478
Emoji	735,023	275,861	24,133	26,363
Hashtag+Emoticon	1,741,767	379,571	152,005	104,120
Hashtag+Emoji	786,594	303,490	48,069	62,052
Emoticon+Emoji	2,419,383	625,398	152,277	94,765
All	2,465,884	650,771	175,923	130,220

Label Type	Anger	Fear	Disgust	Trust
Hashtag	31,109	25,066	25,724	30,501
Emoticon	101,939	128,287	101,842	454,768
Emoji	196,936	344,978	287,583	847,695
Hashtag+Emoticon	132,736	152,931	127,343	483,781
Hashtag+Emoji	226,565	368,792	312,381	874,633
Emoticon+Emoji	297,888	472,773	388,197	1,298,420
All	327,208	496,160	412,777	1,323,897

repetitions of more than two letters consecutively (e.g. "cuteee", "cuteeeee", etc.) with only two, on the assumption that the number of repeating letters was arbitrary. Because we are interested in the emotions of the authors of tweets, quoted text was excluded as it may represent a retweet or someone else's opinion.

We compiled a training dataset D_t consisting of subsets corresponding to each of the four binary emotion pairs: D_t^1 (joy/sadness), D_t^2 (anticipation/surprise), D_t^3 (anger/fear), D_t^4 (trust/disgust). We used the labels appearing in our emotional lexicon to group tweets from D_k into emotion classes within the appropriate D_t^n set, then removed them so that classification would rely solely on language and non-emotional hashtags. The dataset for each binary emotion pair was normalized so that there were equal numbers of tweets for each member of the pair. As such, the total number of tweets for each emotion pair differed in proportion to the number appearing in the D_k tweet dataset, in which occurrences of JOY far outweigh those for other emotions (see Figure 2). The resulting training set contained approximately three million tweets, with each emotion pair (in equal numbers for each emotion in a pair) represented as shown in Table 2.

4   Experiments

We performed two experiments: (1) a cross-validation of emotional class assignments to the different label types, to investigate their correspondence; and (2) evaluation of binary classifiers trained with various combinations of label types on a small manually-labeled dataset of emotional tweets. In all experiments, classification was performed using Naïve Bayes (NB) and Maximum Entropy (ME)

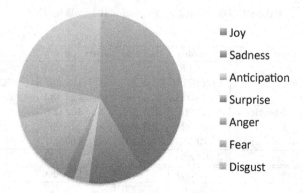

Fig. 2. Emotion proportions based on labels in D_k

Table 2. Distribution of tweets for each of four datasets D_t^n

Training set	Size
D_t^1 (joy/sadness)	1,301,542
D_t^2 (anticipation/surprise)	260,440
D_t^3 (anger/fear)	654,416
D_t^4 (trust/disgust)	825,554
D_t	3,041,952

from the Natural Language Toolkit (NLTK) 2.0.4[9]. Because of the size of the input, all experiments were run using concurrent algorithms on a machine with 158 2.2 GHz AMD Opteron 6174 12-core CPUs. Because of the long running times for training, our experiments include only unigrams as features; however, previous studies [12, 23] have shown that classifiers trained on unigrams outperform those trained on additional phenomena such as bigrams and part-of-speech information.

4.1   Experiment 1: Cross-Validation of Emotional Labels

Our approach relies on the assumption that sets of hashtags, emoticons, and emoji associated with the same emotion are indeed indicative of the same underlying phenomenon. To validate this assumption, we tested the ability of each label to predict the emotion(s) signaled by the other labels. Separate binary classifiers for each label convention were trained on each dataset $D_t^n, 0 < n \leq 4$, using the set of emotion labels for that convention as noisy labels. We evaluated against the same 12 subsets (3 label types, 4 binaries) of D_t^n.

---

[9] http://www.nltk.org

Accuracies are given in Table 3. Experiments used both Naïve Bayes and Maximum Entropy for classification. Ten-fold cross-validation was used for within-label tests. Full sets were used for all other tests. The values in the table show the highest accuracies; with a few exceptions, the accuracies returned by the two different classifiers were no more than a few percentage points apart. All accuracies are significantly higher than chance according to χ^2 tests, except one, shown with strikethrough, which is in fact significantly *lower* than chance. We suspect this is because we have only two emoji for anticipation in our lexicon, although further investigation is needed to determine the actual cause.

The results show that many of the classifiers trained on data labeled using one label type can distinguish classes that were labeled using the other two labels, suggesting that the emotion assignments are relatively reliable, or at least consistent, among the three label types. The clear exception is the lack of ability for hashtags to predict emoticons and emoji for ANTICIPATION; this likely results from the fact that very few emoticons and emoji can be considered indicative of anticipation (which has no obvious facial depiction), whereas hashtags such as #cantwait and #excited unambiguously signal this emotion. ANTICIPATION is one of the two primary emotions that is included in Plutchik's scheme but not in Ekman's[10], the scheme most commonly used in previous studies, which means that there exists no established set of labels for this emotion nor comparative data from other work. We therefore repeated our experiments with a variety of different ANTICIPATION emoji and emoticons, but these variants did not improve our results and in some cases actually worsened them. At the least, our results suggest that hashtags are likely a better source for automatic labeling of this emotion in tweets.

In general, accuracies are more consistently high for JOY and SADNESS than the other emotions. This result is similar to that reported in [23], where cross-label testing for classifiers trained on emoticons and hashtags performed relatively well for distinguishing JOY ("happy"), SADNESS ("sad"), and ANGER as compared with the other three emotions in their study[11], although their accuracies overall were much lower than ours (60-65% range).

4.2   Experiment 2: Classifier Evaluation

Evaluation was performed using a manually labeled set D_m of 420 tweets that is disjoint from either D_e and D_t, consisting of 400 emotional tweets annotated for at least one emotion from at least one emotion pair, and 20 neutral tweets with no emotion from any pair. Because the collection of tweets was random, the distribution of emotion classes in D_m is roughly proportional to their representation in D_k and D_e.

---

[10] The lack of a pictorial representation for ANTICIPATION may in fact account for its absence in Ekman's emotion scheme, which is based on facial expressions.

[11] [23] uses Ekman's six emotions.

Table 3. Highest accuracies for cross-validation of emotional labels on datasets D_t^n. Values in italics used a Naïve Bayes classifier, non-italics used Maximum Entropy.

Test		Train		
Label	Emotion	Hashtag	Emoticon	Emoji
Hashtag	Joy	73.8%	95.2%	94.5%
	Sadness	92.1%	*98.8%*	*99.3%*
	Anticipation	82.8%	*62.8%*	*34.5%*
	Surprise	*86.5%*	81.5%	61.1%
	Anger	*74.7%*	26.3%	*98.6%*
	Fear	78.1%	77.8%	79.8%
	Disgust	*82.2%*	87.0%	92.0%
	Trust	*85.2%*	92.8%	96.1%
Emoticon	Joy	93.3%	78.3%	89.0%
	Sadness	98.5%	*83.0%*	*95.4%*
	Anticipation	15.0%	*81.4%*	92.6%
	Surprise	93.8%	70.7%	88.5%
	Anger	36.9%	*63.6%*	*47.0%*
	Fear	51.7%	89.6%	56.7%
	Disgust	58.9%	*85.2%*	*58.3%*
	Trust	82.4%	82.1%	89.7%
Emoji	Joy	81.3%	85.0%	75.8%
	Sadness	98.0%	*96.9%*	*83.3%*
	Anticipation	*5.8%*	97.5%	80.8%
	Surprise	*90.1%*	59.1%	65.2%
	Anger	86.2%	46.6%	*81.9%*
	Fear	*87.0%*	*42.1%*	70.8%
	Disgust	*83.6%*	*91.4%*	*81.6%*
	Trust	*88.7%*	74.1%	77.5%

Annotation was performed by two annotators. The annotation procedure presented a randomly selected tweet to the annotator together with five annotation options. For example, "joy/sadness" is presented as follows:

```
omg I freaking love sweet potatoes! Literally ate one today!
```

```
[1] joy
[2] sadness
[3] neutral
[4] don't know for this emotion pair
[5] don't know for any pair (leave tweet out of dataset)
```

Options 4 and 5 allow the annotator to identify tweets that are difficult to understand and/or rate, either for a particular emotion or any emotion. Each tweet was annotated for all four emotional binary pairs. In cases where the annotator identified the tweet as "neutral" (option 3) for all four emotion pairs, the tweet was labeled NEUTRAL (non-emotional).

Evaluation was performed for classifiers trained using each label type as well as all possible combinations of labels. The accuracies for this experiment are given in Table 4. All values were found to be significantly different from chance based on χ^2 tests, except one (shown with strikethrough).

Table 4. Evaluation results from Experiment 2. Values in bold are the highest scores for each emotion pair. Strikethrough identifies values that are not statistically significant.

Train	Test					
	Joy/Sadness			Anticipation/Surprise		
	ME	NB	Size	ME	NB	Size
Hashtag	86.3%	73.8%	58,650	66.1%	60.3%	48,016
Emoticon	89.1%	84.8%	705,054	68.8%	69.8%	136,956
Emoji	88.7%	80.1%	551,722	64.0%	67.2%	48,266
Hashtag+Emoticon	**91.0%**	84.0%	759,142	73.0%	**75.7%**	208,240
Hashtag+Emoji	88.7%	80.1%	606,980	72.0%	72.5%	96,138
Emoticon+Emoji	90.2%	83.6%	1,250,796	65.1%	70.9%	189,530
All	90.6%	85.5%	1,301,542	71.4%	**75.7%**	260,440

Train	Test					
	Anger/Fear			Disgust/Trust		
	ME	NB	Size	ME	NB	Size
Hashtag	78.5%	74.6%	50,132	90.6%	86.6%	51,448
Emoticon	58.5%	~~49.2%~~	203,878	85.1%	87.1%	203,684
Emoji	80.8%	78.5%	393,872	90.1%	82.2%	575,166
Hashtag+Emoticon	70.0%	62.3%	265,472	89.1%	88.1%	254,686
Hashtag+Emoji	80.8%	79.2%	453,130	90.6%	84.2%	624,762
Emoticon+Emoji	**84.6%**	80.8%	595,776	89.1%	85.1%	776,394
All	83.1%	82.3%	654,416	**91.1%**	84.7%	825,554

Experiment 2 yields accuracies between 75% and 91%[12] for tests on manually labeled data, which exceed those reported in similar studies [3,23,27]. The results indicate that combining all three label types as distant labels yields the highest accuracies, or accuracies within (roughly) a percentage point of the highest. The remaining values are relatively consistent and reveal no pattern that indicates a particular label combination out-performs the others. The only anomaly in the results is the low accuracies for emoticons on "anger/fear", but this may be due to the difficulty of depicting fear with an emoticon (emoji provide a somewhat better depiction), making that pair particularly difficult to distinguish for emoticons alone. We attribute our stronger results both to the use of binary classifiers, which reduces the complexity of the classification task, and to the inclusion of emoji as well hashtags and emoticons as (noisy) labels for creating the training set. The improvements are likely to come from having more labeled

---

[12] Accuracies fall between 85% and 91% if we eliminate the problematic "anticipation/-surprise" class.

data and because the classifier is less likely to be led down the wrong path by certain correlations with one label type versus another (e.g., if the emoticon : (" were used to indicate surprise by a large portion of writers), thus providing us with something like ensemble noisy labeling.

Experiments 1 and 2 together give us some confidence the various labels actually signal the emotions we are assuming they do. That is, while the results from Experiment 1 verify the cross-label consistency of emotion assignments, they do not provide evidence that the assigned emotions correspond with human judgement. The strong results from Experiment 2, which uses manually labeled data, shows that the emotions associated with the labels are also reasonably consistent with independent human judgements, providing evidence that the associations made in the emotion lexicon are valid.

5   Next Steps

Our goal is to use the binary classifiers for the four emotion pairs to emulate a single multi-way classifier that identifies emotions in tweets. In fact, this combination of classifiers would identify up to four emotions (i.e., at most one from each pair of mutually exclusive emotions) in a tweet, which is appropriate since annotators identified multiple emotions in a large percentage of tweets in our manually labeled dataset. However, we also need to distinguish tweets containing no emotional content (which is the vast majority of tweets) from those containing an emotion from one or more of the four pairs. To address this, we have begun experimenting with four *neutral* binary classifiers, one for each emotion pair, that distinguishes tweets containing either of the emotions in that pair from those that do not, that is, tweets that include any of the six remaining emotions in Plutchik's system or contain no emotion at all. In turn, the combination of a classifier for one of the four emotion pairs with its corresponding neutral classifier would emulate a single three-way classifier that identifies each tweet as containing one of the emotions in the pair or as emotionally neutral; subsequently combining the three-way classifiers for each of the four emotion pairs as shown in Figure 3 emulates a more complex multi-way classifier that identifies *all* of the emotions present in a tweet *or* labels it as non-emotional.

To train a neutral classifier for each emotion pair, we can use the results from the classifiers with the highest accuracies from Experiment 2. Since these classifiers return one emotion of a binary pair for any tweet, even when neither is present, we assume that results with lower probabilities reflect situations where the tweet actually contains neither emotion or contains no emotion at all. Based on this assumption, we determine the optimum cutoff probabilities for each emotion–that is, the value below which probabilities reported by the relevant emotional binary classifier identify tweets that do not contain one emotion from the pair or are emotionally neutral–by iterating over all possible probabilities to determine the one that best predicts the results in the manually labeled dataset. Once this process is complete, the cutoff values with maximum accuracy are retained for classification.

We have so far applied this procedure to create a first set of neutral classifiers for each emotion pair. We performed a two-fold cross-validation of these classifiers using the manually labeled dataset D_m. The accuracies are given in Table 5. Accuracies for JOY and SADNESS, and to a slightly lesser extent ANGER and FEAR, are reasonable, suggesting that it may be possible to develop a reliable multi-way classifier, at least for these emotion pairs.

Table 5. Accuracies for determining neutrals using optimized probabilities

Emotion binary	Accuracy
Joy/Sadness	82.9%
Anticipation/Surprise	44.6%
Anger/Fear	74.7%
Disgust/Trust	61.1%

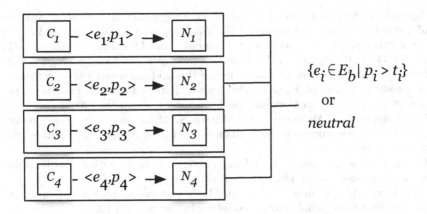

Fig. 3. Combined classifier that returns 1 to 4 emotion(s) from E_b (the set of binary emotions) that are present in a tweet, or *neutral* if the tweet has no emotion, with $e_i \in E_b$, the set of eight binary emotions; p_i the probability for e_i returned by classifier C_i; and t_i the optimal probability threshold for e_i; where $0 < i \le 4$.

Our next steps are to improve the performance of the four binary emotion classifiers as well as the neutral classifiers, and then begin experimenting with the combined classifier configured as shown in Figure 3. Although our initial results for distinguishing neutrals are encouraging, we will need a larger test set with a greater proportion of emotionally neutral tweets to establish more definitive results.

6 Conclusion

The approach outlined in this paper shows that Plutchik's set of four pairs of opposing emotions provides a viable basis for developing binary emotion classifiers for Twitter data that can match or exceed results from previous studies. In addition to emotions and hashtags, which have been used in similar work, we include emoji as emotional labels and show that they may be even more reliable emotion indicators than their pictorial cousins, emoticons. We have shown how the binary emotion classifiers can be combined to emulate a single multi-way classifier, thus avoiding the increased complexity (and corresponding weaker results) of multi-way classification; and how by further combining these classifiers with a combination of binary "neutral" classifiers, we not only emulate a multi-way emotion classifier but also isolate the particular emotions present in a given tweet. Our results on emotion label prediction suggest that our approach can produce reliable classifiers, and we therefore plan to attempt to improve on the work reported here by testing on much larger manually-annotated datasets and experimenting with the combined classifier described in Section 5.

References

1. Agarwal, A., Xie, B., Vovsha, I., Rambow, O., Passonneau, R.: Sentiment Analysis of Twitter Data. In: Proceedings of the Workshop on Language in Social Media (LSM 2011), pp. 30–38. Association for Computational Linguistics, Portland (2011)
2. Alm, C.O., Roth, D., Sproat, R.: Emotions from text: Machine learning for text-based emotion prediction. In: Proceedings of Human Language Technology Conference and Conference on Empirical Methods in Natural Language Processing, pp. 579–586. Association for Computational Linguistics, Vancouver (2005)
3. Balabantaray, R.C., Mohammad, M., Sharma, N.: Multi-class twitter emotion classification: A new approach. International Journal of Applied Information Systems 4(1), 48–53 (2012)
4. Bollen, J., Pepe, A., Mao, H.: Modeling public mood and emotion: Twitter sentiment and socio-economic phenomena. CoRR (2009)
5. Choudhury, M.D., Counts, S., Gamon, M.: Not all moods are created equal! exploring human emotional states in social media. In: ICWSM (2012)
6. Chuang, Z.J., Wu, C.H.: Multi-modal emotion recognition from speech and text. International Journal of Computational Linguistics and Chinese Language Processing 9(2), 45–62 (2004)
7. Danisman, T., Alpkocak, A.: Emotion classification of audio signals using ensemble of support vector machines. In: André, E., Dybkjær, L., Minker, W., Neumann, H., Pieraccini, R., Weber, M. (eds.) PIT 2008. LNCS (LNAI), vol. 5078, pp. 205–216. Springer, Heidelberg (2008)
8. Derks, D., Bos, A.E.R., Von Grumbkow, J.: Emoticons and online message interpretation. Social Science Computer Review 26(3), 379–388 (2008)
9. Dodds, P.S., Harris, K.D., Kloumann, I.M., Bliss, C.A., Danforth, C.M.: Temporal patterns of happiness and information in a global social network: Hedonometrics and twitter. CoRR (2011)
10. Ekman, P.: Universals and cultural differences in facial expressions of emotions. Nebraska Symposium on Motivation 19, 207–283 (1972)

11. Ekman, P.: An argument for basic emotions. Cognition & Emotion 6(3-4), 169–200 (1992)
12. Go, A., Bhayani, R., Huang, L.: Twitter sentiment classification using distant supervision. Processing, 1–6 (2009)
13. González-Ibáñez, R., Muresan, S., Wacholder, N.: Identifying sarcasm in twitter: A closer look. In: Proceedings of the 49th Annual Meeting of the Association for Computational Linguistics: Human Language Technologies, pp. 581–586. Association for Computational Linguistics, Portland (2011)
14. Lee, C.M., Narayanan, S.S.: Toward detecting emotions in spoken dialogs. In: IEEE Transactions on Speech and Audio Processing, pp. 293–303 (2005)
15. Mihalcea, R., Liu, H.: A corpus-based approach to finding happiness. In: AAAI Spring Symposium: Computational Approaches to Analyzing Weblogs, pp. 139–144. AAAI (2006)
16. Mintz, M., Bills, S., Snow, R., Jurafsky, D.: Distant supervision for relation extraction without labeled data. In: Proceedings of the Joint Conference of the 47th Annual Meeting of the ACL and the 4th International Joint Conference on Natural Language Processing of the AFNLP, pp. 1003–1011. Association for Computational Linguistics (2009)
17. Mohammad, S.: From once upon a time to happily ever after: Tracking emotions in novels and fairy tales. In: Proceedings of the 5th ACL-HLT Workshop on Language Technology for Cultural Heritage, Social Sciences, and Humanities, pp. 105–114. Association for Computational Linguistics, Portland (2011)
18. Mohammad, S.: Emotional Tweets. In: SEM 2012: The First Joint Conference on Lexical and Computational Semantics – vol. 1: Proceedings of the main conference and the shared task and vol. 2: Proceedings of the Sixth International Workshop on Semantic Evaluation (SemEval 2012), June 7-8, pp. 246–255. Association for Computational Linguistics, Montréal (2012)
19. Neviarouskaya, A., Prendinger, H., Ishizuka, M.: Analysis of affect expressed through the evolving language of online communication. In: Chin, D.N., Zhou, M.X., Lau, T.A., Puerta, A.R. (eds.) Proceedings of the 12th International Conference on Intelligent User Interfaces, pp. 278–281 (2009)
20. Pak, A., Paroubek, P.: Twitter as a corpus for sentiment analysis and opinion mining. In: Calzolari, N., Choukri, K., Maegaard, B., Mariani, J., Odijk, J., Piperidis, S., Rosner, M., Tapias, D. (eds.) Proceedings of the Seventh International Conference on Language Resources and Evaluation (LREC 2010), European Language Resources Association (ELRA), Valletta (2010)
21. Plutchik, R.: Emotion: Theory, research, and experience. In: Theories of Emotion, vol. 1. Academic Press, New York (1980)
22. Provine, R., Spencer, R., Mandell, D.: Emotional Expression Online: Emoticons punctuate website text messages. Journal of Language and Social Psychology 26(3), 299–307 (2007)
23. Purver, M., Battersby, S.: Experimenting with distant supervision for emotion classification. In: Proceedings of the 13th Conference of the European Chapter of the Association for Computational Linguistics, pp. 482–491. Association for Computational Linguistics, Avignon (2012)
24. Read, J.: Recognising Affect in Text using Pointwise-Mutual Information. Master's thesis, University of Sussex (2004)
25. Read, J.: Using emoticons to reduce dependency in machine learning techniques for sentiment classification. In: Proceedings of the ACL Student Research Workshop, pp. 43–48. Association for Computational Linguistics, Stroudsburg (2005)

26. Roberts, K., Roach, M.A., Johnson, J., Guthrie, J., Harabagiu, S.M.: EmpaTweet: Annotating and Detecting Emotions on Twitter. In: Calzolari, N., Choukri, K., Declerck, T., Do?an, M.U., Maegaard, B., Mariani, J., Odijk, J., Piperidis, S. (eds.) Proceedings of the Eight International Conference on Language Resources and Evaluation (LREC 2012), European Language Resources Association (ELRA), Istanbul (2012)
27. Seol, Y.S., Kim, D.J., Kim, H.W.: Emotion Recognition from Text Using Knowledge-based ANN. In: Proceedings of ITC-CSCC (2009)
28. Tanaka, Y., Takamura, H., Okumura, M.: Extraction and classification of facemarks. In: Proceedings of the 10th International Conference on Intelligent User Interfaces, IUI 2005, pp. 28–34. ACM, New York (2005)
29. Wang, W., Chen, L., Thirunarayan, K., Sheth, A.: Harnessing Twitter Big Data for Automatic Emotion Identification. In: International Conference on Social Computing (2012)

Using Google n-Grams
to Expand Word-Emotion Association Lexicon

Jessica Perrie, Aminul Islam, Evangelos Milios, and Vlado Keselj

Dalhousie University, Faculty of Computer Science
Halifax, NS, Canada B3H 4R2
jssc.perrie@gmail.com,
{islam,eem,vlado}@cs.dal.ca

Abstract. We present an approach to automatically generate a word-emotion lexicon based on a smaller human-annotated lexicon. To identify associated feelings of a target word (a word being considered for inclusion in the lexicon), our proposed approach uses the frequencies, counts or unique words around it within the trigrams from the Google n-gram corpus. The approach was tuned using as training lexicon, a subset of the National Research Council of Canada (NRC) word-emotion association lexicon, and applied to generate new lexicons of 18,000 words. We present six different lexicons generated by different ways using the frequencies, counts, or unique words extracted from the n-gram corpus. Finally, we evaluate our approach by testing each generated lexicon against a human-annotated lexicon to classify feelings from affective text, and demonstrate that the larger generated lexicons perform better than the human-annotated one.

1   Introduction

Problem. Users exchange ideas and opinions by writing blogs, product reviews and comments, producing a massive amount of information. Applications for sentiment and emotion analysis that take advantage of this data to automatically find the feelings conveyed by the word choice, can be used, for example, to track feelings towards a product over time [1].

Consider, for example, the words *delightful* and *gloomy*; according to the National Research Council of Canada (NRC) word-emotion association lexicon, *delightful* is associated with uplifting feelings like anticipation, and joy, while *gloomy* is associated with negative feelings like sadness [1].

While there are hundreds of possible emotions to choose from, many studies have used a small subset of basic emotions. Our study uses emotions as defined by Plutchik: anger, anticipation, disgust, fear, joy, sadness, surprise and trust, because annotating hundreds of emotions would be expensive and difficult, while Plutchik's basic set are well-founded in psychological, physiological and empirical research [1]. They are a superset of the Ekman emotions, which are commonly used in emotion studies [2,3], and are not mostly composed of negative emotions [1]. The sentiments (positive and negative) are also included in our study, but are treated exactly like the emotions. In this paper, both sentiments and emotions are referred together as *feelings*.

A. Gelbukh (Ed.): CICLing 2013, Part II, LNCS 7817, pp. 137–148, 2013.

Sentiment and emotion analysis applications have lexicon- or dictionary-based approaches when they use a general word lexicon as a starting point (and then may refine results with more domain- or feature- specific terms) [4]. Word-emotion lexicons, especially ones created by human-annotators, are essential to evaluate automatic approaches, like the one presented in this paper, that identify emotions associated with additional terms [1].

Motivation. We present an automatic approach to generate word-emotion lexicons using a smaller word-emotion lexicon and the Google n-gram corpus. Automatic approaches, like the one proposed, have many advantages over human-annotated or manual approaches. Although manual approaches tend to be more reliable, automatic approaches require less work and avoid human random error [5]. Furthermore, manually created lexicons are noted for having relatively poor coverage of technical and scientific terms that are essential to analyze research papers [5]. Another major limitation is the additional labour needed to translate the lexicon into each new language [5].

The main advantage of automatic methods is in their creation. Automatic construction approaches expand lexicons by following the smaller lexicon's patterns [5]. Additionally, depending on the similarity of languages and assuming the data needed for that approach is available, the automatic construction can also be applied to generate an emotion lexicon in another language, or plot out the evolution of different words over time [6,7]. Therefore, unlike manual methods, a smaller amount of human work is needed.

Given the advantages of word-emotion lexicons and their use in emotion and sentiment analysis, we developed an approach to generate effective word-emotion association lexicons. Each lexicon was built by comparing the data within the Google n-gram corpus and using a training lexicon of seed words, words where the associated feeling is already known. Training sets in our study are subsets of the NRC lexicon. In Section 4, we present three different methods with two variations of finding the feeling associations of target words in novel ways: the frequency of surrounding feeling associated words, the number of times surrounding feeling associated words occur, and the number of times unique surrounding feeling associated words occur. Finally, in Section 5, the lexicons generated by our methods are evaluated against the testing lexicon in a simple feeling classification task.

2   Related Work

In this section we present a description of related work: sentiment or emotion lexicons that were expanded using automatic methods.

In [5], Turney presented an unsupervised learning algorithm to find synonyms by comparing their Pointwise Mutual Information collected by Information Retrieval (PMI-IR) which measures the association between two terms, a target word and a possibly related word, by finding their probabilities of appearing together within the same document [5]. As the definition of "document" became smaller and meant the two words must appear within ten terms of each other (within a 10-gram), it was observed that the results for matching each synonym improved. In our study we used trigrams. Turney also used PMI-IR to classify the sentiment at document-level of reviews based on the average semantic orientation of their phrases. The orientation for each phrase was found by calculating the mutual information between it and the word "excellent" and "poor" [8]. Most similar to our work, Turney extended this idea further to find the polarity of target words by looking at their statistical association based on their co-occurrence with fourteen positive or negative seed words that kept their polarity no matter the context [9]. To measure co-occurrence, he counted when the target word was within ten words of the polarity word. We extend this idea by only considering appearances within three words and using over 10,000 words as seeds.

For emotion lexicons, automatic approaches have used large corpora from the web. In [10], Yang et al. used weblog corpora and a collocation model to build an emotion lexicon from online articles. Blog data were used because they were timestamped and because blogs can express emotional states of users who may use emoticons to represent their feelings [10]. A training set was used to measure the word's associations with one of forty possible emoticons–each emoticon represented an emotion–by a modified version of Pointwise Mutual Information. This approach had two variations by choosing the top n collocated word-emotion pairs; the first variation had 4,776 entries with 25,000 word sense associations, and the second had 11,243 entries and 50,000 sense pairs [10]. In their comparison of the two lexicons, they observed that the larger one had better performance in classifying emotions.

The use of the NRC word sense lexicon with Google n-grams was briefly touched on in [6] in which is it stated that "[w]ords found in proximity of target entities can be good indicators of emotions associated with the targets." Using Google n-grams frequency data from books scanned up to July 15, 2009, Mohammad placed the n-grams into bins of five years and measured the percentage of different emotion words that appeared in 5-grams with certain target words [6]. This idea is similar to our work, except we expand on it to build a lexicon with emotion and sentiment associations, but do not consider changes of the associations over time, although that is a possible future application.

3   Resources

NRC Word-Emotion Association Lexicon. The NRC word-emotion association lexicon version 0.92 is used in our study to build and test our proposed approach. It contains about 14,200 individual terms and their associations to each of the eight Plutchik basic emotions and two sentiments: anger, anticipation, disgust, fear, joy, negative, positive, sadness, surprise, trust. Each word in the lexicon has ten 2-level values indicating its association for each feeling. For example, a word like *torture* has the values $\langle 1, 1, 1, 1, 0, 1, 0, 1, 0, 0 \rangle$, which indicates there exists an association between *torture* and the feelings anger, anticipation, disgust, fear, negative, and sadness, while there are no associations to feelings of joy, positive, surprise nor trust.

The NRC lexicon was made by dividing the annotation work to a large network of laborers through Mechanical Turk [1]. The NRC lexicon terms were chosen from the most frequent English nouns, verbs, adjectives and adverbs selected from the *Macquarie Thesaurus* and Google n-gram corpus, and from other emotion lexicons like the General Inquirer and the WordNet Affect Lexicon [1].

Google N-Gram Corpus. The Google Web 1T n-gram corpus, contributed by Google Inc., contains English word n-grams (from uni-grams to 5-grams) and their observed frequencies calculated over one trillion words from web page text collected by Google in January 2006. The text was tokenized following the Penn Treebank tokenization, except that hyphenated words, dates, email addresses and URLs are kept as single tokens. The n-grams themselves must appear at least 40 times to be included in the Google n-gram corpus[1].

In October 2009, Google released the Web 1T 5-gram, 10 European Languages Version 1 [11], consisting of word n-grams and their observed frequency for ten European languages: Czech, Dutch, French, German, Italian, Polish, Portuguese, Romanian, Spanish, and Swedish. Thus, it is possible to use our proposed approach to generate lexicons for these languages as well.

Our study uses only trigrams ($n = 3$) from the Google n-gram corpus. Some examples of trigrams provided by the corpus are *he was a* with a supplied frequency 3,683,417; *hehe was a* with 52; and *he was an* with 563,471.

4   The Proposed Approach

The method to develop our proposed approach is shown at high-level in Figure 1. Actions above the second dashed line are explained in this section; actions below the dashed line are explained in the next section where we evaluate our computed lexicon in a feeling classification method.

Description of Approach. To find the feeling associations from each target word, the approach first searches for that word in the n-gram corpus, finds all the n-grams that contain the target word, and, within each n-gram, finds surrounding words from the training lexicon which we call `assoc_word`. It then generates three vectors of size ten (one value for each of the ten feelings) for each target word: `assoc_freq`, `assoc_counts` and `assoc_unique` as defined in Figure 2. To normalize the results, the totals for each of these sums where a feeling is not associated are also detected, respectively as `assoc_not_freq`, `assoc_not_counts`, `assoc_not_unique`.

Each value in each of the three vectors is normalized by taking it over the sum of itself and its inverse (e.g., normalized `assoc_freq[joy]` = `assoc_freq[joy]` / (`assoc_freq[joy]` + `assoc_not_freq[joy]`)) and is farther referred to as a "feeling association strength". In our approach three different methods are used for each of the three normalized vectors. If the feeling association strength of a certain feeling for a target word is higher than a tuned parameter, threshold-1 (as defined by our variations) then that target word is classified as having an association to that feeling. Alternately, if that feeling association strength is below another threshold-2, then that target word is identified as *not* having an

---

[1] Details can be found at www.ldc.upenn.edu/Catalog/docs/LDC2006T13/readme.txt

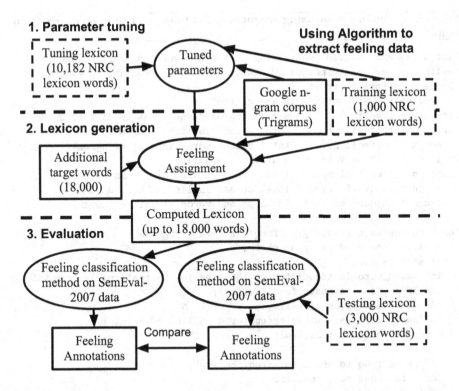

Fig. 1. High-level overview of proposed approach and evaluation. Parameter tuning and lexicon generation are carried out using three different methods in two variations; therefore, six different lexicons are produced. Each lexicon is evaluated against the Testing lexicon by measuring their performance on the SemEval-2007 text using the same classification method: keyword spotting.

association to that feeling. Feeling associations for a target word may also be classified as "unknown" because of a lack of sufficient data.

Our approach can be used with any type of n-grams–bigram, trigram, 4-gram or 5-gram; however, from our experiments, we found that trigrams produced the best results–a greater difference in feeling association strengths between target words with an association and target words without an association. Therefore, we believe that 4-grams and 5-grams are less suitable because they included too much noise in the form of surrounding words that were not indicative of the target word's associated feelings. These results also suggest that bigrams don't contain enough surrounding words to classify the target word.

Method 1: Feeling association strength: normalized **assoc_freq**. This idea follows the idea that frequencies of surrounding words in close proximity to a target word are indicative of its associated feelings [6]. Surrounding words that have a high frequency of occurring with the target word are assumed to share their associated feelings more strongly.

Extracting feeling data using frequencies, counts, and unique words from Google n-grams

```
list_of_ngrams: n-grams containing both assoc_word and target_word
feeling: anger, anticipation, disgust, fear, joy, negative, positive,
  sadness, surprise, trust
Derived over assoc_words with assoc. feeling
  assoc_freq[feeling]:          sum of n-gram frequencies in list_of_ngrams
  assoc_counts[feeling]:        total counts in list_of_ngrams
  assoc_unique[feeling]:        total unique words in list_of_ngrams
Derived over assoc_words without assoc. feeling
  assoc_not_freq[feeling]:    total freq. in list_of_ngrams
  assoc_not_counts[feeling]: total counts in list_of_ngrams
  assoc_not_unique[feeling]: total unique words in list_of_ngrams


for each (ngram_phrase, ngram_freq) in list_of_ngrams
  for each assoc_word in ngram_phrase
   for each feeling
     if (assoc_word is associated to this feeling in training lexicon)
       add ngram_freq to assoc_freq[feeling];
       add 1 to assoc_counts[feeling];
       if (assoc_word wasn't yet encountered in list_of_ngrams)
         add 1 to assoc_unique[feeling];
     else
       add ngram_freq to assoc_not_freq[feeling];
       add 1 to assoc_not_counts[feeling];
       if (assoc_word wasn't yet encountered in list_of_ngrams)
         add 1 to assoc_not_unique[feeling];
```

Fig. 2. Given a target word, use the training lexicon to find the total frequency, the total counts and the total unique words of the feelings (emotions and sentiments) of surrounding words in the Google n-gram corpus.

Method 2: Feeling association strength: normalized **assoc_counts**. Method 2 measures the variety of words in different n-grams listed in the trigram corpus.

Method 3: Feeling association strength: normalized **assoc_unique**. The idea comes from observing the data, and assuming that if a greater number of different surrounding words convey the same feeling, then that feeling is more strongly associated with the target word.

Validation. The first challenge with our approach was dealing with scarcity of data within the n-gram corpus [5]. Furthermore, this step is needed because we found removing target words with scarce data reduced the number of falsely detected associations in the tuning lexicon. Consider the relatively obscure word *obi*, which, according to the NRC lexicon, has associations with *disgust*, *fear*, and *negative*. Within the trigram corpus, no surrounding words of *obi* with associations to *disgust* are spotted, which incorrectly suggests that *obi* is not associated with *disgust*. Therefore, we need a baseline validation to ensure that enough data is available before we can classify an associated feeling for better and more precise results.

Thus, for each possible feeling of each target word, if its `assoc_unique` was smaller than $10 \log_{10}$ of the number of words with that feeling in the training lexicon, than that word-feeling association was declared unknown. However, this specific value is arbitrary because while other thresholds work better for some feelings, they do not work better for all and the amount of improvement in each result depended on the feeling type. Future work could be done in identifying this threshold more specifically.

Tuned Parameters. Each method has two variations with bounds based off the two different tuned parameter sets. The first variation's goal is to maximize the number of true values found between the computed lexicons and the human-annotated lexicon. The second variation's goal is to maximize the precision and recall of the lexicons produced when compared to the human-annotated lexicon. For each feeling, it was observed that there was a range where feeling association strengths of the tuning lexicon words with an association, and the words without an association, would overlap. The second variation works by declaring most word-feeling associations with feeling association strengths within this range as being unknown, which produces a smaller number of true values in the computed lexicon.

Variation 1: Threshold: [0.1, 0.1). If the feeling association strength for a certain feeling of a target word is ≥ 0.1, the target word is classified as having an association to that feeling; else, the target word was classified as not associated to that feeling. This value is arbitrary, because other thresholds produce similar results; however, after observing the different tuning lexicon words, most feeling association strengths with an association were over this threshold, while most feeling association strengths without an association were below.

Variation 2: Threshold: (0.05, 0.15). We expand the threshold by 0.05 to reduce the number of falsely classified associations. If the feeling association strength for a certain feeling of a target word is ≥ 0.15, the word is classified with an association to that feeling. If the feeling association strength is ≤ 0.05, then the word is classified with not having an association to that feeling. Finally, if the feeling association strength is between 0.05 and 0.15, the association of the target word to that feeling remains unknown.

Results of Comparing Human-annotated Feelings with Computed Feelings on Tuning Lexicon Words. The results of comparing the tuning lexicon with the computed lexicon built using the same words with each method at each variation are presented in Table 1. We measured the precision (p)–the number of true and detected word-feeling associations over the number of detected word-feeling associations; the recall (r)–the number of true and detected word-feeling associations over the number of true word-feeling associations in the tuning lexicon; the f-measure (f)–an average of the precision and recall as outlined in the first equation in Eq. 1; and the accuracy (a)–a measurement involving detected true associations (TP) and no associations (TN), and the number of falsely detected associations (FP) and no associations (FN), as shown in the second equation in Eq. 1.

$$f = \frac{2*p*r}{p+r} \qquad a = \frac{\text{TP} + \text{TN}}{\text{TP} + \text{TN} + \text{FP} + \text{FN}} \tag{1}$$

Table 1. Matching of the hand-annotated feelings of the words in the tuning lexicon with computed feelings of the same words based on the training lexicon and Google n-grams for the selected parameters using each method (M). In each variation (Var.), for each feeling, **boldface** values are the highest of their type: f-measure (f) or accuracy (a). The interval next to each variation name indicates the "middle area" where feeling association strengths (as measured by the type of method) are ignored if they fall within this interval, or indicate an association if greater, or no association if lower.

| | Var.1: [0.1, 0.1) | | | | | | Var.2: (0.05, 0.15) | | | | | |
| | M1:freq | | M2:count | | M3:uniq | | M1:freq | | M2:count | | M3:uniq | |
Feeling	f	a	f	a	f	a	f	a	f	a	f	a
anger	0.53	0.84	0.57	0.86	**0.60**	**0.87**	0.64	0.91	0.68	0.93	**0.81**	**0.94**
anticipation	0.20	0.72	0.24	0.77	**0.28**	**0.85**	0.28	0.76	**0.40**	**0.85**	0.33	0.71
disgust	0.39	0.84	0.50	0.88	**0.56**	**0.90**	0.39	0.90	0.46	0.95	**0.67**	**0.98**
fear	0.49	0.78	0.47	0.79	**0.52**	**0.79**	0.50	0.81	0.59	**0.84**	**0.61**	0.71
joy	0.30	0.78	**0.33**	0.81	0.33	**0.88**	0.33	0.83	0.45	0.89	**0.71**	**0.97**
negative	**0.58**	**0.66**	0.58	0.60	0.50	0.40	0.68	**0.67**	**0.70**	0.58	0.65	0.48
positive	**0.33**	**0.32**	0.31	0.21	0.30	0.18	**0.40**	**0.27**	0.36	0.22	0.32	0.19
sadness	0.36	0.78	0.39	0.81	**0.47**	**0.85**	0.50	0.85	0.61	0.91	**0.75**	**0.93**
surprise	**0.20**	0.90	0.15	0.92	0.12	**0.93**	**0.17**	0.94	0.13	**0.96**	0.00	0.95
trust	**0.30**	**0.57**	0.29	0.49	0.28	0.44	0.36	**0.54**	0.45	0.36	**0.48**	0.32

Discussion. In all variations and methods, negative feelings like *anger, disgust, fear, negative* and *sadness* tend to have higher f-measures than positive or neutral feelings like *anticipation, joy, positive, surprise* and *trust*. From observing the data, positive feeling association strengths between the words with associations and words without associations were less different. This result suggests that most word-feeling combinations in trigrams are related to expressing negative emotions [2]. Additionally, the poor results for positive feelings may be because the training lexicon has fewer words with associations to them, and thus, did not have enough positive feeling words to spot. It is also possible that words surrounding positive target words in trigrams don't reflect positive feelings.

With the exception of *trust*, the sentiments *negative* and *positive* have lower values of f-measure and accuracy, suggesting that polarities may act different than emotions, and thus, should be treated differently.

From the results for Variation 1 and Variation 2, Method 3 produces the highest results, which supports its design. Graphs of the feeling association strengths for Method 1 and Method 2 did not have as great a difference between words with an association and words without an association.

Feeling Assignment. To test our methods, we created a lexicon for each method with each variation using the 3,000 target words from the testing lexicon and 15,000 words commonly used within the Google unigram corpus that were not included in the NRC lexicon. The number of word-feeling associations of each lexicon is presented in Table 2.

Table 2. Summary of the number of words in each of the generated lexicons as classified by the different methods (M) and variations (Var.). Variation 2 has more words with only unknown associations, because it assigns most word-feeling associations with feeling association strength within an overlapping area as unknown. The intervals associated with the variations are explained in Table 1.

Feelings	Var.1: [0.1, 0.1)			Var.2: (0.05, 0.15)		
	M1:freq	M2:count	M3:uniq	M1:freq	M2:count	M3:uniq
anger	808	824	794	467	375	208
anticipation	1696	1528	1039	735	318	34
disgust	332	305	198	150	90	16
fear	1545	1524	1692	893	727	454
joy	1227	1194	778	612	424	87
negative	4854	5656	7382	3160	3356	4156
positive	10788	12529	13137	7966	9601	12090
sadness	844	771	628	442	292	100
surprise	143	59	7	54	8	0
trust	4084	4982	5775	2109	1861	1135
only unknown	4391	4391	4391	6378	6149	4988

5   Evaluation

We tested each lexicon against a baseline, the NRC testing lexicon. Because we are only interested in testing the generated lexicon and not the classification method, each lexicon was put through the same naive lexicon-based feeling classifier, keyword spotting, using data from the SemEval-2007 Task 14: Affective Text as shown from Figure 1. In future, we will consider a more accurate lexicon-based method.

SemEval-2007 Task 14: Affective Text data is a collection of news titles (which are often written to provoke readers' emotions) from newspapers and news websites like Google news and CNN [3]. All 1,250 headlines are human-annotated with measures of six emotions–anger, disgust, fear, joy, sadness and surprise–and a sentiment–negative, positive or neutral. The agreement using the Pearson correlation measure among the annotators for each feeling varied, but was lowest for *disgust* and *surprise*. Because emotions *anticipation* and *trust* are not included in this dataset, they are not included in the evaluation.

The human-annotated measurements of feelings are mapped to 1 (meaning there is an association between the feeling and headline) or 0 (meaning there is not an association), in accordance with the coarse-grained evaluation in the SemEval task. In our evaluation, we use all 1250 sentences from this dataset.

Approach. The emotion and sentiment classification method used in this evaluation was keyword spotting as shown in Figure 3.

Keyword spotting procedure to classify feelings in headlines

```
Preprocessing headline (transform to lowercase, remove punctuation, and
  tokenize words)
for each feeling
  for each word in headline
    if (lexicon entry for word-feeling association is not unknown)
      add 1 to count[feeling]
      if (word is associated to feeling)
        add 1 to temp[feeling]
      else add 0 to temp[feeling]
  if (temp[feeling]/count[feeling] is greater or equal to 0.5)
    headline is associated to feeling
  else headline is not associated to feeling
```

Fig. 3. Given a headline, use lexicon to find if associated feeling exists

Results of Evaluation. The results of using the testing lexicon and our generated lexicons are displayed in Table 3. The result formulas are like the equa-
tions Eq. 1, except we're considering headline-feeling associations instead of
word-feeling associations. (Software to recreate these results may be found at
http://www.CICLing.org/2013/data/138)

While the results of the testing lexicon may seem too low to properly judge the
lexicons, both the f-measures and accuracies of the emotions are within 0.0170 of
the lower bounds for the results of the systems that participated in the SemEval
task. For all emotions over those systems, the average f-measure was 0.0993, the
average accuracy was 0.8791, the highest f-measure was 0.3038, and the highest
accuracy was 0.9730 [3].

Table 3. Results from the feeling classification performance of the computed lexicons
from each variation (Var.) and method (M), and the human-annotated testing lexicon
on the SemEval-2007 Task 14 data set. If a value is **boldface** under any of the Method
lexicons, that value is greater than the Testing lexicon (Test.) for either f-measure (f)
or accuracy (a).

| | | | Var.1: [0.1, 0.1) | | | | | | Var.2: (0.05, 0.15) | | | | | |
| | Test. | | M1:freq | | M2:count | | M3:uniq | | M1:freq | | M2:count | | M3:uniq | |
Feeling	f	a	f	a	f	a	f	a	f	a	f	a	f	a
anger	0.06	0.93	**0.11**	0.86	**0.12**	0.87	**0.11**	0.89	**0.12**	0.89	**0.11**	0.91	**0.07**	**0.96**
disgust	0.00	0.97	**0.21**	0.95	**0.02**	0.93	**0.05**	0.97	0.00	0.96	0.00	0.97	0.00	**0.98**
fear	0.17	0.87	**0.27**	0.77	**0.25**	0.77	**0.25**	0.74	**0.25**	0.81	**0.19**	0.82	**0.21**	**0.87**
joy	0.12	0.85	**0.14**	0.80	0.10	0.81	0.06	**0.87**	0.08	0.83	0.07	**0.87**	0.00	**0.88**
sadness	0.08	0.86	**0.21**	0.80	**0.22**	0.82	**0.17**	0.86	**0.12**	**0.86**	**0.13**	0.85	0.00	**0.88**
surprise	0.05	0.93	**0.07**	0.93	0.00	**0.95**	0.00	**0.96**	0.00	**0.95**	0.00	**0.96**	0.00	**0.96**
positive	0.15	0.81	**0.22**	0.43	**0.22**	0.38	**0.22**	0.37	**0.22**	0.51	**0.22**	0.46	**0.22**	0.38
negative	0.19	0.74	**0.39**	0.60	**0.40**	0.56	**0.39**	0.48	**0.34**	0.66	**0.37**	0.67	**0.38**	0.56
Average	0.10	0.87	**0.20**	0.77	**0.17**	0.76	**0.16**	0.77	**0.14**	0.81	**0.14**	0.81	**0.11**	0.81
Highest	0.19	0.97	**0.39**	0.95	**0.40**	0.95	**0.39**	0.97	**0.34**	0.96	**0.37**	0.97	**0.38**	0.98

Discussion. From the results in Table 3, the larger lexicons generated in Variation 1 give higher f-measures than the lexicons in Variation 2, while the smaller more accurate lexicons in Variation 2 give higher accuracies than the lexicons in Variation 1. This latter result likely occurs because most of the human-annotated headlines are not associated to a feeling (when using the 1 or 0 mapping). Because smaller lexicons would only classify a smaller number of words within the headlines, a larger number of these likely-not-associated-to-a-feeling headlines would remain, by default, not associated to any feelings, and thus increase the accuracy. Compared to the testing lexicon, Variations 2's results are still notable because its lexicons are larger than in the testing set (so fewer headlines are left by default with no associations), and yet, it still produces higher accuracies. Variation 1's better performance in f-measures suggest that larger lexicons, with their greater coverage of possible words, increase the precision and recall of feeling classification tasks to find if associations exist, but are less accurate when finding when associations do not exist.

Compared to the testing lexicon, the lexicon computed with Method 1, Variation 1, has the highest f-measures, despite having some of the lowest f-measures in Table 1, which suggests that other tests are needed besides f-measure and accuracy to find the best approach to generate a lexicon. Because Method 1 was based on frequencies, our results add credibility to other frequency-based automatic approaches like Pointwise Mutual Information. Method 3, Variation 2 has the highest accuracies; however, Method 3 does not have as high f-measures, which does not help in identifying if headlines have feeling associations.

For both variations, the computed lexicons performed better for negative emotions like *anger*, *fear* and *sadness*, which farther suggests negative emotions are expressed more in the trigram corpus [2]. The poor performance of *disgust* and *surprise* may result because they had the lowest agreement between the human annotators of the SemEval Affective Text, and, as shown in Table 2, all generated lexicons had a relatively smaller number of word-feeling associations for these feelings, suggesting less accuracy to correctly identify them.

Overall, these results suggest that larger lexicons created using automatic methods can perform feeling classification tasks better than smaller human-annotated ones in terms of f-measure and accuracy. Large lexicons created with less accurate methods (Variation 1) tend to have better f-measure, while smaller lexicons (but still larger than the human-annotated lexicon) with better f-measures (Variation 2) tend to have better accuracy.

6 Conclusion

We proposed a new approach to generate a lexicon by automatic means using data provided by the Google n-grams corpus and NRC lexicon. Our approach consists of using the frequencies of n-grams, the counts of surrounding words or the unique counts of surrounding words at two different variations of tuned parameters to produce lexicons with a relatively large or small number of word-feeling associations. The larger lexicons had more words, but less accuracy than the smaller lexicons. From our evaluation of these computed lexicons against the testing lexicon, we provide evidence that suggests larger lexicons generated with less accurate methods perform better, and that more measurements, in conjunction with precision, recall and accuracy, are needed to find an approach to generate an effective lexicon.

In addition to the future work mentioned in previous sections, we intend to look into using the n-grams farther by searching for the context around each target word and then searching for an identical context where a word from the training lexicon is used. We will also look into handling target words differently depending on how they are used or their parts of speech.

References

1. Mohammad, S., Turney, P.: Crowdsourcing a word–emotion association lexicon. Computational Intelligence (2012)
2. Kozareva, Z., Navarro, B., Vázquez, S., Montoyo, A.: Ua-zbsa: a headline emotion classification through web information. In: Proceedings of the 4th International Workshop on Semantic Evaluations, SemEval 2007, pp. 334–337. Association for Computational Linguistics, Stroudsburg (2007)
3. Strapparava, C., Mihalcea, R.: Semeval-2007 task 14: affective text. In: Proceedings of the 4th International Workshop on Semantic Evaluations, SemEval 2007, pp. 70–74. Association for Computational Linguistics, Stroudsburg (2007)
4. Lu, Y., Castellanos, M., Dayal, U., Zhai, C.: Automatic construction of a context-aware sentiment lexicon: an optimization approach. In: Proceedings of the 20th International Conference on World Wide Web, WWW 2011, pp. 347–356. ACM, New York (2011)
5. Turney, P.D.: Mining the web for synonyms: PMI-IR versus LSA on TOEFL. In: Flach, P.A., De Raedt, L. (eds.) ECML 2001. LNCS (LNAI), vol. 2167, pp. 491–502. Springer, Heidelberg (2001)
6. Mohammad, S.: From once upon a time to happily ever after: tracking emotions in novels and fairy tales. In: Proceedings of the 5th ACL-HLT Workshop on Language Technology for Cultural Heritage, Social Sciences, and Humanities, LaTeCH 2011, pp. 105–114. Association for Computational Linguistics, Stroudsburg (2011)
7. Amiri, H., Chua, T.S.: Mining sentiment terminology through time. In: Proceedings of the 21st ACM International Conference on Information and Knowledge Management, CIKM 2012, pp. 2060–2064. ACM, New York (2012)
8. Turney, P.D.: Thumbs up or thumbs down?: semantic orientation applied to unsupervised classification of reviews. In: Proceedings of the 40th Annual Meeting on Association for Computational Linguistics, ACL 2002, pp. 417–424. Association for Computational Linguistics, Stroudsburg (2002)
9. Turney, P.D., Littman, M.L.: Measuring praise and criticism: Inference of semantic orientation from association. ACM Trans. Inf. Syst. 21, 315–346 (2003)
10. Yang, C., Lin, K.H.Y., Chen, H.H.: Building emotion lexicon from weblog corpora. In: Proceedings of the 45th Annual Meeting of the ACL on Interactive Poster and Demonstration Sessions. ACL 2007, pp. 133–136. Association for Computational Linguistics, Stroudsburg (2007)
11. Brants, T., Franz, A.: Web 1t 5-gram, 10 european languages version 1. In: Linguistic Data Consortium, Philadelphia, PA, USA (2009)

A Joint Prediction Model
for Multiple Emotions Analysis in Sentences

Yunong Wu, Kenji Kita, Kazuyuki Matsumoto, and Xin Kang

Faculty of Engineering
University of Tokushima
wuyunong@iss.tokushima-u.ac.jp,
{kita,matumoto,kangxin}@is.tokushima-u.ac.jp

Abstract. In this study, we propose a scheme for recognizing people's multiple emotions from Chinese sentence. Compared to the previous studies which focused on the single emotion analysis through texts, our work can better reflect people's inner thoughts by predicting all the possible emotions. We first predict the multiple emotions of words from a CRF model, which avoids the restrictions from traditional emotion lexicons with limited resources and restricted context information. Instead of voting emotions directly, we perform a probabilistic merge of the output words' multi-emotion distributions to jointly predict the sentence emotions, under the assumption that the emotions from the contained words and a sentence are statistically consistent. As a comparison, we also employ the SVM and LGR classifiers to predict each entry of the multiple emotions through a problem-transformation method. Finally, we combine the joint probabilities of the multiple emotions of sentence generated from the CRF-based merge model and the transformed LGR model, which is proved to be the best recognition for sentence multiple emotions in our experiment.

Keywords: Multiple emotions, Joint prediction, CRF, LGR.

1 Introduction

Affective information computing is drawing more attention in the recent studies of artificial intelligence [1]. Identifying people's inner emotion states is really a challenging issue compared to the traditional sentiment analysis, since people's emotion states are very private and often change with high frequency. And predicting emotions from the texts has become a common method for emotion analysis, because the textual information such as blogs is relatively easy to be extracted from the Internet, which have embedded the rich emotional states in people's daily lives.

The studies of affective information computing are categorized by two levels including the coarse-level which focuses on the sentiment polarity analysis and the fine-level which studies the human emotions. Specifically, the sentiment polarity analysis would classify the texts (especially the product reviews and the Twitter messages) into the positive, the negative, and the neutral categories. And the emotion analysis finds more subtle human emotions.

A. Gelbukh (Ed.): CICLing 2013, Part II, LNCS 7817, pp. 149–160, 2013.

For emotion analysis, the previous studies have made strong assumptions on the text emotion distribution, by restricting a single emotion label for a piece of text. However, as we know, the real emotional states of human beings are more complicated than a single emotion label that can be represented.

As we have observed through a large amount of Blog articles, the text emotions often fall into multiple emotion categories. This phenomenon becomes more common in the long texts such as sentences and documents. Another important observation is that the emotions are accumulative. In the study of [2], it showed that document emotions were composed of accumulated word emotions.

In this study, we propose a joint prediction model for multiple emotions analysis in the sentence level. The emotion categories include Expect, Joy, Love, Surprise, Anxiety, Sorrow, Anger, and Hate. We first extract the multiple emotions of words in sentences with a context-sensitive Conditional Random Field (CRF) model [3]. The CRF model could generate reliable probabilistic predictions on the multi-emotion label sequences in the sentences. Then we perform a probabilistic merge of the words' multi-emotion predictions to get the multi-emotion distributions in sentences. The main assumption under the probability merge is that the sentence emotion probabilities are statistically consistent with the embedded word emotion probabilities, and the probability volumes could be accumulated just like the word emotions could be accumulated in the sentence emotions.

We learn another probabilistic model on sentence emotions directly from the word distributions with a transformed Logistic Regression (LGR) model [4]. For each emotion category, a binary classifier is trained with the emotion-related and the emotion-unrelated sentences. And the binary prediction results are combined to predict the multi-emotions through a problem-transformation method. The transformed LGR model for multi-emotion analysis has specific drawbacks compared to the CRF-based merge model, in that emotions are supposed to be independently distributed, which however is a very strong assumption in the text emotion analysis. Nevertheless, as the two models are distinguished by the different assumptions and the completely different emotional features, they should support each other in predicting the multi-emotion distributions and jointly predict sentence emotions with better performance.

We combine the joint probabilities of the multiple emotions of sentence from the CRF-based merge model and the transformed LGR model, to generate the final sentence emotion probabilities. As a comparison, we also employ the binary SVM classifiers to predict the multi-emotions of sentences in the same fashion as in the transformed LGR model, except that Support Vector Machine (SVM) classifier [5] generates 1-0 results for the emotion and non-emotion prediction, while the LGR classifier produces probability results.

The results of the transformed SVM model, the transformed LGR model, and the CRF-based merge model are regarded as the baseline.

The rest of this paper is organized as follows: Section 2 introduces the related works in recent years. Section 3 describes joint prediction model for multiple emotion analysis. Section 4 illustrates the experiment. Finally, section 5 concludes this paper.

2   Related Works

Recent studies on affective information computing focused on coarse and fine-grained analysis. The coarse-grained affect studies conducted experiments on the sentiment polarity classification on the product reviews and the messages of Micro-Blogs like Twitter [6]. In [7][8], they carried out experiments on the sentiment classification using the movie review dataset. [9] and [10] collected a corpus from Twitter and classified them into the three categories of positive, negative and neutral respectively. The fine-grained affect studies worked on the emotion classification in more subtle emotion categories. [11] obtained a lot of emotion-provoking events from web, and conducted classification tasks in the coarse-grained and fine-grained emotions separately. [12] explored the emotion prediction problem on the emotional sentences from some Children's fairy tales. [13] proposed an automatic identification of six emotions on text based on knowledge and corpus. However, all these researches were focused on the single emotion analysis. Seldom studies worked on multiple emotions analysis except [14], in which a Hierarchical Bayesian Network was employed to analyze the complex emotions of words. As we know, in the large text pieces, such as sentences and documents, single emotion labels can't exactly express the real emotional states of the writers. This is the reason why we study the prediction on multiple emotions, which would better capture emotions of writers.

3   Joint Prediction Model of Multiple Emotions

We explore the multiple emotions of sentence through a joint prediction model composed of a CRF-based merge model and a LGR model. Both models generate the sentence emotion predictions in the K-dimensional probability vectors, in which K is the number of emotion categories.

3.1   CRF-Based Merge Model for Sentence Emotion Prediction

An important observation about the emotion distribution in different levels of texts from the Blog articles is that the emotions are accumulative. The emotions of higher level texts are statistically consistent with the emotions of lower level texts. Therefore, it would be reasonable to predict sentence emotions from the embraced word emotions.

3.1.1 Word Emotion Recognition
We make use of a CRF model to predict a sequence of the multi-emotion labels for the words in a sentence, for considering the rich context information. This model could generate reliable probabilities on the multi-emotions of each word in the sentence, by marginalization over the joint output probability.

Specifically, we first train a CRF model on a training set selected from the Ren-CECps, which incorporates the context information such as the N-gram words, the degree, the negative and the conjunction modifications as the word emotion features. The candidate word labels in the CRF model include the eight emotion categories for the emotional words as well as a No-emotion category for the non-emotional words. The output of the CRF model is the joint probability of the multi-emotion labels together with a No-emotion label for a sequence of words in a sentence. Next, we calculate the marginal emotion probability on each word:

$$\emptyset\left(y_i^{(j)}\right) = \sum_{y_i^{(j')} \neq y_i^{(j)}} p\left(y_i^{(1)} \dots y_i^{(n_i)} \middle| x_i\right) \tag{1}$$

where $p\left(y_i^{(1)} \dots y_i^{(n_j)} \middle| x_i\right)$ is the prediction given by CRF for sentence x_i. In the formula (1), $\emptyset\left(y_i^{(j)}\right)$ is a $K+1$ dimensional probability vector of the j_{th} word emotion in i_{th} sentence, which consists of the probability values in the corresponding emotion categories as well as in the No-emotion category.

3.1.2 Sentence Emotion Recognition

In this part, we predict the sentence multi-emotions through a probabilistic merge process, in which the emotion probabilities of the embraced words are accumulated in the sentence emotion probability. It has to be noticed that in word emotion recognition, each word gets probabilities for the emotional labels and the No-emotion label at the same time. For the true emotional words, their emotion probabilities are effective factors in the sentence probability, while for the non-emotional words, the probabilistic volume is almost monopolized by the No-emotion label, and the other emotion labels get low and even meaningless probability volumes. Therefore, to get precise accumulated sentence emotions, we have to avoid the effect from the No-emotion words. This is done by selecting a threshold of the No-emotion probability, and removing the words whose No-emotion label has a higher probability than threshold.

Threshold Selection

From a series of candidate values, the threshold of No-emotion label is selected by examining the biggest F_{score} of the No-emotion word classification on a validation set. The detail is shown in Algorithm 1.

Algorithm 1. Calculate the threshold

```
(t_best, F1_best) ← (0.001, 0)
for t = 0.001 to 0.999 do
  (tp, fp, tn, fn) ← (0, 0, 0, 0)
for i = 1 to D do
  if p_i > t and y_i = 1 then
     tp ← tp+1
  else if p_i > t and y_i = 0 then
     fp ← fp+1
  else if p_i ≤ t and y_i = 0 then
     tn ← tn+1
```

```
else if p_i ⩽ t and y_i = 1 then
    fn ← fn+1
  end if
end for
F1 = 2*tp/(2*tp+fp+fn)
if F1 > F1_best then
  t_best = t
  F1_best = F1
end if
end for
return t_best
```

By removing the No-emotion words, we can further restrict the probability vectors of the rest emotion words from $K + 1$ dimensions into K dimensions. The normalization of word emotion probability guarantees that the emotion probabilities of different words are comparable.

Factor Product of Emotion Probability Vectors

Factors are used to generally represent the joint distribution of several variables, and the emotion probability vector can be viewed as the factor over the emotion variable. To combine the distributions in different factors, a factor product is often performed by multiplying the values of the same entries in all the factors:

$$\emptyset^{(1)} = \left(\emptyset_1^{(1)} \dots \emptyset_k^{(1)}\right)$$
$$\emptyset^{(2)} = \left(\emptyset_1^{(2)} \dots \emptyset_k^{(2)}\right) \tag{2}$$
$$\emptyset^{(1)}\emptyset^{(2)} = \left(\emptyset_1^{(1)}\emptyset_1^{(2)} \dots \emptyset_k^{(1)}\emptyset_k^{(2)}\right)$$

in which $\emptyset^{(1)}$ and $\emptyset^{(2)}$ are two factors of length k, and $\emptyset^{(1)}\emptyset^{(2)}$ is the factor product. For the factor of sentence emotion probability, we have the calculations as follows:

$$p^c(y_i|x_i) \propto \emptyset(y_i|x_i) = \prod_{j=1}^{n_i} \emptyset\left(y_i^{(j)}\middle|x_i\right) \tag{3}$$

in which we multiply all the word emotion factors $\emptyset\left(y_i^{(j)}\middle|x_i\right)$ to get the sentence emotion factor $\emptyset(y_i|x_i)$. The factor product generates a vector of emotion probabilities, with the volume in each entry proportional to the corresponding sentence emotion probability. We perform the normalization to the sentence emotion factor to make sure the probability values in all the entries in $p^c(y_i|x_i)$ should sum to 1, and all the sentence emotion probability vectors are comparable.

It has to be noticed that through the probabilistic merge, the output of the CRF-based merge model is a probability vector for a multinomial distributed emotion variable, with the restriction of $\sum_{y_i} p^c(y_i|x_i) = 1$. And the different emotions in prediction have to depend on each other.

3.2   Transformed LGR Model for Sentence Emotion Prediction

In contrast to the CRF-based merge model, in which the sentence emotions are learned from the word emotions, we propose another probabilistic model for sentence emotion prediction which directly learns the sentence emotions from the word distributions with a Logistic Regression (LGR) model. We call this the transformed LGR model, because the multi-emotions of sentences are transformed from K binary classification results. Specifically, we separately train K binary-classifiers h_k^G on each emotion category, and the k_{th} binary-classifier generates a probability p_k^G representing the possibility of having the k_{th} emotion in this sentence. All the prediction results from the K binary-classifiers are combined to predict the sentence multi-emotions.

$$y_{ik} = h_k^G(x_i) = p_k^G(y_i|x_i) \tag{4}$$

Like the CRF-based merge model, the transformed LGR model generates the sentence multi-emotion prediction in a K-dimensional probability vector, in which K is the number of emotion categories. However, in contrast to the CRF-based merge model, the output of LGR model is a vector of independent probability values. Because each binary-classifier is trained separately for a specific emotion category, and each entry of the output emotion probability vector separately evaluates the possibility of existence of certain an emotion in the sentence.

3.3   Joint Prediction of CRF and LGR

The CRF-based merge model and the transformed LGR model are essentially two different models. In the CRF-based merge model, the main assumption is a statistical consistency between the word emotion probability and the sentence emotion probability. Therefore, the sentence multi-emotion is predicted through a probabilistic merge by employing the word emotions as features. In the CRF-based merge model, the contents of the words are only considered in the prior process of word emotion recognition and get ignored in the sentence emotion prediction. On the contrary, the transformed LGR model assumes a direct relationship between the sentence emotions and the word contents. By fitting a logistic function on the feature of words for each emotion category, the LGR classifier could generate the probability for a specific emotion label in the sentence.

Another important difference between the two models is the meaning of outputs. The CRF-based merge model outputs a K-dimensional probability vector for a multinomial distributed emotion variable. The output vector as a whole indicates a sort on the emotion labels, in which the larger probability volume in an entry of the output vector suggests that the sentence is more probable to express the corresponding emotion than the others. And different emotions depends on each other with the restriction of $\sum_{y_i} p^c(y_i|x_i) = 1$. For the transformed LGR model, the output is a collection of K distinct probabilities, each of which specifically evaluate the possibility of existence of a particular emotion in the sentence. Different emotions are independent on each other.

Because the two models predict the sentence emotions under different assumptions and specify the multi-emotion from separate aspects, their results would suggest the sentence emotions in different aspects. We intend to acquire the multiple sentence emotions through the joint prediction composed of the CRF-based merge model and the transformed LGR model. The probability vector of the sentence emotion, in which each element corresponds to the probability of a particular emotion existence, is combined by the factor product as depicted in formula (5).

$$p^{CG}(y_i) \propto p^C(y_i|x_i)p^G(y_i|x_i) \qquad (5)$$

The probability on each emotion category is generated by multiplying the corresponding output entries from the CRF-based merge model p^C and the transformed LGR model p^G respectively. We get the joint prediction of the sentence emotions p^{CG}_{joint} by normalizing the factor product using formula (6).

$$p^{CG}_{joint} = \frac{p^C(y_i|x_i)p^G(y_i|x_i)}{\sum_{y_i'} p^C(y_i'|x_i)p^G(y_i'|x_i)} \qquad (6)$$

The CRF-LGR joint model generates K-dimensional probability vectors for the sentence emotions. To confirm the existence of each emotion, we need to select the thresholds for each emotion category from the validation set, using the same method as depicted in Algorithm 1. Specifically, we select each threshold from the candidate set by examining the biggest F_{score} of the corresponding single emotion classification.

4   Experiment

4.1   Experimental Method

We study the multiple emotion prediction in sentences from Ren-CECps[1], which is a well annotated emotion corpus on Chinese Blog articles. Each sentence in this corpus is labeled with several emotions in the eight basic emotion categories of Expect, Joy, Love, Surprise, Anxiety, Sorrow, Anger, and Hate. Because most Blog articles are written in arbitrary styles, we have to filter some extremely short and meaningless sentences, such as a series of punctuations.

The sentences are divided into a training set of 18,630 sentences, a validation set and a test set of 6214 sentences in each separately. We extract the words and the word-POS pairs in the sentences, as the candidate features for emotion prediction. Besides the CRF and LGR models, we also employ the SVM classifier in a similar transformation process as in the LGR to make further comparison.

4.2   Baseline Methods

The experiment results from the single models, including transformed SVM, transformed LGR, and CRF-based merge models, are regarded as the baseline for the sentence multiple emotion prediction.

---

[1] Ren-CECps is a Chinese emotion corpus containing 1,487 manually annotated documents, which can be found at http://a1-www.is.tokushima-u.ac.jp/member/ren/ Ren-CECps1.0/DocumentforRen-CECps1.0.html

The SVM and LGR are binary classification algorithms. We have to train binary-classifiers on each emotion category from the training set, in a one-against-all fashion, to build models for the multi-emotion classifiers. Specifically, the transformed SVM model would generate k binary results for a sentence, indicating the existence of each emotion. The transformed LGR model predicts the sentence emotions by calculating k probabilities, indicating the confidence of having these emotions in the sentence. And we select the confidence threshold in the validation set for each emotion label, to confirm the existence of this emotion in the LGR output.

For the CRF model, we recognize the emotional words and their emotions, with a probabilistic output for the word emotion distribution. We merge the word emotions through a factor product of the word emotion probability vectors, and normalize the result factor to get the multi-emotion probabilities. We select the thresholds on each emotion category as the transformed LGR model, to confirm the existence of emotions in the output.

4.3   Evaluation

The multiple emotion analysis in sentences can be viewed as a multi-label classification problem. We employ six multi-label evaluation methods, including Hamming Loss [15], Accuracy, Precision, Recall [16], $MicroF_{score}$ and $MacroF_{score}$, to thoroughly analyze the emotion classification results. The details of the evaluation methods are illustrated below.

Hamming Loss: the average percentage of misclassified labels.

$$\text{hoss}(H) = \frac{1}{D}\sum_{i=1}^{D}\frac{|y_i \text{ XOR } z_i|}{K} \tag{7}$$

where H could be SVM, LGR, CRF, or the joint model. y_i is the predicted emotion labels for the i_{th} sentence, z_i is the corresponding true emotion labels, and $y_i \text{ XOR } z_i$ is the number of different entries in y_i and z_i. $K = 8$ indicates the number of entries in the multiple emotions, and D is the size of the data set.

Accuracy: the average percentage of correctly classified labels among all the correctly and incorrectly classified labels.

$$\text{Accuracy}(H) = \frac{1}{D}\sum_{i=1}^{D}\frac{|y_i \cap z_i|}{|y_i \cup z_i|} \tag{8}$$

Precision: the average percentage of correctly classified labels among all the predicted labels.

$$\text{Precision}(H) = \frac{1}{D}\sum_{i=1}^{D}\frac{|y_i \cap z_i|}{|y_i|} \tag{9}$$

Recall: the average percentage of correctly classified labels among all the true labels.

$$\text{Recall}(H) = \frac{1}{D}\sum_{i=1}^{D}\frac{|y_i \cap z_i|}{|z_i|} \tag{10}$$

$MicroF_{score}$ and $MacroF_{score}$: the averaged measure of precision and recall, for multiple emotions analysis.

When calculating F_{score} for 2-label (binary) classification problems, we need to count number of correctly predicted positive labels (tp), the number of incorrectly predicted positive labels (fp), the number of correctly predicted negative labels (tn), and the number of incorrectly predicted negative labels (fn). And the formula for the F_{score} is

$$F_{score} = \frac{2*tp}{2*tp+fp+fn} \tag{11}$$

When calculating $MicroF_{score}$ and $MacroF_{score}$ for the multi-label classification problems, we have for each label k a set of counts as (tp_k, fp_k, tn_k, fn_k). The formula for $MicroF_{score}$ is

$$MicroF_{score} = \frac{2*tp^{Micro}}{2*tp^{Micro}+fp^{Micro}+fn^{Micro}} \tag{12}$$

where

$$tp^{Micro} = \sum_{k=1}^{K} tp_k \quad , \quad fp^{Micro} = \sum_{k=1}^{K} fp_k$$
$$tn^{Micro} = \sum_{k=1}^{K} tn_k \quad , \quad fn^{Micro} = \sum_{k=1}^{K} fn_k \tag{13}$$

The $MicroF_{score}$ evaluates the multi-label classification results by summing all the correctly predicted positive results as the true positive count (tp), and summing all the incorrectly predicted positive results as the false positive count (fp), and the same treatment on the true negative count (tn) and the false negative count (fn). In $MicroF_{score}$, the different labels are not explicitly distinguished, and the score evaluates the overall correctness and completeness of the result. And the formula for $MacroF_{score}$ is

$$MacroF_{score} = \frac{1}{k}\sum_{k=1}^{K} F_{score}^{(k)} \tag{14}$$

where, $F_{score}^{(k)}$ is the F_{score} of the k_{th} emotion type. The $MacroF_{score}$ evaluates the mean of the F_{score} on all the categories in the result.

4.4   Results and Discussion

We evaluate the results of multiple emotions from the CRF-based merge model, the transformed LGR and SVM models, and the joint prediction model, on the test set with different features. The details are shown in table 1.

Table 1. HamLoss, Accuracy, Precision, Recall, MicroF and MacroF of different methods

	SVM. w	SVM. wp	LGR. w	LGR. wp	CRF	CG. w	CG. wp
HamLoss	17. 36	18. 12	**17. 20**	17. 29	22. 20	**22. 35**	22. 95
Accuracy	34. 70	34. 25	34. 84	34. 19	**39. 87**	**42. 36**	41. 57
Precision	42. 65	42. 09	42. 80	42. 30	**47. 41**	**48. 64**	47. 86
Recall	41. 22	41. 51	41. 31	40. 27	**56. 45**	67. 01	**67. 22**
MicroF	**46. 50**	45. 45	46. 64	45. 86	47. 41	**51. 45**	50. 91
MacroF	**38. 33**	37. 49	37. 82	37. 22	38. 09	**43. 36**	42. 58

In table 1, SVM.w indicates the SVM model with word feature, and CG.wp corresponds to the joint model of the CRF and LGR model, with the word-POS feature. We choose Hamming Loss, MicroF and MacroF as our primary evaluation indicators. For the single models, LGR.w ranks the best Hamming Loss of 17.20%, while SVM.w and CRF achieve the highest MicroF and MacroF scores respectively. The Precision, Recall, together with the primary evaluations suggest that LGR generally performs better than SVM model, while CRF-based merge mode outperforms the transformed models. For the joint prediction models, the CG.w and CG.wp achieve much better results than all the single models.

The results also suggest that for the problem of sentence multi-emotion prediction, the word emotion features construct a better pattern for the sentence emotion classification than the word and word-POS pair features. Also, the result comparisons of LGR.w v.s. LGR.wp and SVM.w v.s. SVM.wp indicate that the words in a sentence could better reflect sentence emotions than the word-POS pairs.

To further analysis the results of our multi-emotion classification models, we examine the outputs in all the single emotion categories. We use the F_{score} to evaluate the models' performance. Fig. 1 shows the details.

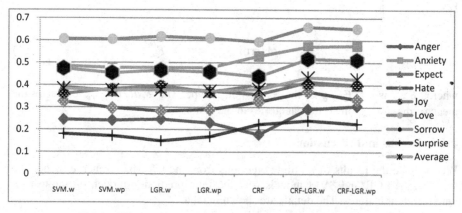

Fig. 1. F_{score}'s in each emotion category from different models

Emotions of Love, Anxiety and Sorrow are well predicted by all the single models and the joint prediction models, indicating that these emotions are relatively easy to recognize. Expect, Joy, and Hate get the relative medially ranked F_{score}'s. The results also suggest that Surprise and Anger are hard to predict, even with the joint prediction models.

By examining the training data and incorrectly predicted cases, we find the major problem in the predicting emotions of Surprise and Anger is the lack of the words annotated with these emotions. In other words, some emotion such as Surprise and Anger requires more effective emotional features to be classified.

5   Conclusions

In this study, we propose a joint prediction model, composed of a CRF-based merge model and a transformed LGR model, for predicting the multiple emotions in sentences. We explore the word emotion features from a context-sensitive CRF model, and merge the probabilistic outputs of word emotions to predict the sentence emotion probabilities. The CRF-based merge model generates a probability vector for the multinomial-distributed sentence emotion. We also explore the word and word-POS features for the sentence multi-emotion prediction, with the transformed LGR and SVM models. Binary LGR and SVM classifiers are trained on each emotion category separately, and their predictions are directly combined to classify the multi-emotions. Like the CRF-based merge model, the transformed LGR model also generates a K-dimensional probability vector. However, the two models are completely different in their basic assumptions and the investigated emotional features, which inspired the joint prediction model to analyze the sentence multi-emotions in different aspects.

We employ the multi-label evaluation methods to examine our models in sentence multi-emotion classification. All the evaluation indicators suggest that the joint prediction model achieves the most promising results, compared to the other base models. We also examine the F_{score} for in each single emotion category from the different models. The results indicate that among the eight emotion categories, some emotions like Love, Anxiety, and Sorrow are easy to predict with current models, while some emotions, such as Surprise and Anger, are hard to classify and might require more effective emotional features. We regard exploring effective emotional features for the multi-emotion classification as a future direction, and expect to develop appropriate models to recognize these emotional patterns in texts.

Acknowledgements. This research has been partially supported by the Grant-in-Aid for Scientific Research (B), 21300036, and Grant-in-Aid for Young Scientists (B), 23700252 from the Japan Society for the Promotion of Science.

Reference

1. Picard, R.W.: Affective Computing. MIT Press, Cambridge (1997)
2. Wu, Y., Kita, K., Ren, F., Matsumoto, K., Kang, X.: Exploring EmotionalWords for Chinese Document Chief Emotion Analysis. In: Proc. of 25th PACLIC, pp. 597–606 (2011)

3. Lafferty, J., McCallum, A., Pereira, F.: Conditional random fields: Probabilistic models for segmenting and labeling sequence data. In: Proceedings of ICML, pp. 282–289 (2001)
4. Menard, S.: Applied Logistic Regression Analysis (Sage University Paper Series 07-106). Sage Publications, Thousand Oaks (1995)
5. Suykens, J.A.K., Vandewalle, J.: Least squares support vector machine classifiers. Neural Process. Lett. 9(3), 293–300 (1999)
6. Bermingham, A., Ghose, A., Smeaton, A.F.: Classifying sentiment in microblogs: Is brevity an advantage? In: Proc. of CIKM, pp. 1833–1836. ACM (2010)
7. Pang, B., Lee, L., Vaithyanathan, S.: Thumbs up? Sentiment Classification using Machine Learning. In: Proc. of the Conference on EMNLP, pp. 79–86 (2002)
8. Turney, P.: Thumbs up or thumbs down? Semantic orientation applied to unsupervised classification of reviews. In: Proc. of the ACL, pp. 417–424 (July 2002)
9. Pak, A., Paroubek, P.: Twitter as a corpus for sentiment analysis and opinion mining. In: Proc. of the LREC 2010 (ELRA), Valletta, Malta, pp. 19–21 (May 2010)
10. Go, A., Bhayani, R., Huang, L.: Twitter sentiment classification using distant supervision. Technical report, Stanford Digital Library Technologies Project (2009)
11. Tokuhisa, R., Inui, K., Matsumoto, Y.: Emotion Classification Using Massive Examples Extracted from the Web. In: Proc. of Coling 2008, pp. 881–888 (2008)
12. Alm, C.O., Roth, D., Sproat, R.: Emotions from text: Machine learning for text-based emotion prediction. In: Proc. of HLT/EMNLP (2005)
13. Strapparava, C., Mihalcea, R.: Learning to identify emotions in text. In: Proceedings of the 2008 ACM symposium on Applied computing, NewYork, pp. 1556–1560 (2008)
14. Kang, X., Ren, F.: Predicting Complex Word Emotions and Topics through a Hierarchical Bayesian Network. journal of China Communications 9(3), 99–109 (2012)
15. Schapire, R.E., Singer, Y.: BoosTexter: A Boosting-based System for Text Categorization. In: Machine Learning, vol. 39, pp. 135–168 (2000)
16. Godbole, S., Sarawagi, S.: Discriminative Methods for Multi-labeled Classification. In: Dai, H., Srikant, R., Zhang, C. (eds.) PAKDD 2004. LNCS (LNAI), vol. 3056, pp. 22–30. Springer, Heidelberg (2004)

Evaluating the Impact of Syntax and Semantics on Emotion Recognition from Text

Gözde Özbal[1] and Daniele Pighin[2],*

[1] FBK-irst, Trento, Italy
gozbalde@gmail.com
[2] UPC, Barcelona, Spain
daniele.pighin@gmail.com

Abstract. In this paper, we systematically analyze the effect of incorporating different levels of syntactic and semantic information on the accuracy of emotion recognition from text. We carry out the evaluation in a supervised learning framework, and employ tree kernel functions as an intuitive and effective way to generate different feature spaces based on structured representations of the input data. We compare three different formalisms to encode syntactic information enriched with semantic features. These features are obtained from hand-annotated resources as well as distributional models. For the experiments, we use three datasets annotated according to the same set of emotions. Our analysis indicates that shallow syntactic information can positively interact with semantic features. In addition, we show how the three datasets can hardly be combined to learn more robust models, due to inherent differences in the linguistic properties of the texts or in the annotation.

1 Introduction

Automatically recognizing the emotion conveyed in a piece of text is a challenging and recently popular topic in computational linguistics. The common goal of all the studies that have been conducted in this area is developing systems which can detect the emotions of users and express various types of emotions [1]. A possible solution to this problem has many potential applicative scenarios in opinion mining, market analysis [2], affective interfaces for computer-mediated communication and human-computer interaction, personality modeling and profiling, consumer feedback analysis, and text-to-speech synthesis [3].

Both knowledge and corpus based approaches have been used to recognize emotions from text at various levels of granularity, from word level to document level. While the first type relies on linguistic models or prior knowledge to identify the dominant emotion in a piece of text, the second applies statistical language modeling techniques. According to the current state-of-the-art, the second type tends to give better results mainly due to its capability to adapt to different domains [1].

---

* Daniele Pighin is currently a postdoc at Google in Zurich, Switzerland.

A. Gelbukh (Ed.): CICLing 2013, Part II, LNCS 7817, pp. 161–173, 2013.

In this paper, we adopt a machine learning approach to categorize texts from different domains (news headlines, stories and blogs) into six basic emotion classes defined by [4], to explore if and how syntactic and semantic features contribute to the accuracy of classification. To this end, we employ Support Vectors Machines (SVM) [5,6] using tree kernel [7] based models and compare the results obtained by using various data representations. The first tree representation is an artificial tree inspired by the work of [8], which deals with the task of sentiment classification of Twitter data into positive, negative and neutral categories. Differently from the representation suggested by this study, we enrich artificial trees with semantic features obtained from WordNet-Affect [9] and SentiWordNet [10]. In addition, we add other semantic features obtained with Latent Semantic Analysis (LSA) method. Furthermore, we experiment with dependency and constituency parse trees with or without the addition of semantic features.

The rest of the paper is structured as follows. In Section 2, we briefly introduce tree kernel functions and explain how they can be employed to effectively design structured features for linguistic tasks. In Section 3, we review the state-of-the-art relevant to the task of emotion recognition from text. In Section 4, we describe the data that we experiment with, the preparation of this data and the additional resources that we utilize. In Section 5, we present the design of the structured features that we use. In Section 6, we explain our evaluation method and discuss the results of the experiments that we have conducted. Finally, in Section 7, we draw conclusions and outline ideas for possible future work.

2   Tree Kernel Functions

A kernel function [11] defines pairwise object similarity as an implicit dot product carried out in some high dimensional space, making it possible to effectively leverage huge feature spaces in common supervised learning frameworks such as the Perceptron or Support Vector Machines. [12] introduced a special class of kernel defined over pairs of trees, and named it a *Tree Kernel* (TK). A TK is a special case of a convolution kernel [13] that measures pairwise tree similarity as the number of substructures shared by two trees. In the TK feature space, each admissible substructure (or *fragment*) constitutes a feature, i.e. a dimension of the feature space. Different classes of TKs allow for different substructures to be considered, thus yielding different results. For the scope of this paper, we consider a very general TK variant, the Partial Tree Kernel (PTK) introduced by [14], for which a valid fragment is any connected substructure of a tree.

One of the most exploited feature of TKs is their ability to generate a great number of structured features, and to assign them a weight in the implicit fragment space. As such, they are often used to prototype novel features via structured representations. [15] employ a PTK to build a TK driven model for question answering. Sequences (with gaps) of words or part-of-speech (POS) tags, which could be modeled using string kernels [16,17], are here evaluated by a PTK on pairs of ad-hoc engineered trees. A fake syntax is used as a container for the sequences of words/POS tags, and to allow for the computation of the TK. [18] employ TKs to model all the stages of a semantic role labeling process including argument boundary recognition, role classification and reranking of complete predicate argument structures. Specific structured features are designed for each of these subtasks. [8] design an artificial tree for sentiment analysis on Twitter data and build models to classify tweets into positive, negative and neutral sentiment categories. Models using unigrams, feature vectors and tree kernels are compared. The TK is applied to artificial trees especially designed for these experiments. The TK based model outperforms the other two models by a significant margin and the most important features are found to be the prior polarity of words along with POS tags for the classification. In the present paper we adopt a very similar formalism to design structured representations of the data combining semantic features with a minimal amount of syntactic clues.

3   Related Work

Many techniques have been proposed for the task of recognizing emotions from text. This section will address the related studies and give brief information about their methodologies.

[9] base their research on the idea that a potential affective meaning is conveyed through every word. Accordingly, an affective lexical resource called WordNet-Affect is created as an extension of WordNet. In this resource, synsets are associated with words directly referring to emotional states (e.g. "joy" or "fear"). The similarity between a term and affective categories is calculated by using cosine similarity.

[2] propose several algorithms for the "SEMEVAL 2007 task on Affective learning". The approach is mainly based on exploiting the co-occurrence of words with ones having explicit affective meaning. The classification in WordNet Affect is utilized to collect six lists of affective words by using the labeled synsets in its dataset. An emotion is represented in three different ways: the first one is the vector of the word that denotes the emotion itself (shortly named as *LSA single word*); the second represents the synset of the emotion (shortly named as *LSA emotion synset*); the last also adds the words in all the synsets which are labeled with the emotion in question, in addition to the previous set (shortly named as *LSA all emotion words*). The similarity calculations are made in the same way with [9]. *LSA all emotion words* model provides the highest recall and F-measure, in terms of coarse grained evaluations. The baseline system which identifies the emotions in a text according the presence of words from WordNet-Affect achieves the best precision. The best results in terms of fine-grained evaluations are obtained by UPAR-7 [19]. This rule based system exploits dependency graphs to understand what is said about the main subject. The emotion of each word is determined using both WordNet-Affect and SentiWordNet, then the main subject rating is boosted. The system called SWAT [20], which uses a unigram model trained to annotate emotional content and conducts synonym expansion on the emotion label words with the help of Roget Thesaurus, and UA [21], which uses Pointwise Mutual Information for emotion scoring and applies statistical methods on the data retrieved by three search engines cannot not beat the other systems in either fine-grained or coarse-grained evaluations.

[3] experiment on data obtained from blogs by using corpus based unigram features, the emotion lexicon obtained from *Roget's Thesaurus* and features from WordNet-Affect. The combination of these features in an SVM-based learning environment results in F-measure values significantly better than the baseline for all emotion categories.

[1] is a recent study where a heterogeneous emotion-annotated dataset combining news headlines, fairy tales and blogs is used in an SVM learning environment with bag-of-words, n-grams and lexical emotion features obtained from WordNet-Affect. Using feature sets from WordNet-Affect does not improve the accuracy of the classifier and a general result cannot be drawn regarding the other two feature sets.

[22] extract emotional expressions at the word and phrase level from English blog sentences and assign six basic emotion tags together with their intensity types using an SVM based supervised framework. According to the feature selection mechanism applied to various linguistic and syntactic features, the emotion word, POS, intensifier and direct dependency features help to improve the extraction of emotional expressions and the identification of emotions and intensities of sentences. As another observation, transitive dependency relations, causal verbs and discourse markers play an effective role in sentential emotion tagging.

4   Data and Pre-processing

We use three different datasets for our experiments. In this section, we will describe each of them and explain the pre-processing phase that we apply together with the resources that we utilize.

The first dataset was prepared for the "SEMEVAL 2007 task on Affective Text" and it consists of news headlines from major newspapers and Google News search engine. The development set consists of 250 headlines, and the test set consists of 1,000 headlines. Each headline is annotated with one of Ekman's [4] six basic emotions (anger, disgust, fear, joy, sadness and surprise) together with the degree of the emotional load. The interval for the emotion annotations is determined between 0 and 100, where 0 represents no emotional load and 100 means maximum emotional load. For our experiments, we use the emotion with the highest load as the sentence label, and we only consider the emotions having a score greater than 50, which was specified by the organizers of the task for the coarse-grained evaluation scheme.

The second dataset [23] consists of blog post sentences which were collected from the web. Each sentence is labeled by four annotators with one of Ekman's 6 emotion classes or determined to be neutral. For our experiments, we use 4090 sentences for which the annotators agreed on the emotion category.

The third data is collected as part of a dissertation research [24] and it includes sentences from stories. Unlike the above two datasets, the sentences here are annotated with one of the 5 emotion classes: angry-disgusted, fearful, happy, sad and surprised. The creators of this dataset decided to merge anger and disgust into one single class due to data sparsity and related semantics between

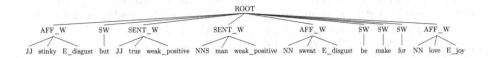

Fig. 1. An artificial tree

each other. We only use 1200 of the sentences with high agreement among the annotators (i.e. sentences with four identical affective labels). It should be noted that we use only the examples with high agreement from the blog and story datasets, since we do not want to introduce any arbitrariness in deciding how to handle the cases with low or no agreement.

To build our tree-kernel model, we first pre-process each dataset. We use TreeTagger [25] for POS tagging, Stanford tokenizer for tokenizing and Stanford parser [26] for parsing sentences. We identify stop words with the help of an online resource[1]. For the sentences in the blog dataset, we assign each emoticon a positive, negative or neutral label based on the emoticon list provided by Wikipedia[2].

As explained in Section 5, to build our structured features we need to pre-calculate the valence and emotion class of all content words in the datasets. We use SentiWordNet [10] to assign a *positive/negative* valence label to words, and WordNet-Affect to determine the prior affective category of single words. As an alternative to WordNet-Affect, we assign an affective label to a word by calculating its similarity with the emotion categories by means of Latent Semantic Analysis (LSA) [27]. To achieve that, we use a vector space induced from ~100 million words of the British National Corpus[3].

5 Structured Feature Design

We employ three different tree representations conveying different amounts of syntactic information, and we measure pairwise similarity between such structures with PTK [28].

The first representation, which we call an *artificial tree*, is inspired by [8], and it conveys a compact representation of a sentence with minimal amount of syntactic information. The only non-trivial difference between the formalism of [8] and the one that we employ is that we enrich the trees with semantic features coming from different sources due to the distinction between the tasks. In order to convert a sentence into an artificial tree representation, we first initalize the tree to the root node (i.e. *(ROOT)*). Afterwards, we tokenize, lemmatize, and POS tag the sentence. Then, for each token we apply the following algorithm.

---

[1] http://www.lextek.com/manuals/onix/stopwords1.html
[2] http://en.wikipedia.org/wiki/List_of_emoticons
[3] http://www.hcu.ox.ac.uk/bnc/

If the token is an exclamation mark, a question mark or an ellipsis, or it represents a negation with any of *no, neither, not, n't,* or *never,* we add a leaf node to the root node with the corresponding tag (*(EXC_MARK), (QUEST_MARK), (ELLIPSIS), (NEG_W)* respectively). We only keep the punctuation marks previously mentioned and ignore the rest since we believe only this subset has the potential of conveying an emotion. Instead, if the token is a stop word, we add the subtree *(SW (token))* to the root.

If the token does not conform to any of the conditions stated above, we check whether the lemma belongs to the WordNet hierarchy and if so, we investigate whether it has an affective meaning. If this lemma is found to have any kind of affective connotation, we add the subtree *(AFF_W (POS) (lemma) (Affective_Category)).* Otherwise, we add *(WORD_NO_MEANING (token))* to the root node. To obtain the affective connotation, we adopt two alternative strategies. These two strategies correspond to two variants of the artificial tree structure. In the first method, we check if the lemma is present in one of the emotion categories in WordNet Affect and we add the information about the POS, lemma and WordNet Affect class to the root as *(AFF_W (POS) (lemma) (WordNet_Affect_Category)).* In the second method, we measure the LSA similarity between the vector representation of the lemma and of each emotion synset. Similarity calculation is carried out with a method similar to *LSA emotion synset* [2]. We create 6 vectors representing the synset of each emotion by summing up the related synonym vectors. Then, to determine the emotion of a specific word, we measure the LSA similarity of that word with each emotion vector and consider the emotion with maximum similarity. If the LSA similarity is higher than a threshold value[4], we add the substructure *(AFF_W (POS) (lemma) (LSA_Affect_Category))* to the root as before. If the token is found not to be related to any affective class, we look up the term in SentiWordNet to measure its valence and if it occurs in the resource, we add the node *(SENT_W (POS) (lemma) (Sentiment_Class)),* and if not (i.e. if the word is neutral in terms of sentiment), we add *(SENT_W (POS) (lemma) (SENTIMENT_neutral))* to the root.

An excerpt of the artificial tree obtained from the sentence "Stinky but true: men's sweat is made for love" is shown in Figure 1. Though very simple, the artificial tree can effectively capture the relative position of words, their grammatical function, the distance between words and the presence of revealing tokens (e.g., exclamation marks or negations) in specific contexts where they can be determinant.

In addition, we consider dependency and constituency parses of a sentence as produced by the Stanford parser. In both cases, we also generate two variants of the plain parse trees by enriching them with semantic nodes. We use the same set of rules explained in the first tree representation to inject the syntactic trees with information about the emotional content of words. In this way, for a constituency parse the node encoding the POS tag of an affective word is replaced with the

---

[4] The value of the threshold has been empirically set to 0.6 by inspecting a small sample of data.

Table 1. F1 measure, leave-one-out on separate datasets

Train/Test	Class	Baseline	Artificial tree		Constituency			Dependency		
			LSA	WNA	Plain	LSA	WNA	Plain	LSA	WNA
Story	Ang+Disg	43.53	46.78	**59.03**	45.23	45.36	56.24	50.94	38.58	47.21
	Fear	*59.34*	37.10	**56.69**	40.21	35.32	49.70	52.05	33.50	45.33
	Joy	*67.29*	56.05	60.39	55.67	54.32	59.16	**62.11**	54.70	57.62
	Sadness	48.00	45.38	**50.45**	45.00	42.24	48.34	48.83	39.08	44.60
	Surprise	19.39	31.11	36.08	40.00	37.86	39.62	**41.38**	38.25	37.14
Blog	Anger	35.97	38.06	**38.92**	26.92	26.20	35.51	37.78	31.4	38.62
	Disgust	21.21	46.64	**49.35**	34.74	40.34	46.25	46.48	40.87	45.05
	Fear	*49.75*	35.90	47.57	37.42	31.08	40.72	45.61	33.98	**48.09**
	Joy	38.70	47.95	58.22	42.87	42.14	53.55	**58.83**	45.34	53.6
	Sadness	30.03	42.07	46.53	31.94	34.48	42.14	40.57	40.54	**47.59**
	Surprise	17.46	26.58	**37.57**	28.57	15.39	25.85	33.54	15.51	29.11
SEMEVAL	Anger	*13.33*	5.88	5.71	6.66	**11.77**	5.26	0	0	0
	Disgust	0.00	34.78	33.33	43.48	**44.44**	33.33	41.67	34.78	34.78
	Fear	29.26	**46.58**	43.68	39.36	32.13	36.58	44.58	42.73	45.41
	Joy	17.46	35.50	33.71	29.87	27.32	31.14	**36.77**	33.01	33.94
	Sadness	6.49	51.39	45.70	43.89	37.42	42.57	**53.08**	52.20	50.31
	Surprise	5.63	35.29	34.57	32.35	29.03	26.02	35.48	**36.37**	**36.37**

subtree *(AFF_W (POS) (lemma) (Affective_Category))*. For the dependency trees, we add the semantic information between a relation node and the modifier lemma, e.g., *(head (relation (AFF_W (Affective_Category (modifier)))))*. As in the case of the artificial tree representation, the semantic features are obtained either via WordNet Affect or LSA.

6   Evaluation and Results

In this section, we report the results that we obtained by training and testing a support vector machine using a Partial Tree Kernel to measure pairwise similarity[5]. To learn the TK models we used the software package SVM-Light-TK[6], which extends an SVM optimizer with support for a wide range of structural kernel families.

For the evaluation we replicated the coarse-grained evaluation scheme used for the SEMEVAL 2007 task on affective text [2]. We did so to be able to carry out a fair comparison of our models against those participating in the evaluation campaign on the same sub-task. The decision of not considering the fine-grained evaluation has both practical and methodological reasons. Of the three datasets considered, one (Story) does not have emotion intensities. The others (Semeval and Blog) use different scales to represent intensity ([0,100] vs. low/med/high). By only considering emotion classes, we are able to provide a unified framework

---

[5] All the experiments were also run using the Syntactic Tree Kernel [12], which is expected to outperform the PTK on constituency parsed data. Instead, we observed that PTK consistently produced the best results.

[6] http://disi.unitn.it/moschitti/Tree-Kernel.htm

Table 2. F1 measure, training on Blog(all) + Story(all) + SEMEVAL(train), test on SEMEVAL(test)

Class	Baseline	Artificial tree		Constituency			Dependency		
		LSA	WNA	Plain	LSA	WNA	Plain	LSA	WNA
Anger	6.06	**8.33**	7.27	0.00	0.00	7.70	8.00	6.55	5.79
Disgust	0.00	0.00	0.00	0.00	0.00	0.00	0.00	0.00	0.00
Fear	3.33	30.33	**37.84**	18.51	27.15	34.85	23.40	23.78	29.70
Joy	1.10	28.57	**33.75**	16.56	23.78	31.63	26.90	32.00	30.36
Sadness	6.61	29.54	33.96	15.95	**40.98**	33.61	9.76	38.42	34.55
Surprise	6.90	8.17	**11.54**	4.55	0.00	7.85	8.34	7.02	10.35

Table 3. F1 measure, training on Blog(all), test on Story(all)

Class	Baseline	Artificial tree		Constituency			Dependency		
		LSA	WNA	Plain	LSA	WNA	Plain	LSA	WNA
Ang+Disg	*43.53*	31.34	42.73	30.89	29.37	39.89	32.87	29.87	**42.93**
Fear	*59.34*	5.78	33.17	2.42	2.40	31.52	3.55	3.51	**33.65**
Joy	*67.29*	54.38	63.95	51.42	56.16	**66.67**	51.34	55.93	65.22
Sadness	*48.00*	34.36	44.26	10.75	22.23	32.26	25.00	32.57	**45.34**
Surprise	*19.39*	3.42	13.84	12.21	3.39	**14.17**	9.45	0.00	12.50

for all the experiments across all the datasets, and we do not introduce any bias by forcing an arbitrary mapping between the two different scales.

To calculate the baseline for each emotion in each dataset, we adopt the approach of [2] and use the six lists of affective words collected from WordNet-Affect based on the synsets labeled with the six emotions. We count how many words in each test sentence are associated with each emotion, and classify the sentence based on the most represented emotion.

To make the most of the limited amount of data available, we first conducted leave-one-out evaluation on each dataset. The results of this experiment are shown in Table 1. Each group of rows is related to a different dataset. For each class and dataset, we show the F1 measure obtained with the different data representations that we adopt: (1) the artificial tree with semantic enrichment via LSA or WordNet-Affect (WNA), and (2) the constituency and (3) dependency parses without semantic enrichment (Plain) or with semantic enrichment via LSA or WNA. In each row, the best configuration is highlighted in bold. The baseline is shown in italic when it performs better than any of the models.

On the Story and Blog corpora (first two blocks of rows) the baseline is better than any other configuration only in three cases. It can be seen how semantic features coming from WordNet-Affect in combination with artificial trees and constituency parses produce very good results. By comparing the columns for Constituency/Plain and Constituency/LSA, we can also observe how the addition of LSA features generally leads to lower performance than using only

Table 4. F1 measure comparison on the SEMEVAL test data

Model	Anger	Disg.	Fear	Joy	Sadn.	Surp.
LSA-SW	11.43	**4.68**	22.80	25.88	21.20	12.23
LSA-ES	13.45	3.00	22.00	30.55	23.06	13.38
LSA-AEW	11.58	3.87	21.91	30.83	20.61	14.10
NB/blogs	16.77	-	5.63	32.87	21.43	2.63
SWAT	7.06	-	18.27	14.91	17.44	11.78
UA	**16.03**	-	20.06	4.21	1.76	**15.00**
UPAR7	3.02	-	4.72	11.87	30.38	2.27
AT+WNA	7.27	-	**37.84**	**33.75**	**33.96**	11.54

syntactic data. The same effect can be observed with dependency parses. In addition, the artificial tree and the dependency parses constantly outperform the constituency parses, showing that the task can benefit from minimal amounts of syntactic information such as information about word order or word-to-word relations. We can also observe that the baseline for the story dataset is especially high and difficult to beat. A possible explanation is that since this dataset consists of fairy tales, the language used is not only simple but also strongly characterized. This is in line with the findings of [29], who showed that fairy tales tend to have very high emotion densities, measured as the number of affective words observed for every fixed number of words. For this reason, there are few cases in which the combination of syntactic and semantic clues can help classification. In other cases, the syntactic features only have the effect of making the semantic information sparser and more difficult to learn from.

The third block of rows (labeled as SEMEVAL) shows results which contrast with the previous ones. In fact, while they confirm the fact that artificial trees and dependency parses are more adequate formalisms for the task, these results clearly point to LSA (in four cases out of six) as the best source for semantic features. The reason for this difference may lie in the fact that the *LSA emotion synset* method that we employ was specifically designed for the SEMEVAL task, and optimized on the available development data. In addition, it is worth noting that the inter-annotator agreement reported for this task is not very high [2], and the annotation process for obtaining more training data is found to be significantly difficult due to the ambiguity in the definition of the task [20][7].

To further explore the differences between the datasets and the possibility of using them jointly, we carried out two more experiments: In the first one, we trained on the whole Blog and Story corpus together with the development set of the SEMEVAL task. We then conducted the test on the 250 test examples of the SEMEVAL test set[8]. The results of this experiment are shown in Table 2.

---

[7] One notable difficulty reported is determining whether to label the emotion experienced by the reader or by the subject of the headline. As another difficulty, many headlines can be annotated in different ways depending on the viewpoint of the annotator (e.g. Italy defeats France in World Cup Final) [20].

[8] The baseline in this case is different since it is calculated only on the test examples.

This model is trained on approximately 6 times more data than the SEMEVAL leave-one-out model (third block of rows in Table 1), while still including all the training data for the task. Therefore, we would expect it to perform much better. Instead, with this training setup classification accuracy drops quite dramatically for all six classes when the best result for each class among all models is considered. This fact suggests that the three datasets are quite heterogeneous, and that combining them to obtain more training data may not be a good decision. Interestingly, with this configuration WNA features result to be more determinant, even though the test is conducted on the SEMEVAL corpus.

As another experiment, we removed the SEMEVAL dataset from the equation and classified the Story dataset by using a model learned on the Blog corpus [9]. The outcome of this experiment is shown in Table 3. Also in this case, the results seem to suggest that the resources are quite heterogeneous. At the same time, the combination of dependency trees with WNA features generally produces the best results, with the artificial tree coming very close. This confirms the results obtained with the leave-one-out evaluation and the findings of [30], who observe that affective lexicon features have a positive impact on generalization for emotion recognition. Conversely, these results appear to contradict the findings of [1]. According to them, features extracted from WordNet-Affect do not produce notable improvements in accuracy. In our case, the positive effect may be triggered by the subtle interaction between the semantic features with the very shallow syntax encoded by the two kinds of structures.

Finally, the comparison of our results with those obtained by the systems participating in the SEMEVAL 2007 task on affective computing is shown in Table 4. The first four models (*LSA single word*, *LSA emotion synset*, *LSA all emotion words* and *NB trained on blogs* represented as *LSA-SW*, *LSA-ES*, *LSA-AEW* and *NB/blogs* respectively) are proposed by the organizers of this task. The first three of them use different vector representations of emotions to calculate LSA similarity, while the fourth exploits a Naïve Bayes classifier trained on a corpus of blog posts. The other systems listed are SWAT [20], UA [21] and UPAR7 [19] respectively. As previously mentioned in Section 3, SWAT uses a unigram model, while UA is based on statistics gathered from three search engines and UPAR7 is a rule based system using a linguistic approach. The last line shows the results of our approach using artificial trees and WNA features after training on Blog, Story and SEMEVAL development data and testing on SEMEVAL test data. This is the only configuration with the same test split that we can fairly compare against the other systems. With this configuration, the Artificial tree + WNA model outperforms all the other models in classifying three of the six classes, and especially for fear and sadness it produces much better results than any other system. The macro-average of the F-1 measure for this model is 20.72, which is 3 points higher than 17.57, the macro-average of the best model reported in the SEMEVAL 2007 task.

---

[9] This decision is motivated by the Blog corpus being larger, and hence more suitable for learning.

7 Conclusions and Future Work

We have presented a systematic study aimed at understanding the extent to which syntactic and semantic information can improve emotion recognition accuracy. We ran all the experiments in a supervised learning framework, exploiting the ability of tree kernel functions to discover relevant features in very high dimensional spaces to streamline the feature engineering stage. We selected three datasets annotated according to the same taxonomy of emotions, and we ran a large set of experiments to validate several hypotheses.

The first question that we tried to answer is whether syntax plays a relevant role for this task. The results that we obtained by comparing very simple artificial trees to constituency and dependency parses strongly suggest that only local and shallow syntactic features, such as information about word sequences or POS tags, improve the accuracy of classification. The inclusion of too much syntactic information only increases the sparsity of the problem with a negative effect on the final accuracy.

Second, we compared the extent to which semantic features obtained with LSA or WordNet-Affect can contribute to the classification accuracy. The experiments show that the latter is generally a better alternative, while the former produces better results only when using a specific configuration of training and test data. In relation to the previous point, we also observe that to effectively exploit the semantic features, syntactic overhead should be kept to a minimum.

Lastly, we investigated the possibility of combining different datasets to learn more reliable and accurate models by using larger amounts of data. The results suggest that the existing resources are too heterogeneous to be combined successfully. This fact may be explained with the fact that the three datasets pertain to three different domains. On the other hand, they may also be related to the lack of a unified annotation framework, resulting in annotation biases specific to each dataset.

In the future, we will continue investigating this topic by incorporating data from other resources and experimenting with other combinations of syntactic and semantic information. In particular, we will focus on the design of novel kernel functions embedding the calculation of semantic similarity and affective connotation, similarly to [31]. In this way, the ability of the kernel to generalize from the semantic layer will be decoupled from the syntactic properties of the structured representations employed.

References

1. Chaffar, S., Inkpen, D.: Using a heterogeneous dataset for emotion analysis in text. In: Butz, C., Lingras, P. (eds.) Canadian AI 2011. LNCS, vol. 6657, pp. 62–67. Springer, Heidelberg (2011)
2. Strapparava, C., Mihalcea, R.: Learning to identify emotions in text. In: SAC 2008: Proceedings of the ACM symposium on Applied computing, pp. 1556–1560. ACM, New York (2008)

3. Aman, S., Szpakowicz, S.: Using Roget's thesaurus for fine-grained emotion recognition. In: Proceedings of IJCNLP, pp. 296–302 (2008)
4. Ekman, P.: Facial expression and emotion. American Psychologist 48, 384–392 (1993)
5. Boser, B.E., Guyon, I., Vapnik, V.: A Training Algorithm for Optimal Margin Classifiers. In: Proceedings of the 5th Annual Workshop on Computational Learning Theory (1992)
6. Vapnik, V.N.: Statistical Learning Theory. Wiley-Interscience (1998)
7. Collins, M., Duffy, N.: New ranking algorithms for parsing and tagging: kernels over discrete structures, and the voted perceptron. In: Proceedings of ACL 2002, Stroudsburg, PA, USA, pp. 263–270 (2002)
8. Agarwal, A., Xie, B., Vovsha, I., Rambow, O., Passonneau, R.: Sentiment Analysis of Twitter Data. In: Proceedings of the Workshop on Language in Social Media, pp. 30–38. Association for Computational Linguistics (2011)
9. Strapparava, C., Valitutti, A.: WordNet-Affect: an affective extension of WordNet. In: Proceedings of LREC, vol. 4, pp. 1083–1086 (2004)
10. Esuli, A., Sebastiani, F.: Sentiwordnet: A publicly available lexical resource for opinion mining. In: Proceedings of LREC, pp. 417–422 (2006)
11. Aizerman, M., Braverman, E., Rozonoer, L.: Theoretical foundations of the potential function method in pattern recognition learning. Automation and Remote Control 25, 821–837 (1964)
12. Collins, M., Duffy, N.: Convolution kernels for natural language. In: Advances in Neural Information Processing Systems, vol. 14, pp. 625–632. MIT Press (2001)
13. Haussler, D.: Convolution kernels on discrete structures. Technical report, Dept. of Computer Science, University of California at Santa Cruz (1999)
14. Moschitti, A.: Efficient Convolution Kernels for Dependency and Constituent Syntactic Trees. In: Fürnkranz, J., Scheffer, T., Spiliopoulou, M. (eds.) ECML 2006. LNCS (LNAI), vol. 4212, pp. 318–329. Springer, Heidelberg (2006)
15. Moschitti, A., Quarteroni, S., Basili, R., Manandhar, S.: Exploiting Syntactic and Shallow Semantic Kernels for Question/Answer Classification. In: Proceedings of ACL 2007 (2007)
16. Lodhi, H., Saunders, C., Shawe-Taylor, J., Cristianini, N., Watkins, C., Scholkopf, B.: Text classification using string kernels. Journal of Machine Learning Research 2, 563–569 (2002)
17. Cancedda, N., Gaussier, E., Goutte, C., Renders, J.M.: Word sequence kernels. Journal of Machine Learning Research 3, 1059–1082 (2003)
18. Moschitti, A., Pighin, D., Basili, R.: Tree Kernels for Semantic Role Labeling. Computational Linguistics 34, 193–224 (2008)
19. Chaumartin, F.R.: Upar7: a knowledge-based system for headline sentiment tagging. In: Proceedings of SemEval 2007, pp. 422–425. Association for Computational Linguistics, Stroudsburg (2007)
20. Katz, P., Singleton, M., Wicentowski, R.: Swat-mp:the semeval-2007 systems for task 5 and task 14. In: Proceedings of the SemEval 2007, pp. 308–313. Association for Computational Linguistics, Prague (2007)
21. Kozareva, Z., Navarro, B., Vazquez, S., Montoyo, A.: Ua-zbsa: A headline emotion classification through web information. In: Proceedings of SemEval 2007, pp. 334–337. Association for Computational Linguistics, Prague (2007)
22. Das, D., Bandyopadhyay, S.: Identifying emotional expressions, intensities and sentence level emotion tags using a supervised framework. Emotion 1, 95–104 (2010)

23. Aman, S., Szpakowicz, S.: Identifying expressions of emotion in text. In: Matoušek, V., Mautner, P. (eds.) TSD 2007. LNCS (LNAI), vol. 4629, pp. 196–205. Springer, Heidelberg (2007)
24. Alm, E.C.O.: Affect in Text and Speech. PhD thesis, University of Illinois at Urbana-Champaign (2008)
25. Schmid, H.: Probabilistic part-of-speech tagging using decision trees. In: Proceedings of the International Conference on New Methods in Language Processing, Manchester, UK (1994)
26. Klein, D., Manning, C.D.: Accurate unlexicalized parsing. In: Proceedings of ACL 2003, Stroudsburg, PA, USA, pp. 423–430 (2003)
27. Deerwester, S., Dumais, S.T., Furnas, G.W., Landauer, T., Harshman, R.: Indexing by latent semantic analysis. Journal of the American Society for Information Science 41, 391–407 (1990)
28. Moschitti, A.: Efficient convolution kernels for dependency and constituent syntactic trees. In: Fürnkranz, J., Scheffer, T., Spiliopoulou, M. (eds.) ECML 2006. LNCS (LNAI), vol. 4212, pp. 318–329. Springer, Heidelberg (2006)
29. Mohammad, S.M.: From once upon a time to happily ever after: Tracking emotions in mail and books. Decision Support Systems 53, 730–741 (2012)
30. Mohammad, S.: Portable features for classifying emotional text. In: Proceedings of NAACL HLT 2012, Stroudsburg, PA, USA, pp. 587–591 (2012)
31. Bloehdorn, S., Basili, R., Cammisa, M., Moschitti, A.: Semantic Kernels for Text Classification based on Topological Measures of Feature Similarity. In: Proceedings of ICDM, Hong Kong (2006)

Chinese Emotion Lexicon Developing via Multi-lingual Lexical Resources Integration

Jun Xu[1], Ruifeng Xu[1,*], Yanzhen Zheng[1], Qin Lu[2],
Kai-Fai Wong[3,4], and Xiaolong Wang[1]

[1] Key Laboratory of Network Oriented Intelligent Computation,
Shenzhen Graduate School, Harbin Institute of Technology, Shenzhen, China
[2] Department of Computing, The Hong Kong Polytechnic University, Hong Kong
[3] Department of SEEM, The Chinese University of Hong Kong, Hong Kong
[4] Key Laboratory of High Confidence Software Technologies, Ministry of Education, China
{xujun,xuruifeng}.hitsz.edu.cn, csluqin@comp.polyu.edu.hk,
kfwong@se.cuhk.edu.cn, xlwang@insun.hit.edu.cn

Abstract. This paper proposes an automatic approach to build Chinese emotion lexicon based on WordNet-Affect which is a widely-used English emotion lexicon resource developed on WordNet. The approach consists of three steps, namely translation, filtering and extension. Initially, all English words in WordNet-Affect synsets are translated into Chinese words. Thereafter, with the help of Chinese synonyms dictionary (Tongyici Cilin), we build a bilingual undirected graph for each emotion category and propose a graph based algorithm to filter all non-emotion words introduced by translation procedure. Finally, the Chinese emotion lexicons are obtained by expanding their synonym words representing the similar emotion. The results show that the generated-lexicons is a reliable source for analyzing the emotions in Chinese text.

Keywords: Emotion lexicon development, Emotion analysis, Multi-lingual.

1 Introduction

Sentiment analysis studies how to identify and extract the subjective information in text. It may be divided into two aspects which are opinion and emotion, respectively. Opinion, in general, is the judgment or evaluation of a speaker or a writer with respect to some topic, such as negative/positive, pros/cons. While emotion is a strong human feeling, the emotional state of a person, such as joy, anger, sadness, fear, etc.

With the popularity of the social network, social media plays an important role in the information release and dissemination. It has been a new and a good media platform for fast and wide spread of information nowadays. To measure and recognize the emotional changes of population in large scale is one of the most important areas in social sciences [1] and economics studies [2]. Therefore, text emotion analysis research attracts much attention. Many emotion analysis approaches essentially rely

---

* Corresponding author.

A. Gelbukh (Ed.): CICLing 2013, Part II, LNCS 7817, pp. 174–182, 2013.

on emotion lexical resources containing words and their associated emotions. Thus, the establishing of emotion lexicon is recognized as the foundation in the research of emotion analysis.

However, the emotion lexicon is still not easily available for Chinese or other resource poor languages. This paper focuses on the automatic construction of a Chinese emotion lexicon, starting from WordNet-Affect which is a widely used emotion lexicon in English, through the translation and integration with multi-lingual lexical resources.

The rest of this paper is organized as follows. Section 2 gives a brief review on the construction of emotion lexical resource. Section 3 presents a brief description for two lexical resources which will be used in this study. Section 4 introduces our proposed three steps automatic Chinese emotion lexicon construction approach. Section 5 gives the performance evaluation and Section 6 concludes.

2 Emotion Lexicon Construction: State of the Art

The emotion lexicons can be used as semantic knowledge base for emotion analysis. For the development of emotion lexicons, there are two questions faced: which word can be used to express emotions? and what kind of emotion or set of emotions that the words convey? In the previous studies on emotion lexicon construction, there are two major approaches adopted: extension from semantic lexical resource and corpus-based extraction with heuristic rule.

2.1 Extension with Semantic Lexical Resource

To create an emotion lexicon automatically, the existing lexical resource may be a good starting point. Starting with WordNet, Strapparava and Valitutti developed WordNet-Affect [3].Several seed emotion words are manually chose and then the correlation of relations defined in WordNet (e.g., causes, entailment and so on), emotional tags and domain tags are used to expand. Finally, WordNet-Affect, the collection of emotion synsets are obtained by exploiting associated affection. With WordNet-Affect, many researchers attempt to expand it to other languages for developing multi-lingual emotion lexicons. Sokolova and Bobicev [4] translated every word in WordNet-Affect to Romanian and Russian. They used three machine learning methods to classify the emotions of these words which are represented by the word spelling and word form. Torii et al. [5] constructed a Japanese WordNet-Affect directly according to WordNet-Affect's SyssetID by making use of Japanese WordNet.

2.2 Corpus-Based Extraction with Heuristic Rule

Using emoticons (such as ":)" and ": o") in the blogs as the clues, Yang et al. [6] exploited co-occurrence based algorithm in collocations to extract emotion words from blog corpora. Xu L. et al. [7] built a Chinese affective lexical ontology. The emotions of the ontology are hierarchical categorized into 7 categories on first level and 20 categories on second level. They annotated the emotion label and intensity for each emotion word, which are manually collected from related semantic lexicons. In contrast to the method of expanding from semantic lexicon, they finally labeled the emotion category and compute the intensity for all candidate words automatically based on mutual information on a large corpus. Xu G. et al. [8] proposed a graph-based approach to identify the emotion label of a word. They computed the similarity between the candidate words and seed words with different similarity metrics by leveraging un-annotated corpora, lexicon resources, heuristic rules and so on. Thereafter, they built the word similarity matrices after integration to label each candidate word iteratively based on their proposed graph-based algorithm. Quan and Ren [10] identified the emotion words by training a Maximum Entropy based classifier on an emotional labeling corpus, Ren-CECps [9] with semantic features.

3   Lexical Resources Used in this Research

3.1   English Emotion Lexicon – WordNet-Affect

The English WordNet-Affect is a widely used emotion lexical resource with affective annotation. It was developed on WordNet based on Ekman's six emotion types (anger, disgust, fear, joy, sadness, surprise) theory. WordNet-Affect is a subset of WordNet which contains the essential knowledge related to emotion analysis.

WordNet-Affect is provided in six files named by the six emotions, respectively. Each file lists the synsets they contain per line. Following line is an example synset entry in WordNet-Affect.

```
n#05588822 umbrage offense
```

In this line, the first letter gives the part of speech (POS) of this entry and it is followed by the synset ID, and then the synonyms in this synset.

3.2   Chinese Synonym Dictionary – Tonngyici Cilin

Tongyici Cilin (in short Cilin) is a well-used Chinese synonyms dictionary, which was published in 1983. It contains about 50 thousands of Chinese words. Three-level conceptual categories are adopted to cluster synonyms according to their semantic relationships. The top level category consists of 12 main classes. The second level category consists of 94 classes. While the third level concepts are classified into 1428 classes.

In this study, we use HIT IR-Lab Tongyici Cilin[1] (Extended)(EClin for short) as Chinese lexical resources. ECilin extended the three-level categories of original Tongyici Cilin to five levels while the rare and unusual words are filtered out. At the same time, some new words are added in. In ECilin, a capital letter is used to label the fourth level, which is the concept clusters. The deepest one stands for atomic concepts, in which words are nearly synonyms. There are three tags: "=" stands for the same sense; "#" for antonyms and "@" for enclosure which means the word has no synonyms. An example entry of ECilin is given below. All words "欢腾 欢跃 手舞

---

[1] http://ir.hit.edu.cn

足蹈 欢呼雀跃" (clam happy) of the entry are followed by a sense code "Ga01A04", the five-level categories. The words having the same sense code can be regarded as of similar meaning.

Ga01A04=欢腾 欢跃 手舞足蹈 欢呼雀跃

4   Our Approach

In this section, our approach for constructing a Chinese emotion lexicon is presented. This approach contains three steps: translation, filtering and extension.

4.1   Translation

The goal of translation is to translate English emotion words in WordNet-Affect to Chinese as the emotion word candidates as much as possible for the following proce-dure. To translate each word in the WordNet-Affect synsets from English to Chinese, two online machine translation systems are used, i.e., BaiDu Translator[2] and YouDao Translator[3]. Both of them are well-known and widely used in Chinese-English transla-tion area, as well as they support free API. The YouDao Translator outputs all transla-tions with corresponding part-of-speech (POS) tags. In this study, the target translated words whose POS match the source word's POS are returned for following procedure.

Table 1. An example of WordNet-Affect synsets translation

WordNet-Affect Synset	n#05588321 wrath	n#05588822 umbrage	offense
A, Translated Re-sults from BaiDu	NULL	阴影,树荫;簇叶,愤怒;生气	罪过;犯法;过错冒犯;触怒,引起反感的事物
B, Translated Re-sults from YouDao	愤怒; 激怒	不快;生气,树荫,怀疑	犯罪;过错,进攻,触怒,反感的事物
A∪B	愤怒; 激怒	阴影;树荫;簇叶;愤怒;生气;不快,怀疑	冒犯;引起反感的事物;犯法;犯罪;罪过;触怒;过错;进攻

Table 1 demonstrates an example of the synset translation procedure. In Table 1, "NULL" denotes that there is no returned translations since there are some words in the English synset can't be translated to Chinese words. To ensure the integrity of translation procedure, the union of all outputs from different machine translation sys-tems, i.e., A∪B is admitted.

---

[2] http://fanyi.baidu.com/
[3] http://fanyi.youdao.com

4.2   Filtering

The original English words in a synset of WordNet-Affect have similar meanings, while their corresponding Chinese translations have much ambiguity as there is no way to obtain accurate equivalent words during the translation. In the translation step, all possible translations for all of their senses are provided. It means that many noisy or irrelevant words are introduced. For instance, the Chinese words "树荫(shade of tree)", "簇叶(foliage)" in the translations of synset "n#05588822". Such kind of words should be filtered. Therefore, we propose a bilingual undirected graph based filtering algorithm for automatic sense disambiguation. The Chinese words in the translated synsets which convey the same emotion will be figured out.

For each emotion category, the bilingual graph G is constructed as follows:

Step 1 Create a R-vertex graph with no edges. R is a starting vertex.

Step 2 Let $S = \{s_1, s_2 \dots s_n\}$ denotes the set of synsets, s_i is synset ID. For each $s_i \in S$, add s_i as a synset vertex and add an edge (R, s_i) between R and a synset s_i.

Step 3 Let $E = \{e_1, e_2 \dots e_m\}$ denotes the synonyms in S, For each $e_j \in E$, add e_j as an English word vertex and add an edge (e_j, s_i) if and only if e_j belongs to synset s_i.

Step 4 Lets $C = \{c_1, c_2 \dots c_l\}$ denotes all translated Chinese words, For each $c_k \in C$, add c_k as a Chinese word vertex and add an edge (c_k, e_j) if and only if c_k is in e_j 's translation results.

Step 5 For all Chinese word vertices, if two words are synonyms, then add an edge between them.

ECilin is utilized in this research for synonym judgment.

Figure 1 shows a partial bilingual graph after adding edges to link the synonyms. As shown in Figure 1, each simple path[4] between R and a Chinese word vertex include a synset and an English word vertex. If a Chinese word vertex has at least two simple paths to reach the vertex R, and these paths go through different synset and English vertex, the Chinese word can be treated as emotion word. It means that such Chinese word may share the same emotion sense in different English synsets. For example, "愤怒", "激怒", "不快" , "生气", "触怒", "冒犯" vertices in Figure 1 are classified as members of emotion lexicon of "anger" in our filtering algorithm.

The pseudo code of the proposed bilingual graph based filtering algorithm is given below. The Chinese word c is treated as a terminal vertex. We use depth first search to detect all simple paths between the start vertex R and c firstly (line 2). If there are at least two paths which do not have same synset and English word vertices, c can be annotated as an emotion word with a corresponding label. (line 3-5).

---

[4] A simple path is a path with no repeated vertices.

4.3 Extension

With the above procedures, six emotion lexicons are obtained corresponding to each emotion category, respectively. As shown in Table 2, the words for each emotion are very few in number. There are 220 unique words in "Anger" category, 58 words in "Disgust", 152 words in "Fear", 516 words in "Joy", 200 words in "Sadness", and 67 words in "Surprise". Obviously, current emotion lexicon is not efficient enough for a practical application of Chinese emotion analysis. Naturally, our aim is further extend current lexicon.

In this study, ECilin is utilized here for lexicon extension. For all words in current emotion lexicons, if it is found in ECiliin, all of the words with the same sense code are added to the corresponding emotion lexicon. Generally speaking, the words with the same sense code which have higher hit frequency have more relevance to the corresponding emotion, and thus they may be added to this emotion lexicon. After the extension procedure, the lexicon of each category has a great increment in number as Table 2 shows.

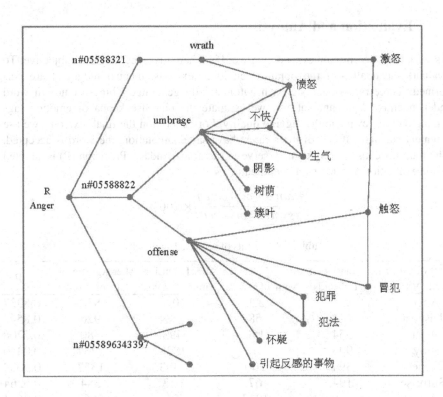

Fig. 1. A partial bilingual graph of "Anger"

Algorithm 1. The bilingual graph based filtering algorithm

```
Input: G, the constructed bilingual graph
       S, set of synset vertices
       E, set of English word vertices
       C, set of Chinese word vertices
Output: O, set of Chinese emotion words
1: for each word c ∈ C do
2:     Use the Depth First Search to find the paths set
           p = {p(R,c)|all simple paths between R and c};
3:     if ∃ p₁ ∈ P and ∃ p₂ ∈ P, while p₁ ≠ p₂, p₁ ∩ p₂ ∩ S = ∅
           and p₁ ∩ p₂ ∩ E = ∅
       then
4:             Add c to O;
5:     end if
6: End for
```

5   Evaluation and Analysis

As we know, the emotion carried by a word is inherently uncertain and subjective. To evaluate the quality of the obtained emotion lexicons, manual judgment are performed. Three raters annotate each automatically generated Chinese emotion word independently. After annotations, we estimate the pairwise kappa of emotion tags among them to evaluate the tagging quality. For generation the final lexicons, we use a lenient standard. If two of three raters have same annotation, the word is accepted. The final lexicons after agreement serve as the gold standard. Precision (P) is adopted as the evaluation metric and it is computed as

$$\frac{\#human\_corrected}{\#system\_propsed} \times 100\%$$

Table 2. All results of proposed approach

Emotion Category	After Translation (Num. of Words)	After Filtering (Num. of Words)	After Extension (Num. of Words)	After Agreement (Num. of Words)	P
Anger	525	220	1022	852	0.8337
Disgust	144	58	1084	926	0.8542
Fear	354	152	493	380	0.7708
Joy	993	516	1838	1737	0.9450
Sadness	394	200	1493	1357	0.9089
Surprise	194	67	613	384	0.6264
Total	2604	1213	6543	5667	0.8614

Table 2 shows the number of words for each emotion category after agreement by raters as well as the precision. It is observed that our proposed approach achieves a good precision. It is also observed that the lexicon generation for some specific emotion, such as "Surprise", achieves lower precision. As we know, a word or phrase may express more than one emotion. For example, the idiom "惊慌失措 (dismayed)" expresses both "surprise" and "fear". Statistics on Ren-CECps also shows that about 15.1% Chinese emotion words are multi-emotion ones which express complex feelings in its usage [10]. After agreement, 218 multi-emotions words are kept.

Table 3. Inter-rater agreements by category on generated lexicons

Emotion Category	Num. of Words	Averaged Kappa(K)
Anger	1022	0.6398
Disgust	1084	0.5772
Fear	493	0.6864
Joy	1838	0.3865
Sadness	1493	0.6399
Surprise	613	0.5475
Macro-averaged		0.5796

Table 3 shows the averaged value of the kappa coefficient for each emotion category, respectively. The values vary from 0.5475 to 0.6864 except for the lexicon of "Joy". Though agreements vary within categories, the macro-averaged kappa value is near to 0.6. This value is considered as good performance which indicates a good level of agreement. It also states that the final obtained emotion lexicon after agreement is reliable. For the "Joy" lexicon, the Kappa value is lowest (i.e., 0.3865). However, as shown in Table 2, the "Joy" lexicon has a high precision (0.9450). This happens when Kappa deviates from the normal distribution. It ignores the high inter-observer agreement of the annotation result.

Compared to work of Xu, G. et al. in [8], the proposed approach achieved a much larger lexicon with good precision. Compared to the work reported in [7], our approach saves human labor in great deal. Furthermore, the construction process of our proposed approach is easy to repeat.

6   Conclusion

In this paper, we presented an approach for developing Chinese emotion lexicon by using a English emotion lexicon and a Chinese thesaurus. This lexicon is developed starting from a English emotion lexicon, WordNet-Affect, through the translation, filtering and extension. We translated the WordNet-Affect synsets into Chinese, and afterwards integrated with another Chinese thesaurus, Tongyici Cilin to filter irrelevant words and also to expand it. The obtained Chinese emotion lexicon is freely available at http://icrc.hitsz.edu.cn/emotion_lexcions.rar.

In the future, we will continue to enrich this resource to make it useful in affective computing and emotion-based human inter-action applications.

Acknowledgement. This work is supported by the China Postdoctoral Science Foundation (No. 2011M500670), National Natural Science Foundation of China (No. 61203378 and No. 61272383), Shenzhen Foundational Research Funding (NO. JCYJ2012 0613152557576) and General Research Fund of Hong Kong (No. 417112).

References

1. Dodds, P.S., Danforth, C.M.: Measuring the Happiness of Large Scale Written Expression: Songs, Blogs, and Presidents. Joural of Happiness Studies 11(4), 441–456 (2009)
2. Bollen, J., Mao, H., Zeng, X.-J.: Twitter Mood Predicts the Stock Market. Journal of Computational Science 2(1), 1–8 (2011)
3. Strapparava, C., Valitutti, A.: WordNet-Affect: An Affective Extension of WordNet. In: Proceedings of the 4th International Conference on Language Resources and Evaluation, pp. 1083–1086 (2004)
4. Sokolova, M., Bobicev, V.: Classification of Emotion Words in Russian and Romanian Languages. In: Proceedings of the International Conference RANLP 2009, pp. 416–420 (2009)
5. Torii, Y., Das, D., Bandyopadhyay, S., Okumura, M.: Developing Japanese Word-Net Affect for Analyzing Emotions. In: Proceedings of the 2nd Workshop on Computational Approaches to Subjectivity and Sentiment Analysis, pp. 80–86 (2011)
6. Yang, C., Lin, K.H.-Y., Chen, H.-H.: Building Emotion Lexicon from Weblog Corpora. In: Proceedings of the 45th Annual Meeting of the Association for Computational Linguistics Companion Volume: Proceedings of the Demo and Poster Sessions, pp. 133–136 (2007)
7. Xu, L., Lin, H., Pan, Y., Ren, H., Chen, J.: Constructing the Affective Lexicon Ontology. Journal of the China Society for Scientific and Technical Information 27(2), 180–185 (2008)
8. Xu, G., Meng, X., Wang, H.: Build Chinese Emotion Lexicons Using a Graph Based Algorithm and Multiple Resources. In: Proceedings of the 23rd International Conference on Computational Linguistics, pp. 1209–1217 (2010)
9. Quan, C., Ren, F.: Construction of a Blog Emotion Corpus for Chinese Emotional Expression Analysis. In: Proceedings of the 2009 Conference on Empirical Methods in Natural Language Processing, pp. 1446–1454 (2009)
10. Quan, C., Ren, F.: An Exploration of Features for Recognizing Word Emotion. In: Proceedings of the 23rd International Conference on Computational Linguistics, pp. 922–930 (2010)

N-Gram-Based Recognition of Threatening Tweets

Nelleke Oostdijk and Hans van Halteren

Radboud University Nijmegen, CLS – Dept. of Linguistics / CLST
{N.Oostdijk,hvh}@let.ru.nl

Abstract. In this paper, we investigate to what degree it is possible to recognize threats in Dutch tweets. We attempt threat recognition on the basis of only the single tweet (without further context) and using only very simple recognition features, namely n-grams. We present two different methods of n-gram-based recognition, one based on manually constructed n-gram patterns and the other on machine learned patterns. Our evaluation is not restricted to precision and recall scores, but also looks into the difference in yield of the two methods, considering either combination or means that may help refine both methods individually.

Keywords: social media, text mining, text classification, manually constructed rules, machine learning.

1 Introduction

In recent years the microblogging service Twitter has gained immense popularity. Estimates are that in the Netherlands alone each day over 3 million tweets are posted. The very short 140-character messages are primarily used for sharing information on what is going on right there and then. However, as journalists, policy makers, businesses, marketing agencies etc. have been quick to discover, the collective information has also great potential when it comes to finding out about things that are about to happen or that have only just taken place, and what the prevailing sentiments are. For searching and retrieving information and for sentiment mining, existing NLP techniques are being deployed rather successfully.

However, there is also a dark side of the internet as in the perceived anonymity of the medium people are being bullied, harassed, and even threatened with violence. As acts of intimidation, harassment, and other forms of threatening are criminal offences punishable by law, law enforcement agencies are under pressure to develop a policy for dealing with these phenomena.[1] A possible course of action could be to monitor the internet so that immediate action can be taken when a threat is made. Such a task becomes only feasible when tools are available that will support it.

In the present study we investigate, for Dutch tweets, whether on the basis of the content of a single tweet (without further context) we can detect automatically whether

---

[1] See also `http://www.rcmp-grc.gc.ca/qc/pub/cybercrime/cybercrime-eng.htm`

A. Gelbukh (Ed.): CICLing 2013, Part II, LNCS 7817, pp. 183–196, 2013.

it contains a threat. This task is quite hard, as threats cannot be detected simply by means of a set of keywords or phrases.

For the present study we adopt the following working definition of what constitutes a threat:

> A threat is a declaration of an intention to cause death or bodily harm to a person or persons, to damage or destroy their personal property, or to kill or injure an animal that is the property of a person.[2]

Under this definition tweets that are intended to annoy, alarm or otherwise cause emotional distress to another person are not considered to hold a threat. Also verbal abuse of another person or persons does not by itself constitute a threat.

The recognition of what constitutes a real threat is especially difficult as there are numerous tweets containing riddles or jokes, or where people are being sarcastic or ironic (so that it would immediately be clear to someone that what was being said was not to be taken seriously). Other tweets where a threat is not normally taken seriously is where a tweet clearly refers to for example a game setting, a movie, or a soap series.

Recognizing a threat is all the more difficult as in a language like Dutch there are numerous expressions which hint at harm or violence, but which are generally understood as figures of speech (e.g. *je kunt doodvallen* ('drop dead'), *op sterven na dood* ('almost dead'), *rijp voor de sloop* ('ready to be demolished': 'written off')). Moreover, many words are ambiguous and only point towards a threat in particular contexts. For example, a word like *maken* ('to make') is mostly neutral, also when it occurs as part of a separable verb (e.g. *opmaken* ('to format'), *doormaken* ('to go through')). However, when it occurs as part of the verb *afmaken* it may be neutral (as in *huiswerk afmaken* ('to finish homework')) or threatening (as in *jou afmaken* ('to finish you off')).

In the present paper we investigate two approaches that might be employed for the task of automatically detecting threats in Dutch tweets. In the first approach we attempt to manually construct a set of n-grams that should detect threats. In the second approach, we use machine learning to discover which (surface) features characterize threats. The task is defined as a classification task in which the two approaches each attempt to classify tweets as either threatening or non-threatening, depending on whether or not they contain a threat. The approaches are evaluated and compared for efficacy but also so as to see how one approach might advance the other.

The structure of the paper is as follows. In Section 2 we describe the data used for development and testing. A description of the manual construction of the n-grams is given in Section 3, while in Section 4 the machine learning approach is described. A quantitative analysis of the test results is given in Section 5. The two approaches are compared qualitatively in Section 6. Section 7 concludes this paper.

---

[2]   Note that we are not looking for a legal definition, but rather for a definition that can be operationalized when attempting to identify what constitutes a threat when dealing with tweets. The definition is rather loosely based on that given in Black's Law Dictionary [1] and the Canadian Criminal Code [2].

2   Experimental Data

For our experiments we need data representing threats and also data representing non-threats. Although with Twitter large amounts of data are available, we do not know which tweets are threatening. Therefore we decided to use large random samples of data as background corpus for development and for measuring precision. For the positive examples (used for development and measuring coverage) specialized collections are needed.

2.1   Collection of Dutch Twitter Threats

Threatening tweets were obtained from the website www.doodsbedreiging.nl, a site which allegedly wants to raise a public debate on the phenomenon of threats made through Twitter.[3] Over the past two years or so the site has published over 5,000 threats that were posted on Twitter. We downloaded two data sets, viz. one that we used as development set and the other that we held apart and used as test set. As we found that not all downloaded tweets answered to our definition of what constitutes a threatening tweet, all data was checked manually and non-threats were removed. As a result in the (threat) development set (henceforth TDS) 4,564 tweets remain, while the (threat) test set (TTS) comprises 583 tweets. The TTS fully postdates and has no overlap with the TDS.

Data clean-up for both data sets involved the removal of collection artifacts such as the hash tag #doodsbedreiging, retweet markers (rt, RT etc.), time stamps and user names (@username). Moreover, in the development set proper names and URLs were anonymized so as to avoid recognizing regular targets (such as the controversial politician Geert Wilders) rather than the threat itself. Subsequently all data were tokenized: punctuation marks were separated from the word tokens and all upper case characters were converted to lower case. Complexes of punctuation marks and symbols, probably meant as emoticons, were not broken up into parts.

2.2   Samples of Dutch Twitter in General

For a large random sample of general tweets to be used as development set, we extracted some 2.3 million tweets, viz. the tweets from a single day in 2011, from a much larger set of Dutch Twitter data collected through the Dutch e-science centre [3]. As in the collection process a language filter was applied, the data contains virtually no dialect or street language, which we do find in the data from www.doodsbedreiging.nl. As test set, a random set of 1 million tweets was sampled from the same collection, with time stamps between October 2011 and September 2012. In what follows we refer to the general development set as the GDS and to the general test set as the GTS.

---

[3] Cf. the editorial on www.doodsbedreiging.nl

3   Manually Constructed Recognition Patterns

In the first of two approaches we want to compare, we use a set of manually constructed recognition patterns. Here we rely on our (linguistic) intuition as native speakers of Dutch. In the process, the development sets (TDS and GDS) are used for further inspiration and for obtaining more objective information as to how frequently certain patterns occur and with what senses.

The set of patterns consists of (token)[4] n-grams, more specifically positive and negative unigrams, bigrams, trigrams, and skipgrams (bigrams and trigrams). By definition, the tokens in bigrams and trigrams are adjacent while in skip bigrams they are non-adjacent. In a skip trigram, however, one of three situations may arise: (1) the first two tokens are adjacent, while the third is non-adjacent to the second, (2) the last two tokens are adjacent, while the first is non-adjacent to the second, or (3) the three tokens are all non-adjacent. There is no differentiation in pattern strength.

The total number of base n-grams[5] is 16,190. Of these 3,129 are positive and 13,061 negative. The distribution over the different n-gram types is given in Table 1.

Table 1. Characterization of the base n-gram set: distribution of n-gram types. The labels used are as follows: <NG1>=unigram, <NG2>=bigram, <NG3>=trigram, <SG2>=skip bigram, <SG3as>=skip trigram with only the first two tokens adjacent, <SG3sa>=skip trigram with only the last two token adjacent, <SG3ss>=skip trigram with only non-adjacent tokens.

n-gram type	positive	negative
<NG1>	304	--
<NG2>	831	1190
<NG3>	519	2875
<SG2>	709	201
<SG3as>	277	2944
<SG3sa>	299	2938
<SG3ss>	190	2913

3.1   N-Grams Expected in Threatening Tweets

The manual patterns focus on the recognition of phrasings that overtly express a threat. Therefore, most positive n-grams contain an action verb that is indicative of some violent action. Examples are *doden* ('to kill'), *(neer)steken* ('to stab'), *vermoorden* ('to murder') and *(neer/af/dood)schieten* ('to shoot'). As threats typically refer to something happening in the near or not too distant future - such as that the sender of the tweet is going to inflict harm upon the receiver or, put differently, the receiver is

---

[4]   Tokens are words, numbers, punctuation marks, or symbols.

[5]   Base n-grams are expressed using conventional spelling, with the exception of spelling variants involving different spacing in words (cf. note 5). See also Section 4.3 which describes how spelling variation is handled.

going to experience something bad happening to him - the verb form commonly is first or second person present tense or future.[6] Examples can be found in the unigrams <snijd> ('cut'), <schiet> (shoot) and <djoek> ('kill') and the bigrams <ik vermoord> ('I kill') and <gaat sterven> ('are going to die').

As the n-grams are token-based and no part-of-speech information can be brought to bear to disambiguate between homographs of, for example, a noun and a verb (*dood*, 'death'/'kill'), or a present tense verb form and a past participle (*vermoord*, 'kill'/'killed'), the unigrams are likely to overgenerate. Therefore, in many such cases we have opted to use a (skip) bigram rather than a unigram (<ik dood> ('I kill') and <ik vermoord> ('I murder')).[7]

The large proportion of n-grams that are not unigrams can further be explained by the fact that in Dutch there are many separable verbs (e.g. *doodsteken* ('to stab to death'), for which the first person present tense is *steek dood*) and there is a frequent use of subject-verb inversion (so that apart from the bigram <ik vermoord> we also need to specify the inverse <vermoord ik>).

3.2   N-Grams Inhibiting Erroneous Recognition

Negative n-grams are brought into play in order to delimit the extent to which the positive n-grams are overgenerating. Thus where the unigram <aanval> ('attack') will yield a great many false accepts including *hart aanval* ('heart attack'), *paniek aanval* ('panic attack'), *schijn aanval* ('mock attack'), the inclusion of such instances as negative n-grams effectively cancels them out.[8]

While there are quite a few cases where it suffices to identify an adjacent item that 'disarms' the otherwise threatening wording, there are also many cases where it is only clear from the wider semantic context that there is actually no threat. When we look once more at the word *aanval* we find that it is more commonly used in non-threatening contexts, for example in a sports context (soccer, basketball, tennis, etc.) or when talking about politics (politicians 'attacking' each other in a polical debate). Negative skip bigrams in which we include domain-specific words (for example, in the case of *aanval* words from the sports context like *doelpunt* ('goal'), *middenveld* ('centre field'), *rechterflank* ('right wing'), *wedstrijd* ('match'), *bal* ('ball'), *beker* ('cup'), and *finale* ('final')) cancel out positive matches in non-threatening contexts and contribute to reducing the proportion of false accepts.

Virtually all negative skip trigrams are directed at canceling out positive matches that are the result of skip bigrams applying across clause boundaries. For example, the

---

[6]   The expression of future time in Dutch requires the use of an auxiliary such as *gaan* ('go') or *zullen* ('shall') with the infinitive form of the verb.

[7]   The proportion of unigrams is still fairly substantial. This is due to the fact that they also include some proclitic forms (such as *kschiet* ('I shoot') and *ksteek* ('I stab')), and contracted forms such as *ikwurg* ('I strangle') and *iksla* ('I hit') where there is no space between the word tokens where there normally would be.

[8]   All of these are compounds which normally in Dutch are written as single words. However, in tweets we find that they are frequently written as separate words.

skip bigram <maak af> (from the separable verb *afmaken* ('to finish off')) finds a match in the tweet

> *maak jij nog 3 screenshots met 3 zinnen er onder? moet maandag af x*
> [Eng: will you make 3 screenshots with 3 sentences below them? must be
>   ready by Monday x]

where the tokens *maak* and *af* occur in different clauses and therefore are completely unrelated items. The negative skip trigram <maak ? af> identifies the match as a false accept and cancels it. We included the following tokens as clause boundary markers: ., : ; ? ! *en of* (punctuation, 'and' and 'or').

3.3   Spelling Variation

As there is a great deal of spelling variation in tweets, we can expect to miss out on many threatening tweets if we employ the n-grams in their base form, i.e. using essentially conventional spelling. We therefore automatically expanded the set of n-grams by including possible spelling variants of the word tokens.[9] To this end we used data from previous work on spelling variation [3], where spelling variants were clustered and represented by means of a normal form. The spelling suggestions were manually checked and where necessary removed.[10] Where on the basis of the development set we were aware of variants that did not occur among the suggestions, such variants were added. This was the case for some word tokens that are typical of Dutch street language (e.g. *deade* for Dutch *dood* ('dead') and *joeke* for *djoeke*, i.e. Dutch *doden* ('to kill')). After expansion the n-gram set comprised some 11.3 million n-grams (see also Section 6.3).

3.4   Limitations of the Present n-Grams

With the present n-grams there are clearly limitations to what can be expressed and the amount of control one may have over a pattern:
- The n-grams are (on occasion too) limited in size: max n=3;
- The length of the skip cannot be defined;
- Negative n-grams are applied independently of the positive n-gram they have been designed to cancel out;
- As the base n-grams are expanded, spelling variants are introduced for individual word tokens in isolation, i.e. not in the context of the n-gram.

---

[9] We refrained from expanding the negative bigrams.

[10] Items that were removed include items that had inadvertently been associated with a particular cluster (as for example *bloedband* ('blood tie'), one of the suggested variants for *bloedbad* ('blood-bath')), but also items that were at odds with what the pattern is attempting to match such as third person verb forms where the pattern is directed at first person: in Dutch the morpheme –*t* marks the third person singular form (cf. *snijdt* (3rd person singular of *snijden* ('to cut')) vs *snijd* (1st person singular)); while we do want to include *snij* as variant for *snijd*, we want to exclude *snijdt*.

4   Machine Learning of Recognition Patterns

The second approach we test for recognizing threatening tweets is machine learning. Now, a machine learning system rather than a human expert attempts to identify those n-grams that are indicative of threats. Because of computational complexity, it cannot make use of skip trigrams, but unigrams, bigrams, trigrams en skip bigrams are all available. As training material, the machine learner has access to the development sets also used in manually constructing patterns (TDS and GDS). In order to maintain optimal comparability with the first approach, we will set the acceptance threshold for the machine learning system in such a way that, on the GDS, it will accept the same amount of the tweets, about 0.8%.

4.1   Machine Learning System

Our machine learning system will have to decide whether or not a tweet is threatening or not, purely on the basis of the text in the tweet. This task is very similar to other text classification tasks, but differs in the amount of text that is available. We have decided to base our system on the Linguistic Profiling (LP) system [5]. However, it is necessary to change this system because of the shortness of tweets. Where LP bases its judgements on both overuse and underuse of n-grams, underuse cannot be used here. In the on average ten words present in tweets, practically all n-grams will be underused. Overuse will also have to be treated differently. In a text of about a thousand words, an n-gram may be overused more or less, but in a tweet one can only sensibly use presence or absence and LP's weighting based on the frequency in the test text should therefore not be used. On the other hand, the degree of overuse in the training material can still be used fruitfully.

Therefore, we use the following procedure. During training we determine which n-grams occur more frequently in the set of tweets known to be threatening (TDS) than in a background corpus of tweets (GDS), and to which degree. To determine this degree we split the TDS and GDS into blocks of 100 tweets (comparable to the texts of about one thousand words that LP has been used for in other tasks). On the GDS, we calculate the means and the standard deviations for the frequencies per block of the various n-grams. Then, on the TDS, we calculate for each block how many standard deviations the occurring n-grams are overused. The average of this value over the blocks is taken to be the degree of overuse. During testing, every presence of an overused n-gram yields a contribution to the recognition score equal to the degree of overuse, raised to the power determined by a hyperparameter P_O. The hyperparameter is set automatically during the training process.

However, when we simply add the scores for all n-grams, longer tweets can be expected to get higher scores than shorter tweets. We need to introduce some kind of correction for the text length. We have chosen to divide the score by the number of tokens in the tweet, raised to the power determined by a second hyperparameter P_L, again set during training. Finally, the corrected score is compared to a threshold to determine acceptance.

4.2   The Training Process

During training, the system learns the degree of overuse of all n-grams and the optimal settings of the two hyperparameters and the threshold. To find the optimal settings, we go through a full training-test sequence, applying ten-fold cross-validation on the TDS, as it is rather small. In this process, we try various settings for the hyperparameters, using a rough grid in a first cycle and a finer grid in a second cycle. The best values found after the second cycle are used when the system is actually applied. We determine the best values by measuring how many tweets from the background corpus are accepted when the threshold is set in such a way that the false reject rate on the TDS is kept under a specified percentage (here 5%) and choosing the values where this accept rate is lowest.

Rather than a single recognizer, using the full GDS as its background corpus, we built three recognizers which each filter out non-threats.[11] The first is trained using the full GDS as background corpus, the second using only those GDS tweets accepted by the first recognizer and the third using only those accepted by the second recognizer. For each of the three training processes, we allowed the system to falsely reject 5% of the full TDS. As we wanted the system to accept the same amount of tweets as the manual patterns, we needed to reduce the final number slightly, which we did by adjusting the threshold for the third recognizer. The eventual three filters will reduce the GTS from 1M to 47,684 (-95.2%), to 17,001 (-64.4%) and finally to 9,188 (-46.0%).

4.3   Types of N-Grams Playing a Role

Where, in the manual construction of patterns, n-grams are chosen on semantic grounds, the machine learner has no notion of meaning and works purely with statistics. It selects those n-grams which systematically occur more often in threatening tweets than in randomly selected tweets. On the basis of the approximately 80,000 tokens in the TDS, the machine learner selects 337,084 n-grams (7,674 unigrams, 34,080 bigrams, 51,361 trigrams and 243,969 skip bigrams).

If we examine these n-grams, we can identify a number of clear groups. First of all, there are the references to the planned violence that were also targeted in the manual construction of patterns. These include action words like *vermoorden* ('to murder') and *aanslag* ('attack'), but also weapons like *bom* ('bomb') or *kraspen* ('scratching pen'), and targeted body parts like *kop* ('head') or *strot* ('throat'). Secondly, there are the intended targets themselves, which can be people (individual persons, groups of people, institutions/organizations) and/or their possessions, but also parts of the infrastructure, buildings, etc. For example, *jeugdzorg* ('child welfare organization'), *politiebureau* ('police station'), and *school* ('school'). With individual persons particularly there is lot of name calling (e.g. *hoer* ('whore') and *mongool* ('Downie', i.e. person suffering from Down syndrome)) and frequent use of abusive forms of address. Examples of the latter frequently involve the use of adjectives like *vuile*, *vieze*, *gore* or *smerige* (all various degrees of 'dirty'). Next we find interjections, such as *wollah* (street language 'I swear', 'truly') or *kanker* (originally 'cancer'). Then

---

[11] On the development sets, this sequential set-up outscored the single recognizer by 2%.

there are words expressing that we are talking about a future event (*morgen* ('tomorrow')), possibly containing a warning (*wacht maar* ('just wait')) The next group are the pronouns one might expect to be more prevalent in threats, such as *ik* ('I'), *je* ('you'). Finally, we also see very general words which we cannot link directly to threats, such as *en* ('and') and *de* ('the'). As these even occur as unigrams, this may well just be caused by statistical coincidence.

The coincidence hypothesis is possibly confirmed by the observation that n-grams are not used in all three recognizers. For example, the unigram *de* is only used in the third one. On the other hand, the differences between recognizers sometimes also have a reason. The bigram *ik ga* ('I go', 'I will'), for instance, is active in the first two recognizers, but no longer in the third one. Apparently, the fact that something is announced appears to be handled at the start of the filtering process and is no longer significant in the third phase.

Of the 3,125 positive n-grams in the manually constructed patterns (before spelling expansion), 477 (15%) are also selected by the machine learner. Interestingly, even though the training set is not that large, a further 210 overlapping n-grams are found containing spelling variation.[12] As could be expected, most of the overlapping n-grams (641 out of 667) are active in all three recognizers.

5   Test Results: Quantitative Evaluation

We tested the two systems by applying them to the general and threat test sets (i.e. the GTS and the TTS resp.). We then examined all tweets from the GTS that were accepted by either system (15,312 tweets) and marked those which we deemed to be threats as described above (1,134 tweets).[13] The resulting data was used in the subsequent evaluation.

5.1   Overall Recall and Precision

The recall and precision scores of the various systems on the test sets are summarized in Table 2.

The manually constructed patterns recognize 84.8% (3871/4564) of the TDS, 84.7% (494/583) of the TTS and 79.9% (906/1134) of the threats we found in the GTS. The machine learner, with a threshold accepting the same amount of tweets on the GDS, recognizes 90.0% (4108/4564), 90.1% (525/583) and 55.8% (633/1134) respectively. However, for the machine learner we can vary the threshold, which leads to the recall scores shown in Figure 1. We see that, for both systems, there is hardly any difference between the recall on the TDS and TTS. Recall on the randomly selected tweets (from the GTS) is lower, though, for the machine learner scores quite a lot lower.

---

[12] These are not just idiosyncratic n-grams from the training data as 62 of the 210 (30%) are also found in the 1M tweets of the GTS, versus 271 of the 477 (57%).

[13] As also described above, this task is a difficult one and we have to assume that we missed some threats. Furthermore, there will of course also be threats that were not caught by the systems. As a result, the recall figures below can be taken to be (reasonably accurate) overestimates, but the precision figures will be underestimates.

Table 2. Recall and precision scores of various systems on various data sets.MP represents the manually constructed patterns. MP- = MP without spelling variation, MP+ = MP with spelling variation. ML represents the machine learner. The last two columns show (simple) combinations, in which MP is used with the spelling variation active.

	MP-	MP+	ML	ML or MP+	ML and MP+
Recall TDS	81.8%	84.8%	90.0%	95.5%	79.3%
Recall TTS	82.5%	84.7%	90.1%	95.5%	79.2%
Recall threats in GTS	75.5%	79.9%	55.8%	100.0%[14]	35.7%
Precision threats in GTS	12.2%	12.1%	6.9%	7.4%	30.1%

Table 3. Number of recognition patterns used by the various systems.MP represents the manually constructed patterns. MP- = MP without spelling variation, MP+ = MP with spelling variation. ML represents the machine learner. POS refers to positive n-grams and NEG to negative n-grams.

	MP- (POS/NEG)	MP+ (POS/NEG)	ML
# patterns in total	3125/13056	~7.09M/~4.25M	337,084
# patterns used on GTS	589/795	918/917	162,071
# patterns used for accepted tweets	578/83	876/102	83,917
# patterns used for correctly accepted tweets	268/13	357/15	20,141

If we examine the various threat sets (TDS, TTS, and threats in GTS) more closely, we observe that the tweets extracted from www.doodsbedreiging.nl form a rather biased sample. These are the threats that someone apparently found to be of particular interest, e.g. when they target well-known people or institutions such as schools. They also have a certain level of seriousness. The bulk of threats in the random sample, however, concern potentially violent disagreements between individuals, and are often likely to be bluster rather than real intent. We also have the impression that the language use in the two sets differs. The manually constructed patterns suffer somewhat from the differences between the data sets, but not very much. The machine learner, however, suffers greatly from the shift in data type. In order to reach the same kind of recall as seen on the threat sets, we would need to collect a training set at least as large as our TDS.

---

[14] Remember that we only checked tweets accepted by one of the two systems. There are probably more threatening tweets among the one million in the GTS.

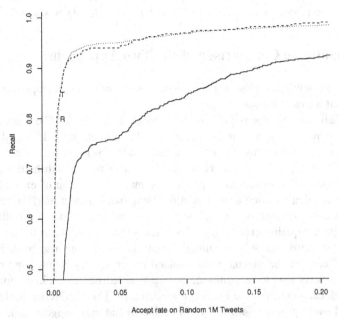

Fig. 1. The recall on the various known threat sets as a function of the accept rate on the GDS. The lines represent the machine learning recalls for the TDS (dotted), the TTS (dashed) and marked threats from the GTS (full). The markers T and R represent the manual pattern recalls for the two threat sets (TDS and TTS, represented by T) and GTS (R).

5.2 Effectiveness of Negative N-Grams

As we saw in Section 4,[15] the number of negative n-grams in the manually constructed set was far larger than the number of positive n-grams, while the machine learner could only use positive n-grams. When we look into the effectiveness of the negative n-grams we find that on the GTS they boost the precision of the manual patterns from 10.1% to 12.1% (+19%) as they prevent 1,569 tweets from being falsely accepted. There is very little loss of recall: on the TDS 30 threats are missed (-0.8%), on TTS 4 (-0.8%) and on GTS 5 (-0.5%).

5.3 Effectiveness of Modeling Spelling Variation

Modeling spelling variation increases the recall measured on all sets (Table 2). Where the gain for the threat sets, TDS and TTS with 3.7 and 2.7% respectively, is already worthwhile, the gain for the GTS is as much as 5.8%. Precision, on the other hand, is decreased much less, about 0.8%. Apart from variants where letters are repeated any number of times as for example in the various variants for *gaat dood* ('will die', which include *gaaaat dood , gaaat dooood, gaat dooود*), a very frequent but more systematic type of spelling variant involves leaving out the final *−n* with infinitive

---

[15] See also Table 3.

forms (e.g. *aanvalle(n)* ('to attack'), *afschiete(n)* ('to shoot'), *djoeke(n)* ('to kill'), *murdere(n)* ('to murder'), *gooie(n)* ('to throw'), *neerknalle(n)* ('to shoot down')).

6   Qualitative Comparison of the Two Approaches

Apart from presenting a general evaluation, we can now also compare the two approaches that we used for our recognition task.

First of all, we can observe that both approaches are viable. The machine learner appears to score a bit better for the already available threat sets (TDS and TTS) and the human expert's patterns do better on the random selection of tweets, but both produce quite acceptable results. However, both also need a substantial amount of work, be it manual construction of patterns or manual selection of examples for the learner. We see that an often used reason for using machine learning, the reduction of labour by reusing apparently compatible data sets and annotations, is an illusion here as the recognition quality greatly degrades when we move to differently sampled data.

The two systems operate in a quite different manner as can also be deduced from Table 3. Where, for the manually constructed patterns, only a few n-grams activate and almost always lead to recognition, the machine learner uses a large amount of n-grams which each contribute a bit to the recognition. This difference leads to a relatively small overlap in recognized tweets (Table 2) and may suggest some manner of combination. However, union or intersection do not appear to be very useful, as we can see in Table 2, unless we are dealing with a task where either precision or recall is less important. And a voting technique is useless since the patterns provide only a yes/no decision (barring the rather low number of tweets where more than one pattern is present). This means that we should rather examine whether and how one approach can help improve the other.

6.1   Lessons for the Machine Learner

In order to see how the machine learner might be improved, we took the threatening tweets in the GTS which were recognized by the manually constructed patterns, but not by the machine learner, and examined which n-grams were apparently missed by the machine learner. For these 501 tweets, there were 249 different patterns active (in total 558 matches). 142 of these (403 matches) were also known to and used by the machine learner, but the threshold was not reached. 9 n-grams (25 matches) were used in some but not all the three recognizers (1 only in the first filter, 8 in the first two). In only 9 of the 25 matches, the tweet was rejected by the filter missing the pattern, but it is not clear if the presence of the n-gram would have helped. More interesting is the set of 99 n-grams (133 matches) which the machine learner missed altogether. 11 of these (12 matches) concern skip trigrams, an n-gram type which the machine learner does not use at all. The number does not appear high enough to introduce skip trigrams, given the concomitant computational cost. For 44 n-grams (46 matches), some also skip trigrams, there is some kind of non-standard spelling. This would imply that we should look into the possibility of handling spelling variation for the machine learner too. We fear that the method used here for the manually constructed patterns is far too liberal and that we should rather attempt to normalize training and test material in some way [3]. The remaining n-grams (48, of which 22 unigrams, 6 bigrams, 5 trigrams and 15 skip bigrams) have simply not been seen in the training material. They sometimes concern more rare types of violence, like *stenigen* ('to stone', but equally present are far more normal types, like *afknallen* ('to shoot down') and *vechten met* ('to fight with'). If we want to hold on to a pure machine learning approach, the solution here is to collect more training data, probably also more geared towards the type of threatening tweets that we want to find. The manually constructed patterns can of course be useful here in filtering tweets for this collection process.

6.2   Lessons for the Manual Construction of Patterns

Conversely, in order to see how the manually constructed patterns might be improved, we took the threatening tweets in the GTS which were only recognized by the machine learner. Again we listed the n-grams, this time those which were active in the machine learner's recognition. In this case, we might consider adding new patterns to our collection, copied directly from this list. However, as we have already seen, the machine learner uses very large amounts of n-grams, also ones that are innocent by themselves but correlated with threats. As a result, the 228 tweets in question yield a list of 9,630 n-grams so far unrepresented in the patterns. Most of these have no place in our patterns as they seem to have no direct bearing on threats. All in all it is doubtful whether examining the list is more fruitful than simply examining the set of additionally accepted tweets. However, we should keep in mind that this set was only constructed through a large amount of work, viz. the inspection of more than 15,000 tweets.

7   Conclusion

We have attempted to recognize threatening tweets, on the one hand using manually constructed recognition patterns and on the other hand machine learning. Both methods used token n-grams as a handle on the meaning of the tweets and both had access to the same development data and the same test data.

An evaluation on unseen data showed that both methods led to good results (85% or more recall when accepting less than 1% of the input data) when tested on unseen data that has been collected in the same way as the training data, with the machine learner having a slight edge. However, when testing on data collected in a different way, the recall of the manually constructed patterns dropped slightly, but that of the machine learner significantly.

We conclude that, for this kind of data and task, both methods require a substantial investment of labour before they can reach an acceptable level of quality, be it the construction of patterns or the collection of training material. For machine learning, there is the possibility of the shortcut of reusing existing data sets, but this shortcut proves effective only if the existing data set and annotation are very close to the target data set and task.

As for recognizing threats, we deem that both methods do provide a good start but also show room for improvement. Each method can help to some degree in improving the other, but the current precision levels are still rather low and significant amounts of manual intervention will probably be needed. We expect that progress can be made faster by investing in more information-rich methods instead of approximating meaning by way of surface features like n-grams.

References

1. The Law Dictionary. Featuring Black's Law Dictionary Free Online Legal Dictionary, 2nd edn., http://thelawdictionary.org/search2/?cx=partner-pub-4620319 056007131%3A7293005414&cof=FORID%3A11&ie=UTF-&q=threat&x=6&y=6
2. Canadian Criminal Code, http://www.rcmp-grc.gc.ca/qc/pub/cybercrime/cybercrime-eng.htm
3. Tjong Kim Sang, E.: Het Gebruik van Twitter voor Taalkundig Onderzoek. TABU: Bulletin Voor Taalwetenschap 39(1/2), 62–72 (2011)
4. van Halteren, H., Oostdijk, N.: Towards Identifying Normal Forms for Various Word Form Spellings on Twitter. CLIN Journal 2, 2–22 (2012), http://www.clinjournal.org/sites/default/files/1VanHalteren2012_0.pdf
5. van Halteren, H.: Linguistic Profiling for Author Recognition and Verification. In: Scott, D., Daelemans, W., Walker, M.A. (eds.) Proceedings of the 42nd Annual Meeting of the Association for Computational Linguistics, Barcelona, Spain, July 21-26. ACL, Barcelona (2004)

Distinguishing the Popularity between Topics: A System for Up-to-Date Opinion Retrieval and Mining in the Web

Nikolaos Pappas[1,2], Georgios Katsimpras[2], and Efstathios Stamatatos[2]

[1] Idiap Research Institute, Rue Marconi 19, 1920 Martigny, Switzerland
nikolaos.pappas@idiap.ch
http://www.idiap.ch/
[2] University of the Aegean,
Dep. of Information and Communication Systems Engineering,
Karlovassi 83200, Samos, Greece
gkatsimpras@gmail.com, stamatatos@aegean.gr
http://www.icsd.aegean.gr/

Abstract. The constantly increasing amount of opinionated texts found in the Web had a significant impact in the development of sentiment analysis. So far, the majority of the comparative studies in this field focus on analyzing fixed (offline) collections from certain domains, genres, or topics. In this paper, we present an online system for opinion mining and retrieval that is able to discover up-to-date web pages on given topics using focused crawling agents, extract opinionated textual parts from web pages, and estimate their polarity using opinion mining agents. The evaluation of the system on real-world case studies, demonstrates that is appropriate for opinion comparison between topics, since it provides useful indications on the popularity based on a relatively small amount of web pages. Moreover, it can produce genre-aware results of opinion retrieval, a valuable option for decision-makers.

Keywords: Opinion Retrieval, Text Mining, Sentiment Analysis, Information Extraction, Utility-Based Agents.

1 Introduction

A huge number of user-generated content on various topics is created every day in social networks, news media, blogs, discussion forums and other sources in the Web. This content oftenly expresses opinions of users about certain products, people, services, etc. and therefore the need of computational treatment of opinion, sentiment, and subjectivity in text has become crucial [12]. Many applications, such as brand analysis, measuring marketing effectiveness, influence network analysis and many more, exploit the existing opinionated information.

During the last decade, considerable progress has been achieved in opinionated document retrieval. Most of the published studies are targeting blogs (TREC) [7, 10] and can be roughly categorized into two categories: lexicon-based [9,21] and classification-based [4,22]. The former utilize subjective dictionaries and decide whether the occurrences of these words suggest an opinionated

A. Gelbukh (Ed.): CICLing 2013, Part II, LNCS 7817, pp. 197–209, 2013.

document. The latter, develop subjectivity classifiers, based machine learning on opinionated and non-opinionated text. The proposed approaches are using fixed and offline collections of texts, taken from certain domains (e.g. blogs, movie reviews, message boards) or certain corpora.

In addition, opinion mining conclusions can differ according to the examined web genres (e.g. certain products may have good promotion articles but poor comments in blogs). So far, the task of collecting online domain-independent opinionated texts from various web sources in order to be used for opinion mining applications, has not been studied thoroughly. Moreover research on focused crawling usually deals with the more general task of collecting any kind of documents about a certain topic (e.g., [1,3,11]). However, opinion mining applications require the discovery of certain web genres that mostly comprise opinionated texts. Moreover, it is not yet possible to estimate the number of opinionated texts needed to extract reliable conclusions on the total polarity of opinions about particular topics.

In this paper, we present an online system for opinion retrieval and mining which handles the above subjects together: it discovers up-to-date topic-related documents dynamically from web sources using focused crawling techniques by targeting to specific genres (news, blogs, discussions) which are highly likely to contain opinionated texts; detects user-generated content regions inside the related pages by using web segmentation and noise removal techniques; computes a confidence score which quantifies the relatedness of the page to the given topic; and lastly performs automatic subjectivity and polarity detection on the sentences of the detected regions.

The main contribution of this paper is four-fold: (a) a unified framework for the discovery of topic-related opinionated texts in web pages, (b) a genre-based analysis of topic popularity[1], (c) a sentiment score estimation of opinionated regions of web pages, and (d) an efficient approach to estimate the sentiment polarity of topics using a few hundred documents.

The rest of this paper is organized as follows. Section 2 reviews the related research work. Sections 3, 4 and 5 provide an overview of the system and its components, whereas Section 6 describes the examined case studies. Finally, Section 7 summarizes the conclusions drawn from this study.

2   Related Work

There is a large body of research conducted for opinion retrieval and mining since TREC Blog was introduced in 2006 [10]. Most of these approaches are performing in a two-stage retrieval model. Firstly, one of the standard Information Retrieval methods is applied for locating topically relevant documents and secondly, various opinion mining/sentiment analysis algorithms are used to discover and identify opinionated texts within the documents.

---

[1] We refer to popularity using the definition i.e. 'well-liked, admired by the people'. The detected positive and negative opinions of the people are used as indications for their admiration degree for a given topic.

The aforementioned approaches focus on detecting the subjectivity for each document, using various opinion mining methods such as subjectivity word/phrase dictionaries [9, 20, 21], machine learning algorithms [22] or proximity and phrase matching [19]. In [9], is presented a system which consists of three major modules: a fact-oriented information retrieval, dictionary-based opinion mining method and spam filtering. The information retrieval module in [20] utilizes proximity and phrase matching while the opinion module integrates a number of factors, such as frequency-based heuristics, special pronoun patterns and adjective/adverb-based heuristics. Zhang et al. [22] perform a concept-base information retrieval [5], machine learning opinion detection and a ranking algorithm for filtering the irrelevant information.

Many other related works utilize machine learning techniques such as SVMs [4] or focus on subjective/polarity classification [16–18]. In [4], SVM is used to classify sentences as opinionated or non opinionated, then decide whether the sentences are topic-specific and lastly compute a total document score by summing the SVM scores of the examined sentences. In [17], subjective language features are identified, such as low-frequency words, word collocations, adjectives and verbs, from corpora and used them in the subjectivity classification. In a more recent approach [2], Gelani et al. proposed a probabilistic model using proximity information of opinionated terms.

3 Overview of the System

The architecture of the proposed system is displayed in Fig. 1. The two major components are the Crawling Module and the Mining Module. The first is responsible for gathering relevant documents to a specific topic, while the second extracts and identifies opinionated documents. Both components are operating asynchronously using the Messaging Module to communicate[2], which provides scalability and robustness. The code for the system is available online[3].

Based on given topic query, the first task is to find a set of appropriate seed pages to guide the crawling procedure. To this end, the query is sent via Seeding Module to major search engines (e.g., Google, Yahoo, etc.) and the top results of each search engine, form the list of seed web pages. These results are stored in a distributed object memory and forwarded to the Crawling Module which initializes n Focused Crawler Agents (FCAs), each one using an equally-sized chunk of seeds while the crawled URLs are stored in a distributed database[4].

At the same time, n Opinion Miner Agents (OMAs) are initialized to process the web pages discovered by each FCA. The OMAs are responsible to segment the page into textual parts and filter out the non-informative parts (i.e., non-opinionated texts or texts irrelevant to the query) and then decide about the subjectivity and the polarity of each opinionated text.

---

[2] http://www.rabbitmq.com/
[3] https://github.com/nik0spapp/icrawler
[4] http://www.mongodb.org/

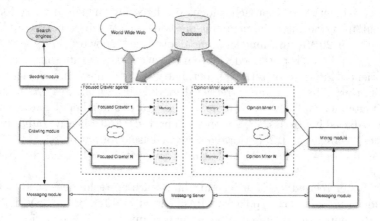

Fig. 1. The basic architecture and the components of the system

4   Discovery of Topic-Related Web Documents

The information retrieval component of the system is a state-of-art focused crawl-ing procedure. The idea is that, given a query, up-to-date relevant documents can be retrieved from various domains and web-genres by following the path of a focused crawler, but also in a real-time manner. For the purposes of our sys-tem, [13] is especially suitable. It is an agent-based focused crawling framework that is able to retrieve topic- and genre-related web documents in an automated and real-time manner.

The focused crawler agents displayed in Fig. 1 are making use of a utility function that weights an unvisited URL p and consists of two components: one for the topic relevance and one for the genre relevance.

$$Linkscore(p) = w_T * Linkscore_T(p) + w_G * Linkscore_G(p) \qquad (1)$$

The $Linkscore_T$ and $Linkscore_G$ are relevance scores based on topic and genre accordingly; and they are computed by using link analysis techniques (see [13]). For our experiments we used equally weighted these two scores ($w_T = w_G = 0.5$) since it has been shown that it leads to both topic and genre related document discovery. In addition, we selected the news, blogs and discussions genres for seed URLs and for weighting the genre component in the above equation, since these genres are more likely to contain opinionated texts. For the implementation we used Scrapy, a python-based crawling framework[5].

5   Opinion Retrieval and Mining

The Mining Module is responsible for the extraction of the opinionated textual parts from web pages and the estimation of their sentiment polarity. An OMA

---

[5] http://scrapy.org/

performs web page segmentation, assigns a confidence score which indicated the relevancy of the document being processed and estimates the sentiment subjectivity and polarity of the page. It learns from its previous experience with a page and uses this knowledge for solving more accurately and the sentiment analysis problem in future processing (Section 5.3).

In Fig. 2 the page processing by the OMA is displayed. Initially, it receives a message from an FCA to perform a task, connects to the corresponding database and retrieves all the relevant pages. Then, for each page, three basic procedures are executed; web page segmentation, page filtering, and sentiment analysis.

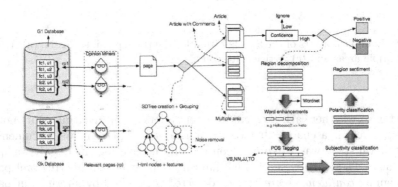

Fig. 2. The processing steps of an OMA are displayed for a given page: (a) web page segmentation (b) page filtering and (c) sentiment analysis

5.1   Web Page Segmentation

For this task a mechanism is needed to segment a web page into semantically-coherent parts that correspond to the basic textual components of the web page. Moreover, it is convenient that the noisy segments (i.e., ads, banners, etc.) are removed. A very recent approach that handles the above issues in an efficient manner, is presented in [14]. It exploits visual and non-visual characteristics of a web page encapsulated in a DOM Tree with additional features, called SD-Tree, and performs the layout classification and extraction using SD-algorithm.

We adopted this method because it provides robust identification of informative textual parts and it yields promising results as a web page type classifier in a realistic web setting. The output of this processing is a set of informative annotated regions in to three possible classes (Article, Multiple areas and Article with comments). Output examples are displayed in Fig. 3.

5.2   Page Filtering

The web page segmentation mechanism provides a set of segments with informative text of user-generated content; a source of potential opinions. However, it is

Fig. 3. Example outputs of the SD algorithm for the three possible classes: (a) Article, (b) Multiple areas and (c) Article with comments

not yet clear whether each extracted segment refers to the given query or another subject. There is a chance that the existence of the query in the document at the retrieval stage was not present on the informative regions (e.g. it was part of the ads). Therefore, we need a mechanism to filter out all the irrelevant pages by assigning *confidence score* to each detected region and by filtering out pages with low score (i.e. unlikely to refer to the given query).

The *confidence score* for a page i is calculated by the weighted combination of the presence of the topic in the detected regions, the URL and the title:

$$Confidence_i = w_1 * ArticleContextScore(i)$$
$$+ w_2 * CommentsContextScore(i)$$
$$+ w_3 * MultipleContextScore(i)$$
$$+ w_4 * UrlScore(i)$$
$$+ w_5 * TitleScore(i)$$

Regarding the type of the document, some context scores of the above formulation may be equal to 0. The weights can be learned from an annotated corpus of region class and relevance value pairs. For our experiments we used the weights below which yielded good results for each of the classes: (a) Article: ($w_1 = 0.4$, $w_4 = 0.3$, $w_5 = 0.3$), (b) Article with comments: ($w_1 = 0.2$, $w_2 = 0.2$, $w_4 = 0.2$, $w_5 = 0.4$) and (c) Multiple areas: ($w_3 = 0.4$, $w_4 = 0.3$, $w_5 = 0.3$). For example, given the query *Audi*, if it is present in the title and URL the confidence would be: $0.4 * 0 + 0.3 * 1 + 0.3 * 1 = 0.6$.

In the case some of the non-zero weighted regions are missing from the page, their weights are distributed equally to the rest of the coefficients. To this end, we select the documents with high confidence scores based on a threshold t. The threshold values range from 0 to 1. The closer to 1 the threshold is, the greater the confidence about the topic. For the experiments we used the value of $t = 0.6$.

5.3 Sentiment Analysis

The confidence mechanism provides related documents to a given topic. The next step is to to detect whether a given document contains subjective information or not. In order to learn dynamically the domain knowledge for a given query we use self-trained machine learning algorithms (see [6,15]). Initially, the filtered regions are decomposed into sentences (Fig. 2). The sentences are then pre-processed in three steps: (a) tokenization, (b) spell-checking based on WordNet and (c) part-of-speech (POS) tagging. Next, the set of sentences in the text area is given as input to our subjectivity classifier. Each sentence is classified as subjective or not. All sentences that are labeled as subjective are then forwarded to our polarity classifier. And thus, the sentiment for each sentence is determined.

Subjectivity classification. We adopted the method presented in [15] which is a bootstrapping process that learns linguistically rich extraction patterns for subjective expressions. High-precision classifiers using a subjectivity lexicon (MPQA[6]), label unannotated data to create a large training set, which is given to an extraction pattern learning algorithm. The learned patterns are then used to identify more subjective sentences. The bootstrapping process learns many subjective patterns and increases recall while maintaining high precision. To make the learning algorithm tractable in an online setting, we activate only the n-most frequent patterns at each learning step.

Polarity classification. Similarly, we adopted a bootstrapping method presented in [6]. The method follows three steps: (a) rule-based polarity classification with high precision [18], (b) training of an SVM classifier[7] using as input data the high scored instances from the rule-based classifier and (c) classification with the self-trained SVM classifier. The rule-based polarity classifier makes use of a subjectivity lexicon (MPQA[4]) and proceeds as follows: preprocessing, feature extraction, polar expression marking, negation modeling, intensifier marking, heuristic weighting and classification. Since we target on web text, we further extended the MPQA lexicon with informal and swear words as well as a great amount of emoticons. Lastly, for tractability reasons, we trained the SVM for a given query in a first short run and then we use it online in a second longer run.

Total Sentiment Estimation. Let D be a set of topic-related documents, r_{ij} the i-th region of document dj, and $Score(r_{ij})$ the sentiment score of r_{ij}. Then, the total *sentiment score* is defined as follows:

$$TotalScore(D) = \sum_{d_j \in D} \left(\sum_{r_{ij} \in d_j} Score(r_{ij}) \right) \in R \qquad (2)$$

Unlike the Eq. 2 where the detected regions are treated equally, the *normalized sentiment score* weighs them based on the region length as follows:

$$NormalizedScore(D) = \sum_{d_j \in D} \left(\sum_{r_i \in d_j} \frac{Score(r_{ij})}{|r_{ij}|} \right) \in R \qquad (3)$$

---

[6] http://mpqa.cs.pitt.edu/lexicons/subj_lexicon/
[7] http://pyml.sourceforge.net/tutorial.html#svms

where $|r_{ij}|$ is the length of the region r_{ij} in words. Lastly, given a the set of regions with positive sentiment score r_{pos} in D and r_{neg} with negative sentiment score accordingly, we compute the *sentiment ratio* as follows:

$$SentimentRatio(D) = \frac{|r_{pos}|}{|r_{pos}| + |r_{neg}|} \in [0, 1] \tag{4}$$

6   Experiments

In this section we examine the overall effectiveness of the proposed system to estimate the total sentiment polarity of the retrieved opinions for a given topic query in the Web. The study focuses on the system's ability to provide structured sentiment analysis results as well as on the number of pages required to form a reliable calculation of the sentiment. Since the system is designed to run in the Web (web pages not yet necessarily indexed by search engines), it is more appropriate to evaluate in real-world case studies rather than offline collections. The selected case studies concern well-known subjects that enable us to properly validate the produced results and were performed in October 2011.

6.1   Case Study 1: Distinguishing the Popularity between Topics

In the first case study we examined queries on two well-known political concepts: *democracy* and *fascism*. The presented system was used to discover a predefined number of relevant web pages for each query (1,000 relevant pages), extract the opinionated texts from them and calculate their sentiment polarity.

Figure 4 depicts the distribution of the detected relevant text regions over three major types (articles, multiple areas, and comments), the total sentiment score, and the total normalized sentiment score for both queries. As expected, the *democracy* query has a far more positive sentiment score in all three region types. In addition, the *fascism* query has negative sentiment scores in two types of pages (articles and multiple areas). Interestingly, the relatively high sentiment score of the *fascism* query for comments indicates an increased use from people with far-right radical political opinions.

The normalized sentiment score seems to be able to better represent the differences in sentiment polarity since it takes into account the length of the extracted text regions. For example, in articles usually there are a lot of long sentences with neutral polarity so the overall sentiment score tends to be lower. On the other hand, the normalized sentiment score indicates the intensity of the positive or the negative sentiment polarity.

A more detailed look in the distribution of sentiment polarity with respect to the three region types is given in Fig. 5 for *democracy* and *fascism* queries. In the former case, the positive sentiment is dominant in all region types with more emphasis in articles. Despite the increased percentage of neutral polarity in multiple areas and in comments, the positive opinions are in all cases greater than the negative opinions with an average difference of 20%. In the latter case, the

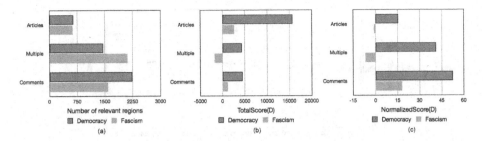

Fig. 4. Overall results for *democracy* and *fascism*: (a) number of relevant regions, (b) total sentiment scores and (c) total normalized sentiment scores per region type

negative polarity is greater than the positive one in most of the regions (articles and multiple areas). The difference of the positive versus negative polarity is not so intense in the comment regions.

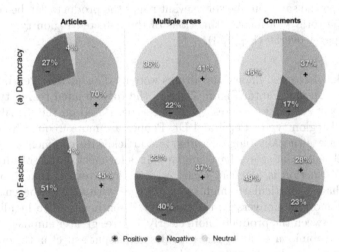

Fig. 5. Percentage of sentiments per region type for (a) *democracy* and (b) *fascism*

Finally, Fig. 6 shows the sentiment ratio (Eq. 4) for both queries (y-axis) during the process of discovering relevant web pages (x-axis) on these topics. The sentiment ratio remains practically stable after a few hundred pages have been examined. Moreover, there is a notable difference between the sentiment scores of the two queries indicating a much more positive polarity for *democracy* in comparison to *fascism*. This means that we can reliably decide about the sentiment polarity in short time.

6.2   Case Study 2: Ranking of Competitive Products

This experiment focuses on the examination of the system when it deals with a set of queries on competitive products in the same thematic area. In this case, it

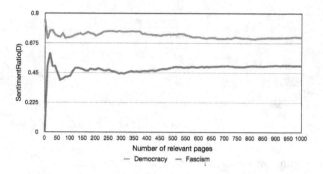

Fig. 6. Sentiment ratio curves for *democracy* and *fascism* queries

is crucial to provide comparative sentiment results and decide about a general ranking of the products according to the opinions found on web pages. We used a threshold of 300 relevant pages to be discovered for each of the product queries in the set. Given the same number of relevant pages the products can be compared based on the total sentiment estimations of the detected region types and the discovered pages overall (Eq. 2, 3).

Soft Drinks. Five well-known soft drinks were used as queries: Pepsi, Dr. Pepper, Sprite, 7up and Fanta. Figure 7 shows the topic-related region types, the total sentiment estimation scores per region type. Based on the distribution of the detected region types, Pepsi and Dr. Pepper are more frequently discussed in multiple areas (usually blogs, forums) and article with comments.

The total sentiment scores have similar values for most of the soft drinks and sentiment distinction is not very clear. A closer look reveals that 7up, Sprite, and Fanta have a particularly high score in pages with articles, potentially the result of promotion. Conversely, the normalized sentiment score highlights the differences between the products more clearly; it gives greater emphasis to pages with multiple opinionated areas and provides a different aspect in the evaluation of opinions (potentially of end users) about the products.

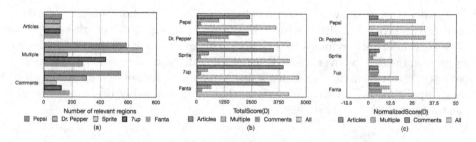

Fig. 7. Overall results for soft drinks: (a) number of relevant regions per type, (b) total sentiment score and (c) total normalized sentiment score per soft drink

Lastly, we compared the ranking based on the total sentiment estimation (Eq. 2, 3) to the ranking of the soft drinks based on social media metrics (number of likes, number of people talking) of their major groups on Facebook (Table 1). The ranking based on the normalized sentiment score matches closely to the one obtained based on the social media metrics. Sprite is probably low ranked due to the neutral or negative opinions found. Also, it has the smallest number of talking people from all the soft drinks.

Table 1. List of soft drinks and IM clients ranked by the social media metrics and the rankings based on total sentiment score and total normalized sentiment score

Rank	Soft drink	Likes	Talking	Both	TotalScore	NormalizedScore
1_{st}	Dr. Pepper	12,093,912	187,011	12,280,923	7up	Dr. Pepper
2_{nd}	Pepsi	11,835,244	236,105	12,071,349	Dr. Pepper	Pepsi
3_{rd}	Sprite	8,574,563	50,192	8,624,755	Sprite	Fanta
4_{th}	Fanta	2,650,072	84,080	2,734,152	Fanta	7up
5_{th}	7up	785,967	75,996	861,963	Pepsi	Sprite
	IM Client	**Followers**	-	-	TotalScore	NormalizedScore
1_{st}	Google Talk	405,818	-	-	Google Talk	Google Talk
2_{nd}	Skype	367,385	-	-	Skype	Skype
3_{rd}	MSN	82,896	-	-	MSN	MSN
4_{th}	AOL	14,431	-	-	AOL	ICQ
5_{th}	ICQ	14,138	-	-	ICQ	AOL
				NDCG:	0.841	0.993

Instant Messaging (IM) Clients. Similarly, some well-known IM clients were also used: Google talk, Skype, MSN messenger, AOL messenger and ICQ. We compared the ranking based on the total sentiment estimation (Eq. 2, 3) to the ranking of them based on their followers in Twitter[8]. In this case, the ranking based on each of the estimation scores matched almost perfectly the ranking based on the social media metrics. AOL and ICQ were ranked falsely based on the normalized score but they were not clearly distinguishable either based on the number of followers (14,431 and 14,138 accordingly).

Finally, we computed the average normalized cumulative gain (NDCG) [8] for both soft drinks and IM clients. In Table 1, we can observe that the normalized sentiment score performed better than the simple one in the examined queries. The long subjective sentences seem to be less important than shorter ones in the total estimation over the text regions.

7   Conclusions and Future Work

We presented an online system for topic-based opinion retrieval and mining in the Web. Rather than making use of static well-defined document collections, we

---

[8] Some of the IM clients' official groups were missing from Facebook (e.g. Google talk).

acquire dynamic collections in real-time from the Web. Such collections targeted to certain web genres, can provide up-to-date sources of opinionated text about a given topic. The opinion mining agents are able to extract opinionated textual parts from web pages and estimate their sentiment polarity while ignoring irrelevant and noisy regions. Useful conclusions can then be drawn based on the distribution of positive and negative opinions over the detected regions.

A series of experiments demonstrated that the system can provide a total estimation about the popularity of certain topics as well as comparative results for competitive topics. The genre-aware output of the sentiment results, can be of crucial importance for decision-makers since they can estimate the result of promotion as well as the potential difference in the opinion between the general population and some influential people. In addition, the system provides efficient results since a few hundred web pages are usually enough to estimate the total sentiment polarity about a given query.

A dimension of the system that could be further explored concerns the date that each opinionated text was created. This temporal information can be used to express the change of sentiment polarity about a certain topic over time and to provide an in-depth analysis for a certain time period.

Acknowledgements. The work described in this article was supported by the European Union through the inEvent project FP7-ICT n. 287872 (see http://www.inevent-project.eu).

References

1. Chen, X., Zhang, X.: Hawk: A focused crawler with content and link analysis. In: Proc. of the International Conference on e-Business Engineering (ICEBE), Xi'an Jiaotong Xian, China (2008)
2. Gerani, S., Carman, M.J., Crestani, F.: Proximity-based opinion retrieval. In: 33rd International Conference on Research and Development in Information Retrieval (SIGIR), Geneva, Switzerland (2010)
3. Hati, D., Sahoo, B., Kumar, A.: Adaptive focused crawling based on link analysis. In: Proc. of 2nd International Conference on Education Technology and Computer (ICETC), Shanghai, China (2010)
4. Jia, L., Yu, C., Zhang, W.: Uic at trec 2008 blog track. In: Proc. of The 17th Text Retrieval Conference (TREC), Gaithersburg, USA (2008)
5. Liu, S., Liu, F., Yu, C., Meng, W.: An effective approach to document retrieval via utilizing wordnet and recognizing phrases. In: Proc. of the 27th International Conference on Research and Development in Information Retrieval (SIGIR), Sheffield, United Kingdom (2004)
6. Wiegand, D.K.M.: Bootstrapping supervised machine-learning polarity classifiers with rule-based classification. In: Proceedings of the 1st ECAI-Workshop on Computational Approaches to Subjectivity and Sentiment Analysis (WASSA), Lisbon, Portugal (2009)
7. Macdonald, C., Ounis, I., Soboroff, I.: Overview of trec-2009 blog track. In: Proc. of The 17th Text Retrieval Conference (TREC), Gaithersburg, USA (2009)

8. Manning, C.D., Raghavan, P., Schtze, H.: Introduction to Information Retrieval. Cambridge University Press (2008)
9. Mishne, G.: Multiple ranking strategies for opinion retrieval in blogs. In: Proc. of the 15th Text Retrieval Conference (TREC), Gaithersburg, USA (2006)
10. Ounis, I., de Rijke, M., Macdonald, C., Mishne, G., Soboroff, I.: Overview of the trec-2006 blog track. In: Proc. of TREC, Gaithersburg, USA (2006)
11. Pal, A., Tomar, D.S., Shrivastava, S.C.: Effective Focused Crawling Based on Content and Link Structure Analysis. Journal of Computer Science 2(1) (2009)
12. Pang, B., Lee, L.: Opinion mining and sentiment analysis. Foundations and Trends in Information Retrieval 2(1-2), 1–135 (2008)
13. Pappas, N., Katsimpras, G., Stamatatos, E.: An agent-based focused crawling framework for topic- and genre-related web document discovery. In: Proc. of the 24th IEEE International Conference on Tools with Artificial Intelligence (ICTAI), Athens, Greece (2012)
14. Pappas, N., Katsimpras, G., Stamatatos, E.: Extracting informative textual parts from web pages containing user-generated content. In: Proc. of the 12th International Conference on Knowledge Management and Knowledge Technologies (i-Know), Graz, Austria (2012)
15. Riloff, E., Wiebe, J.: Learning extraction patterns for subjective expressions. In: Proc. of the International Conference on Empirical Methods in Natural Language Processing (EMNLP), Sapporo, Japan (2003)
16. Turney, P.D.: Thumbs up or thumbs down?: semantic orientation applied to unsupervised classification of reviews. In: Proc. of the 40th Annual Meeting on Association for Computational Linguistics (ACL), Philadelphia, USA (2002)
17. Wiebe, J., Wilson, T., Bruce, R., Bell, M., Martin, M.: Learning subjective language. Computational Linguistics 30(3), 277–308 (2004)
18. Wilson, T., Wiebe, J., Hoffmann, P.: Recognizing contextual polarity in phrase-level sentiment analysis. In: International Conference on Human Language Technology and Empirical Methods in Natural Language Processing (HLT/EMNLP), Vancouver, Canada (2005)
19. Yang, K.: Widit in trec 2008 blog track: Leveraging multiple sources of opinion evidence. In: Proc. of The 17th Text Retrieval Conference (TREC), Gaithersburg, USA (2009)
20. Yang, K., Yu, N., Valerio, R., Zhang, H.: Widit in trec-2006 blog track. In: Proc. of The 14th Text Retrieval Conference (TREC), Gaithersburg, USA (2006)
21. Zhang, M., Ye, X.: A generation model to unify topic relevance and lexicon-based sentiment for opinion retrieval. In: Proc. of the 31st International Conference on Research and Development in Information Retrieval (SIGIR), Singapore (2008)
22. Zhang, W., Yu, C., Meng, W.: Opinion retrieval from blogs. In: Proc. of the 16th International Conference on Information and Knowledge Management (CIKM), Lisbon, Portugal (2007)

No Free Lunch in Factored Phrase-Based Machine Translation [*]

Aleš Tamchyna and Ondřej Bojar

Institute of Formal and Applied Linguistics,
Malostranské náměstí 25, 11800 Praha
{tamchyna,bojar}@ufal.mff.cuni.cz

Abstract. Factored models have been successfully used in many language pairs to improve translation quality in various aspects. In this work, we analyze this paradigm in an attempt at automating the search for well-performing machine translation systems. We examine the space of possible factored systems, concluding that a fully automatic search for good configurations is not feasible. We demonstrate that even if results of automatic evaluation are available, guiding the search is difficult due to small differences between systems, which are further blurred by randomness in tuning. We describe a heuristic for estimating the complexity of factored models. Finally, we discuss the possibilities of a "semi-automatic" exploration of the space in several directions and evaluate the obtained systems.

1 Introduction

Phrase-based statistical machine translation [1] is probably the most popular approach to MT today. However, its models use no linguistic information for translating—words are treated as mere strings, no internal structure is considered. As such, phrase-based models suffer from certain inherent limitations that some linguistic insight might help to overcome. Factored models are an extension of phrase-based translation. They were introduced by [2] with the aim to reduce several problems of the paradigm, centered around the inability to handle linguistic description beyond surface forms. In a factored model, the system no longer translates words. Instead, each word is represented by a *vector of factors* that can contain the surface form, but also lemma, word class, morphological characteristics or any other information relevant for translation.

Factored models can employ various types of additional information to improve translation quality on many language pairs in various aspects like morphological coherence [3–8], grammatical coherence [9], compound handling [10] or domain adaptation [11, 12].

In factored translation, decoding is decomposed into a series of mapping steps: translation steps map source factors to target factors, generation steps operate

---

[*] This work was supported by the Czech Science Foundation grant P406/11/1499 and the EU project MosesCore (FP7-ICT-2011-7-288487).

A. Gelbukh (Ed.): CICLing 2013, Part II, LNCS 7817, pp. 210–223, 2013.

solely on the target side. There are many ways of defining a factored system. We can vary the set of source and target factors, but also the mapping steps and the order of their application.

Factored systems are mainly designed based on linguistic intuition, yet there may exist interesting configurations which lack a straightforward linguistic interpretation. The aim of this work is to analyze whether factored systems could be generated *automatically*, i.e. whether we can create an algorithm to decide, given a language pair and possible factors, which configuration will produce the best translations.

2   Factored Phrase-Based Translation

As in phrase-based translation, the main source of data for training a factored model is a parallel corpus. In this case, the corpus can be factored; each word can be annotated with arbitrary linguistic information.

In factored models, translation consists of applying *translation* and *generation* steps that gradually fill in the target-side factors and produce a final translation.

Translation steps (T) operate on phrases, they map a defined subset of source factors to a defined subset of target factors. The translation proceeds similarly as in the phrase-based scenario, it operates on phrases, i.e. contiguous sequences of words regardless of any syntactic structure.

Generation steps (G) operate on the target side, their input is a subset of factors (already generated, e.g. by a previous translation step) and they give at output another subset of target factors. Generation steps operate on single target words, so no word alignment is necessary. In fact, additional monolingual data can be used in their training.

The example in Figure 1 shows a scenario with two translation steps and one generation step. Source lemmas are translated to target lemmas, similarly for tags (translation). The joint information is then used on the target side to generate final surface forms (generation); for each word, the step generates its surface form based on lemma and tag (factors that were filled in by the previous translation steps). Note that factored models used in practice are *synchronous*—the same segmentation into phrases is used for all translation steps.

2.1   Translation Options in Factored Models

Factored models, especially the more complex setups, can dramatically increase the computational cost—the combination of translation options of various steps

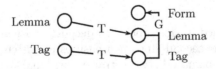

Fig. 1. Factored translation. An example of translation and generation steps

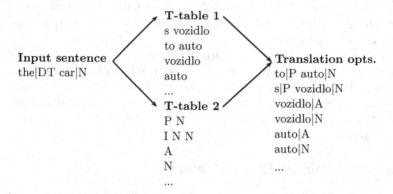

Fig. 2. Phrase expansion in factored models. Options can be used multiple times, such as "DT N"→"N", or completely discarded if they are inconsistent, such as "DT N"→"I N N".

can cause a combinatorial explosion. Generating all of them is costly in terms of computational time and memory. During decoding, pruning will likely discard good hypotheses, as stacks will be filled with too many factor combinations.

Consider the example shown in Figure 1. This particular translation system uses two translation tables (lemma→lemma, tag→tag) and one generation table (target lemma|tag→form). For each source phrase, the decoder generates all possible translations of the lemmas. Then it combines each lemma with all consistent translations of the tags (resulting in a subset of Cartesian product of the lemma/tag options). Finally, each combination generates zero, one or more (phrases of) target forms. The first two expansions are illustrated in Figure 2.

An expansion is considered *consistent* if the target side has the same length (we are filling in additional factors of a given target phrase) and if the shared factors match.

If the steps share some of the output factors, the order of application of mapping step plays a significant role. In this case, only consistent translation options can be generated during expansion. This restriction has two effects for phrase expansion. First, it limits the number of translation options generated from the existing options. Second, it discards those partial options for which no consistent expansion exists.

For example, suppose that we define two separate translation steps:

1. lemma→lemma
2. tag→lemma

If the steps are applied in this order, the decoder will first generate possible lexical translations. The second step then ensures consistency with the source morphology (e.g. disambiguate between translating English words as nouns or verbs). If we invert the order, the tags will be "translated" first, resulting in an explosion of translation options (the decoder has to produce all lemmas that the source tag can be mapped to).

2.2   Factors

We process our data with Treex,[1] a modular framework for natural language processing. We use tagging and shallow and deep parsing on both sides (English and Czech), enabling us to work with a wide range of linguistic information. Detailed documentation of the discussed factors can be found in PDT[2] and PCEDT[3].

From the morphological layer, we extract the *lemma* and *morphological tag* of each word. Czech lemmas are disambiguated. English tags come from the Penn Treebank tagset [13], Czech tags use the positional system of the Prague Dependency Treebank 2.0 [14]. This tagset is much richer than the English counterpart—about a half of the 4000 possible tags were actually seen in a corpus.

On the surface-syntactic (so-called analytical) layer, words are annotated with their *analytical function*. Examples of analytical functions include Sb for subject or Pred for predicate.

The tectogrammatical layer describes the deep syntactic structure of sentences. It contains annotation of phenomena that border on the syntax and semantics, such as semantic roles, (grammatical) coreference or valency. We draw a number of factors directly from the annotation:

t-lemma Tectogrammatical lemma, i.e. the deep-syntactic lemma.

functor Describes syntactic-semantic relation of a node to its parent node. Its possible values include ACT (actor), PAT (patient) or ADDR (addressee).

grammatemes A set of factors that describe meaning-bearing morphological properties of t-nodes. We extracted the following categories:

gender Grammatical gender.

number Grammatical number.

sempos Semantic part of speech. This factor classifies autosemantic words into 4 classes: nouns, adjectives, adverbs and verbs (with their respective subcategories).

tense This attribute specifies the tense of verbs.

verbmod This factor indicates the verb mood.

negation Indicator of negation.

formeme Contains a projection of some morpho-syntactic information from the morphological and analytical layers.

2.3   Software

We use a common set of tools for statistical MT: GIZA++ [15] for computing word alignments, SRILM [16] for creating language models and the Moses toolkit [17] for decoding.

---

[1] http://ufal.mff.cuni.cz/treex/
[2] http://ufal.mff.cuni.cz/pdt2.0/doc/pdt-guide/en/html/
[3] http://ufal.mff.cuni.cz/pcedt2.0/en/

3   Space of Factored Configurations

In this section, we describe the space of possible factored configurations. A taxonomy of factored systems was proposed by [18]. From this perspective, our work considers Direct (one translation step) and Single-Step (multiple mapping steps within a single search) factored setups.

3.1   Enumeration of Possible Configurations

We can partially order factored setups by the number of mapping steps and explore them in a canonical order (T, TT, TG, TTT,...). Each of these setups can use many combinations of factors and mappings.

Even for one mapping step (this must be a translation step), there are many possible configurations: on the source side, it must use at least one of the lexical factors, but it can also include any number of additional factors, leading to an exponential number of possibilities.[4] The situation on the target side is similar. An exhaustive evaluation is thus intractable even with one translation step.

When multiple mapping steps are involved, the number of configurations explodes further. We analyzed configurations of two mapping steps and the number of factors on each side restricted to 2. Let the first factor (denoted by 0) be the surface form on both sides.

Table 1 shows the viable configurations. For each combination, we provide an example of a potentially good translation system to demonstrate that these combinations warrant exploration. The last column contains our estimate of the number of possible combinations of factored values, given our setting: 12 factors on top of the surface forms, two of which are lexically informative (lemma, tlemma).

We found 13 possible factored scenarios for two mapping steps and estimate that 1142 systems would have to be evaluated if our goal was to explore the space exhaustively. These results demonstrate that an exhaustive search is unrealistic even in this extremely restricted setting. If we hope to find good configurations in this space automatically, we have to guide our search somehow.

4   Evaluation of Factored Configurations

In order to navigate in this space, ideally, we would hope to find a heuristic that would help us predict the translation quality without much computation. But let us back off to a simpler question—can we even reliably compare two factored systems?

The simplest way of evaluating two MT systems is to translate a test set using both of them and compare the achieved BLEU scores [19]. This procedure however disregards the fact that model tuning is randomized. Factored systems can have many parameters (usually 5 for each translation step, 2 for generation

---

[4] The number of configurations is proportional to the size of the *power set* of the set of source factors S, i.e. $2^{|S|}$.

Table 1. Enumeration of configurations with two mapping steps

| Mapping Steps | | Sample Plausible Setup | | Estimated |
First	Second	First Mapping Step	Second Mapping Step	Combinations
0→0	1→0	form→form	tag→form	12
0→1	1→0,1	form→POS	lemma→form|POS	48
1→0	0→0	lemma→form	form→form	2
1→0	0→0,1	lemma→form	form→form|tag	24
1→1	0→0,1	tag→tag	form→form|tag	144
1→0,1	0→0	lemma→form|POS	form→form	24
1→0,1	0→1	lemma→form|POS	form→POS	24
0→0,1	1→0	form→form|tag	lemma→form	144
0→0,1	1→1	form→form|tag	tag→tag	144
0,1→0	0→0,1	form|tag→form	form→form|tag	144
0,1→0	1→0,1	form|lemma→form	lemma→form|tag	144
0,1→1	0→0,1	form|tag→lemma	form→form|lemma	144
0,1→1	1→0,1	form|lemma→lemma	lemma→form|lemma	144

steps), adding dimensions to the weight space and thus increasing the effects of randomness.

Our task also requires us to compare systems which are very close in performance. Can we distinguish the random variance in tuning from a true difference between systems?

We evaluated two algorithms for tuning, minimum error rate training [20] and pairwise-ranked optimization [21]. MERT uses random starting points to avoid reaching local optima. PRO samples its training examples randomly (pairs of translations with high differences in BLEU), but unlike MERT, it is empirically very stable.

In these experiments, we used CzEng 0.9 [22], a richly annotated parallel Czech-English corpus. We trained on a random subset of 200 thousand sentences, development a test data were random 1000-sentence samples from the respective sections of the same corpus.

We used two alternative decoding paths, one that translated form|factor → form and another that only mapped form → form (as a back-off). Each of these paths represents five weights that need to be optimized.

Table 2 shows the evaluated factors. We ran MERT for each factor three times. We can see that differences in BLEU scores in MERT runs are often as high as 0.5 absolute point, which is roughly the same as the improvement we expect from incorporating a useful factor in the system.

Furthermore, if we disregard statistical significance and look simply at the BLEU scores, we might draw very different conclusions depending on which MERT run we consider. We can even entirely invert the ordering of some factors:

- tag (25.07) > functor (25.03) > sempos (25.01) > baseline (24.66)
- baseline (25.16) > sempos (25.01) > functor (24.99) > tag (24.61)

Table 2. BLEU scores achieved by multiple MERT runs and PRO

Factor	BLEU (3 runs)	Mean	StDev	BLEU-PRO
child(0)→tlemma	24.75, 25.12, 25.43	**25.10**	0.28	24.82
functor	24.99, 25.03, 25.26	**25.09**	0.12	24.56
—	24.66, 25.15, 25.16	24.99	0.23	24.84
formeme	24.58, 25.08, 25.09	24.92	0.24	24.79
sempos	24.75, 25.00, 25.01	24.92	0.12	24.90
tag	24.61, 24.74, 25.07	24.81	0.19	24.90
lemma	24.34, 24.80, 24.88	24.67	0.24	24.81

Moreover, if we use just one MERT run and do a statistical significance test, specifically the bootstrap resampling as introduced by [23], the confidence intervals are so wide that we cannot consider any two systems to be significantly different.[5]

Regarding PRO, our experiments confirmed the stability of the algorithm. However, notice that the order of factors achieved by MERT and PRO is very different. Also, even though MERT is much less stable, it often finds a better set of weights than PRO.

We therefore decided to evaluate all of our experiments by running MERT several (3) times and calculating mean and standard deviation. However we cannot rely on these scores to guide a fully automatic search.

5   Estimating Complexity of Factored Setups

We developed a tool that estimates the number of partial translation options (i.e. translation with factors partially filled in) generated by each step. This estimation is done without decoding and only uses small sample phrase tables. An automatic search for configurations can use this estimate of complexity to prevent training of unrealistic setups. The estimates for individual steps can provide further insights for analysis.

If we estimate the average number of options for a single step, we cannot use the arithmetic mean because extracted phrases obey the power law in a sense: phrases that occur only once have only one translation in the phrase table. These phrases actually make up most of the phrase table but in fact they are almost never used. On the other hand, very frequent phrases tend to have a large number of translations. We therefore use a frequency-weighted average (t_i denotes the number of translations and f_i is the source phrase frequency):

$$avg = \frac{\sum_i f_i \cdot t_i}{\sum_i f_i} \tag{1}$$

---

[5] Recently, pair-wise significance tests that sample from multiple runs of the optimizer have been suggested [24].

When multiple steps are used, the decoder first generates partial options according to the first step and then expands them in the following steps. Each expansion must be consistent. An example of an expansion was shown in Figure 2.

To approximate this procedure of expansion, we factor each source phrase according to the length of translations and the values of fixed target factors. So each source phrase effectively becomes several source phrases. We then count their translation options separately.

So far we have discussed how to approximate the number of translation options for translation steps. Generation steps are slightly different as generation is done word-by-word. This implies that for a phrase of length k, there will be about avg^k translation options. Instead of k we use the average phrase length according to the first translation table.

When combining the translation and generation steps to obtain an estimate of the number of full translation options, we simply multiply the individual estimates. For each step, we also account for the observed difference in the average number of translation options between tables trained on the full data set and our sample tables (this only needs to be computed once). In our case the ratio was roughly 1.3.

We did not find a way to estimate the effect of *implicit* pruning: for example, we might have a step that translates tag $\rightarrow$ tag and a following translation step form $\rightarrow$ form|tag. Some of the previously generated tags will be discarded (if the second step did not generate them) and some of the expansions as well (if their tag was not generated by the previous step). This is the primary source of errors in our estimates, especially for generation steps.

5.1   Evaluation

We evaluated the estimation accuracy for several factored systems. We modified Moses to emit the average number of translation options and compared the results obtained when translating a test set with our prediction. Table 3 shows the results ("t:" and "g:" distinguish translation and generation steps).

As we progress to more complicated setups, the results start to suffer from the deficiency of the heuristic (as discussed above). However, while the absolute values are wrong, the ordering of the setups is correct. This allows us to use the heuristic to pinpoint difficult configurations and the problematic steps in them. For example, the last setup (with functors) ran many times longer than the identical configuration with tags (despite the fact that there are far more tags than functors). This difference is correctly discovered by the heuristic.

6   Experiments

In this section, we describe the conducted experiments. Because of the discussed difficulties—the absence of a reliable method for evaluation, the small and insignificant differences in BLEU and the enormous number of possible configurations—we did not carry out a fully automatic exploration of the space of factored setups. Instead, we conducted several sets of experiments in a few targeted research directions; given a small set of factors, a fixed setting and the predictor of setup complexity, we were able to carry out a "semi-automatic" search.

The main source of data for our experiments is CzEng in its latest release 1.0 [25]. It is a richly annotated Czech-English parallel corpus with over 15 million parallel sentences from 7 different domains. We do not use the whole CzEng in the experiments (otherwise the duration of experiments would prohibit *any* search), we limit ourselves to the news domain as the source of both parallel data for translation model training and target-side monolingual data for language modeling.

Our development data (for system tuning) are the test set for WMT11 translation task [26]. For final evaluation of each system, we use WMT test set for 2012. The evaluation data for WMT are news articles, hence the choice of training data. Table 4 shows basic statistics of the data.

6.1   Additional Source Factor

We evaluated the usefulness of all additional factors in combination with the translation of surface forms. The setup was the following:

1. form|*extra* → form
2. (form → form)

All factors were evaluated with and without the alternative path. Results are summarized in Table 5. Baseline system is denoted by '—'. The ± sign denotes the standard deviation over 3 runs of the optimizer. MERT was used for tuning of the systems.

We still see only very little improvements over the baseline BLEU, complicated by variance that makes most of the differences insignificant. Even so, several factors stand out in both scenarios as potentially valuable for modeling the English-Czech translation.

Table 3. Estimation of the number of translation options per phrase

Mapping Steps	Estimation	Moses Avg.
t:form→form	$1.3 \cdot 5.38 \doteq 7$	**12**
t:tag→tag +	$1.3 \cdot 11.28 \cdot$	
+ t:form→form|tag	$1.3 \cdot 1.28 \doteq 24$	**85**
t:lemma→lemma +	$1.3 \cdot 5.23 \cdot$	
+ t:tag→tag +	$1.3 \cdot 57.25 \cdot$	**173**
+ g:lemma|tag→form	$1.3 \cdot 1.13 \doteq 655$	
t:lemma→lemma +	$1.3 \cdot 5.19 \cdot$	
+ t:functor→functor +	$1.3 \cdot 52.48 \cdot$	**5153**
+ g:lemma|functor→form	$1.3 \cdot 16.54 \doteq 9903$	

In the first column, factors that lead to data sparsity were penalized due to the absence of a back-off. Formeme stands out as the most prominent example, with the BLEU score 2.41 and almost no deviation; all MERT runs converged in a few iterations. Adding this factor diluted the data so much that translation became impossible. Factors that achieved high scores in this column can be (relatively) safely added to translation systems: they do not make the data much sparser and increase translation quality. The best factors are highlighted: analytical function, negation, tense. Grammatical number and tag are also potentially useful.

Analytical function provides roles of English words (subject, predicate etc.) which help disambiguate target-side morphology—in Czech, subjects are almost always in nominative case while objects frequently appear in accusative or dative case.

Tense helped disambiguate verb forms mainly when the predicate contained an auxilliary verb specifying future or past tense. Our annotation assigns this tense also to the main verb (e.g. "will|post go|post") making its translation easier even when it is translated independently (as a one-word phrase).

We suspect that the benefit of the negation attribute is more due to the annotation rules—nouns are (almost always) assigned an empty value, while verbs, adjectives and adverbs are assigned either "neg0" or "neg1". Thus the negation attribute provides a coarse-level PoS tagging useful for modelling the overall sentence structure.

In the second column, even factors that introduce some degree of data sparsity can achieve high scores—they may help in modeling some rare but difficult phenomena. In the situations where the additional information is not helpful, the alternative path maintains good quality of translation. Functor, analytical function and tense appear to be the most promising factors according to this column.

We used the results to create a combination of factors that we then evaluated separately. As it is not clear which back-offs should be used when multiple factors are combined, we evaluated several approaches; the results are summarized in Table 6 and demonstrate quite clearly that the simplest back-off (just translating surface forms) works best—the overall BLEU score is the highest and this setup was also the most stable one.

6.2 Multiple Mapping Steps

In this section, we evaluated a typical factored scenario with several factors. The scenario consists of two consecutive translation steps: lemma → lemma and

Table 4. Statistics of the data used in experiments

Data Set	Data Source	Sentences	En Words	Cs Words
Training	CzEng 1.0 news	197053	4641026	4193078
Development	WMT11 test set	3003	74822	65602
Test	WMT12 test set	3003	72955	65306

Table 5. BLEU scores of configurations with 1 translation step

Factor	Single Path	+Alternative
—	9.93±0.03	—
afun	**10.08±0.08**	**10.11±0.09**
formeme	2.41±0.01	9.95±0.02
functor	9.08±0.08	**10.07±0.08**
gender	9.70±0.05	9.87±0.06
lemma	9.93±0.08	9.66±0.30
negation	**10.05±0.03**	9.99±0.02
number	10.00±0.03	9.96±0.08
person	9.92±0.03	9.79±0.18
sempos	9.93±0.06	9.95±0.16
tag	10.00±0.07	9.95±0.11
tense	**10.06±0.05**	**10.05±0.06**
tlemma	8.62±0.06	9.99±0.15
verbmod	9.56±0.04	9.94±0.10

one additional factor to its counterpart. This is followed by a generation step that takes the lemma and the additional factor and generates surface form on the target side. All of the factors have a language model on the target side. An alternative path maps surface form directly to all three target factors.

This setup has been used with tags in the past and improvements have been reported on similarly small datasets. Our results are shown in Table 7. Systems without a score ran for too long (one MERT iteration took over a day); this was correctly predicted by our complexity heuristic.

We achieved a large gain in BLEU (roughly 1.1 point absolute) when we used morphological tag as the additional factor, which confirms previous findings. However, no other factor was beneficial in this scenario.

Table 6. Back-off strategies and achieved BLEU scores

Translation Steps	BLEU
form|afun → form : : form|functor → form : : form|tense → form	10.00±0.29
form|afun|functor|tense → form : : form|afun → form	10.08±0.10
form|afun|functor → form : : form|tense → form	10.10±0.08
form|afun|functor|tense → form : : form → form	**10.24±0.02**

Table 7. BLEU scores of systems with 2 translation and 1 generation steps

Factor	BLEU	Prediction of Complexity
formeme	9.91±0.05	4573
—	9.93±0.03	7
tag	**11.05±0.03**	655
functor	—	9903
sempos	—	38412
tense	—	13607

7 Discussion

7.1 Experimental Results

We were able to improve translation performance (0.3 BLEU absolute) when using a single translation step by combining well-performing factors on the source side. We showed that analytical function, tense and functors as used in the PCEDT annotation are the most useful from a wide range of attributes for modeling factored phrase-based transfer of English into Czech.

We also evaluated a scenario that consists of multiple mapping steps. Unfortunately, similarly to the previous set of experiments, we were unable to identify any new useful factors, so even though our improvement in BLEU score is quite large (over 1.1 points), our findings are not new.

7.2 Search for Factored Configurations

It seems that finding the correct combination of steps and factors is not a task that an algorithm can solve, especially not by brute force—the number of possibilities explodes no matter which direction of exploration we take. A clever automatic search in the space of configurations does not seem feasible due to the low reliability of automatic MT evaluation and frequent large variance in scores across different optimization runs.

We believe it is possible to search for factored configurations semi-automatically given a particular research goal—the methods and tools that we developed can assist in selecting the most suitable factored setup from a limited number of possibilities.

8 Conclusion

We provided an analysis of the paradigm of factored machine translation. We described the complexity of the space of configurations. We proposed a heuristic that can successfully predict which factored setups are too complex to be feasible. We carried out a "semi-automatic" search for factored configurations in several directions and evaluated the results.

In the future, we would like to apply the developed machinery to more complex setups and richer sets of factors but obviously with a manual guidance. We would also like to improve the precision of the heuristic for complexity estimation.

References

1. Koehn, P., Och, F.J., Marcu, D.: Statistical phrase-based translation. In: HLT/NAACL (2003)
2. Koehn, P., Hoang, H.: Factored translation models. In: EMNLP-CoNLL, pp. 868–876. ACL (2007)
3. Bojar, O.: English-to-Czech Factored Machine Translation. In: Proc. of ACL WMT, Prague, Czech Republic, pp. 232–239. ACL (2007)
4. Avramidis, E., Koehn, P.: Enriching morphologically poor languages for statistical machine translation. In: Proc. of ACL/HLT, Columbus, Ohio, pp. 763–770. ACL (2008)
5. Badr, I., Zbib, R., Glass, J.: Segmentation for English-to-Arabic statistical machine translation. In: Proc. of ACL/HLT Short Papers, Columbus, Ohio, pp. 153–156. ACL (2008)
6. Ramanathan, A., Choudhary, H., Ghosh, A., Bhattacharyya, P.: Case markers and morphology: addressing the crux of the fluency problem in English-Hindi SMT. In: Proc. of ACL/IJCNLP, Suntec, Singapore, vol. 2, pp. 800–808. ACL (2009)
7. Koehn, P., Haddow, B., Williams, P., Hoang, H.: More linguistic annotation for statistical machine translation. In: Proc. of WMT and MetricsMATR, Uppsala, Sweden, pp. 115–120. ACL (2010)
8. Yeniterzi, R., Oflazer, K.: Syntax-to-Morphology Mapping in Factored Phrase-Based Statistical Machine Translation from English to Turkish. In: Proc. of ACL, Uppsala, Sweden, pp. 454–464. ACL (2010)
9. Birch, A., Osborne, M., Koehn, P.: CCG Supertags in Factored Statistical Machine Translation. In: Proc. of ACL WMT, Prague, Czech Republic, pp. 9–16. ACL (2007)
10. Stymne, S.: German Compounds in Factored Statistical Machine Translation. In: Nordström, B., Ranta, A. (eds.) GoTAL 2008. LNCS (LNAI), vol. 5221, pp. 464–475. Springer, Heidelberg (2008)
11. Koehn, P., Schroeder, J.: Experiments in domain adaptation for statistical machine translation. In: Proc. of ACL WMT, Prague, Czech Republic, pp. 224–227. ACL (2007)
12. Niehues, J., Waibel, A.: Domain adaptation in statistical machine translation using factored translation models. In: EAMT (2010)
13. Santorini, B.: Part-of-Speech Tagging Guidelines for the Penn Treebank Project. In: University of Pennsylvania, School of Engineering and Applied Science, Dept. of Computer and Information Science, Philadelphia (1990)
14. Hajič, J., Panevová, J., Hajičová, E., Sgall, P., Pajas, P., Štěpánek, J., Havelka, J., Mikulová, M., Žabokrtský, Z., Ševčíková Razímová, M.: Prague Dependency Treebank 2.0. LDC2006T01 (2006) ISBN: 1-58563-370-4
15. Och, F.J., Ney, H.: Improved statistical alignment models. In: ACL. ACL (2000)
16. Stolcke, A.: SRILM - an extensible language modeling toolkit. In: Proc. of IC-SLP2002 - INTERSPEECH. ISCA, Denver (2002)

17. Koehn, P., Hoang, H., Birch, A., Callison-Burch, C., Federico, M., Bertoldi, N., Cowan, B., Shen, W., Moran, C., Zens, R., Dyer, C., Bojar, O., Constantin, A., Herbst, E.: Moses: Open Source Toolkit for Statistical Machine Translation. In: Proc. of ACL: Demo and Poster Sessions, Prague, Czech Republic, pp. 177–180. ACL (June 2007)
18. Bojar, O., Jawaid, B., Kamran, A.: Probes in a Taxonomy of Factored Phrase-Based Models. In: Proc. of ACL WMT, Montréal, Canada, pp. 253–260. ACL (2012)
19. Papineni, K., Roukos, S., Ward, T., Zhu, W.J.: BLEU: A method for automatic evaluation of machine translation. In: ACL, pp. 311–318. ACL (2002)
20. Och, F.J.: Minimum error rate training in statistical machine translation. In: Proc. of ACL, Sapporo, Japan, pp. 160–167. ACL (2003)
21. Hopkins, M., May, J.: Tuning as ranking. In: EMNLP, pp. 1352–1362. ACL (2011)
22. Bojar, O., Žabokrtský, Z.: CzEng0.9: Large Parallel Treebank with Rich Annotation. Prague Bulletin of Mathematical Linguistics 92 (2009)
23. Koehn, P.: Statistical Significance Tests for Machine Translation Evaluation. In: Proc. of EMNLP, Barcelona, Spain (2004)
24. Clark, J.H., Dyer, C., Lavie, A., Smith, N.A.: Better hypothesis testing for statistical machine translation: Controlling for optimizer instability. In: Proc. of ACL (Short Papers), pp. 176–181. ACL (2011)
25. Bojar, O., Žabokrtský, Z., Dušek, O., Galuščáková, P., Majliš, M., Mareček, D., Maršík, J., Novák, M., Popel, M., Tamchyna, A.: The Joy of Parallelism with CzEng 1.0. In: Proc. of LREC, İstanbul, Turkey, pp. 3921–3928. ELRA (2012)
26. Callison-Burch, C., Koehn, P., Monz, C., Zaidan, O.: Findings of the 2011 Workshop on Statistical Machine Translation. In: Proc. of ACL WMT, Edinburgh, Scotland, pp. 22–64. ACL (2011)

Domain Adaptation in Statistical Machine Translation Using Comparable Corpora: Case Study for English-Latvian IT Localisation

Mārcis Pinnis, Inguna Skadiņa, and Andrejs Vasiļjevs

Tilde
{marcis.pinnis,inguna.skadina,andrejs}@tilde.lv

Abstract. In the recent years, statistical machine translation (SMT) has received much attention from language technology researchers and it is more and more applied not only to widely used language pairs, but also to under-resourced languages. However, under-resourced languages and narrow domains face the problem of insufficient parallel data for building SMT systems of reasonable quality for practical applications. In this paper we show how broad domain SMT systems can be successfully tailored to narrow domains using data extracted from strongly comparable corpora. We describe our experiments on adaptation of a baseline English-Latvian SMT system trained on publicly available parallel data (mostly legal texts) to the information technology domain by adding data extracted from in-domain comparable corpora. In addition to comparative human evaluation the adapted SMT system was also evaluated in a real life localisation scenario. Application of comparable corpora provides significant improvements increasing human translation productivity by 13.6% while maintaining an acceptable quality of translation.

Keywords: comparable corpus, statistical machine translation, software localisation, under-resourced languages, Latvian, narrow domain.

1 Introduction

In the recent years, SMT has become the dominant paradigm not only for widely-used languages, but also for under-resourced languages. However, lack of sufficiently large parallel corpora limits the building of reasonably good quality machine translation (MT) solutions for these languages. Because of this reason there is a growing interest in research of comparable corpora as a source for extracting data useful for training MT systems.

In this paper we describe our research on using comparable corpora for adaptation of an SMT system for translation from English into the under-resourced language: Latvian. The Latvian language belongs to the Baltic language group of the Indo-European language family, with less than 2.5 million speakers worldwide. It is a morphology rich language with a rather free word order. Since there is a relatively small number of Latvian speakers, content in Latvian is also limited. Only few bi/multilingual parallel corpora contain Latvian, among them the largest are JRC-Acquis [21], DGT-TM [22], and Opus [24].

A. Gelbukh (Ed.): CICLing 2013, Part II, LNCS 7817, pp. 224–235, 2013.

These corpora have sufficient data only for building legal domain SMT systems with high BLEU scores when evaluated on in-domain texts [18]. However, these systems are not suitable for other domains, e.g., automotive or information technology (IT).

Although quality of MT systems has been criticized a lot, due to a growing pressure on efficiency and cost reduction, MT receives more and more interest from the localisation industry. Localization companies have to increase volume of translation and decrease costs of services in order to remain competitive in the market.

In this paper we address both these challenges. We show that, for language pairs and domains where there is not enough parallel data available (1) in-domain comparable corpora can be used to increase translation quality and (2) if comparable corpora are large enough and can be classified as strongly comparable (i.e., have many similar text fragments, sentence pairs or phrases overlapping between the different languages) then the trained SMT systems applied in the localisation process increase productivity of human translators.

In the next chapters we present our work on English-Latvian SMT system adaptation to the IT domain: building a comparable corpus, extracting semi-parallel sentences and terminological units from the comparable corpus, and adapting the SMT system to the IT domain with the help of the extracted data. We describe evaluation results demonstrating that data extracted from comparable corpora can significantly increase BLEU score over a baseline system. Results from the application of the adapted SMT system in a real life localisation task are presented showing that SMT usage increased the productivity of human translators by 13.6%.

2   Related Work

2.1   Comparable Corpora in Machine Translation

Applicability of comparable corpora for MT is a relatively new field of research. While methods on how to use parallel corpora in MT are well studied (e.g. [6]), methods and techniques for comparable corpora have not been thoroughly investigated.

The latest research has shown that adding extracted parallel lexical data from comparable corpora to the training data of a SMT system improves the system's performance in view of word coverage [5]. It has been also demonstrated that language pairs with little parallel data can benefit the most from exploitation of comparable corpora [8]. Munteanu and Marcu [9] achieved significant performance improvements from large comparable corpora of news feeds for English, Arabic and Chinese over a baseline MT system, trained on existing available parallel data.

However, most of such experiments are performed with widely used language pairs, such as French-English [1, 2], Arabic-English [2] or English-German [23], while for under-resourced languages (e.g., Latvian), possible exploitation of comparable corpora for machine translation tasks is less studied [17].

2.2   Machine Translation in Localisation

Different aspects of post-editing and machine translatability have been researched since the 90-ies (a comprehensive overview has been provided by O'Brien [11]). Recently several productivity tests have been performed in translation and localisation industry settings at Microsoft [16], Adobe [4], Autodesk [15] and Tilde [19]. In all these tests authors report productivity increase. However, in many cases they also indicate on significant performance differences in the various translation tasks. Also increase of the error score for translated texts is reported.

As the localization industry experiences a growing pressure on efficiency and performance, some developers have already integrated MT in their computer-assisted translation (CAT) products, e.g. SDL Trados, ESTeam TRANSLATOR and Kilgrey memoQ.

3   Collecting and Processing Comparable Corpus

3.1   Comparable Corpus

For our experiment we used an English-Latvian comparable corpus containing texts from the IT domain: software manuals and Web crawled data (consisting of IT product information, IT news, reviews, blogs, user support texts including also software manuals, etc.). The corpus was acquired in an artificial fashion in order to simulate a strongly comparable narrow domain corpus (that is, a corpus containing overlapping content in a significant proportion).

To get more data for our experiments we used two different approaches in creation of comparable corpus. Thus the corpus consists of two parts. The first part contains documents acquired from different versions of software manuals of a productivity software suite split in chunks of less than 100 paragraphs per document and aligned at document level with *DictMetric* tool [20]. As a very large number of alignments was produced, we filtered document pairs so that for each source and target language document there were no more than the top three alignments (for both languages separately) included.

The second part consists of an artificially created strongly comparable corpus from parallel data that is enriched with Web crawled non-comparable and weakly comparable data. The parallel data was split in random chunks from 40 to 70 sentences per document and randomly polluted with sentences from the Web crawled data from 0 to 210 sentences. The Web corpus sentences were injected in random positions in English and Latvian documents separately, thus heavily polluting the documents with non-comparable data. The Web crawled data was collected using the *Focussed Monolingual Crawler* (FMC) from the ACCURAT Toolkit [12]. The Web corpus consists of 232,665 unique English and 96,573 unique Latvian sentences. The parallel data before pollution contained 1'257,142 sentence pairs.

The statistics of the English-Latvian comparable corpus are given in Table 1. Note that the second part of the corpus accounts for 22,498 document pairs.

Table 1. Comparable corpus statistics

English documents	Latvian documents	Number of aligned document pairs	Number of aligned document pairs after filtering
27,698	27,734	385,574	45,897

Although, this comparable corpus has been artificially created, the whole process chain of system adaptation described in the following sections is the same for any comparable corpus, e.g., it can be applied to corpora automatically acquired from the Web.

3.2   Extraction of Semi-parallel Sentence Pairs

The parallel sentence extractor *LEXACC* [23] was used to extract semi-parallel sentences from the comparable corpus. Before extraction, texts were pre-processed – split into sentences (one sentence per line) and tokenized (tokens separated by a space).

Because the two parts of our corpus differ in terms of comparable data distribution and the comparability level, different confidence score thresholds were applied for extraction. The threshold was selected by manual inspection of extracted sentences so that most (more than 90%) of the extracted sentence pairs would be strongly comparable or parallel.

Table 2 shows information about data extracted from both parts of the corpus using the selected thresholds.

Table 2. Extracted semi-parallel sentence pairs

Corpus part	Threshold	Unique sentence pairs
First part	0.6	9,720
Second part	0.35	561,994

3.3   Extraction of Bilingual Term Pairs

We applied the ACCURAT Toolkit to acquire in-domain bilingual term pairs from the comparable corpus following the process thoroughly described in [13], which then were used in the SMT adaptation process. At first, the comparable corpus was monolingually tagged with terms and then terms were bilingually mapped. Term pairs with the confidence score of mapping below the selected threshold were filtered out. In order to achieve a precision of about 90%, we selected the confidence score threshold of 0.7. The statistics of both the monolingually extracted terms and the mapped terms are given in Table 3.

Table 3. Term tagging and mapping statistics

Corpus part	Unique monolingual terms		Mapped term pairs	
	English	Latvian	Before filtering	After filtering
First part	127,416	271,427	847	689
Second part	415,401	2,566,891	3,501	3,393

The term pairs were further filtered so that for each Latvian term only those English terms having the highest mapping confidence scores would be preserved. We used Latvian term to filter term pairs, because Latvian is a morphologically richer language and multiple inflective forms of a word in most cases correspond to a single English word form (although this is a "rude" filter, it increases the precision of term mapping to well over 90%).

As can be seen in Table 3, only a small part of the monolingual terms were mapped. However, this amount of mapped terms was sufficient for SMT system adaptation as described in the following sections. It should also be noted that in our adaptation scenario translated single-word terms are more important than multi-word terms as the adaptation process of single-word terms partially covers also the multi-word pairs that have been missed by the mapping process.

4   Building SMT Systems

We used the LetsMT! platform [25] based on the Moses tools [7] to build three SMT systems: the baseline SMT system (trained on publicly available parallel corpora), the intermediate adapted SMT system (in addition data extracted from comparable corpus was used) and the final adapted SMT system (in-domain terms integrated). All SMT systems have been tuned with minimum error rate training (MERT) [3] using in-domain (IT domain) randomly selected tuning data containing 1,837 unique sentence pairs.

4.1   Baseline SMT System

For the English-Latvian baseline system, the DGT-TM parallel corpora of two releases (2007 and 2011) were used. The corpora were cleaned in order to remove corrupt sentence pairs and duplicates. As a result, for training of the baseline system a total of 1'828,317 unique parallel sentence pairs were used for translation model training and a total of 1'736,384 unique Latvian sentences were used for language model training.

4.2   Domain Adaptation through Integration of Data Extracted from Comparable Corpora

In order to adapt the SMT system for the IT domain, the extracted in-domain semi-parallel data (both sentence pairs and term pairs) were added to the parallel corpus used for baseline SMT system training. The whole parallel corpus was then cleaned and filtered with the same techniques as for the baseline system. The statistics of the filtered corpora used in SMT training of the adapted systems (intermediate and final) are shown in Table 4.

Table 4. Training data for adapted SMT systems

	Parallel corpus (unique pairs)	Monolingual corpus
DGT-TM (2007 and 2011) sentences	1'828,317	1'576,623
Sentences from comparable corpus	558,168	1'317,298
Terms form comparable corpus	3,594	3,565

Table 4 shows that there was some sentence pair overlap between the DGT-TM corpora and the comparable corpora content. This was expected as DGT-TM covers a broad domain and may contain documents related to the IT domain. For language modelling, however, the sentences that overlap in general domain and in-domain monolingual corpora have been filtered out from the general domain monolingual corpus. Therefore, the DGT-TM monolingual corpus statistics between the baseline system and the adapted system do not match.

After filtering, a translation model was trained from all available parallel data and two separate language models were trained from the monolingual corpora:

- Latvian sentences from the DGT-TM corpora were used to build the general domain language model;
- The Latvian part of extracted semi-parallel sentences from in-domain comparable corpus were used to build the in-domain language model.

4.3 Domain Adaptation through Terminology Integration

To make in-domain translation candidates distinguishable from general domain translation candidates, the phrase table of the domain adapted SMT system was further transformed to a term-aware phrase table [14] by adding a sixth feature to the default five features used in Moses phrase tables. The following values were assigned to this sixth feature:

- "2" if a phrase in both languages contained a term pair from the list of extracted term pairs.
- "1" if a phrase in both languages did not contain any extracted term pair; if a phrase contained a term only in one language, but not in both, it received "1" as this case indicates of possible out-of-domain (wrong) translation candidates;

In order to find out whether a phrase contained a given term or not, every word in the phrase and the term itself was stemmed. Finally, the transformed phrase table was integrated back into the adapted SMT system.

5   Automatic and Comparative Evaluation

5.1   Automatic Evaluation

The evaluation of the baseline and both adapted systems was performed with four different automatic evaluation metrics: BLEU, NIST, TER and METEOR on 926 unique IT domain sentence pairs. Both, case sensitive and case insensitive, evaluations were performed. The results are given in Table 5.

Table 5. Automatic evaluation results

System	Case sensitive?	BLEU	NIST	TER	METEOR
Baseline	*No*	*11.41*	*4.0005*	*85.68*	*0.1711*
	Yes	10.97	3.8617	86.62	0.1203
Intermediate	*No*	*56.28*	*9.1805*	*43.23*	*0.3998*
adapted system	Yes	54.81	8.9349	45.04	0.3499
Final adapted	*No*	**56.66**	**9.1966**	**43.08**	**0.4012**
system	Yes	**55.20**	**8.9674**	**44.74**	**0.3514**

The automatic evaluation shows a significant performance increase of the improved systems over the baseline system in all evaluation metrics. Comparing two adapted systems, we can see that making the phrase table term-aware (*Final adapted system*) yields further improvement over intermediate results after just adding data extracted from comparable corpora (*Intermediate adapted system*). This is due to better terminology selection in the fully adapted system. As terms comprise only a certain part of texts, the improvement is limited.

5.2   Comparative Evaluation

For the system comparison we used the same test corpus as for automatic evaluation and compared the baseline system against the adapted system. Figure 1 summarizes the human evaluation results using the evaluation method described in [18]. From 697 evaluated sentences, in 490 cases (70.30±3.39%) output of the improved SMT system was chosen as a better translation, while in 207 cases (29.70±3.39%) users preferred the translation of the baseline system. This allows us to conclude that for IT domain texts the adapted SMT system provides better translations as the baseline system.

System 1 total (A):	207	Params	P ± err	Lower	Upper
System 2 total (B):	490	N = A+B	29.70 ± 3.39	26.31	33.09
Total:	697	K = A			
		N = A+B	70.30 ± 3.39	66.91	73.69
		K = B			

26%	7%	67%

Fig. 1. System comparison by total points (System 1 – baseline, System 2 – adapted system)

Figure 2 illustrates the evaluation on sentence level: for 35 sentences we can reliably say that the adapted SMT system provides a better translation, while only for 3 sentences users preferred the translation of the baseline system. It must be noted that in this figure we present the results only for those sentences for which there was a statistically significant preference to the first or second system by the evaluators.

System 1 (A):	3	Params	P ± err	Lower	Upper
Tie (C):	6	N = A+B+C	6.82 ± 7.45	0.00	14.27
System 2: (B)	35	K = A			
Undefined:	882	N = A+B+C	79.55 ± 11.92	67.63	91.46
Total:	926	K = B			
		Params	P ± err	Lower	Upper
		N = A+B	7.89 ± 8.57	0.00	16.47
		K = A			
		N = A+B	92.11 ± 8.57	83.53	100.00
		K = B			

7%	14%	12%	68%

17%	84%

Fig. 2. System comparison by count of the best sentences (System 1 – baseline, System 2 – adapted system)

6   Evaluation in Localisation Task

The main goal of this evaluation task was to evaluate whether integration of the adapted SMT system in the localisation process allows increasing the output of translators in comparison to the efficiency of manual translation. We compared productivity (words translated per hour) in two real life localisation scenarios:

- Translation using translation memories (TM's) only.
- Translation using suggestions of TM's and the SMT system that is enriched with data from comparable corpus.

6.1   Evaluation Setup

For tests 30 documents from the IT domain were used. Each document was split into two parts. The length of each part of a document was 250 to 260 adjusted words on average, resulting in 2 sets of documents with about 7,700 words in each set.

Three translators with different levels of experience and average performance were involved in the evaluation cycle. Each of them translated 10 documents without SMT support and 10 documents with integrated SMT support. The SDL Trados translation tool was used in both cases.

The results were analysed by editors who had no information about techniques used to assist the translators. They analysed average values for translation performance (translated words per hour) and calculated an error score for translated texts. Individual productivity of each translator was measured and compared against his or her own productivity. An error score was calculated for every translation task by counting errors identified by an editor and applying a weighted multiplier based on the severity of the error type (1):

$$ErrorScore = \frac{1000}{n} \sum_i w_i e_i \tag{1}$$

where n is a number of words in translated text, e_i is a number of errors of type i, w_i is a coefficient (weight) indicating the severity of type i errors. Depending on the error score the translation gets a translation quality grade (*Superior, Good, Mediocre, Poor* or *Very poor*) assigned (Table 6).

Table 6. Quality grades based on error scores

Superior	Good	Mediocre	Poor	Very poor
0...9	10...29	30...49	50...69	>70

6.2   Results

Usage of MT suggestions in addition to TM's increased the productivity of the translators on average from 503 to 572 words per hour (13.6% improvement). There were significant differences in the results of different translators from performance increase by 35.4% to decreased performance by 5.9% for one of the translators (see Table 7). Analysis of these differences requires further studies but most likely they are caused by working patterns and the skills of individual translators. The average productivity for all the translators has been calculated using the formula (2).

$$Productivity\ (scenario) = \frac{\sum_{Text=1}^{N} Adjusted\ words(Text,scenario)}{\sum_{Text=1}^{N} Actual\ time(Text,scenario)} \qquad (2)$$

Table 7. Results of productivity evaluation

Translator	Scenario	Actual productivity	Productivity increase or decrease	Standard deviation of productivity
Translator 1	TM	493.2	35.39%	110.7
	TM+MT	667.7		121.8
Translator 2	TM	380.7	13.02%	34.2
	TM+MT	430.3		38.9
Translator 3	TM	756.9	-5.89%	113.8
	TM+MT	712.3		172.0
Average	TM	503.2	13.63%	186.8
	TM+MT	571.9		184.0

According to the standard deviation of productivity in both scenarios (without MT support 186.8 and with MT support 184.0) there were no significant performance differences in the overall evaluation (see Table 8). However, each translator separately showed higher differences in translation performance when using the MT translation scenario.

The overall error score (shown in Table 8) increased for one out of three translators. Although the total increase in the error score for all translators combined was from 24.9 to 26.0 points, it still remained at the quality evaluation grade "Good".

Table 8. Localisation task error score results

Translator	Scenario	Accuracy	Language quality	Style	Terminology	Total error score
Translator 1	TM	6.8	8.0	6.8	1.6	23.3
	TM+MT	9.9	14.4	7.8	4.1	36.3
Translator 2	TM	8.2	10.1	11.7	0.0	30.0
	TM+MT	3.8	11.7	7.6	1.5	24.6
Translator 3	TM	4.6	9.5	7.3	0.0	21.4
	TM+MT	3.0	8.3	6.0	0.8	18.1
Average	TM	6.5	9.3	8.6	0.5	24.9
	TM+MT	5.4	11.4	7.1	2.1	26.0

7   Conclusion

The results of our experiment demonstrate that it is feasible to adapt SMT systems for a particular domain with the help of comparable data and integrate such SMT systems for highly inflected under-resourced languages into the localisation process.

The use of the English->Latvian domain adapted SMT suggestions (trained on comparable data) in addition to the translation memories lead to the increase of translation performance by 13.6% while maintaining an acceptable ("*Good*") quality of the translation. However, our experiments also showed a relatively high difference in translator performance changes (from -5.89% to +35.39%), which suggests that for more justified results the experiment should be carried out with more participants. It would also be useful to further analyse correlation between the regular productivity of translator and the impact on productivity by adding MT support.

Error rate analysis shows that overall usage of MT suggestions decreased the quality of translation in two error categories (language quality and terminology). At the same time this degradation is not critical and the result is still acceptable for production purposes.

To our knowledge, this is the first evaluation of usability of SMT systems enriched with comparable data for translation into a less-resourced highly inflected language. This is also one of the first evaluation of SMT for an under-resourced highly inflected language in the localisation environment.

Acknowledgements. The research leading to these results has received funding from the research project "2.6. Multilingual Machine Translation" of EU Structural funds, contract nr. L-KC-11-0003 signed between ICT Competence Centre and Investment and Development Agency of Latvia. Many thanks to our colleagues Juris Celmiņš, Elita Kalniņa and Artūrs Pudulis for participation in the localisation experiments and Ieva Dātava for her support and comments.

References

1. Abdul-Rauf, S., Schwenk, H.: On the use of comparable corpora to improve SMT performance. In: Proceedings of the 12th Conference of the European Chapter of the Association for Computational Linguistics, Athens, Greece, pp. 16–23 (2009)
2. Abdul-Rauf, S., Schwenk, H.: Parallel sentence generation from comparable corpora for improved SMT. Machine Translation 25(4), 341–375 (2011)
3. Bertoldi, N., Haddow, B., Fouet, J.B.: Improved Minimum Error Rate Training in Moses. The Prague Bulletin of Mathematical Linguistics 91, 7–16 (2009)
4. Flournoy, R., Duran, C.: Machine translation and document localization at Adobe: From pilot to production. In: MT Summit XII: Proceedings of the Twelfth Machine Translation Summit, Ottawa, Canada (2009)
5. Hewavitharana, S., Vogel, S.: Enhancing a Statistical Machine Translation System by using an Automatically Extracted Parallel Corpus from Comparable Sources. In: Proceedings of the Workshop on Comparable Corpora, LREC 2008, pp. 7–10 (2008)
6. Koehn, P.: Statistical Machine Translation. Cambridge University Press (2010)

7. Koehn, P., Federico, M., Cowan, B., Zens, R., Duer, C., Bojar, O., Constantin, A., Herbst, E.: Moses: Open Source Toolkit for Statistical Machine Translation. In: Proceedings of the ACL 2007 Demo and Poster Sessions, Prague, pp. 177–180 (2007)

8. Lu, B., Jiang, T., Chow, K., Tsou, B.K.: Building a large English-Chinese parallel corpus from comparable patents and its experimental application to SMT. In: Proceedings of the 3rd Workshop on Building and using Comparable Corpora: from Parallel to Non-Parallel Corpora, Valletta, Malta, pp. 42–48 (2010)

9. Munteanu, D., Marcu, D.: Extracting parallel sub-sentential fragments from nonparallel corpora. In: Proceedings of the 21st International Conference on Computational Linguistics and the 44th Annual Meeting of the Association for Computational Linguistics, Morristown, NJ, USA, pp. 81–88 (2006)

10. Munteanu, D., Marcu, D.: Improving Machine Translation Performance by Exploiting Non-Parallel Corpora. Computational Linguistics 31(4), 477–504 (2006)

11. O'Brien, S.: Methodologies for measuring the correlations between post-editing effort and machine translatability. Machine Translation 19(1), 37–58 (2005)

12. Pinnis, M., Ion, R., Ştefănescu, D., Su, F., Skadiņa, I., Vasiļjevs, A., Babych, B.: ACCURAT Toolkit for Multi-Level Alignment and Information Extraction from Comparable Corpora. In: Proceedings of System Demonstrations Track of ACL 2012, Jeju Island, Republic of Korea (2012)

13. Pinnis, M., Ljubešić, N., Ştefănescu, D., Skadiņa, I., Tadić, M., Gornostay, T.: Term Extraction, Tagging and Mapping Tools for Under-Resourced Languages. In: Proceedings of the 10th Conference on Terminology and Knowledge Engineering, Madrid, Spain (2012)

14. Pinnis, M., Skadiņš, R.: MT Adaptation for Under-Resourced Domains – What Works and What Not. In: Proceedings of the Fifth International Conference Baltic HLT 2012, pp. 176–184. IOS Press, Tartu (2012)

15. Plitt, M., Masselot, F.: A Productivity Test of Statistical Machine Translation Post-Editing in a Typical Localisation Context. The Prague Bulletin of Mathematical Linguistics 93, 7–16 (2010)

16. Schmidtke, D.: Microsoft office localization: use of language and translation technology (2008), http://www.tm-europe.org/files/resources/TM-Europe2008-Dag-Schmidtke-Microsoft.pdf

17. Skadiņa, I., Aker, A., Mastropavlos, N., Su, F., Tufiş, D., Verlič, M., Vasiļjevs, A., Babych, B., Clough, P., Gaizauskas, R., Glaros, N., Paramita, M.L., Pinnis, M.: Collecting and Using Comparable Corpora for Statistical Machine Translation. In: Proceedings of LREC 2012, Istanbul, Turkey, May 21-27, pp. 438–445 (2012)

18. Skadiņš, R., Goba, K., Šics, V.: Improving SMT for Baltic Languages with Factored Models. In: Proceedings of the Fourth International Conference Baltic HLT 2010, Riga, Latvia, October 7-8, pp. 125–132 (2010)

19. Skadiņš, R., Puriņš, M., Skadiņa, I., Vasiļjevs, A.: Evaluation of SMT in localization to under-resourced inflected language. In: Proceedings of the 15th International Conference of the European Association for Machine Translation EAMT 2011, Leuven, Belgium, May 30-31, pp. 35–40 (2011)

20. Su, F., Babych, B.: Measuring Comparability of Documents in Non-Parallel Corpora for Efficient Extraction of (Semi-) Parallel Translation Equivalents. In: Proceedings of the EACL 2012 Joint Workshop on Exploiting Synergies between Information Retrieval and Machine Translation (ESIRMT) and Hybrid Approaches to Machine Translation (HyTra), Avignon, France, April 23-27, pp. 10–19 (2012)

21. Steinberger, R., Eisele, A., Klocek, S., Pilos, S., Schlüter, P.: DGT-TM: A freely Available Translation Memory in 22 Languages. In: Proceedings of LREC 2012, Istanbul, Turkey, pp. 454–459 (2012)
22. Steinberger, R., Pouliquen, B., Widiger, A., Ignat, C., Erjavec, T., Tufis, D., Varga, D.: The JRC-Acquis: A multilingual aligned parallel corpus with 20+ languages. In: Proceedings of LREC 2006, Genoa, Italy, pp. 2142–2147 (2006)
23. Ştefănescu, D., Ion, R., Hunsicker, S.: Hybrid Parallel Sentence Mining from Comparable Corpora. In: Proceedings of the 16th Conference of the European Association for Machine Translation (EAMT 2012), Trento, Italy, May 28-30, pp. 137–144 (2012)
24. Tiedemann, J.: News from OPUS - A Collection of Multilingual Parallel Corpora with Tools and Interfaces. In: Recent Advances in Natural Language Processing, vol. V, pp. 237–248 (2009)
25. Vasiļjevs, A., Skadiņš, R., Tiedemann, J.: LetsMT!: a cloud-based platform for do-it-yourself machine translation. In: Proceedings of the 50th Annual Meeting of the Association for Computational Linguistics (ACL 2012), Jeju, Republic of Korea, July 10. System Demonstrations, pp. 43–48 (2012)

Assessing the Accuracy of Discourse Connective Translations: Validation of an Automatic Metric

Najeh Hajlaoui and Andrei Popescu-Belis

Idiap Research Institute, Rue Marconi 19, 1920 Martigny, Switzerland
{Najeh.Hajlaoui,Andrei.Popescu-Belis}@idiap.ch

Abstract. Automatic metrics for the evaluation of machine translation (MT) compute scores that characterize globally certain aspects of MT quality such as adequacy and fluency. This paper introduces a reference-based metric that is focused on a particular class of function words, namely discourse connectives, of particular importance for text structuring, and rather challenging for MT. To measure the accuracy of connective translation (ACT), the metric relies on automatic word-level alignment between a source sentence and respectively the reference and candidate translations, along with other heuristics for comparing translations of discourse connectives. Using a dictionary of equivalents, the translations are scored automatically, or, for better precision, semi-automatically. The precision of the ACT metric is assessed by human judges on sample data for English/French and English/Arabic translations: the ACT scores are on average within 2% of human scores. The ACT metric is then applied to several commercial and research MT systems, providing an assessment of their performance on discourse connectives.

Keywords: Machine translation, MT evaluation, discourse connectives.

1 Introduction

The evaluation of machine translation (MT) output has been revolutionized, in the past decade, by the advent of reference-based metrics. While not entirely substitutable to human judges, these metrics have been particularly beneficial as a training criterion for statistical MT models, leading to substantial improvements in quality, as measured by a variety of criteria. Reference-based metrics such as BLEU [13], ROUGE [5] or METEOR [1] rely on a distance measure between a candidate translation and one or more reference translations to compute a quality score. However, such metrics work best when averaging over large amounts of test data, and are therefore a reflection of global text quality and MT performance, rather than measuring a specific ability to correctly translate a given linguistic phenomenon. At best, large classes of linguistic phenomena can be assessed, e.g. by restrictions of METEOR or using the method proposed by [15].

Recent extensions of statistical MT algorithms to text-level or discourse-level phenomena deal with problems that are relatively sparse in texts, though they are crucial to the understanding of text structure. Examples include the translation of

A. Gelbukh (Ed.): CICLing 2013, Part II, LNCS 7817, pp. 236–247, 2013.

discourse connectives [7] and pronouns [9]. Evaluating the performance of MT systems on such phenomena cannot be done with the above metrics, and often such studies resort to manual counts of correct vs. incorrect translations.

In this paper, we introduce a reference-based metric for one type of discourse-level items, namely discourse connectives. These are lexical items (individual words or multi-word expressions) that signal the type of rhetorical relation that holds between two clauses, such as contrast, concession, cause, or a temporal relation such as synchrony or sequence. We define a method, called ACT for Accuracy of Connective Translation, which uses word-level alignment together with other features to determine the reference and candidate translations of a given source-language connective, and then to compute a score based on their comparison. Moreover, ACT identifies a subset of occurrences for which manual scoring is useful for a more accurate judgment. We focus on a small number of English connectives, and evaluate their translation into French and Arabic by a baseline and by a connective-aware SMT system. We show first that ACT matches closely the human judgments of correction, and then provide benchmark scores for connective translation.

The paper is organized as follows. In Section 2, we define the ACT metric, first using dictionary-based features, and then using word-alignment information. In Section 3, we validate the ACT metric by comparing it to human judgments, compare it briefly to previous proposals, and show how it can be generalized from English/French to English/Arabic translation. Finally, in Section 4, we provide results on three systems, giving an idea of current capabilities.

2   Definition of the ACT Metric for Discourse Connectives

The translation of an English connective to French may vary depending on the type of discourse (or rhetorical) relation that is conveyed. There are several theories of discourse structure, but the largest manually annotated corpus to date, in English, is Penn Discourse Treebank (PDTB) [14]. Discourse relations can be explicit, i.e. marked by connectives, or implicit. In the first case, the relation is equated with the "sense" of the connective. Four top-level senses (these are: temporal, contingency, comparison, expansion) are distinguished, with 16 sub-senses on a second level and 23 on a third level. The PDTB thus provides a discourse-layer annotation over the Wall Street Journal Corpus, with 18,459 explicit relations (marked by connectives) and 16,053 implicit ones.

To consider the example of a frequent discourse connective, the English *"while"* can have three senses:

- A contrast sense (French: *alors que, tandis que, mais,* etc.)
- A temporal sense (French: *tout en, tant que, quand, pendant que,* etc.)
- A concessive sense (French: *cependant, bien que, même si,* etc.)

Similarly, the English connective *since*, often signals a temporal relation, which can be translated to French by *depuis (que), dès que,* etc., but can also signal a causal relation, which can be translated into French by *comme, puisque, étant donné que,* etc.

Consequently, the evaluation of the accuracy of connective translation should ideally consider if the sense conveyed by the target connective is identical to (or at least compatible with, e.g. more general) the sense of the source connective. If sense labels were available for connectives (as in the PDTB annotation) for both source and target texts, including MT output, then evaluation would amount at identifying the connectives and comparing their senses. However, this is not the case, and therefore an evaluation metric for connectives must do without the sense labels.

2.1   ACT: Accuracy of Connective Translation

The idea of the proposed evaluation metric, named ACT for Accuracy of Connective Translation is the following. For each discourse connective in the source text that must be evaluated (typically an ambiguous connective), the metric first attempts to identify its translation in a human reference translation (as used by BLEU) and its candidate translation. Then, these are compared and scored. The specification of these two procedures appears in this section and the following ones.

To identify translations, ACT uses in a first step a dictionary of possible translations of each discourse connective type, collected from training data and validated by humans. If a reference or a candidate translation contains more than one possible translation of the source connective, then we use alignment information to detect the correct connective translation. If we have irrelevant alignment information (not equal to a connective), then we compare the word position (word index) between the source connective alignment in the translation sentence (candidate or reference) and the set of candidate connectives to disambiguate the connective's translation, and we take the nearest one to the alignment.

The ACT evaluation algorithm is given below using the following notations, and we suppose that there is a connective in the source sentence (at least one).

- Src: the source sentence
- Ref: the reference translation
- Cand: the candidate translation
- C: Connective in Src
- T(C): list or dictionary of possible translations of C (made manually)
- Cref: Connective translation of C in Ref
- Ccand: Connective translation of C in Cand

Table 1 shows 6 different possible cases when comparing a candidate translation with a reference one. We firstly check if the reference translation contains one of the possible translations of this connective, listed in our dictionary ($T(C) \cap Ref \neq \emptyset$). After that, we similarly check if the candidate translation contains a possible translation of this connective or not ($T(C) \cap Cand \neq \emptyset$). Finally, we check if the reference connective found above is equal (case 1), synonym (case 2) or incompatible (case 3) to the candidate connective (Cref=Ccand). Because discourse relations can be implicit, correct translations might also appear in cases 4–6 which are for non translated connectives. In general, they are due to a valid drop [17] and in a small number of cases to missing translations in our dictionary (not introduced to avoid interference with other cases).

Table 1. Basic evaluation method without alignment information

T(C) ∩ Ref ≠ Φ	T(C)∩Cand≠Φ	Cref=Ccand	Decision	
1	1	1	"Same connective in Ref and Cand ==>likely ok !"	1
		~	"Synonym connectives in Ref and Cand ==>likely ok !"	2
		0	"Incompatible connectives"	3
	0		"Not translated in Cand ==> likely not ok"	4
0	1		"Not translated in Ref but translated in Cand ==> indecided, to check by Human"	5
	0		"Not translated in Ref nor in Cand ==> indecided"	6

In total, these different combinations can be represented by 6 cases. For each one, the evaluation script prints an output message corresponding to the translation situation (Table 1). These 6 cases are:

- Case 1: same connective in the reference (Ref) and candidate translation (Cand).
- Case 2: synonymous connective in Ref and Cand.
- Case 3: incompatible connective in Ref and Cand.
- Case 4: source connective translated in Ref but not in Cand.
- Case 5: source connective translated in Cand but not in Ref.
- Case 6: the source connective neither translated in Ref nor in Cand.

For case 1 (identical translations) and case 2 (equivalent translations), ACT counts one point, otherwise zero (for cases 3-6). We thus use a dictionary of equivalents to rate as correct the use of synonyms of connectives classified by senses (case 2), as opposed to identity only. (A semi-automatic method based on word alignment of large corpora can be used to builds the dictionary of equivalents. We describe it more in detail in section 3.3 for the English-Arabic pair.)

One cannot automatically decide for case 5 if the candidate translation is correct, given the absence of a reference translation. We advise then to check manually these candidate translations by one or more human evaluators. Similarly, for case 6, it is not possible to determine automatically the correctness of each sentence. Therefore, we count them as wrong to adopt a strict scoring procedure (to avoid giving credit for wrong translations), or we check them manually as with the ACTm score.

ACT generates as output a general report, with scores of each case and sentences classified by cases. The following example illustrates case 2, "synonymous connectives". The candidate translation keeps the same sense (*concession*) as the reference translation by using a synonym connective (*Ccand* = *bien que* and *Cref* = *même si*) as a translation for the source connective (*Csrc* = *although*).

```
Csrc=although (althoughCONCESSION)   Cref=même si   Ccand=bien que
SOURCE: although traditionally considered to be non-justiciable , these
fundamental principles have been applied in a number of cases .
REFERENCE: même si ils sont traditionnellement considérés comme non
justiciables , ces principes fondamentaux ont été appliqués à plusieurs
reprises .
CANDIDATE: bien que toujours considéré comme non-justiciable , ces
principes fondamentaux ont été appliquées dans un certain nombre de cas
```

The total ACT score is the ratio of the total number of points to the number of source connectives. Three versions of the score are shown in Equations (1)–(3) below. A strict but fully automatic version is ACTa, which counts only Cases 1 and 2 as correct and all others as wrong. A more lenient automatic version excludes Case 5 from the counts and is called ACTa5. Finally, ACTm also considers the correct translations found by manual scoring of Case 5 (their number is noted |Case5corr|).

$$ACTa = (|case1| + |case2|)/\sum\nolimits_{i=1}^{6} |casei| \tag{1}$$

$$ACTa5 = (|case1| + |case2|)/\sum\nolimits_{i=1}^{4} |casei| + |case6| \tag{2}$$

$$ACTm = (|case1| + |case2| + |case5corr|)/\sum\nolimits_{i=1}^{6} |casei| \tag{3}$$

where |caseN| is the total number of discourse connectives classified in caseN.

In order to improve ACT and to limit errors, we describe in the next two sections the use of alignment information and numeric position information to improve the detection of the correct connectives when more than one possible connective translation is detected by simple dictionary lookup.

2.2   ACT Improved by Alignment Information

In order to reduce the number of errors due to the existence of more than one connective in a given sentence, we need to match correctly the source connective with the reference and the candidate connectives, respectively in the reference translation and in the candidate translation.

In the example below, both the reference and the candidate translation contain three potential connectives: *mais* (literally: *but*), *pas encore* (literally: *not yet*), and *encore* (literally: *again*). The question is then how we can get the third *encore* as a translation of *yet* and not the other ones. Let us add the following notations:

- CR = alignment(Src,Ref,C), CR is the reference connective in the reference sentence as a result of the alignment with the source connective C.
- CC = alignment(Src,Cand,C), CC is the candidate connective in the candidate sentence as a result of the alignment with the candidate connective C.

To resolve the ambiguity, we firstly propose to use the alignment information as disambiguation module. Theoretically, several cases can be observed depending on the alignment result (CR and CC) and on its intersection with the list of possible translations of a given connective C noted T(C), knowing that alignment information can be sometimes wrong. We now use alignment information to make an automatic disambiguation improving the 6 cases of Table 1. We check if CR (respectively CC) is a possible translation of the source connective (CR∈ T(C)) (respectively (CC∈ T(C))). If yes, Cref (respectively Ccand) will be replaced by CR (respectively CC) as shows the following example.

```
SENTENCE 13 Csrc:yet {} CR:
SENTENCE 13 Csrc:yet {20} CC:encore
SENTENCE 13: Csrc = yet (yetADVERB)   Cref = pas encore      Ccand =
encore ==> case 2: Synonym connectives in Ref and Cand ==>likely ok !
```

SOURCE 13: he intends to donate this money to charity , but hasn 't
decided which yet .
REFERENCE 13: il compte en faire don à des œuvres de bienfaisance , **mais**
il n' a **pas encore** concrètement décidé lesquelles .
CANDIDATE 13: il a l ' intention de donner cet argent de la charité ,
mais qui n ' a **pas encore** décidé .

The source connective (Csrc) is *yet*, of which there is more than one possible
translation in the candidate sentence (*mais* and *pas encore*). CR is empty, Cref (*mais*)
is then replaced by the nearest connective (*pas encore*) to the source one comparing
numeric positions (see 2.3). In general, if CR (respectively CC) is not a possible
translation of the source connective, two procedures based on the calculation of the
numeric position are used depending on the value of CR (respectively CC) (empty or
not). The following section shows how we proceed to detect the right connective.

2.3   ACT Improved by Numeric Position Information

For many reasons, the alignment of the source connective with the target sentence
might not result in a connective. This could be due to the result of a misalignment or
an error-alignment but it can be also because the source connective is translated
implicitly. Two main cases are distinguished: (1) the alignment information (CR in
Ref respectively CC in Cand) is empty. We then take the nearest connective to the
source connective comparing numeric positions. (2) The alignment information is not
empty but contains a non-connective: we then take the nearest connective to the
alignment comparing numeric positions.

Formally, we can summarize the translational and alignment situation by the
following notations and conditions. If the two following conditions are true:

- We have more than one possible translation of (C) in Ref, let's say n (n>1).
- CR is not a possible translation of (C), that is, CR is not a connective.

Then we apply the first heuristic (1) if CR (respectively CC) is empty, if not we apply
the second heuristic (2).

The following example shows another example of disambiguation, which makes
ACT more accurate. Before disambiguation, the sentence is classified in case 1 since
the same connective *si* (literally: *if*) is detected in the reference and in the candidate,
but it is a false case 1. After disambiguation, this sentence will be classified in the
correct case (case 2) since it contains a synonym connective (*bien que* and *même si*).

BEOFRE DISAMBIGUATION: Csrc = although Cref=Ccand = si
AFTER DISAMBIGUATION: Csrc = although (althoughCONCESSION) Cref = bien
que   Ccand = même si==> case 2: Synonym connectives in Ref and Cand

*SOURCE 5: we did not have it so bad in ireland this time **although** we have had many serious wind storms on the atlantic .*
*REFERENCE 5: cette fois-ci en irlande , ce n' était pas **si** grave , **bien que** de nombreuses tempêtes violentes aient sévi dans l' atlantique .*
*CANDIDATE 5: nous n' était pas **si** mauvaise en irlande , cette fois , **même si** nous avons eu vent de nombreuses graves tempêtes sur les deux rives de l' atlantique .*

3   Evaluation of the ACT Metric

3.1   Comparison with Related Work

The METEOR metric [1] uses a monolingual alignment between two translations to be compared: a system translation and a reference one. METEOR performs a mapping between unigrams: every unigram in each translation maps to zero or one unigram in the other translation. Unlike METEOR, the ACT metric uses a bilingual alignment (between the source and the reference sentences and between the source and the candidate sentences) and the word position information as additional modules to disambiguate the connective situation in case there is more than one connective in the target (reference or candidate) sentence. ACT may work without these modules.

The evaluation metric described in [6] indicates for each individual source word which systems (among two or more systems or system versions) correctly translated it according to some reference translation(s). This allows carrying out detailed contrastive analyses at the word level, or at the level of any word class (e.g. part of speech, homonymous words, highly ambiguous words relative to the training corpus, etc.). The ACT metric relies on the independent comparison of one system's hypothesis with a reference.

An automatic diagnostics of machine translation and based on linguistic checkpoints [16] and [10] constitute a different approach from our ACT metric. The approach essentially uses the BLEU score to separately evaluate translations of a set of predefined linguistic checkpoints such as specific parts of speech, types of phrases (e.g. noun phrases) or phrases with a certain function word.

A different approach was proposed by [15] to study the distribution of errors over five categories (inflectional errors, reordering errors, missing words, extra words, incorrect lexical choices) and to examine the number of errors in each category. This proposal was based on the calculation of Word Error Rate (WER) and Position-independent word Error Rate (PER), combined with different types of linguistic knowledge (base forms, part-of-speech tags, name entity tags, compound words, suffixes, prefixes). This approach does not allow checking synonym words having the same meaning like the case of discourse connectives.

3.2   Error Rate of the ACT Metric

In order to estimate the accuracy of ACT and the improvements explained above, we manually evaluated it on a subset of 200 sentences taken from the UN EN/FR corpus with 207 occurrences of the seven English discourse connectives (*although, though, even though, while, meanwhile, since, yet*). We counted for each of the six cases the number of occurrences that have been correctly vs. incorrectly scored by ACT (each correct translation scores one point). The results were, for case 1: 64/0, case 2: 64/3, case 3: 33/4, case 4: 1/0, and for case 6: 0/0. Among the 38 sentences in case 5, 21 were in fact correct translations. The ACT error scores by case are 0% for case 1, case 4 and case 6, case 2: 4.2%, and case 3: 10%.

Therefore, the ACTa score was about 10% lower than reality (lower than the score computed by humans), while ACTa5 and ACTm were both about only 0.5% lower. Without using the disambiguation module, ACTa error score is more or less the same, while ACTa5 and ACTm were both about 2% than reality, word alignment thus improves the accuracy of the ACT metric.

A strict interpretation of the observed ACT errors would conclude that ACT differences are significant only above 4%, but in fact, as ACT errors tend to be systematic, we believe that even smaller variations (especially for ACTa) are relevant.

Two (opposite) limitations of ACT must be mentioned. On the one hand, while trying to consider acceptable (or "equivalent") translation variants, ACT is still penalized, as is BLEU, by the use of only one reference translation. On the other hand, the effect on the human reader of correctly vs. wrongly translated connectives is likely more important than for many other words.

3.3   Towards a Multilingual ACT Metric

The main resource needed to port the ACT metric to another language pair is the dictionary of connectives matching possible synonyms and classifying connectives by sense. In order to find the possible translations of the seven ambiguous English connectives and based on a large corpus analysis of translations of English discourse connectives into Arabic, we used an automatic method based on alignment between sentences at the word level using GIZA++ [11] and [12]. We experimented with the large UN parallel corpus to find out the Arabic connectives that are aligned to English ones. It is a corpus of journal articles and news:

- English: 1.2 GB of data, with 7.1 million sentences and 182 million words.
- Arabic: 1.7 GB of data, with 7.1 million of sentences and 154 million words.

Table 2. Translations of the 386 occurrences of 'while' with explicit alignments (out of 1,002)

Arabic translations of 'while'			
Buckwalter	Arabic	N.	%
bynmA	بينما	139	36.0%
w+	و	110	28.5%
Hyn	حين	66	17.1%
mE	مع	54	14.0%
w+ bynmA	وبينما	6	1.6%
w+ mE	ومع	5	1.3%
w+ Hyn	وحين	6	1.6%
Total		**386**	**100%**

For the alignment task, the data was tokenized and lowercased for English, and transliterated and segmented using MADA [2] for Arabic. Table 2 shows the correspondences between the one of the seven English connective "*while*" and Arabic translations detected automatically using the annotation projection from English sentences to Arabic ones.

Starting from that table (similarly for the other six English connectives), we cleaned firstly the Arabic vocabulary by merging several translations into one entry and checking also sentences to correct the alignment information. Secondly, we added other possible (known) translations to complete the dictionary. Thirdly, in order to classify the dictionary by sense, we checked manually the meaning of each connective based on a small number of sentences (10 to 50 sentences). For instance, the Arabic possible translations of "while" can be classified along three senses, Contrast, Concession and Temporal, as follows.

$whileCONTRAST="mE An ثا|mE مع|lAn بلا|lkn لكن";

$whileCONCESSION="Alrgm الرغم|rgm رغم|A*A اذا|A* اذ";

$whileTEMPORAL="bynmA بينما|Ely Hyn على حين|fy Hyn فى حين";

From lack of space, we list only one example of English connective. This research was recently published [3] and the same technique will be adapted and adopted to extend ACT in two ways: by adapting it to a new language pair and by adapting it to find out the correspondences and the sense of more connectives. Additional research is needed to assess the variability and sensitivity of the measure. Once we had the dictionary of synonyms connectives classified by sense, we adapted ACT metric to English-Arabic language pair.

We performed a similar evaluation for the English-Arabic version of ACT taking 200 sentences from the UN EN/AR corpus with 205 occurrences of the seven discourse connectives. Results are as follows (correctly vs. incorrectly): for case 1: 43/4, case 2: 73/2, case 3: 27/4, case 4: 19/2, and for case 6: 5/1. Among the 25 sentences in case 5, 9 were in fact correct translations. The error scores by case are then case 1: 8.5%, case 2: 2.6%, case 3:13%, case 4: 9.5%, and case 6: 16%.

Therefore, the ACTa score was about 5% lower than score computed by human, while ACTa5 and ACTm were both about 0.5% lower.

4   Benchmark ACT Scores for the Translation of Connectives

4.1   Configuration of ACT

As shown in Fig. 1, ACT can be configured and used with two main versions: with or without disambiguation module. Two subversions of the disambiguation version can be used: (1) without saving alignment model using just GIZA++ as alignment tool. (2) with training and saving an alignment model using MGIZA++ (a multithreaded version of GIZA++) which is trained in a first step on the Europarl corpus [4] giving an alignment model to be applied on the new data (Source, Reference) and (Source, Candidate). In the following experimentation, we will use the 3 configurations of ACT: ACT without disambiguation, ACT without saving the alignment model, and ACT with saving the alignment model.

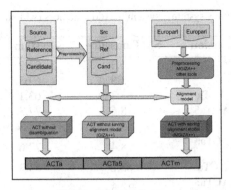

Fig. 1. ACT architecture

4.2   Data

In all the following experiments, we made use of a set of 2100 sentences taken from the UN EN/FR corpus, with 2207 occurrences of the seven discourse connectives mentioned above (at least 300 occurrences for each one). We used 3 MT systems to translate from English to French. Since our objective is to observe a range of state-of-the-art (benchmark) scores for translation of connectives, we study the accuracy of three systems: an SMT baseline system trained on the Europarl corpus and two commercial systems (anonymized as system1 and system2) to test the ACT metric.

4.3   Experiments and Results

BLEU is computed here on detokenized, lowercased text, by using the NIST Mteval script (version 11b, available from www.itl.nist.gov/iad/mig/tools/).   ACT is computed on tokenized and lowercased text. ACT includes a pre-processing step in order to normalize French connectives. For example, we might find *lorsqu'* and *lorsque* as translations of the connective *while* respectively in the reference sentence and in the candidate sentence.

Table 3 contains BLEU, NIST and ACT scores respectively for the SMT baseline system, system1 and system2. The 3 configurations of ACT are all used giving each one 2 scores (ACTa, ACTa5). ACTm is not provided because we did not check manually how many translations in case 5 were actually correct. As shown in section 3 there were approximately 30-50% of correct translations among the total number of instance of case5.

For each system and for this test set, ACT scores are more or less stable, which shows that any version of ACT is useful. If we compare the 3 systems based on BLEU and NIST scores, the classification is the same as the one based on the ACT scores but ACT is a more sensitive indicator specific of the accuracy of connective translation.

Table 3. SMT baseline, system1, system2, 2100 sentences (without checking case 5)

Metric	Version	SMT baseline	System1	System2
BLEU		26.3	24.2	20.3
NIST		6.88	6.63	5.97
ACT without	ACTa	63.7	63.1	61.7
disambiguation	ACTa5	78.6	77.3	75.3
ACT without	ACTa	63.7	63.3	61.6
saving	ACTa5			
alignment		78.4	77.6	75.2
ACT with	ACTa	63.6	63.3	61.6
saving	ACTa5			
alignment		78.3	77.5	75.2

5   Conclusion and Perspectives

We proposed a new distance-based metric to measure the accuracy of connective translation, ACT. This measure is intended to capture the improvement of an MT system that can deal specifically with discourse connectives. Such models have been shown to perform with BLEU score gains of up to +0.60 points, but the semi-automated evaluation metric ACT shows improvements of up to 8% in the translation of connectives. We measured the variation of ACT scores comparing to the variation to distance-based metrics (BLEU/NIST metric).

Our second goal is to work towards a multilingual metric by adapting the developed metric (initially for English to French) to other pairs of languages (English-Arabic, Arabic-French, etc), focusing on connectives. We are working on 2 news target languages (Italian and German). In a second step, we will extend ACT to other words (mainly verbs and pronouns).

We have also presented here a semi-automatic method to find out Arabic possible translations functionally equivalent to English connectives. It consists to project connectives detected on the English side to the Arabic side of a large corpus using alignment information between sentences at the word level. Starting from the result of this method, we built a dictionary of English-Arabic connectives classified by senses. This successful technique based on large parallel corpus will be adopted to adapt ACT to other new language pair.

Acknowledgments. We are grateful to the Swiss National Science Foundation for its support through the COMTIS Sinergia Project, n. CRSI22_127510 (see www.idiap.ch/comtis/).

References

1. Denkowski, M., Lavie, A.: METEOR-NEXT and the METEOR Paraphrase Tables: Improved Evaluation Support for Five Target Languages. In: Proc. of the ACL 2010 Joint Workshop on Statistical Machine Translation and Metrics MATR, Uppsala (2010)

2. Habash, N., Rambow, O.: Arabic tokenization, part-of-speech tagging and morphological disambiguation in one fell swoop. In: Proc. of ACL 2010, Ann Arbor, MI, pp. 573–580 (2005)
3. Hajlaoui, N., Popescu-Belis, A.: Translating English Discourse Connectives into Arabic: a Corpus-based Analysis and an Evaluation Metric. In: Proc. of the CAASL4 Workshop at AMTA 2012 (Fourth Workshop on Computational Approaches to Arabic Script-based Languages), San Diego, CA, p. 8 (2012)
4. Koehn, P.: Europarl: A Parallel Corpus for Statistical Machine Translation. In: Proc. of the Tenth Machine Translation Summit, Phuket, pp. 79–86 (2005)
5. Lin, C.-Y., Och, F.J.: Automatic Evaluation of Machine Translation Quality Using Longest Common Subsequence and Skip-Bigram Statistics. In: Proc. of the ACL, Barcelona (2004)
6. Max, A., Crego, J.M., Yvon, F.: Contrastive Lexical Evaluation of Machine Translation. In: Proc. of the International Conference on Language Resources and Evaluation (LREC), Valletta, Malta (2010)
7. Meyer, T., Popescu-Belis, A.: Using sense-labeled discourse connectives for statistical machine translation. In: Proc. of the EACL 2012 Joint Workshop on Exploiting Synergies between IR and MT and Hybrid Approaches to MT (ESIRMT-HyTra), Avignon, pp. 129–138 (2012)
8. Meyer, T., Popescu-Belis, A., Hajlaoui, N., Gesmundo, A.: Machine Translation of Labeled Discourse Connectives. In: Proc. of AMTA 2012 (10th Conference of the Association for Machine Translation in the Americas), San Diego, CA, p. 10 (2012)
9. Nagard, R.L., Koehn, P.: Aiding pronoun translation with co-reference resolution. In: Proc. of the Joint 5th Workshop on Statistical Machine Translation and Metrics (MATR), Uppsala, pp. 258–267 (2010)
10. Naskar, S.K., Toral, A., Gaspari, F., Way, A.: A framework for diagnostic evaluation of MT based on linguistic checkpoints. In: Proc. of MT Summit XIII, Xiamen, China (2011)
11. Och, F.J., Ney, H.: Improved Statistical Alignment Models. In: Proc. of the ACL, Hong-Kong, China, pp. 440–447 (2000)
12. Och, F.J., Ney, H.: A systematic comparison of various statistical alignment models. Computational Linguistics 29(1), 19–51 (2003)
13. Papineni, K., Roukos, S., Ward, T., Zhu, W.J.: BLEU: a method for automatic evaluation of machine translation. In: Proc. of ACL, Philadelphia, PA, pp. 311–318 (2002)
14. Prasad, R., Dinesh, N., Lee, A., Miltsakaki, E., Robaldo, L., Joshi, A., Webber, B.: The Penn Discourse Treebank 2.0. In: Proc. of 6th International Conference on Language Resources and Evaluation (LREC), Marrakech, Morocco, pp. 2961–2968 (2008)
15. Popovic, M., Ney, H.: Towards automatic error analysis of machine translation output. Computational Linguistics 37(4), 657–688 (2011)
16. Zhou, M., Wang, B., Liu, S., Li, M., Zhang, D., Zhao, T.: Diagnostic evaluation of machine translation systems using automatically constructed linguistic check-points. In: Proc. of COLING, Manchester, UK, pp. 1121–1128 (2008)
17. Zufferey, S., Cartoni, B.: English and French causal connectives in contrast. Languages in Contrast 12(2), 232–250 (2012)

An Empirical Study on Word Segmentation for Chinese Machine Translation*

Hai Zhao[1,2,**], Masao Utiyama[3], Eiichiro Sumita[3], and Bao-Liang Lu[1,2]

[1] MOE-Microsoft Key Laboratory of Intelligent Computing and Intelligent System
[2] Department of Computer Science and Engineering,
Shanghai Jiao Tong University, #800 Dongchuan Road, Shanghai, China, 200240
[3] Multilingual Translation Laboratory, MASTAR Project
National Institute of Information and Communications Technology
3-5 Hikaridai, Keihanna Science City, Kyoto, 619-0289, Japan
zhaohai@cs.sjtu.edu.cn, {mutiyama,eiichiro.sumita}@nict.go.jp,
bllu@sjtu.edu.cn

Abstract. Word segmentation has been shown helpful for Chinese-to-English machine translation (MT), yet the way different segmentation strategies affect MT is poorly understood. In this paper, we focus on comparing different segmentation strategies in terms of machine translation quality. Our empirical study covers both English-to-Chinese and Chinese-to-English translation for the first time. Our results show the necessity of word segmentation depends on the translation direction. After comparing two types of segmentation strategies with associated linguistic resources, we demonstrate that optimizing segmentation itself does not guarantee better MT performance, and segmentation strategy choice is not the key to improve MT. Instead, we discover that linguistical resources such as segmented corpora or the dictionaries that segmentation tools rely on actually determine how word segmentation affects machine translation. Based on these findings, we propose an empirical approach that directly optimize dictionary with respect to the MT task for word segmenter, providing a BLEU score improvement of 1.30.

1 Introduction

Word segmentation is regarded as a primary step for Chinese natural language processing, as Chinese words are not naturally defined with spaces appearing between words. Word segmentation is usually helpful for better understanding Chinese meaning though it is not always necessary. In this decade, researchers have developed quite a lot of techniques to seriously improve the segmentation performance, work motivated by a series of shared tasks on Chinese word

---

* This work was partially supported by the National Natural Science Foundation of China (Grant No. 60903119, Grant No. 61170114, and Grant No. 61272248), and the National Basic Research Program of China (Grant No. 2009CB320901 and Grant No.2013CB329401).
** This work was partially done as the first author was at NICT with support of MASTAR project.

A. Gelbukh (Ed.): CICLing 2013, Part II, LNCS 7817, pp. 248–263, 2013.

segmentation, SIGHAN Bakeoff, has given especially satisfactory segmentation results for various further application in Chinese processing [1–3]. Typically, a segmenter has to be trained on a segmentation corpus subject to a predefined segmentation standard. A segmenter that is based on statistical learning can give a F-score of more than 95% in word segmentation performance evaluation.

However, researchers have realized that different natural language processing tasks may have quite different requirements for the segmentation task, which is often beyond the issues of segmentation performance or standards [4, 5]. A typical example of this concern is from Chinese related machine translation (MT). Basically, we try to answer two questions about the role of Chinese word segmentation in machine translation,

(1) Is word segmentation necessary?

(2) If it is, then which segmentation strategy should we adopt for better machine translation performance?

To the first question, our answer will be a NO, or more precisely, word segmentation strategies should be carefully selected so that it can really outperform a character aligning system. In theory, the current phrase-based alignment MT system is supposed to discover a phrase table at last, which right performs a similar operation over sentences as a word segmenter does. However, existing empirical works show that word segmentation can help an MT system work better than a system without word segmentation [6]. Later in this paper, we will actually show that word segmentation does not always lead to better machine translation performance.

To the second question, a number of empirical studies have been conducted [7, 8], and various improved segmentation strategies proposed. In this work, we continue the empirical study by expanding on the contents of existing work. What is the most different between previous work and this one is that various segmentation strategies in this paper are examined and compared by considering the affect of both linguistic resources and approach characteristics. In addition, we also consider both Chinese-to-English and English-to-Chinese translation tasks, while the latter translation task was seldom considered in existing work.

2   Related Work

All word segmentation strategies that are applied to machine translation can be put into two categories. One is the joint model, which is integrated into the aligning or decoding procedure of machine translation, and the other is the independent model, which may be flexibly used independent of an MT system. Independent models can be further split into two sub-classes, statistical and rule-based. The latter is sometimes called the dictionary (lexicon or vocabulary) based approach as a word list is specified aforehand for segmentation. If we distinguish word segmenters according to their data set sources, then we may also put them into two categories, monolingual-motivated and bilingual-motivated.

According to our knowledge, Xu et al. [6] is the first work on the use of word segmentation in MT, and their results showed that segmentation generated by word alignments may achieve competitive results compared to using monolingual segmenters with a predefined third-party dictionary.

Later Xu et al. [9] proposed a joint segmentation model that uses word lattice decoding in phrase-based MT. This work was generalized to hierarchical MT systems and other language pairs in the work of Dyer et al. [10]. Both of these methods need a specific monolingual segmentation to generate the final word lattices.

Xu et al. [11] proposed a Bayesian semi-supervised Chinese word segmentation model which uses both monolingual and bilingual information to derive segmentation suitable for MT. Their approach models the source-to-null alignment and has been shown to be a special case of the model in the work of Nguyen et al. [12]. Both Xu et al. [11] and Nguyen et al. [12] belong to joint models and used Gibbs sampling for inference.

Ma and Way [13] proposed a bilingually motivated segmentation approach for MT. Their approach first uses the output from an existing statistical word aligner to obtain a set of candidate "words", then according to a metric, the co-occurrence frequencies, the segmentation of the respective sentences in the parallel corpus will be iteratively modified. These modified sentences will be fed back to the word aligner, which produces new alignments.

For other improvement about monolingual word segmenters, Chang et al. [7] suggested that tuning granularity of Chinese "words" given by segmenters can enhance machine translation. Zhang et al. [8] proposed that concatenating various corpora regardless of their different specifications can help producing a better segmenter for MT.

Though word segmentation is a concern especially for Chinese machine translation, it is also a consideration for other non-Chinese language pairs, Koehn et al. [14] and Habash and Sadat [15] showed that data-driven methods for splitting and preprocessing can improve Arabic-English and German-English MT, and Paul et al. [16] and Nguyen et al. [12] proposed a language independent segmentation strategy to improve MT for different language pairs.

3 Word Segmenters

We will try to evaluate the two main word segmentation approaches, statistical and dictionary-based (rule-based), in this paper. For the statistical approach, a segmentation corpus should be available for segmenter training. Character-based tagging has been shown as an effective strategy for corpus-based segmentation information acquisition according to results of the SIGHAN Bakeoff shared tasks [17–20]. This approach was initially proposed in the work of Xue and Shen [21] and it needs to define the position of character inside a word. Traditionally, the four tags, b, m, e, and s stand, respectively, for the beginning, middle, end of a word, and a single character word since then [21]. Later Zhao et al. [19] furthermore introduced two tags, b_2 and b_3, for the second and third character in a word and demonstrated better performance. The following n-gram features from [19] were used as basic features,

Table 1. Corpus statistics

Corpus		PKU2	MSRA2	CTB3
training set	#word	1.1	2.37	0.51
(M)	#char	1.83	3.9	0.83
test set	#word	104	107	155
(K)	#char	173	188	257

(a) $C_n(n = -1, 0, 1)$,
(b) $C_n C_{n+1}(n = -1, 0)$,
(c) $C_{-1} C_1$,

where C stands for a character and the subscripts for the relative order to the current character C_0.

Conditional random fields (CRF) has become popular for word segmentation since it provides better performance than other sequence labeling tools [22], and it will be adopted as our machine learning tool.

From the first to the third SIGHAN bakeoff, each time organizers provided four data sets for evaluation, in which two sets are traditional Chinese and the other two simplified Chinese. As our parallel corpus for MT is simplified Chinese, we consider adopting all six simplified data sets from Bakeoff 1,2 and 3. These six data sets are noted as CTB1, PKU1, MSRA2, PKU2, CTB3, and MSRA3. However, for the training set, CTB1 is a subset of CTB3, MSRA3 is a subset of MSRA2, and PKU1 and PKU2 are identical. Thus we only need to adopt three data sets, PKU2, MSRA2, and CTB3 to train our segmenters. Corpus size information is in Table 1.

For dictionary based segmentation strategy, a predefined dictionary should be available for segmentation use. Following the category of the work of Zhao and Kit [23], and assuming the availability of a list of word candidates or words (the dictionary) each associated with a goodness for how likely it is to be a true word. Let $W = \{\{w_i, g(w_i)\}_{i=1,...,n}\}$ be such a list, where w_i is a word candidate and $g(w_i)$ its goodness function that is usually to set to word length. Dictionary based segmentation strategies can apply two types of decoding algorithms.

The first decoding algorithm is the traditional maximal-matching one. It works on a given plain text T to output the best current word w^* repeatedly with $T = t^*$ for the next round as follows,

$$\{w^*, t^*\} = \underset{wt = T}{\operatorname{argmax}}\ g(w) \tag{1}$$

with each $\{w, g(w)\} \in W$. This above algorithm is more precisely referred to as the forward maximal matching (FMM) algorithm Symmetrically, it has an inverse version that works the other way around, and it is referred to backward maximal matching (BMM) algorithm.

The second decoding algorithm is a Viterbi-style one to search for the best segmentation S^* for a text T, as follows:

$$S^* = \underset{w_1 \cdots w_i \cdots w_n = T}{\operatorname{argmax}} \sum_{i=1}^{n} g(w_i), \tag{2}$$

with all $\{w_i, g(w_i)\} \in W$. However, this algorithm subject to the above equation will not work as the goodness function is set to word length, and as the sum of all word lengths will be always the length of the given plain text T. Instead, a so-called shortest path (SP) algorithm will be applied for this case by searching the best segmentation with respect to the following equation,

$$S^* = \underset{w_1 \cdots w_i \cdots w_n = T}{\operatorname{argmin}} n. \tag{3}$$

As it finds a segmentation with minimal number of words, it is named the shortest path.

Traditionally, word segmentation performance is measured by F-score ($F = 2RP/(R + P)$), where the recall (R) and precision (P) are the proportions of the correctly segmented words to all words in, respectively, the gold-standard segmentation and a segmenter's output.

4   Experimental Settings

The MT data set for this study is from the Chinese-to-English patent machine translation subtask of the NTCIR-9 shared task [24]. Both the development and test sets are with single reference. All the data are extracted from patent documents, so it will not be biased towards any existing word segmentation specification that is mostly from the news domain.

The MT training data contains one million sentence pairs; on the Chinese side there are 63.2 million characters, and the English sentences have 35.6 million words. Both the development and test corpora include two thousand sentence pairs. Five-gram language models are trained for both Chinese-to-English and English-to-Chinese translation tasks over the target language data set. No other resources are involved.

The MT system used in this paper is a recent version of Moses[1][25]. We build phrase translations by first acquiring bidirectional GIZA++ alignments [26], the maximal phrase length is set to the default value 7, and using Moses' *grow-diag-final-and* alignment symmetrization heuristic[2]. During decoding, we incorporate the standard eight feature functions of Moses with the lexicalized reordering model. The parameters of these features are tuned with Minimum Error Rate Training (MERT) [26] on the standard development and test sets that were provided by the NTCIR-9 organizers. In addition, we set the maximum

---

[1] http://www.statmt.org/moses/
[2] According to our explorative experiments, this heuristic always outperformed the default setting, *grow-diag-final*.

Table 2. Correlation between F-score and BLEU (%)

Segmenter		CTB3	MSRA2	PKU2
CRF	*F*-score	94.6	97.2	95.1
	BLEU	31.26	31.82	31.74
FMM	*F*-score	82.8	86.9	93.3
	BLEU	31.20	31.32	31.70

distortion limit to 11, as in our experiments this setting always produces better performance. We report the MT performance using the BLEU metric on the standard test corpus with the default scorer *multi-bleu.perl* [27]. All BLEU scores in this paper are uncased if English is the target language.

5   Chinese to English Translation

We check multiple assumptions about how word segmentation affects machine translation.

5.1   Segmentation Performance

Existing work has shown that there is no strong correlation between segmentation F-score and BLEU score [8, 7]. We will confirm this observation again.

The F-scores and BLEU scores are listed in parallel in Table 2. Note that it is meaningless to compare performance between different segmentation conventions. For FMM segmenters, their dictionaries are extracted from the respective CRF segmenter outputs on MT training corpora. We may focus on FMM and CRF segmenters for the same convention, the F-score and BLEU score are separated for different corpus, and it is easy to observe that two types of segmenters output similar results though CRF segmenter slightly outperforms the corresponding FMM segmenter if the latter adopts the dictionary whose words are extracted from the segmentation outputs of the former. The F-score was evaluated over the SIGHAN bakeoff test data set. The CRF segmenters output much higher F-scores, but their corresponding BLEU scores are only slightly higher than FMM segmenters. Thus we have shown again that the F-score and BLEU score correlate insignificantly.

5.2   Segmentation Inconsistence

There is a theory about segmentation inconsistency for machine translation, which is that a segmenter that outputs different segmentation outputs for the same input substring between training corpus and development/test corpus or even for the same corpus easily leads to a poor performance on machine translation. This has been well analyzed in the work of Chang et al. [7] and an empirical metric, conditional entropy, has been proposed to measure segmentation inconsistency inside one segmented corpus. This metric partially may explain why a

Table 3. Correlation between differences of F-score and BLEU (%)

Corpus	CTB3	MSRA2	PKU2
F-score	78.6	84.6	82.8
1-F	21.4	15.4	17.2
BLEU diff(%)	0.2	1.6	0.1

dictionary-based segmentation strategy like FMM sometimes outperforms CRF segmenters.

Here, we introduce more experimental facts that may reflect how segmentation strategies vary over machine translation quality.

First, we compare the difference between outputs from FMM and CRF segmenters. For each segmentation convention, the FMM segmenter will still use the dictionary in which words are extracted from the CRF segmenter's output over the MT training corpus. Regarding the segmentation results of the CRF segmenter as the gold standard, an F-score can be calculated over the FMM segmenter' outputs. We will take the F-score as the quantity consistence between two segmentation outputs and that 100% minus the F-score may correspondingly represent the difference between the two outputs. Table 3 shows comparisons between the 1-F and BLEU score relative differences between the FMM and CRF segmenters. This comparison in Table 3 actually discloses that although two types of segmenters, FMM and CRF, output quite different word segmentation results, their MT results are quite close. Such facts suggest that an MT system may accept quite different segmentation inputs for a degree of translation quality and using similar or related linguistic resource, different segmenters may lead to close MT performance. Meanwhile, this also means that we cannot effectively predict BLEU differences only from segmentation difference.

Second, we check if it is sensitive if we apply different segmentation strategies between the MT training set and developement/test sets. Table 4 shows MT results as CRF and FMM segmenters are respectively applied to the training and development/test sets. In the table, segmentation consistency F-scores are calculated on the training corpus, and the BLEU loss ratio is calculated between two average scores as the same and different segmenters are applied to the training and development/test corpora. An obvious BLEU score loss have been observed from the results, and the magnitude of BLEU score change is kept at a similar level as segmentation difference.

For all tree dictionary based segmentation strategies, FMM, BMM and SP, we also do a similar check. Their segmentation differences are in Table 5 as the dictionary is extracted from output of the CRF segmenter on CTB3 convention. The BLEU scores are in Table 6. The results show that even using the same dictionary, segmentation strategy differences cause quite different BLEU scores.

Based on the above two observations: MT quality is sensitive to segmentation strategy choice if the training set and development/test set adopt different segmentation strategies, though apart from this condition, the current MT system is not so sensitive to segmentation strategy choice if the support linguistic

Table 4. BLEU scores as different segmenters for training and dev/test sets(%)

training	dev/test	CTB3	MSRA2	PKU2
CRF	CRF	31.26	31.82	31.74
FMM	FMM	31.20	31.32	31.70
FMM	CRF	27.75	27.11	28.72
CRF	FMM	25.91	26.39	26.99
BLEU loss ratio		14.1	15.3	12.2
1-F		21.4	15.4	17.2

Table 5. Segmentation differences of dictionary based segmenters(%)

	FMM BMM	BMM SP	SP FMM
F-score	78.0	80.9	95.6
1-F	22.0	19.1	4.4

resource is kept unchanged. We then may cautiously conclude that segmentation strategy itself becomes a factor on segmentation consistence analysis, that is, segmentation consistency for MT evaluation should be only measured among the segmentation output given by the same segmentation strategies.

5.3 Different Dictionary Sources

So far, we only adopt dictionaries that are extracted from CRF segmenter outputs for all dictionary-based segmenters. However, for dictionary sources, we may have more choices than segmented corpora for CRF segmenters. All segmented corpora for CRF segmenters are from the SIGHAN Bakeoff shared task and independent of our MT corpus; therefore, they belong to the out-of-domain resources for the MT task. Intuitively, in-domain linguistic resources are always preferable due to it usually bringing about better performance. Compared to building an in-domain segmented corpus for MT tasks, it is much easier to construct an in-domain dictionary.

We then consider two strategies for generating dictionaries from an MT corpus. One is based on unsupervised segmentation over a monolingual corpus, i.e., the Chinese side of the parallel corpus, and the other is based on the alignment model.

Unsupervised segmentation has been empirically studied in the work of Zhao and Kit [23]. According to the empirical results of this work, Accessor Variety (AV) has shown the best goodness function for unsupervised segmentation incorporated with a Viterbi-style decoding algorithm according to equation 3. AV was proposed in [28] as a statistical criterion to measure how likely a substring is a true word. The AV of a substring $x_{i..j}$ is given as follows:

$$AV(x_{i..j}) = \min\{L_{av}(x_{i..j}), R_{av}(x_{i..j})\} \qquad (4)$$

Table 6. BLEU scores as using different segmenters for training and dev/test sets(%)

training	FMM	FMM	FMM	BMM	BMM	BMM	SP	SP	SP
dev/test	FMM	BMM	SP	FMM	BMM	SP	FMM	BMM	SP
BLEU	31.20	27.08	30.16	27.62	30.47	28.05	30.42	28.06	31.25

Table 7. Dictionary size(K)

AV	ALIGN	ALIGN$_{>1}$	CRF-CTB3	CRF-MSRA2	CRF-PKU2
316	417	142	503	460	465

where the left and right accessor variety $L_{av}(x_{i..j})$ and $R_{av}(x_{i..j})$ are the number of distinct predecessor and successor characters, respectively. In practice, the logarithm of AV is actually used as a goodness measure in equation 3.

Note that AV scores should be calculated for possible character n-grams, which would yield too large of a dictionary. Thus, we first use the Viterbi decoding algorithm with all n-gram AV scores to segment the Chinese MT training corpus, then we build a much smaller dictionary by only collecting all words from the segmented text.

Xu et al. [6] proposed a heuristic rule to generate a dictionary from alignment outputs. Firstly, each Chinese character in the corpus is segmented as a word, then an aligner like GIZA++ is used to train an alignment model with this trivially segmented Chinese text. According to alignment outputs, if two or more successive Chinese characters are translated to one English word, then these Chinese characters will be regarded as a word. This word collection strategy may lead to a large dictionary with remarkable noise. Therefore, we introduce a filtering rule by counting aligning times. For example, only if aligning is observed more than once, will those concerned continuous characters be collected as a word. This strategy (it will be noted as ALIGN$_{>1}$ afterwards.) helps us generate a much smaller dictionary.

Table 7 gives size information for different dictionaries. Numbers of word types generated by CRF segmenters are also given for comparison. All three dictionary-based segmentation approaches, FMM, BMM and SP, are used on all these dictionaries, and the results are in Table 8. *char-seg* in the table means that each character in the corpus will be segmented into a word. The results show that all segmentation strategies may outperform *char-seg*, but their results are not better than those given by every CRF segmenter. However, we also show that the dictionary pruning according to the alignment model can effectively enhance machine translation.

5.4   Segmentation Granularity for Dictionary Approach

Observing that MT is sensitive to segmentation granularity, Chang et al. [7] introduced a novel feature to tune the granularity in the outputs of CRF segmenters. Wang et al. [29] also made the similar observation and proposed using a third-party dictionary to modify a segmented corpus. In this part, we try to

Table 8. BLEU scores of dictionary based segmenters(%)

char-seg	30.14		
dict. / segmenters	FMM	BMM	SP
AV	30.46	30.76	30.62
ALIGN	30.73	30.94	30.90
ALIGN$_{>1}$	31.26	31.55	31.25

Table 9. BLEU scores over different segmentation granularity(%)

dict. / length	Full	5	4	3	2
CRF-CTB3	31.20	31.06	30.81	31.01	**31.22**
CRF-MSRA2	31.32	31.65	**31.73**	31.36	31.66
CRF-PKU2	31.70	31.30	31.31	**30.72**	31.03
AV	30.46	30.50	30.30	30.64	**30.71**
ALIGN$_{>1}$	31.26	31.34	31.43	**31.62**	31.04

verify this observation for dictionary segmenters. FMM is adopted as the decoding algorithm and word length is limited to 2,3,4 and 5 characters, respectively[3]. The results in Table 9 show that such granularity tuning is not too significant for dictionary-based segmentation strategies and the improvement sometimes depends on which dictionary source is adopted.

6   English-to-Chinese Translation

English-to-Chinese translation seems like a simple translation direction reversion, but it may follow quite difference principles. As to our best knowledge, few research endeavors have been done on this topic and this work, is the first attempt that comprehensively explores how word segmentation affects English-to-Chinese translation.

As the target language needs word segmentation and none of standard segmentations are available for evaluation, we will have to report the two types of BLEU scores, one is based on character sequences, the other based on word sequences. All results with different segmenters are given in Table 10, 11 and 12,

As the result of *char-seg* is from the direct optimization on character sequences during aligning and decoding, it is not surprising that it receives a character based BLEU score as high as 40.72, which is much better than any other regular word segmenters.

For further comparison, we re-segment the translation output text of *char-seg* and test corpus with the same segmenter, and the word-based BLEU score will be calculated between these two texts. From the results in Table 10, we see that for two of three segmenters, the trivial segmentation strategy, *char-seg*, outperforms CRF segmenters even in terms of word BLEU score, which is quite

---

[3] Note that most Chinese words are about two-characters long and few Chinese words are longer than five-characters.

Table 10. BLEU scores of English-to-Chinese translation(%): CRF segmenters

Segmenter	BLEU type	CTB3	MSRA2	PKU2
CRF	char	33.16	33.54	32.85
	word	26.11	27.25	25.55
char-seg	word	26.27	21.16	26.27
	char		40.72	

Table 11. BLEU scores of English-to-Chinese translation(%): dictionary segmenters with CRF segmenter generated dictionary

Segmenter	BLEU type	CTB3	MSRA2	PKU2
FMM	char	32.98	33.39	32.49
	word	23.65	24.79	23.90
char-seg	word	22.48	22.87	23.07
	char		40.72	

different from the case of Chinese-to-English translation. These results cast an obvious suspicion on the necessity of word segmentation for English-to-Chinese translation.

As we turn to compare the results of dictionary segmenters, another problem will be disclosed. From the results of Tables 11 and 12, we indeed observe that all dictionary segmenters give higher word BLEU scores than *char-seg*. However, this is not because dictionary segmenters really produce higher word BLEU scores, but that converted word BLEU scores of *char-seg* drop. This case in Table 12 is more serious, where all the converted BLEU scores are only around 10%. Manual observation on $\text{ALIGN}_{>1}$ dictionary shows that too many "words" in it are actually irregular character combinations, not true words. Therefore, this series of experiment results actually show that word BLEU scores may be seriously biased by the low-quality dictionary and it cannot be taken as a reliable metric for English-to-Chinese translation.

Continuing along this train of thought, if we have to take character BLEU as a unique metric for evaluating English-to-Chinese translation, then we will naturally draw a conclusion that word segmentation is not in fact necessary for this type of machine translation task.

7   Finding an Optimal Dictionary

From a linguistic resources perspective, dictionary or segmented corpus, there is not a solid borderline between statistical and dictionary-based segmentation strategies. They can be converted to each other easily. We have let a dictionary segmenter adopt a dictionary collected from the segmentation output of CRF segmenters. Conversely, dictionary segmenters can be used to segment a given text to generate a segmented corpus for training CRF segmenters as well. Our empirical study also shows that when using correlative linguistic resources, either a statistical segmenter or dictionary segmenter gives similar results, and in

Table 12. BLEU scores of English-to-Chinese translation(%): dictionary segmenters with ALIGN$_{>1}$ dictionary

Segmenter	BLEU type	FMM	BMM	SP
ALIGN$_{>1}$	char	33.74	32.45	33.23
	word	20.24	19.27	19.99
char-seg	word	10.10	9.96	10.29
	char		40.72	

this case, none of the segmentation strategies work significantly better than the others. In other words, to optimize a word segmenter, we have to optimize the linguistic resources that it relies on.

Here, we propose an empirical dictionary optimization (more precisely, pruning) algorithm to improve the related dictionary-based segmenters. The algorithm is mostly motivated by the empirical observation that most words in a given dictionary actually provide poor information for aligning and decoding in a specific MT task. As a dictionary with n words is given, our task of dictionary optimization is to find a subset of the dictionary to maximize the machine translation performance. However, we will have to examine 2^n subsets without guidance of any priori knowledge, which is computationally intractable. A solution to this difficulty is introducing a metric to assess how much a word is beneficial for machine translation and guide the later dictionary subset selection. So far, no priori metric has been found to measure how good a segmenter is for machine translation. Thus, most related studies have to directly adopt aligner outputs or even BLEU scores to choose a good segmenter. We will exploit both alignment model and BLEU scores given by MERT on the development set, and aligning counter is adopted as the metric to evaluate how well a word inside the dictionary individually contributes to machine translation[4]. This algorithm is given in Algorithm 1. There are two layers of loops in the algorithm, but in practice, this algorithm usually ends after running the MT routine less than 15 times. In addition, against existing dictionary optimization approaches [13, 31], the proposed one is actually non-parametric, which is more convenient and practical for use.

We consider three different dictionaries for the inputs of the proposed dictionary optimization algorithm and FMM is chosen as the decoding algorithm for the Chinese-to-English translation task, and the results on the test set are given in Table 13. All input dictionaries give higher BLEU scores after optimization. The most improvement, a 1.3% BLEU score, is from ALIGN$_{>1}$, which suggests that an in-domain bilingually motivated dictionary source can bring about better performance.

---

[4] Actually, we have considered various rank metrics in our early exploration. Ma and Way [13] argued that the co-occurrence frequency (COOC) that was proposed in [30] could be better for ranking words, however, our empirical study shows that COOC may lead to unstable performance for quite a lot of dictionary sources.

Algorithm 1. Dictionary optimization

1: **INPUT** An initial dictionary, D
2: **while do**
3:     Segment the MT corpus with D.
4:     Run GIZA++ for alignment model M.
5:     Run MERT and receive BLEU score(on the dev set) b.
6:     Rank all words in D according to aligning times.
7:     Let $counter=0$ and $n=2$
8:     **while** $counter <2$ **do**
9:         Extract top $1/n$ words from D according to aligning times to build dictionary D_n.
10:        Run GIZA++, MERT and receive BLEU score b_n.
11:        **if** $b_n < b_{n-1}$ **then**
12:            $counter = counter + 1$.
13:        **end if**
14:        $n = n + 1$
15:    **end while**
16:    **if** max $\{b_i\} < b$ **then**
17:        **return** D
18:    **end if**
19:    Let $D_0 = $ arg max$_{D_i}$ b_i and $b=$ max $\{b_i\}$
20:    Let $D' = D - D_0$
21:    According to aligning times in M, divide D' into $2n$ dictionaries, $D'_1,...,D'_n, ...,$ D'_{2n}.
22:    **for** top n most-aligned dictionaries, $D'_i, i = 1,...,n$ **do**
23:        Segment the MT corpus with $D_0+D'_i$.
24:        Run GIZA++, MERT and receive BLEU score b'_i.
25:    **end for**
26:    **if** max $\{b'_i\} < b$ **then**
27:        **return** D_0
28:    **end if**
29:    Let $D = $ arg max$_{D_0+D'_i}$ b'_i and $b=$ max $\{b'_i\}$
30: **end while**

Table 13. BLEU scores of segmenters with optimized dictionary (%)

char-seg	30.14		
Dictionary sources	CTB3	AV	ALIGN$_{>1}$
before opti.	31.20	30.46	31.26
after opti.	31.73	31.50	32.56
#running MT routines	6	9	15

Table 14 gives dictionary size information before and after optimization. The results demonstrate that all dictionaries are heavily pruned. The pruning result from dictionary ALIGN$_{>1}$ is especially unusual, as the dictionary with only 7K words that provides the most MT performance improvement among three

Table 14. Dictionary size before and after optimization (K)

Dictionary sources	CTB3	AV	ALIGN$_{>1}$
before opti.	503	316	142
after opti.	35	32	7

dictionary sources is at last obtained through the proposed algorithm, while most previous work often reports that dictionaries with tens of thousands of words at least are required [8, 31].

8   Conclusions

As word segmentation has been shown helpful for Chinese-to-English machine translation, we investigate what type of segmentation strategy can help machine translation work better. First, our empirical study shows that word segmentation is a necessity for Chinese-to-English translation, but not for the case of English-to-Chinese translation. Second, both statistical and dictionary-based word segmentation strategies are examined. We actually show for better machine translation, the key is not the segmentation strategy choice, but the linguistic resources for supporting segmenters. Third, an easy-implemented dictionary optimization algorithm is proposed to improve segmentation for machine translation. Our experiment results show that this approach is effective for different dictionary sources; however,better results come from a domain adaptive and bilingually motivated dictionary, which gives the most improvement with a BLEU score as high as 1.30 %.

References

1. Sproat, R., Emerson, T.: The first international chinese word segmentation bakeoff. In: Proceedings of the Second SIGHAN Workshop on Chinese Language Processing, Sapporo, Japan, pp. 133–143 (2003)
2. Emerson, T.: The second international chinese word segmentation bakeoff. In: Proceedings of the Fourth SIGHAN Workshop on Chinese Language Processing, Jeju Island, Korea, pp. 123–133 (2005)
3. Levow, G.A.: The third international chinese language processing bakeoff: Word segmentation and named entity recognition. In: Proceedings of the Fifth SIGHAN Workshop on Chinese Language Processing, Sydney, Australia, pp. 108–117 (2006)
4. Gao, J., Li, M., Wu, A., Huang, C.N.: Chinese word segmentation and named entity recognition: A pragmatic approach. Computational Linguistics 31, 531–574 (2005)
5. Li, M., Zong, C., Ng, H.T.: Automatic evaluation of chinese translation output: word-level or character-level? In: Proceedings of the 49th Annual Meeting of the Association for Computational Linguistics: Human Language Technologies: Short Papers, HLT 2011, June 19-24, vol. 2, pp. 159–164. Association for Computational Linguistics, Portland (2011)

6. Xu, J., Zens, R., Ney, H.: Do we need chinese word segmentation for statistical machine translation. In: Proceedings of the Third SIGHAN Workshop on Chinese Language Learning, Barcelona, Spain, pp. 122–128 (2004)

7. Chang, P.C., Galley, M., Manning, C.D.: Optimizing Chinese word segmentation for machine translation performance. In: Proceedings of the Third Workshop on Statistical Machine Translation, Columbus, Ohio, USA, pp. 224–232 (2008)

8. Zhang, R., Yasuda, K., Sumita, E.: Improved statistical machine translation by multiple chinese word segmentation. In: Proceedings of the Third Workshop on Statistical Machine Translation, pp. 216–223. Association for Computational Linguistics, Columbus (2008)

9. Xu, J., Matusov, E., Zens, R., Ney, H.: Integrated chinese word segmentation in statistical machine translation. In: Proceedings of IWSLT, Pittsburgh, PA, pp. 141–147 (2005)

10. Dyer, C., Muresan, S., Resnik, P.: Generalizing word lattice translation. In: Proceedings of the 46th Annual Meeting of the Association for Computational Linguistics: Human Language Technologies, Columbus, OH, USA, pp. 1012–1020 (2008)

11. Xu, J., Gao, J., Toutanova, K., Ney, H.: Bayesian semi-supervised chineseword segmentation for statistical machine translation. In: Proceedings of COLING 2008, Manchester, UK, pp. 1017–1024 (2008)

12. Nguyen, T., Vogel, S., Smith, N.A.: Nonparametric word segmentation for machine translation. In: Proceedings of COLING 2010, Beijing, China, pp. 815–823 (2010)

13. Ma, Y., Way, A.: Bilingually motivated domain-adapted word segmentation for statistical machine translation. In: Proceedings of the 12th Conference of the European Chapter of the Association for Computational Linguistics, pp. 549–557. Association for Computational Linguistics, Athens (2009)

14. Koehn, P., Knight, K.: Empirical methods for compound splitting. In: Proceedings of the Tenth Conference on European Chapter of the Association for Computational Linguistics, pp. 187–193. Association for Computational Linguistics, Budapest (2003)

15. Habash, N., Sadat, F.: Arabic preprocessing schemes for statistical machine translation. In: Proceedings of the Human Language Technology Conference of the NAACL, pp. 49–52. Association for Computational Linguistics, New York City (2006)

16. Paul, M., Finch, A., Sumita, E.: Integration of multiple bilingually-learned segmentation schemes into statistical machine translation. In: Proceedings of the Joint 5th Workshop on Statistical Machine Translation and MetricsMATR, pp. 400–408. Association for Computational Linguistics, Uppsala (2010)

17. Low, J.K., Ng, H.T., Guo, W.: A maximum entropy approach to Chinese word segmentation. In: Proceedings of the Fourth SIGHAN Workshop on Chinese Language Processing, Jeju Island, Korea, pp. 161–164 (2005)

18. Tseng, H., Chang, P., Andrew, G., Jurafsky, D., Manning, C.: A conditional random field word segmenter for SIGHAN bakeoff 2005. In: Proceedings of the Fourth SIGHAN Workshop on Chinese Language Processing, Jeju Island, Korea, pp. 168–171 (2005)

19. Zhao, H., Huang, C.N., Li, M.: An improved Chinese word segmentation system with conditional random field. In: Proceedings of the Fifth SIGHAN Workshop on Chinese Language Processing, Sydney, Australia, pp. 162–165 (2006)

20. Zhao, H., Kit, C.: Unsupervised segmentation helps supervised learning of character tagging for word segmentation and named entity recognition. In: Proceedings of the Sixth SIGHAN Workshop on Chinese Language Processing, Hyderabad, India, pp. 106–111 (2008)

21. Xue, N., Shen, L.: Chinese word segmentation as LMR tagging. In: Proceedings of the Second SIGHAN Workshop on Chinese Language Processing, in Conjunction with ACL 2003, Sapporo, Japan, pp. 176–179 (2003)
22. Peng, F., Feng, F., McCallum, A.: Chinese segmentation and new word detection using conditional random fields. In: COLING 2004, Geneva, Switzerland, pp. 562–568 (2004)
23. Zhao, H., Kit, C.: An empirical comparison of goodness measures for unsupervised chinese word segmentation with a unified framework. In: The Third International Joint Conference on Natural Language Processing (IJCNLP 2008), Hyderabad, India, pp. 9–16 (2008)
24. Goto, I., Lu, B., Chow, K.P., Sumita, E., Tsou, B.K.: Overview of the patent machine translation task at the ntcir-9 workshop. In: Proceedings of NTCIR-9 Workshop Meeting, Tokyo, Japan, pp. 559–578 (2011)
25. Koehn, P., Och, F.J., Marcu, D.: Statistical phrase-based translation. In: Proceedings of the 2003 Conference of the North American Chapter of the Association for Computational Linguistics on Human Language Technology, NAACL 2003, vol. 1, pp. 48–54. Association for Computational Linguistics, Stroudsburg (2003)
26. Och, F.J., Ney, H.: A systematic comparison of various statistical alignment models. Comput. Linguist. 29, 19–51 (2003)
27. Papineni, K., Roukos, S., Ward, T., Zhu, W.J.: Bleu: a method for automatic evaluation of machine translation. In: Proceedings of the 40th Annual Meeting on Association for Computational Linguistics, ACL 2002, pp. 311–318. Association for Computational Linguistics, Stroudsburg (2002)
28. Feng, H., Chen, K., Deng, X., Zheng, W.: Accessor variety criteria for Chinese word extraction. Computational Linguistics 30, 75–93 (2004)
29. Wang, Y., Uchimoto, K., Kazama, J., Kruengkrai, C., Torisawa, K.: Adapting chinese word segmentation for machine translation based on short units. In: Proceedings of LREC 2010, Malta, pp. 1758–1764 (2010)
30. Melamed, I.D.: Models of translational equivalence among words. Computational Linguistics 26, 221–249 (2000)
31. Ma, J., Matsoukas, S.: BBN's systems for the Chinese-English sub-task of the NTCIR-9 PatentMT evaluation. In: Proceedings of NTCIR-9 Workshop Meeting, Tokyo, Japan, pp. 579–584 (2011)

Class-Based Language Models
for Chinese-English Parallel Corpus*

Junfei Guo[1,2], Juan Liu[1,**], Michael Walsh[2], and Helmut Schmid[2]

[1] School of Computer, Wuhan University, China
[2] Institute for Natural Language Processing, University of Stuttgart, Germany
guojf@ims.uni-stuttgart.de, liujuan@whu.edu.cn,
{michael.walsh,schmit}@ims.uni-stuttgart.de

Abstract. This paper addresses using novel class-based language models on parallel corpora, focusing specifically on English and Chinese languages. We find that the perplexity of Chinese is generally much higher than English and discuss the possible reasons. We demonstrate the relative effectiveness of using class-based models over the modified Kneser-Ney trigram model for our task. We also introduce a rare events clustering and a polynomial discounting mechanism, which is shown to improve results. Our experimental results on parallel corpora indicate that the improvement due to classes are similar for English and Chinese. This suggests that class-based language models should be used for both languages.

1 Introduction

Language modeling is a topic well studied in Natural Language Processing (NLP). It is used in many NLP tasks such as speech recognition, optical character recognition, and statistical machine translation [1]. A language model assigns a probability to a sequence of n words by means of a probability distribution.

Despite the vast amount of work on language modeling, there has been little focus on building language models of Chinese. Downstream tasks, such as machine translation [2] could easily benefit from improved models.

For both Chinese and English, sparseness is an inherent problem even though the training sets are often large. One solution is to use classes-based generalization to estimate the probability. However, most class-based models with a simple interpolation achieve a modest improvement.

In this paper, we do a comparative study using a Chinese and English parallel corpus. The parallel corpus allows us to make a direct comparison of the perplexities of Chinese and English since the content of the two corpora is the same; this means that any difference in perplexity must be a fundamental "linguistic" difference between Chinese and English. Therefore, we investigate the difference in language models' performance on the parallel corpus.

---

* Junfei Guo acknowledge support by Chinese Scholarship Council during the first author's study in University of Stuttgart.

** Corresponding author.

A. Gelbukh (Ed.): CICLing 2013, Part II, LNCS 7817, pp. 264–275, 2013.

We notice that frequent events probability estimates are hard to improve by classes. Therefore, we carry out clustering on rare events. In particular, we apply a rare events class model [3] on a word-based Chinese corpus. Furthermore, we introduce the History-length Interpolation (HI) model that integrates the Kneser-Ney (KN) model with a class model by optimizing the weights to make the method close to a maximum likelihood estimator.

The main contribution of our work is we use different class-based models on a Chinese and English parallel corpus. To our knowledge, this paper presents the first such direct comparison of the perplexities of the two languages. We show that the perplexity of Chinese is much higher than English and discuss the possible reasons in detail.

In addition, we introduce using classes especially based on rare events to carry out a proper evaluation on a language model. A number of experiments were conducted with the clustering on rare events and all events. We show that all the class-based language models which focus on rare events increase performance.

Finally, we present a polynomial discounting mechanism on the parallel corpus. Polynomial discounting is motivated by the fact that HI and class generation need more probabilities allocated to the backoff probability than KN. The result and the analysis show that all the novel models performs well in Chinese and English. The results of this study may be of interest to researchers working on machine translation or other applications with language models in Chinese. We will show that we can use all of those models for Chinese.

The organization of the rest of the paper is as follows. In Section 2, we introduce some properties of the Chinese language and Chinese NLP. In Section 3, we review the KN model, a recursive model and a top-level interpolated model and two new models with a polynomial discounting mechanism that does better than the recursive model. Section 4 presents the experimental setup materials and methods. Section 5 contains a detailed description of our experimental results, as well as presenting a thorough analysis. Related work is summarized in Section 6. Finally, conclusions and work-in-progress are reported in Section 7.

2 Properties of the Chinese Language

Gao et al. [4] indicated that applying statistical language modeling techniques like trigrams to Chinese is challenging. There are a number of idiosyncrasies in Chinese: there is no standard definition of words in Chinese, word boundaries are not marked by spaces, and there is a limited amount of training data. Normally, there is no separation information between two characters in Chinese text. Such as a Chinese sentence is usually written like the first line in (1):

(1) 冰箱里有早餐!

breakfast is in fridge!

We put an English translation under the sentence in 1 and translate the Chinese sentence character by character in (2):

(2) 冰 箱 里 有 早   餐 ！
ice box in have morning food !

There are two ways to tokenise the sentence:

Character-Based

" 冰╷箱╷里╷有╷早╷餐╷！ "

with a space between each two characters in the sentence and the English translation character by character is in (2).

Word-Based

" 冰箱╷里╷有╷早餐╷！ "

This is a word-based segmentation of the sentence. An external information source is used to determine the word boundaries. The first character "冰" means "ice", the second one "箱" means "box". The two characters are not separated by space, because "冰箱" corresponds to one English word "fridge". We segment the characters to words in a sentence, then it's the word-based tokenized sentence. Some of the words are "compositional" with the characters i.e., the meaning of the word has to do with the meaning of the component characters, such as the above example "早餐"(morning food) forming the word "breakfast".

However, other words are non-compositional. Manning et al.[5] give the following example: The characters "和(harmony)" and "尚 (prefer)" form the word "和尚 (monk)", which means that one who prefers harmony is a monk. However, for each character there are still other meanings such as "和(and)" and "尚(still)".

Here is a sentence with the above problem: it has a sequence with "和尚" in it. The sentence "结婚的和尚未结婚的" can be segmented into two possibilities:

(3) 结婚的 和尚 未 结婚的
married monk not married (implicit "and" before "not")
the married monk and not married

The segmentation in (3) is an incorrect segmentation of the text (because monks do not marry). With the help of a segmentation system we get another possibility:

(4) 结婚的 和 尚 未 结婚的
married and still not married
the married and not married

The segmentation in (4) means that "the married and still not married", which makes sense.

For this reason, we do word-based segmentation on the Chinese sentences. We use the Stanford Word Segmenter [6] to do the segmentation. The experiments are conducted with word-based corpus.

3 Language Models

In this section we describe all the language models used in our experiment: the Kneser-Ney(KN) model, the Dupont-Rosenfeld (DR) model, the top-level (TOP) model, the Polynomial Kneser-Ney (POLKN) model and the Polynomial only (POLO) model. The notation we use throughout the paper is shown in Table 1.

Table 1. Notation

symbol	denotation
$\Sigma[[w]]$	Σ_w(sum over all unigrams w)
w_i^j	a segment from word w_i to word w_j
$c(w_i^j)$	count of w_i^j
$n_{1+}(\bullet w_i^j)$	number of distinct w occurring before w_i^j

3.1 Kneser-Ney Model

The modified Kneser-Ney (KN) trigram model proposed by Chen and Goodman [7] is our baseline model.

We estimate the model parameters on the training set as follows.

$$
\begin{aligned}
p_{KN}(w_3|w_1^2) &= \frac{c(w_1^3) - d'''(c(w_1^3))}{\Sigma[[w]]c(w_1^2 w)} + \gamma_3(w_1^2)p_{KN}(w_3|w_2) \\
\gamma_3(w_1^2) &= \frac{\Sigma[[w]]d'''(c(w_1^2 w))}{\Sigma[[w]]c(w_1^2 w)} \\
p_{KN}(w_3|w_2) &= \frac{n_{1+}(\bullet w_2^3) - d''(n_{1+}(\bullet w_2^3))}{\Sigma[[w]]n_{1+}(\bullet w_2 w)} + \gamma_2(w_2)p_{KN}(w_3) \\
\gamma_2(w_2) &= \frac{\Sigma[[w]]d''(n_{1+}(\bullet w_2 w))}{\Sigma[[w]]n_{1+}(\bullet w_2 w)} \\
p_{KN}(w_3) &= \begin{cases} \frac{n_{1+}(\bullet w_3) - d'(n_{1+}(\bullet w_3))}{\Sigma[[w]]n_{1+}(\bullet w)} & \text{if } c(w_3) > 0 \\ \gamma_1 & \text{if } c(w_3) = 0 \end{cases} \\
\gamma_1 &= \frac{\Sigma[[w]]d'(n_{1+}(\bullet w))}{\Sigma[[w]]n_{1+}(\bullet w)}
\end{aligned}
\tag{1}
$$

The parameters d', d'' and d''' are the discounts for unigrams, bigrams and trigrams, respectively, as defined by Chen and Goodman [7].

3.2 Brown Class Model

The Brown class model [8] is a class sequence model. Cluster transition probabilities p_T are computed using add-one smoothing. Emission probabilities p_E

are estimated by maximum likelihood. The cluster model we used is defined as follows:

$$p_B(w_3|w_1^2) = p_T(g(w_3)|g(w_1w_2))p_E(w_3|g(w_3))$$
$$p_B(w_3|w_2) = p_T(g(w_3)|g(w_2))p_E(w_3|g(w_3))$$

(2)

where $g(v)$ is the class of the uni- or bigram v.

3.3 DR Model

A recursive interpolation model following [9] is proposed by Schütze[3]. The key idea of this recursive model is that class generalization ought to play the same role in history interpolated models as the lower-order distributions: they should improve estimates for unseen and rare events. For a trigram model, this means that we interpolate $p_{KN}(w_3|w_2)$, $p_B(w_3|w_1^2)$ on the first backoff level and $p_{KN}(w_3)$, $p_B(w_3|w_2)$ on the second backoff level, where as the Brown class model [8] is interpolated globally. This motivates the following definition of the recursive model:

$$p_{DR}(w_3|w_1^2) = \frac{c(w_1^3) - d'''(c(w_1^3))}{\Sigma[[w]]c(w_1^2w)} +$$
$$\gamma_3(w_1^2)[\beta_1(w_1^2)p_B(w_3|w_1^2) + (1 - \beta_1(w_1^2))p_{DR}(w_3|w_2)]$$
$$p_{DR}(w_3|w_2) = \frac{n_{1+}(\bullet w_2^3) - d''(n_{1+}(\bullet w_2^3))}{\Sigma[[w]]n_{1+}(\bullet w_2w)} +$$
$$\gamma_2(w_2)[\beta_2(w_2)p_B(w_3|w_2) + (1 - \beta_2(w_2))p_{DR}(w_3)]$$
$$p_{DR}(w_3) = p_{KN}(w_3)$$
$$\beta_i(v) = \begin{cases} \alpha_1 & \text{if } v \in B_{2-(i-1)} \\ 0 & \text{otherwise} \end{cases}$$

(3)

where $\beta_i(v)$ is equal to a parameter α_i if the history (w_1^2 or w_2) is part of a cluster and 0 otherwise. B_1(resp. B_2) is the set of unigram (resp. bigram) histories that is covered by the clusters.

3.4 TOP Model

Brown et al. [8] firstly combined the class-based with other models by interpolation. In [3] Schütze interpolated unigrams models, bigrams models and KN models as the top-level model:

$$p_{TOP}(w_3|w_1w_2) = \mu_1(w_1w_2)p_B(w_3|w_1w_2) + \mu_2(w_2)p_B(w_3|w_2) +$$
$$(1 - \mu_1(w_1w_2) - \mu_2(w_2))p_{KN}(w_3|w_1w_2)$$

(4)

where $\mu_1(w_1w_2) = \lambda_1$ if $w_1w_2 \in B_2$ and 0 otherwise, $\mu_2(w_2) = \lambda_2$ if $w_2 \in B_1$ and 0 otherwise, λ_1 and λ_2 are parameters.

3.5   Polynomial Discounting Model

In [3], a new polynomial discounting mechanism is presented. While the incorporation of the additional polynomial discount into KN is straightforward, they used a discount function that is the sum of $d(x)$ and the polynomial:

$$e(x) = d(x) + \begin{cases} \rho x^\gamma & \text{for } x \geq 4 \\ 0 & \text{otherwise} \end{cases} \quad (5)$$

$$Where(e, d) \in (e', d'), (e'', d''), (e''', d''').$$

This so called POLKN model is identical to KN except that $d(x)$ is replaced with $e(x)$.

3.6   Polynomial Discounting Only Model

This model is a second version of the polynomial discount in [3], which replaces the discount in the following way:

$$e(x) = \rho x^\gamma$$

This model is a simple recursive model without using KN discounts.

4   Experimental Setup

The corpus that we use is a special release of the sentence alignment versions of the Zh-En MultiUN data that was made available in August 2011 in order to support evaluation for IWSLT 2011. MultiUN [10] is a release of the multilingual parallel corpus extracted from official documents of the United Nations. It has 8,824,451 parallel sentences and more than 200M words (Chinese words after segmentation: 237,600,044, English words: 246,005,349). We did word segmentation for Chinese and tokenized both languages in the corpus.

We added a symbol "BOS" in front of each sentence in the corpus, which makes it easy to calculate the probability of the first word in the sentence. For the rare events clustering corpus, the experimental setup is the same as for the all events except the cluster corpus. For example, given a raw sentence: $w_1\ w_2\ w_3\ w_4\ w_5\ w_6$ ..., there are four different clusterings for each size of the vocabulary as follows.

All-Event Unigram Clustering. The cluster corpus is the same as the raw corpus.

All-Event Bigram Clustering. We represent a bigram as a hyphenated word in bigram clustering and change every sentence $w_1\ w_2\ w_3\ w_4\ w_5\ w_6$ ... in the training set to two sentences:

$w_1\text{-}w_2\ w_3\text{-}w_4\ w_5\text{-}w_6$ ...

$w_2\text{-}w_3\ w_4\text{-}w_5$ ...

Unique-Event Unigram Clustering. The cluster corpus should have a sequence of two unique unigrams per line. The sequence occurring in the training set is composed as follows (each bigram occurs only once):

w_1 w_2

w_2 w_3

w_3 w_4

w_4 w_5

w_5 w_6

...

Unique-Event Bigram Clustering. The cluster corpus contains unique sequence of bigrams (occurring in the all-event bigram cluster corpus) per line as follows (each bigram occurs only once):

w_1-w_2 w_3-w_4

w_2-w_3 w_4-w_5

w_3-w_4 w_5-w_6

...

We use the SRILM toolkit [11] for clustering and calculate the counts of unigram, bigram and trigram.

We extract the N most frequent words where N is the base set size ($|B_i|$). In our experiment, the base set size is from 10,000 to 40,000, which is the parameter of the class-based model.

Then we use the SRILM toolkit to perform the clustering for unigrams and bigrams. We use vocabularies of different size $|B_i|$. and the cluster corpus as the input to do the clustering by SRILM. The vocabulary file is the same file used in all-events bigram clustering. We did not use any class count file generated with SRILM, we just used the class definition file and calculated the transition counts on the raw training set as we did for all-events clustering.

We settled on a fixed number of clusters $k = 512$. We mapped all the unknown words in the validation set to "unk" in the cluster definition file. We use maximum likelihood to estimate the emission probabilities p_E.

5   Results and Discussion

The experimental results and the parameters are listed in Table 2a, 2b, 3a and 3b, which show the performance of all the models for a range of base set sizes $|B_i|$ and for classes trained on all events or rare events. We use heuristic grid search to find the optimal parameters. All the perplexity values are reported for the validation set.

Table 2a and 2b show perplexities of several models: TOP, DR, POLKN and POLO. All the experiments are on $|B_i|$=40,000.

The experiment results indicate that all models have a perplexity that is lower than KN model (79.78 for Chinese and 55.56 for English), which suggests that classes improve language model performance.

Table 2. Performance of models for $|B_i| = 40,000$ and classes trained on rare events

(a) English models

		α_1/λ_1	α_2/λ_2	ρ	γ	perp.
1	KN					55.56
2	TOP	0.01	0.01			53.00
3	DR	0.10	0.40			52.90
4	POLKN	0.10	0.40	0.70	0.09	52.88
5	POL0	0.10	0.40	0.50	0.70	52.78

(b) Chinese models

		α_1/λ_1	α_2/λ_2	ρ	γ	perp.
1	KN					79.78
2	TOP	0.01	0.01			76.24
3	DR	0.10	0.40			76.11
4	POLKN	0.10	0.40	0.60	0.20	75.84
5	POL0	0.10	0.40	0.40	0.80	75.78

Table 2b shows that we can achieve considerable perplexity reduction for Chinese. The perplexity drops from 76.24 to 75.84 and 75.78, respectively. In all our experiments, the main result is that the DR model performs better than the TOP model, although the POLKN and POLO models do performs better than the DR model. All the Chinese models' performance is similar to English which suggests that we can use all the models in Chinese.

From the different base set sizes in Table 3a and 3b the main findings on the test set is that both the DR model and the TOP model with large vocabulary outperform the smaller vocabulary for rare events. However the difference for the all-events models is small.

When comparing the performance of all events and rare events in Table 3a and 3b, we see that the rare events model does better than the all events model for Chinese. This confirms that this is the right way of using classes. They should not be formed based on all events in the training set, but only based on rare events.

From Table 3a and Table 3b, it can be seen that the perplexity for DR is also lower than TOP. This again is evidence that the the DR model is superior to the top-level model.

Results are similar for the all events experiments with the DR model and the TOP model, in Chinese and English. Only using rare events increases performance, for both Chinese and English. All the Chinese models perform similarly to the English ones which confirms our previous suggestion that we can use all the models in Chinese and English.

Table 3. Performance of different vocabulary size $|B_i|$, classes trained on all events and rare events

(a) English models

	$|B_i|$	Model	all events			rare events		
			α_1/λ_1	α_2/λ_2	perp.	α_1/λ_1	α_2/λ_2	perp.
1	10,000	TOP	0.007	0.02	53.04	0.01	0.01	53.04
2	40,000	TOP	0.01	0.01	53.05	0.01	0.01	53.00
3	10,000	DR	0.08	0.40	53.03	0.10	0.40	52.97
4	40,000	DR	0.08	0.30	53.04	0.10	0.40	52.90

(b) Chinese models

	$|B_i|$	Model	all events			rare events		
			α_1/λ_1	α_2/λ_2	perp.	α_1/λ_1	α_2/λ_2	perp.
1	10,000	TOP	0.009	0.02	76.29	0.01	0.01	76.28
2	40,000	TOP	0.02	0.01	76.30	0.01	0.01	76.24
3	10,000	DR	0.05	0.30	76.29	0.10	0.40	76.21
4	40,000	DR	0.08	0.30	76.30	0.10	0.40	76.11

We present statistics for the word counts and word types for the parallel corpus in Table 4. We can see that the entropy of Chinese is larger than the entropy of English, (the entropy are computed by unigram model) that is why all the perplexity of Chinese is larger than English in all the models.

Table 4. The entropy of Chinese and English for the training set of parallel corpus

Language	Word Count	Word Type	Entropy
Chinese	203,689,494	585,230	10.15
English	210,618,352	579,936	9.887

For Chinese, we need fewer words to describe the same meaning in the parallel corpus, which shows that Chinese words have more information.

6   Related Work

A large number of studies have proposed class-based models. The Brown model [8] is a class sequence model, in which $p(u|w)$ is computed as the product of a class transition probability and an emission probability. In this work, we use a simplified Brown model which has fewer parameters.

Schütze [3] proposed a recursive model and a top-level interpolated model and two new models with a new polynomial discounting mechanism which we apply to Chinese in our work.

A large volume of work has been published on Chinese language models. An iterative procedure to build a Chinese language model was presented by Luo and Roukos [12].

An empirical study of clustering techniques for Asian language modeling was shown in [13]. Experimental tests are presented for cluster-based trigram models on a Chinese heterogeneous corpus.

Gao et al. [4] proposed a unified approach which automatically gathers training data from the web, creates a high-quality lexicon, segments the training data using this lexicon, and compresses the language model.

Both word-based and character-based models are explored in [14]. It is found that typically word-based modeling outperforms character-based modeling.

Also class-based models have been used in other applications such as part-of-speech tagging [15], speech recognition [16] and question answering [17].

Our work is similar to the combining of word- and class-based language models in [18]. Maltese et al. proposed various class-based language models used in conjunction with a word-based trigram language model by means of linear interpolation. The clustering method [3] is that clusters are formed based on word-based rare events, and that the modified model performs better.

7 Conclusion

In conclusion, we performed a comparative language modeling study on a Chinese and English parallel corpus. To our knowledge, we present the first such direct comparison of the perplexities of the two languages. We find that the entropy of English is smaller than Chinese, which should be the reason why the perplexity of English is smaller.

Additionally, we used novel class-based models to work on the Chinese-English parallel corpus. All results show that class-based models perform better than the KN model. Finally, from the comparison of all the models, we find that the DR model performs better than the TOP model in both languages. We also showed that models based on the DR model with a polynomial discounting mechanism improve results. Finally, we applied the rare-event model for Chinese and English that increases performance further. The results and analysis suggest that all the models also perform well in Chinese.

In the future, we would like to explore further ways to improve Chinese language models, including a more efficient algorithm for parameter optimization. We would also like to extend our work to part-of-speech tagged corpora, as well as comparing character and word-based corpora.

Acknowledgments. This work was funded by China Scholarship council. Junfei Guo wishes to express his gratitude to his supervisor, Prof. Dr. Hinrich Schütze who offered invaluable assistance and guidance during the work and study in University of Stuttgart. We also thank the colleagues in IMS at University of Stuttgart and anonymous reviewers for their comments.

References

1. Manning, C.D., Schütze, H.: Foundations of statistical natural language processing. MIT Press, Cambridge (1999)
2. Koehn, P.: Statistical Machine Translation, 1st edn. Cambridge University Press, New York (2010)
3. Schütze, H.: Integrating history-length interpolation and classes in language modeling. In: Proceedings of the 49th Annual Meeting of the Association for Computational Linguistics: Human Language Technologies, pp. 1516–1525. Association for Computational Linguistics, Portland (2011)
4. Gao, J., Goodman, J., Li, M., Lee, K.F.: Toward a unified approach to statistical language modeling for chinese, vol. 1(1), pp. 3–33 (March 2002)
5. Manning, C.D., Raghavan, P., Schtze, H.: Introduction to Information Retrieval. Cambridge University Press, New York (2008)
6. Chang, P.C., Galley, M., Manning, C.D.: Optimizing chinese word segmentation for machine translation performance. In: Proceedings of the Third Workshop on Statistical Machine Translation, StatMT 2008, pp. 224–232. Association for Computational Linguistics, Stroudsburg (2008)
7. Chen, S.F., Goodman, J.: An empirical study of smoothing techniques for language modeling. In: Proceedings of the 34th Annual Meeting on Association for Computational Linguistics, ACL 1996, pp. 310–318. Association for Computational Linguistics, Stroudsburg (1996)
8. Brown, P.F., deSouza, P.V., Mercer, R.L., Pietra, V.J.D., Lai, J.C.: Class-based n-gram models of natural language. Computational Linguistics 18, 467–479 (1992)
9. Dupont, P., Rosenfeld, R.: Lattice based language models. Technical report (1997)
10. Eisele, A., Chen, Y.: Multiun: A multilingual corpus from united nation documents. In: Chair, N.C.C., Choukri, K., Maegaard, B., Mariani, J., Odijk, J., Piperidis, S., Rosner, M., Tapias, D. (eds.) Proceedings of the Seventh International Conference on Language Resources and Evaluation (LREC 2010). European Language Resources Association (ELRA), Valletta (2010)
11. Stolcke, A.: Srilm - an extensible language modeling toolkit. In: Hansen, J.H.L., Pellom, B.L. (eds.) INTERSPEECH. ISCA (2002)
12. Luo, X., Roukos, S.: An iterative algorithm to build chinese language models. In: Proceedings of the 34th Annual Meeting on Association for Computational Linguistics, ACL 1996, pp. 139–143. Association for Computational Linguistics, Stroudsburg (1996)
13. Gao, J., Goodman, J.T., Miao, J.: The Use of Clustering Techniques for Language Modeling Application to Asian Languages
14. Luo, J., Lamel, L., Gauvain, J.L.: Modeling characters versuswords for mandarin speech recognition. In: Proceedings of the 2009 IEEE International Conference on Acoustics, Speech and Signal Processing, ICASSP 2009, pp. 4325–4328. IEEE Computer Society, Washington, DC (2009)
15. Schütze, H.: Distributional part-of-speech tagging. In: Proceedings of the Seventh Conference on European Chapter of the Association for Computational Linguistics, EACL 1995, pp. 141–148. Morgan Kaufmann Publishers Inc., San Francisco (1995)
16. Yokoyama, T., Shinozaki, T., Iwano, K., Furui, S.: Unsupervised class-based language model adaptation for spontaneous speech recognition. In: Proceedings of 2003 IEEE International Conference on Acoustics, Speech, and Signal Processing (ICASSP 2003), vol. 1, pp. I-236–I-239 (April 2003)

17. Momtazi, S., Klakow, D.: A word clustering approach for language model-based sentence retrieval in question answering systems. In: Proceedings of the 18th ACM Conference on Information and Knowledge Management, CIKM 2009, pp. 1911–1914. ACM, New York (2009)
18. Maltese, G., Bravetti, P., Crépy, H., Grainger, B.J., Herzog, M., Palou, F.: Combining word- and class-based language models: a comparative study in several languages using automatic and manual word-clustering techniques. In: Dalsgaard, P., Lindberg, B., Benner, H., Tan, Z.H. (eds.) INTERSPEECH, pp. 21–24. ISCA (2001)

Building a Bilingual Dictionary
from a Japanese-Chinese Patent Corpus

Keiji Yasuda and Eiichiro Sumita

National Institute of Information and Communications Technology
3-5, Hikaridai, Keihanna Science City, Kyoto,619-0289, Japan
{keiji.yasuda,eiichiro.sumita}@nict.go.jp

Abstract. In this paper, we propose an automatic method to build a bilingual dictionary from a Japanese-Chinese parallel corpus. The proposed method uses character similarity between Japanese and Chinese, and a statistical machine translation (SMT) framework in a cascading manner. The first step extracts word translation pairs from the parallel corpus based on similarity between Japanese kanji characters (Chinese characters used in Japanese writing) and simplified Chinese characters. The second step trains phrase tables using 2 different SMT training tools, then extracts common word translation pairs. The third step trains an SMT system using the word translation pairs obtained by the first and the second steps. According to the experimental results, the proposed method yields 59.3% to 92.1% accuracy in the word translation pairs extracted, depending on the cascading step.

1 Introduction

The number of patent applications in China has been increasing every year, and exceeded half a million in 2011. China has about 1.5 times the number of applications in Japan. As number of applications increase, so does the need for Japanese-Chinese cross language patent surveys, massively increase. Natural language processing (NLP) tools such as machine translation (MT) and cross-lingual information retrieval technologies play very important roles in streamling the patent survey process. For these technologies, high coverage bilingual dictionaries are one of the most important language resources. In this paper, we propose an automatic method for building a bilingual dictionary from an existing Japanese-Chinese parallel corpus.

The proposed method uses the character similarity between Japanese and Chinese and several natural language processing tools, including statistical machine translation (SMT) training toolkits, in a cascading manner.

Section 2 describes the proposed dictionary building method. Section 3 details the experiments extracting a Japanese-Chinese bilingual dictionary from a Japanese-Chinese patent corpus. Section 4 discuss the related research. Section 5 concludes the paper.

A. Gelbukh (Ed.): CICLing 2013, Part II, LNCS 7817, pp. 276–284, 2013.
© Springer-Verlag Berlin Heidelberg 2013

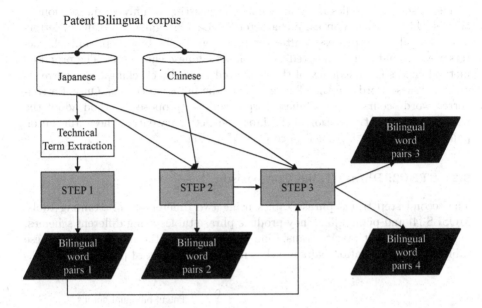

Fig. 1. Overview of the proposed method

2 Proposed Method

The proposed method uses character similarity between Japanese and Chinese and several NLP tools in a cascading manner. As shown in Fig. 1, the proposed method consists of three steps. As preprocessing for the dictionary extraction, Japanese and Chinese parts are first segmented by the POS taggers Chasen [1] and ICTCLAS [2]. Then, the Japanese is given to the automatic tool [3], which extracts technical terms from the given sentences. The proposed method uses the extracted technical terms as dictionary entries. Most of the outputs from the term extractor are composite words which use the Chasen segmentation unit as a component of the composite words. Details on the dictionary extraction steps are shown in Fig. 2 and 3. Each step is detailed in the following subsection.

2.1 STEP1: Character Conversion

The proposed method uses character similarity in Step 1. The writing system in mainland China uses a set of logograms called simplified Chinese characters. Meanwhile, the Japanese writing system uses two sets of phonograms (hiragana and katakana) and one set of logograms ("kanji," or Chinese characters). In this paper, we distinctly use the terms "kanji" and "simplified Chinese characters". They express kanji for the Chinese characters used in Japan, and simplified Chinese characters for the Chinese character used in mainland China, respectively. Most parts of Kanji set and Simplified Chinese character set are assigned to different fields in the Unicode code map, but, all but a few of kanji characters map onto simplified Chinese characters because kanji were borrowed from China.

The first step handles dictionary entries consisting of only kanji. Dictionary entries which contain non-kanji characters such as hiragana, katakana or letters of the alphabet are processed after the first step. In the first step, kanji characters are automatically converted to simplified Chinese characters. The proposed method checks the adequacy of the converted words by checking the occurrence of the Chinese words on the Chinese side of the bilingual corpus. Only if a converted word occurs in the Chinese corpus will the proposed method adopt the word as a Chinese translation of the Japanese dictionary entry. These translation pairs are shown as "Bilingual word pairs 1" in Fig. 2.

2.2 STEP2: Phrase Table Comparison

The second step trains phrase tables using two different SMT training tools, MOSES [4] and pialign [5]. They produce phrase tables using different schemes, and we can extract precise translation pairs by taking the intersection of these tables. These translation pairs are shown as "Bilingual word pairs 2" in Fig. 2.

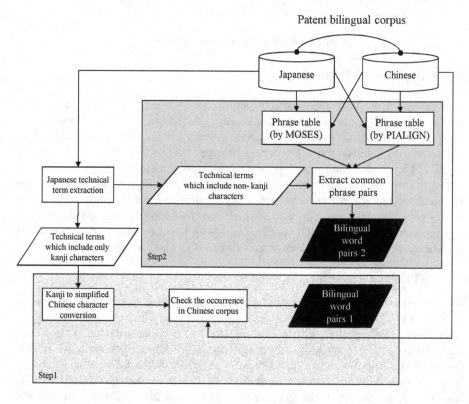

Fig. 2. Framework of the proposed method (Step 1 and 2)

2.3 STEP3: SMT Based Aproach

The third step trains two SMT systems: SMT1 and SMT2. SMT1 is an ordinal monotonic decoding system trained on the original Japanese-Chinese patent corpus. Meanwhile, SMT2 is trained on the word translation pairs obtained in steps 1 and 2. SMT2 regards the composite word unit and component sub-word unit, respectively as the sentence unit and the word unit in the ordinal SMT research.

Both systems monotonically decode segmented Japanese technical terms into Chinese technical terms. Then, one of two outputs is chosen as a Chinese translation of the input Japanese word by the selection scheme shown in Fig. 4. In the selection, SMT2 always has priority and SMT1 is used for supplemental purposes to check the confidence of the extracted translation. There are accuracy differences between Bilingual word pairs 3 and 4, the details of which are explained in section 3.

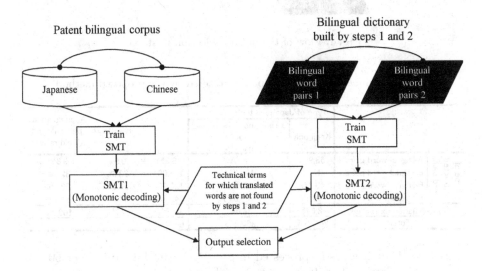

Fig. 3. Framework of the proposed method (Step3)

3 Experiments

In this section, we describe the experiments with the proposed method.

3.1 Experimental Settings

The Japanese-Chinese patent corpus used for the experiments is built by [6], which consists of 993 K bilingual sentence pairs. The preprocessing tool extracts 603K Japanese technical terms. 40% of the technical terms only consist of kanji

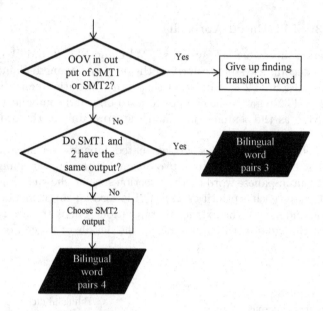

Fig. 4. Flow of the output selection in step3

Table 1. Number of translation pairs extracted by the proposed method

			Type of Japanese techinical term		Total	Percentage	Accumulative percentage (Yield ratio)
			Kanji only	Term including non-kanji character			
Type of extracted word pairs	Bilingual word pairs 1		58,951	N/A	58,951	9.8%	9.8%
	Bilingual word pairs 2		N/A	136,061	136,061	22.6%	32.3%
	Bilingual word pairs 3		43,318	80,471	123,789	20.5%	52.9%
	Bilingual word pairs 4		130,951	109,813	240,764	39.9%	92.8%
No translation extracted			8,675	34,641	43,316	7.2%	100.0%
Total			241,895	360,986	602,881		

characters. They are firstly processed in Step 1. Meanwhile, the other 60% is processed by Step 2. The Japanese technical terms which cannot be paired to the Chinese words in step 1 and 2 are processed in step 3 again.

3.2   Experimental Results

Table 1 shows the number of bilingual translation pairs extracted in each steps. As shown in the table, up to 92.8% of Japanse technical terms can be paired to the Chinese words using the proposed method.

Table 2 shows the precision of the products in each step. For the evaluation, we randomly sampled 200 translation pairs from each products as test sets. If the Japanese technical terms have problems due to automatic term extractor errors, they are removed from the test set. Using the test sets, we carried out a subjective evaluation by a Japanese-Chinese bilingual evaluator. An evaluation

Table 2. Evaluation results of the proposed method (Test set precision)

Type of extracted word pairs	Test set precision
Bilingual word pairs 1	92.1%
Bilingual word pairs 2	79.5%
Bilingual word pairs 3	85.6%
Bilingual word pairs 4	59.3%

criteria is if the meaning of the extracted Chinese translation perfectly matches the Japanese technical term or not.

As shown in the table, "Bilingual word pair 1" gives a high precision of 92.1%, although the yield ratio shown in Table 1 is 9.8%, which is very low. "Bilingual word pair 2" gives a lower precision than the "Bilingual word pair 1", but it is still around 80%. Meanwhile, the precision of "Bilingual word pair 3" even higher than "Bilingual word pair 2". "Bilingual word pair 4" yields the lowest precision, which is 59.3%. There are 10.7 points difference in precision between "Bilingual word pair 3 and 4". Hence, the supplemental usage of SMT1 works favorably.

We also evaluate translation results of SMT 1 by using the "Bilingual word pair 4" test set. The accuracy is 57.8%, which is 1.5 points lower than the results of SMT2 for checking the adequacy of the selection scheme shown in Fig. 4. We can conclude that the selection scheme is adequate.

By using the percentage values in Table 1 and the test set precision in Table 2, we can calculate estimated the precision and recall of connected set as follows:

$$P_1^n = \frac{\sum_{i=1}^n t_i \times w_i}{\sum_{i=1}^n w_i} \tag{1}$$

$$R_1^n = \sum_{i=1}^n t_i \times w_i \tag{2}$$

where P_1^n and R_1^n are the precision and recall of a connected set of "Bilingual word pair 1 to n", respectively. t_i is the test set precision of "Bilingual word pair i" shown in Table 2, and w_i is the percentage shown in Table 1, which is used as weight here.

These estimated values are shown in Table 3. By connecting all of the results ("Bilingual word pair 1 to 4") together, we obtain 68.2% on recall and 73.5% on precision.

3.3 Discussion

To discuss a better selection scheme than the one shown in Fig. 4, we analyze the evaluation results of "Bilingual word pairs 3 and 4". We classify these sets by the input feature (kanji-only term or other), then tally the results for each classification.

As shown in Table 4, input features largely affect precision. For "Bilingual word pair 3", the precision of kanji-only input is close to 90%, which is 7.8

Table 3. Evaluation results of the proposed method (Estimated precision and recall)

Type of extracted word pairs	Estimated precision	Estimated recall
Bilingual word pairs 1	92.1%	9.0%
Bilingual word pairs 1 to 2	83.3%	26.9%
Bilingual word pairs 1 to 3	84.2%	44.5%
Bilingual word pairs 1 to 4	73.5%	68.2%

Table 4. Detailed Evaluation results of Bilingual word pair 3 and 4

Type of Japanese techinical term	Output type	SMT1	SMT2
Kanji only	Bilingual word pairs 3	89.8%	
	Bilingual word pairs 4	53.5%	46.5%
Term including non–kanji character	Bilingual word pairs 3	81.0%	
	Bilingual word pairs 4	60.2%	72.4%

points higher than the precision for the other. In the results of "Bilingual word pair 4", the input feature reverses performance of SMT 1 and SMT 2.

Although these results are on a small data set, it should be possible to improve the selection method by using input features. We will work on an improved selection method by using a large data set in the near future.

4   Related Works

The idea of using the character similarity between kanji and simplified Chinese characters to build a Japanese-Chinese dictionary is not new. Goh et al. used this idea to build a Japanese-Chinese dictionary from manually-built Japanese-Chinese and Chinese-English dictionaries[7].

Tsunakawa et al. also use the idea to build an SMT-based term translator from Japanese-English and Chinese-Japanese dictionaries[8].

Both of these methods use character similarity to build Japanese-Chinese or Chinese-Japanese language resources by pivoting manually made dictionaries in different language pairs. However, the purpose of our research is to build a Japanese-Chinese bilingual dictionary from the parallel corpus, which is very different from this prior research.

Morishita et.al proposed a method foro extracting bilingual dictionary from a parallel corpus [9]. They carried out experiments using a Japanese-English bilingual patent corpus which consisted of 1.8 million sentence pairs[10]. The main idea of their research is similar to ours which is to extract precise word translation pairs from an SMT-based phrase table. They also combined multiple techniques and language resources, including SVM based classifier, manually-built Japanese English dictionary and so on, and their method obtained satisfactory results.

The main difference between Morishita's method and our proposed method is usage of a manually-built bilingual dictionary. The proposed method only uses character conversion knowledge, which is free from copyright. In most cases, manually built bilingual dictionary have usage limitations due to copyright, so a dictionary-free method has a lot of merit.

Additionally, we can use the products of the proposed method in Morishita's method by substituting the manually built dictionary for our product. This combination of the proposed method and Morishita's method may work well to expand and refine our products.

5   Conclusions and Future Works

We propose a method to automatically build a bilingual dictionary from a Japanese-Chinese bilingual corpus.

The proposed method uses several NLP tools and kanji-to-simplified conversion. First, the proposed method automatically extracts technical terms from the Japanese side of the corpus. Then, the method extracts the Chinese translation of the words in three cascading steps. As the required language resources for the proposed method are the bilingual corpus and a simple character conversion table, this method is applicable to any Japanese-Chinese bilingual corpus.

For the experiments, we used 200 randomly sampled translation pairs from the products of the proposed method. Then, we carried out a subjective evaluation. According to the experimental results, the proposed method can produce a Chinese translation up to 92.8% of the Japanese technical terms. The precision of the products is from 59.3% to 92.1%, and these values depend on the cascading steps.

Now, we are manually cleaning the products of the proposed method (all of "Bilingual word pairs 1 and 2") to release a Japanese-Chinese patent dictionary. We believe this language resource will help in the development of patent-related NLP tools in the near future.

References

1. Asahara, M., Matsumoto, Y.: Extended models and tools for high-performance part-of-speech tagger. In: Proceedings of COLING, pp. 21–27 (2000)
2. Zhang, H.P., Yu, H.K., Xiong, D.Y., Liu, Q.: Hhmm-based chinese lexical analyzer ictclas. In: Proceedings of the Second SIGHAN Workshop on Chinese Language Processing, vol. 17, pp. 184–187 (2003)
3. Nakagawa, H., Mori, T.: Automaic term recognition based on statistics of compound nouns and their components. Terminology 9(2), 201–209 (2003)
4. Hoang, H., Koehn, P.: Design of the moses decoder for statistical machine translation. In: Proceedings of ACL Workshop on Software Engineering, Testing and Quality Assurance for NLP, pp. 58–65 (2008)
5. Neubig, G., Watanabe, T., Sumita, E., Mori, S., Kawahara, T.: An unsupervised model for joint phrase alignment and extraction. In: Proceedings of the 49th Annual Meeting of the Association for Computational Linguistics: Human Language Technologies, pp. 632–641. Association for Computational Linguistics, Portland (2011)

6. Lu, B., Tsou, B.K., Jiang, T., Kwong, O.Y., Zhu, J.: Multilingual patents: An english-chinese example and its application to smt. In: Proceedings of the 1st CIPS-SIGHAN Joint Conference on Chinese Language Processing (CLP 2010) (2010)
7. Goh, C.-L., Asahara, M., Matsumoto, Y.: Building a japanese-chinese dictionary using kanji/Hanzi conversion. In: Dale, R., Wong, K.-F., Su, J., Kwong, O.Y. (eds.) IJCNLP 2005. LNCS (LNAI), vol. 3651, pp. 670–681. Springer, Heidelberg (2005)
8. Tsunakawa, T., Okazaki, N., Tsujii, J.: Building a bilingual lexicon using phrase-based statistical machine translation via a pivot language. In: Proceedings of the 22nd International Conference on Computational Linguistics Companion Volume Posters and Demonstrations, pp. 127–130 (2008)
9. Morishita, Y., Bing, L., Utsuro, T., Yamamoto, M.: Estimating translation of technical terms based on phrase translation table and a bilingual lexicon. IEICE Transactions on Information and Systems J-93D(11), 2525–2537 (2010) (in Japanese)
10. Utiyama, M., Isahara, H.: A japanese-english patent parallel corpus. In: Proceedings of MT Summit XI, pp. 475–482 (2007)

A Diagnostic Evaluation Approach
for English to Hindi MT
Using Linguistic Checkpoints and Error Rates

Renu Balyan[1,*], Sudip Kumar Naskar[2], Antonio Toral[2], and Niladri Chatterjee[1]

[1] Indian Institute of Technology Delhi, India
{renu17775,niladri.iitd}@gmail.com
[2] CNGL, School of Computing, Dublin City University, Dublin, Ireland
{snaskar,atoral}@computing.dcu.ie

Abstract. This paper addresses diagnostic evaluation of machine translation (MT) systems for Indian languages, English to Hindi MT in particular, assessing the performance of MT systems on relevant linguistic phenomena (checkpoints). We use the diagnostic evaluation tool DELiC4MT to analyze the performance of MT systems on various PoS categories (e.g. nouns, verbs). The current system supports only word level checkpoints which might not be as helpful in evaluating the translation quality as compared to using checkpoints at phrase level and checkpoints that deal with named entities (NE), inflections, word order, etc. We therefore suggest phrase level checkpoints and NEs as additional checkpoints for DELiC4MT. We further use Hjerson to evaluate checkpoints based on word order and inflections that are relevant for evaluation of MT with Hindi as the target language. The experiments conducted using Hjerson generate overall (document level) error counts and error rates for five error classes (inflectional errors, reordering errors, missing words, extra words, and lexical errors) to take into account the evaluation based on word order and inflections. The effectiveness of the approaches was tested on five English to Hindi MT systems.

Keywords: diagnostic evaluation, automatic evaluation metrics, DELiC4MT, Hjerson, checkpoints, errors.

1 Introduction

The evaluation of MT output is an important but difficult task. Human evaluation, which is still considered to be the most reliable method for evaluating MT systems uses fluency and adequacy, which are frequently measured together on a discrete 5 or 7 point scale, with their average being used as a single score of translation quality [1]. [2] found that inter-annotator agreements between human judgments were low for several tasks; they reported very low kappa values for fluency, adequacy and ranking of full sentences. It

---

* Work done while at CNGL, School of Computing, DCU.

A. Gelbukh (Ed.): CICLing 2013, Part II, LNCS 7817, pp. 285–296, 2013.

was found that inconsistencies occur even when the same annotator is presented the same sentences several times [3]. On top of that, human evaluation is very time consuming and expensive. Hence, cheap and fast automatic MT evaluation metrics are preferred over human evaluation, at least during intermediate steps of development of MT systems (a final human evaluation is still regarded indispensable). Consequently, design of automatic evaluation metrics plays a prominent role in MT research and development.

The state-of-the-art methods for automatic MT evaluation use n-gram based metrics represented by BLEU [4] and closely related NIST [5]. METEOR [6-7] additionally considers stems and synonyms of the words. TER [8] measures the amount of editing required for changing the MT output so that it matches the reference exactly. Globally, these automatic MT evaluation metrics are being studied with great interest for different language pairs. However, their direct applicability to Hindi or other Indian languages needs proper investigation. There have been some efforts in this direction for Indian languages [9-13]. Barring these few exceptions, the subject has not been studied deeply. In addition, most of these approaches cover either human evaluation, or consider modification of existing automatic metrics to make them more suitable for Indian languages. None of these approaches, or their extensions, inform the MT developers or users on the strengths and weaknesses of the MT system concerned and the nature of translation errors made by the system. Consequently, it is difficult for the developers to understand the capability of different modules of an MT system. MT developers need an evaluation approach which can provide useful feedback on the translation quality of MT system in terms of various linguistic features. Similarly, from the user perspective, an evaluation scheme that can assess the strengths and weaknesses of a given MT system is required.

A good metric can be formulated by considering two points. On one hand it should take care of the major mistakes that occur during translation from one language to another. On the other hand, it should take into account how human users evaluate translations in the event of an error. Such a metric should identify these mistakes and their relative importance and measure the translation quality accordingly.

In this paper we consider the relevant errors that may occur during translation from English to Hindi as the main linguistic units. The linguistic units that have been considered for this study are related not only to PoS-based phenomena but also to other types: named entities (NEs), inflections, word order, and phrase based entities. It is important to perform evaluation in terms of these linguistic units and provide feedback about the system performance on these units.

Our final aim is to come up with an approach for diagnostic evaluation of MT that can be adapted to Indian languages. In the present work the experiments have been carried out with the DELiC4MT [14] and Hjerson [15] toolkits. The experiments have been carried out to adapt the tools for Hindi, which can later be extended to other related Indian languages. To the best of our knowledge this is a pioneering work in the direction of diagnostic evaluation with respect to Indian languages.

This paper is an extension of the work reported in [29], which was a first step to-wards the development of diagnostic evaluation measures for Indian languages based on linguistic checkpoints. Our previous work defines a taxonomy of relevant linguistic checkpoints for the English–Hindi language pair (considering both directions) but evaluates only the subset of checkpoints that regard word level phenomena in the English to Hindi direction (presented also here in Table 1). [29] also presented the correlation between commonly used automatic metrics and system level score from DELiC4MT. The current paper extends our previous work by implementing and evaluating phrase level and named entity checkpoints as well as exploring a set of error rates.

The rest of the paper is organized as follows. Related work on diagnostic evaluation is discussed in Section 2. Section 3 and 4 give a brief overview of the diagnostic evaluation tools that have been used for this study: DELiC4MT and Hjerson. Section 5 discusses about linguistic checkpoints. In Section 6 we present the experimental setup and compare the results obtained on the English–Hindi test set using DEL-iC4MT, Hjerson and automatic evaluation metrics. This is followed by conclusions and avenues for future work.

2   Related Work

Diagnostic evaluation of MT has been occasionally addressed in the literature in the last few years. A framework proposed by [16] analyzes the errors manually. The scheme covers five top-level classes: missing words, incorrect words, unknown words, word order and punctuation errors. They identified some important classes of errors for English to Spanish and Chinese to English language pairs using this framework. [17] carried out manual error analysis for Spanish–Catalan and classified errors at orthographic, morphological, lexical, semantic, and syntactic level. However, human error analysis is as time consuming as human evaluation. Automatic methods for error analysis using base forms and PoS tags have been proposed in [18-19]. The proposed methods have been used for estimation of inflectional and reordering errors. The methods use relative differences between the word error rate (WER) and position independent word error rate (PER) for nouns, adjectives and verbs. [20] presents a method for automatic error classification which was found to correlate well with human judgments. [15] describes a tool that classifies errors into five categories based on the hierarchy proposed by [16]. RGBF, a tool for automatic evaluation of MT output based on n-gram precision and recall is described in [21]. The tool calculates the F-score averaged on all n-grams of an arbitrary set of distinct units such as words, morphemes, PoS tags, etc. [22] introduced Addicter, a tool for Automatic Detection and DIsplay of Common Translation errors. The tool allows automatic identification and labeling of translation errors. In [23] translation quality based on the frequencies of different error categories is quantified. [24] used a classifier trained with a set of linguistic features to automatically detect incorrect segments in MT output. [25] proposed diagnostic evaluation of linguistic checkpoints obtained by aligning the parsed source and target language sentences. [26] proposed a framework for diagnostic MT evaluation which offers similar functionality as in [25] but it is language independent. It also provides additional functionality like filtering of noisy checkpoint instances based on PoS tags. The tool however considers only PoS-based linguistic units as checkpoints.

3   DELiC4MT: A Diagnostic MT Evaluation Tool

DELiC4MT [1] (Diagnostic Evaluation using Linguistic Checkpoints for Machine Translation) is an open source tool for diagnostic evaluation of MT. It allows evaluation of MT systems over linguistic phenomena. The various steps involved for diagnostic evaluation using DELiC4MT are: text analysis, word alignment, defining Kybots and evaluation. The evaluation pipeline proceeds as follows:

- The source and target sides of the gold standard (test set) are processed by respective PoS taggers (Treetagger [2] for English and a Hindi PoS tagger for this study) and converted into KYOTO Annotation Format (KAF) [3], to represent textual analysis.
- The test set is word aligned using GIZA++, [4] and identifiers of the aligned tokens are stored.
- Kybot profiles [5] specifying the linguistic checkpoints to be extracted are run on the KAF text, and the matching terms are extracted.
- The evaluation module takes as input the kybot output, the KAF files for source and target languages, the word alignments and the plain output of an MT system. It calculates the performance of the MT system over the linguistic checkpoint(s) considered.

Further details regarding KAF, kybot profiles and calculation of the recall scores can be found in [26].

4   Hjerson

Hjerson implements the edit distance algorithm and identifies the actual words contributing to the standard WER as well as to the recall/precision based PERs called Reference PER (RPER) and Hypothesis PER (HPER) [27]. The RPER errors are defined as the words in the reference which do not appear in the hypothesis, and the HPER errors are the words in the hypothesis which do not appear in the reference. Once the WER, RPER and HPER errors have been identified, the lemmas for each word are added and error classification is performed according to [16] in the following way:

- Inflectional Error: a word in which the full form is marked as RPER/HPER error but the lemmas are the same.
- Reordering Error: a word which occurs both in the reference and in the hypothesis, thus not contributing to RPER or HPER, but is marked as a WER error.

---

[1] http://www.computing.dcu.ie/~atoral/delic4mt
(under the GPL-v3 license).
[2] http://www.ims.unistuttgart.de/projekte/corplex/TreeTagger/
[3] A XML format for text analysis based on representation standards from ISO TC37/SC4.
[4] http://code.google.com/p/giza-pp/
[5] Kybot profiles can be understood as regular expressions over KAF documents,
http://kyoto.let.vu.nl/svn/kyoto/trunk/modules/mining_module/

— Missing Word: a word which occurs as deletion in WER errors and at the same time occurs as RPER error without sharing the lemma with any hypothesis error.
— Extra Word: a word which occurs as insertion in WER errors and at the same time occurs as HPER error without sharing the lemma with any reference error.
— Incorrect Lexical Choice: a word which belongs neither to inflectional errors nor to missing or extra words is considered as lexical error.

Although the method is language independent, availability of lemmas for the particular target language is a requisite. For morphologically rich languages, the error classification should be carried out with base forms, otherwise the morphological errors are not detected and the results would be noisy. The details regarding the errors can be found in [15].

5   Linguistic Checkpoints

A linguistic checkpoint is a linguistically motivated unit; for example, it can be an ambiguous word, a NE, a verb particle construction, a noun-noun compound etc. The level of detail and the specific linguistic phenomena included in the evaluation can vary depending on what the users want to investigate as part of the diagnostic evaluation. The categories that are out of scope for current NLP tools to be recognized have been ignored in this study. The checkpoints included in our evaluation comprise typical linguistic phenomena at word and phrase level.

The DELiC4MT system currently works for word level PoS-based checkpoints only. In practice, any tag used by parsers (e.g. NP, VP, PP, etc.) could be added as a new category easily; though currently these are not implemented in the system. In this study the phrase level checkpoints (e.g. NP and PP) have been extracted indirectly using some of the most frequent PoS patterns for these phrases. The most frequent patterns for NP and PP are identified from the parsed test set. Some of these patterns for noun phrase (NP) are: "a *determiner or a pronoun followed by a noun*" and "*a determiner or a pronoun followed by an adjective and a noun*". Similarly, some of the commonly occurring patterns for prepositional phrase (PP) are; "*'to' followed by a noun*", "*'to' followed by a noun phrase (NP)*", "*a preposition followed by a noun*" and "*a preposition followed by a noun phrase (NP)*". On similar lines, frequent patterns for other phrase level checkpoints (viz. verb phrase (VP), adjective phrase (JJP)) can also be extracted, and the translation quality on these phrase level checkpoints can be examined.

The NE checkpoint has been implemented using a standard NER tool (Stanford NER)[6]. This labels sequences of words in the test set which belong to either of the three classes (person, organization, location). This checkpoint is important as we observed that the existing English to Hindi MT systems do not handle NEs properly. Typically they provide literal translations of the words that constitute an NE, thus leading to poor translation quality.

---

[6] http://nlp.stanford.edu/software/CRF-NER.shtml

The inflectional checkpoint is important for Hindi as adjectives and verbs in a Hindi sentence get inflected according to the gender, number and person (GNP) of the head noun. The verbs get inflected based on the tense, aspect and modality (TAM). Incorrect inflections result in translation errors. Postpositions also play a crucial role in Hindi. They indicate the case markers for nouns. A missing, incorrect or extra postposition can alter the meaning of a sentence completely.

Hindi is a relatively free phrase order language i.e. a sentence can be written correctly in multiple ways by just changing the order of the phrases in a sentence. Although the re-ordered sentences may be semantically equivalent, some of them may be more fluent or frequently used than others. Thus, word order needs to be considered as a checkpoint.

6   Evaluation

6.1   Experimental Setup

The test set considered for this study consists of 1,000 sentences (21,129 words) from the tourism domain. DELiC4MT and Hjerson have been used for diagnostic evaluation of five English to Hindi MT systems: Google Translate[7] (MT1), Bing Translator[8] (MT2), Free-translations[9] (MT3), MaTra2[10] (MT4) and Anusaaraka[11] (MT5). GIZA++ was used for word alignment. Since the test set is far too small to be word aligned using statistical word aligners, an additional parallel corpus comprising of 25,000 sentences (424,595 words) from the same domain was used to avoid data sparseness during alignment. The test set was appended to the additional corpus and the word alignments were generated. Finally the word alignments for the test set sentences were extracted. Treetagger was used to PoS-tag the English dataset, while the Hindi dataset was PoS-tagged using the PoS tagger developed by IIIT, Hyderabad.[12] Regarding linguistic checkpoints, we have considered linguistic units at two levels: word and phrase level. Simple PoS-based checkpoints (noun, verb, adjective, etc.) have been considered at the word level. Four phrase level checkpoints: noun compounds (NCs), verb particle constructions (VPCs), NPs and PPs - have been considered for this study. Other phrase level checkpoints like verb phrase (VP), adverb phrase (RBP) etc. could also be evaluated similarly. Stanford NER was used to extract NE information from the test set. Finally the performance of the system on NEs is evaluated using the usual approach used by DELiC4MT.

Evaluation of checkpoints related to re-ordering, inflections, lexical errors, extra and missing words has been performed using the Hjerson toolkit. A Hindi PoS tagger[9] was used to extract the base forms and PoS tags for the reference and hypothesis translations. In addition to the surface form, PoS tagged forms are also used for evaluation using 'Hjerson'.

---

[7] http://translate.google.com/
[8] http://www.bing.com/translator/
[9] http://www.free-translator.com/
[10] http://www.cdacmumbai.in/matra/
[11] http://anusaaraka.iiit.ac.in/
[12] http://sivareddy.in/downloads

6.2   Results and Discussions

The scores for word level checkpoints across the MT systems using DELiC4MT are shown in Table 1. In addition to the diagnostic evaluation scores, the table also shows the number of instances obtained for each checkpoint. Checkpoint-specific best scores are shown in bold. Checkpoint-specific statistically significant improvements (calculated using paired bootstrap resampling [28]) are reported and shown as superscripts. For representation purposes, we use a, b, c, d and e for MT1, MT2, MT3, MT4 and MT5 respectively. For example, the MT1 score 0.5539^d for the pronoun checkpoint in Table 1 indicates that the improvement provided by MT1 for this checkpoint is statistically significant over MT4 (d).

Table 1. DELiC4MT scores for word level checkpoints for MT systems

Checkpoint	Instances	MT1(a)	MT2(b)	MT3(c)	MT4(d)	MT5(e)
Noun	4538	$\mathbf{0.3792^{b,d,e}}$	$0.3568^{d,e}$	$0.3776^{b,d,e}$	0.2552	0.2925^d
Pronoun	276	$\mathbf{0.5539^d}$	0.5000	$\mathbf{0.5539^d}$	0.4059	0.5490^d
Possessive-Pronoun	184	$\mathbf{0.3464^{d,e}}$	$0.3333^{d,e}$	$\mathbf{0.3464^{d,e}}$	0.0196	0.1699^d
Adjective	1859	$\mathbf{0.3785^{b,d,e}}$	$0.3574^{d,e}$	$0.3772^{b,d,e}$	0.2061	0.2699^d
Adverb	663	$\mathbf{0.4347^d}$	0.4288^d	0.4327^d	0.2402	0.4103^d
verb	2580	$\mathbf{0.2656^{d,e}}$	0.2584^d	$\mathbf{0.2656^{d,e}}$	0.1789	0.2402^d
Preposition	2667	$\mathbf{0.6655^{d,e}}$	$0.6555^{d,e}$	$0.6646^{d,e}$	0.5434	0.6217^d
Modal	128	0.3913	0.3696	0.3913	0.3478	**0.4239**

The following observations were made for evaluation of word level checkpoints:

- MT1 outperforms all the other MT systems in every category except modals.
- MT3 performs almost at par with MT1.
- Verb seems to be the most problematic checkpoint among the word level checkpoints for all the systems except for MT4 and MT5.
- MT5 performs best for the modals category in comparison to the rest of systems.
- All the systems perform poorly on possessive pronouns compared to pronouns in general. All the systems perform best on prepositions followed by the pronouns category.
- MT1, MT2 and MT3 systems perform better for adverbs as compared to modals, whereas MT4 and MT5 perform just in the reverse manner for these categories.

Table 2 shows the scores obtained for phrase level checkpoints using DELiC4MT. We consider the following representation for Table 2: determiner (DT), pronoun (Pro), noun (N), adjective (J) and prepositions (Pr) and prepositional phrase (PP).

Table 2. DELiC4MT scores for phrase level checkpoints for MT systems

Checkpoint	Instance	MT1(a)	MT2(b)	MT3(c)	MT4(d)	MT5(e)
NC	592	**0.2878**^{d,e}	0.2696^{d,e}	0.2863^{d,e}	0.1726	0.2012
VPC	52	0.2000	**0.2545**^{d}	0.2000	0.0364	0.1636
NP	1646	**0.3243**	0.3008	0.3241	0.1755	0. 2284
DT-N	985	**0.4125**^{d,e}	0.3950^{d,e}	0.4120^{d,e}	0.2784	0. 3201
DT-J-N	499	**0.3416**^{d,e}	0.3270^{d,e}	**0.3416**^{d,e}	0.2198	0. 2733^{d}
Pro-N	116	**0.3346**^{d,e}	0.2885^{d,e}	**0.3346**^{d,e}	0.1154	0. 1692
Pro-J-N	46	**0.2083**^{d}	0.1927^{d}	**0.2083**^{d}	0.0885	0. 1510^{d}
PP	1250	**0.3362**	0.3074	0.3361	0.1861	0. 2559
TO-N	27	0.3529	**0.3676**	0.3529	0.2059	0. 2500
Pr-N	447	**0.4178**^{d}	0.3904^{d}	**0.4178**^{d}	0.2664	0. 3970^{d}
TO-DT-N	42	**0.3442**^{d,e}	0.3247^{d,e}	**0.3442**^{d,e}	0.1753	0. 2143
TO-DT-J-N	15	**0.3158**	0.2763	**0.3158**	0. 1184	0. 2368
TO-Pro-N	5	**0.3704**	0.2222	**0.3704**	0.1481	0. 2593
TO-Pro-J-N	1	**0.1667**	**0.1667**	**0.1667**	**0.1667**	0.1389
Pr-DT-N	435	**0.4521**^{d,e}	0.4269^{d,e}	0.4515^{d,e}	0.2837	0. 3588^{d}
Pr-DT-J-N	198	**0.3387**^{b,d,e}	0.3097^{d}	**0.3387**^{d,e}	0.1908	0. 2823^{d}
Pr-Pro-N	51	**0.3762**^{d}	0.3476^{d}	**0.3762**^{d,e}	0.1619	0. 2333
Pr-Pro-J-N	29	0.2271	**0.2415**	0.2271	0.1401	0. 1884

The observations regarding phrase level checkpoints are:

- All the systems perform best for PPs followed by NPs, NCs and VPCs.
- MT2 system performs better than all the other MT systems for VPCs. It is to be noted that MT1 and MT3 produced the best scores for verbs (cf. Table 1).
- All the systems perform worst for VPCs. The results shown here are only for instances of non-separable VPCs. Non-separable VPCs are the VPCs where the particle immediately follows the verb and the verb and particle cannot be separated by any NP. The scores may further degrade if separable VPCs are also taken into consideration.

Scores obtained for the NE checkpoint are shown in Table 3. We have considered location, organization, person and date as NE categories.

Table 3. DELiC4MT scores for NE (location, person, date and organization) checkpoints

Checkpoint	Instances	MT1	MT2	MT3	MT4	MT5
Location	255	**0.2988**	0.2500	**0.2988**	0.1893	0.1422
Organization	315	**0.2851**	0.2654	0.2816	0.1832	0.1186
Person	224	**0.3148**	0.2194	0.3132	0.1765	0.1741
Date	101	0.3249	0.2996	0.3249	0.2318	**0.3502**
NEs (All)	1093	**0.3127**	0.2735	0.3085	0.2033	0.1811

Some of the observations related to the evaluation of NEs are given below:

- MT1 outperforms all the other MT systems in all the NE categories except for the date category where MT5 performs the best. This is expected since RBMT systems are particularly good at translating this kind of expressions.
- MT4 performs better than MT5 for all NE categories except dates, whereas it performs poorly for all the previous checkpoints in comparison to MT5 (cf. Table 1 and 2).

The summary of word level, phrase level, NE and system level scores for all the MT systems are shown in Table 4. The system level scores are calculated by taking the weighted average of word level, phrase level and NE scores. Weighted average is calculated by taking into account the number of instances for each checkpoint. The scores decrease as we move from word level to phrase level checkpoints.

Table 4. Summary for DELiC4MT word, phrase, system level and NE scores

Checkpoint	Instances	MT1	MT2	MT3	MT4	MT5
Word level	12895	**0.4269**	0.4075	0.4262	0.2746	0.3722
Phrase level	3540	**0.2871**	0.2831	0.2866	0.1427	0.2123
NEs	1093	**0.3127**	0.2735	0.3085	0.2033	0.1811
Total / Average Scores	17528	**0.3422**	0.3214	0.3404	0.2069	0.2552
System Scores (Weighted)	-	**0.3915**	0.3740	0.3907	0.2435	0.3280

The document level error rates for all the MT systems as calculated by Hjerson are shown in Table 5.

Table 5. Hjerson Inflectional, Re-ordering, Lexical, Missing and Extra words scores

Types of Errors		MT1	MT2	MT3	MT4	MT5
INFER	ref	5.9	**5.85**	5.9	6.61	6.04
	hypo	6.16	6.08	6.16	7.90	**5.99**
RER	ref	14.77	16.02	14.73	**11.40**	13.54
	hypo	15.42	16.66	15.37	13.63	**13.41**
LEXER	ref	**41.40**	43.23	41.42	48.82	49.71
	hypo	**42.45**	44.30	42.47	55.20	49.20
MISER		8.29	7.90	8.36	15.19	**6.71**
EXTER		5.03	5.06	5.12	**2.4**	7.30

The following observations have been made in terms of different errors for the MT systems:

- MT2 and MT5 produce the minimum number of inflectional errors for reference and hypothesis translations respectively.

- MT4 and MT5 have the least amount of re-ordering errors whereas MT2 has the maximum number of errors for this category.
- MT5 has the minimum number of missing words and MT4 has maximum number of missing words. MT4 has minimum number of extra words but MT5 produces large number of extra words.
- MT1 has minimum and MT4, MT5 have the maximum number of Lexical errors.

The performance of all the MT systems was also evaluated using automatic evaluation metrics: BLEU, NIST, METEOR and TER. According to all the automatic evaluation metrics MT1 performs best followed by MT3, MT2, MT5 and MT4, the same order obtained with system level scores by DELiC4MT (the only exception being MT4 ranked higher than MT5 by TER). However, the point to be noted here is that with automatic evaluation metrics we do not get any additional information about the systems' performance other than the system level scores. The automatic metrics do not provide any information regarding the linguistic features - as to which linguistic units a system translates well or on which ones it performs poorly but DELiC4MT and Hjerson do provide that information.

7   Concluding Remarks and Future Work

This paper has presented a study on diagnostic evaluation of MT for Hindi as the target language. The main objective of the work was to assess the applicability of the state-of-the-art diagnostic evaluation tools DELiC4MT and Hjerson for Indian languages in general, and Hindi in particular. The linguistic checkpoints considered for this study are PoS-based (both word and phrase level), NEs, word order, inflections, missing words, extra words and lexical words. In total 18 checkpoints have been considered for the study. The paper has presented a detailed analysis of the results obtained for five English to Hindi MT systems using DELiC4MT and Hjerson. The translations obtained from these MT systems were also evaluated using some of the most commonly used automatic evaluation metrics. As far as the MT systems are concerned, Google proved to be the best among the 5 systems according to both automatic evaluation metrics and DELiC4MT. The results obtained from Hjerson regarding the linguistic units related to inflections, re-ordering, missing, extra and lexical words indicate that MaTra and Anusaraka perform better for inflection and re-ordering related checkpoints as compared to the other MT systems. The work offers a number of possibilities for future work. There is a need to find which linguistic checkpoints are more important and therefore provide a weight to each checkpoint based on its relative importance. The authors plan to work in this direction in the future.

Acknowledgements. This work has been funded in part by the European Commission through the CoSyne project (FP7-ICT-4-248531) and Science Foundation Ireland (Grant No. 07/CE/I1142) as part of the Centre for Next Generation Localisation (www.cngl.ie) at Dublin City University. The authors would also like to thank Josef Van Genabith, CNGL, Dublin City University, Vineet Chaitanya, IIIT-Hyderabad, Sasi Kumar, CDAC and Siva Reddy for providing support.

References

1. Snover, M., Madnani, N., Dorr, B.J., Schwartz, R.: Fluency, Adequacy, or HTER? Exploring Different Human Judgments with a Tunable MT Metric. In: Proceedings of the 4th EACL Workshop on Statistical Machine Translation, pp. 259–268. Association for Computational Linguistics, Athens (2009)
2. Callison-Burch, C., Fordyce, C., Koehn, P., Monz, C., Schroeder, J.: (Meta-) evaluation of machine translation. In: Proceedings of the Second Workshop on Statistical Machine Translation, Prague, Czech Republic, pp. 136–158 (2007)
3. Stymne, S., Ahrenberg, L.: On the practice of error analysis for machine translation evaluation. In: Proceedings of 8th International Conference on Language Resources and Evaluation (LREC 2012), Istanbul, Turkey, pp. 1785–1790 (2012)
4. Papineni, K., Roukos, S., Ward, T., Zhu, W.: BLEU: A method for automatic evaluation of machine translation. In: Proceedings of 40th Annual Meeting of the ACL, Philadelphia, PA, USA, pp. 311–318 (2002)
5. Doddington, G.: Automatic Evaluation of Machine Translation Quality Using N-gram Co-Occurrence Statistics. In: Proceedings of the Human Language Technology Conference (HLT), San Diego, CA, pp. 128–132 (2002)
6. Banerjee, S., Lavie, A.: METEOR: An automatic metric for MT evaluation with improved correlation with human judgments. In: Proceedings of the ACL 2005 Workshop on Intrinsic and Extrinsic Evaluation Measures for MT and/or Summarization, Ann Arbor, Michigan, pp. 65–72 (2005)
7. Lavie, A., Agarwal, A.: METEOR: An automatic metric for MT evaluation with high levels of correlation with human judgments. In: Proceedings of the Second ACL Workshop on Statistical Machine Translation, Prague, Czech Republic, pp. 228–231 (2007)
8. Snover, M., Dorr, B., Schwartz, R., Micciulla, L., Makhoul, J.: A study of translation edit rate with targeted human annotation. In: Proceedings of Association for Machine Translation in the Americas, AMTA 2006, Cambridge, MA, pp. 223–231 (2006)
9. Chatterjee, N., Balyan, R.: Towards Development of a Suitable Evaluation Metric for English to Hindi Machine Translation. International Journal of Translation 23(1), 7–26 (2011)
10. Gupta, A., Venkatapathy, S., Sangal, R.: METEOR-Hindi: Automatic MT Evaluation Metric for Hindi as a Target Language. In: Proceedings of ICON 2010: 8th International Conference on Natural Language Processing. Macmillan Publishers, India (2010)
11. Ananthakrishnan, R., Bhattacharyya, P., Sasikumar, M., Shah, R.: Some issues in automatic evaluation of English-Hindi MT: More blues for BLEU. In: Proceeding of 5th International Conference on Natural Language Processing (ICON 2007), Hyderabad, India (2007)
12. Chatterjee, N., Johnson, A., Krishna, M.: Some improvements over the BLEU metric for measuring the translation quality for Hindi. In: Proceedings of the International Conference on Computing: Theory and Applications, ICCTA 2007, Kolkata, India, pp. 485–490 (2007)
13. Moona, R.S., Sangal, R., Sharma, D.M.: MTeval: A Evaluation methodolgy for Machine Translation system. In: Proceedings of SIMPLE 2004, Kharagpur, India, pp. 15–19 (2004)
14. Toral, A., Naskar, S.K., Gaspari, F., Groves, D.: DELiC4MT: A Tool for Diagnostic MT Evaluation over User-defined Linguistic Phenomena. The Prague Bulletin of Mathematical Linguistics 98, 121–131 (2012)
15. Popović, M.: Hjerson:An Open Source Tool for Automatic Error Classification of Machine Translation Output. The Prague Bulletin of Mathematical Linguistics 96, 59–68 (2011)

16. Vilar, D., Xu, J., Fernando, L., D'Haro, N.H.: Error analysis of statistical machine translation output. In: Proceedings of Fifth International Conference on Language Resources and Evaluation (LREC 2006), Genoa, Italy, pp. 697–702 (2006)

17. Farrús, M., Costa-jussà, M.R.: Mariño, J. B., Fonollosa, J. A. R.: Linguistic-based evaluation criteria to identify statistical machine translation errors. In: Proceedings of EAMT, Saint Raphaël, France, pp. 52–57 (2010)

18. Popović, M., Ney, H.: Towards automatic error analysis of machine translation output. Computational Linguistics 37(4), 657–688 (2011)

19. Popović, M., Ney, H., Gispert, A.D., Mariño, J.B., Gupta, D., Federico, M., Lambert, P., Banchs, R.: Morpho-syntactic information for automatic error analysis of statistical machine translation output. In: StatMT 2006: Proceedings of the Workshop on Statistical Machine Translation, New York, pp. 1–6 (2006)

20. Popović, M., Burchardt, A.: From human to automatic error classification for machine translation output. In: Proceedings of EAMT 2011, Leuven, Belgium, pp. 265–272 (2011)

21. Popović, M.: rgbF: An Open Source Tool for n-gram Based Automatic Evaluation of Machine Translation Output. The Prague Bulletin of Mathematical Linguistics 98, 99–108 (2012)

22. Zeman, D., Fishel, M., Berka, J., Bojar, O.: Addicter: What Is Wrong with My Translations? The Prague Bulletin of Mathematical Linguistics 96, 79–88 (2011)

23. Fishel, M., Sennrich, R., Popović, M., Bojar, O.: TerrorCat: a translation error categorization-based MT quality metric. In: WMT 2012 Proceedings of the Seventh Workshop on Statistical Machine Translation, Stroudsburg, PA, USA, pp. 64–70 (2012)

24. Xiong, D., Zhang, M., Li, H.: Error detection for statistical machine translation using linguistic features. In: Proceedings of ACL 2010, Uppsala, Sweden, pp. 604–611 (2010)

25. Zhou, M., Wang, B., Liu, S., Li, M., Zhang, D., Zhao, T.: Diagnostic Evaluation of Machine Translation Systems using Automatically Constructed Linguistic Checkpoints. In: Proceedings of 22nd International Conference on Computational Linguistics (COLING 2008), pp. 1121–1128. Manchester (2008)

26. Naskar, S.K., Toral, A., Gaspari, F., Ways, A.: A framework for Diagnostic Evaluation of MT based on Linguistic Checkpoints. In: Proceedings of the 13th Machine Translation Summit, Xiamen,China, pp. 529–536 (2011)

27. Popović, M., Ney, H.: Word Error Rates: Decomposition over POS classes and Applications for Error Analysis. In: Proceedings of the 2nd ACL 2007 Workshop on Statistical MachineTranslation (WMT 2007), Prague, Czech Republic, pp. 48–55 (2007)

28. Koehn, P.: Statistical Significance Tests for Machine Translation Evaluation. In: Proceedings of the Conference on Empirical Methods on Natural Language Processing, EMNLP, pp. 385–395 (2004)

29. Balyan, R., Naskar, S.K., Toral, A., Chatterjee, N.: A Diagnostic Evaluation Approach Targeting MT Systems for Indian Languages. In: Proceedings of the Workshop on Machine Translation and Parsing in Indian Languages (MTPIL-COLING 2012), Mumbai, India, pp. 61–72 (2012)

Leveraging Arabic-English Bilingual Corpora with Crowd Sourcing-Based Annotation for Arabic-Hebrew SMT

Manish Gaurav, Guruprasad Saikumar, Amit Srivastava, Premkumar Natarajan, Shankar Ananthakrishnan, and Spyros Matsoukas

Raytheon BBN Technologies, 10 Moulton Street, Cambridge, MA 02138, USA
{mgaurav,gsaikuma,asrivast,pnataraj,sanantha,smatsouk}@bbn.com

Abstract. Recent studies in Statistical Machine Translation (SMT) paradigm have been focused on developing foreign language to English translation systems. However as SMT systems have matured, there is a lot of demand to translate from one foreign language to another language. Unfortunately, the availability of parallel training corpora for a pair of morphologically complex foreign languages like Arabic and Hebrew is very scarce. This paper uses active learning based data selection and crowd sourcing technique like Amazon Mechanical Turk to create Arabic-Hebrew parallel corpora. It then explores two different techniques to build Arabic-Hebrew SMT system. The first one involves the traditional cascading of two SMT systems using English as a pivot language. The second approach is training a direct Arabic-Hebrew SMT system using sentence pivoting. Finally, we use a phrase generalization approach to further improve our performance.

Keywords: Arabic-Hebrew Statistical Machine Translation, Sentence Pivoting, Amazon Mechanical Turk, Active Learning, Phrase Generalization.

1 Introduction

Over the last decade, Statistical Machine Translation (SMT) systems have shown to produce adequate translation performance for language pairs with large amounts of bilingual training corpora available. Unfortunately, large parallel corpora do not exist for many foreign language pairs like Arabic-Hebrew or Chinese-Arabic etc. The standard annotation process for creating a bilingual corpus for a foreign language pair can be very expensive. Much work has been done to overcome the lack of parallel corpora.

For example, Utiyama and Isahara [6] extract Japanese-English parallel sentences from a noisy-parallel corpus. Resnik and Smith [10] propose mining the web to collect parallel corpora for low-density language pairs. Munteanu and Marcu [2] extract parallel sentences from large Chinese, Arabic, and English non-parallel newspaper corpora. Another approach to overcome the problem of lack of parallel corpora is to use a pivot language approach (Cohn and Lapata, [13]; Utiyama and Isahara, [7]; Wu and Wang [4]; Bertoldi et al. [8]), where a third, more frequent language is leveraged

A. Gelbukh (Ed.): CICLing 2013, Part II, LNCS 7817, pp. 297–310, 2013.

as a pivot language. Utiyama and Isahara [7] use English as a pivot language. Adequate bilingual corpus exists for both source-pivot and pivot-target language pairs. The final translation for the source and target language pair is then obtained by going via the pivot language, either by generating full translations of the source sentence in this pivot language or by bridging the bilingual data to build translation models for the source–target language pair.

One major disadvantage of this approach is that both the translation into the pivot language and the translation into the target language are error-prone and these errors could add up. As a result, on similar training resources, the translation quality of a pivot SMT system could be significantly lower than the translation quality of a direct SMT system.

In this paper, we have explored different techniques in developing a SMT system for Hebrew-Arabic pair of morphologically complex languages. We leverage an active learning approach proposed by Ananthakrishnan et al. [12] to create Arabic-Hebrew parallel corpus by using sentence pivoting technique on Hebrew annotation obtained from Amazon Mechanical Turk starting from an Arabic-English bilingual corpus. We further demonstrate that this corpus can be used to create a preliminary viable SMT system. We have also explored the traditional technique of cascading two SMT systems using a pivot language like English to create Arabic-Hebrew SMT system. We further compare both SMT systems in Arabic-Hebrew domain. Finally, we improve our SMT system by leveraging available bilingual corpus using phrase generalization.

The rest of the paper is organized as follows. Section 2 describes data annotation. Section 3 describes sentence pivoting approach. Section 4 describes cascade approach. Section 5 describes the phrase generalization approach. Section 6 describes MTurk annotation process and translation statistics. This is followed by description of our Experimental paradigm. Subsequently we present the results of our techniques. Finally we conclude and outline future work.

2   Data Annotation

We have used the GALE (Global Autonomous Language Exploitation) Arabic to English bilingual broadcast news corpus. GALE Arabic Blog Parallel Text was prepared by the LDC (Linguistic Data Consortium) and consists of 102K words (222 files) of Arabic blog text and its English translation from thirty-three sources. We initially create a bootstrap English-Hebrew parallel corpus of 1940 sentences with the help of a Hebrew bilingual annotator. We also create a test and tune set by translating English to Hebrew. By leveraging already existing Arabic-English parallel GALE corpora, we map English back to Arabic and thus create a pivoted Arabic-Hebrew bootstrap, test and tune set. One of the problems with creating an Arabic-Hebrew test set in the above mentioned way is duel translation losses that occur during Arabic to English and English to Hebrew translation which reduces the quality of test and dev set. To resolve this problem, we posted source (Arabic) side of the Arabic test set as HITs to be translated to Hebrew on MTurk. We used one of our trusted Turker (who performed a lot of English-Hebrew translations) to create a direct Arabic–Hebrew test set. A schematic diagram of our approach is illustrated in Figure 1.

We leverage active learning approach proposed by Ananthakrishnan et al. [12] to select most optimal English sentences from GALE corpus to translate to Hebrew using MTurk. Ananthakrishnan [12] introduces a novel, fine-grained, error-driven measure of value for candidate sentences obtained by translation error analysis on a domain-relevant held-out development set. Errors identified in translation hypotheses are projected back on to the corresponding source sentences through phrase derivations from the SMT decoder. This projected error is used to obtain a "benefit value" for each source n-gram that serves as a measure of its translation difficulty. Sentence selection is posed as the problem of choosing K sentences from the candidate pool that maximize the sum of the benefit values of n-grams covered by the choice.

The Amazon Mechanical Turk (MTurk) is a crowdsourcing Internet marketplace that enables requesters to co-ordinate the use of human intelligence to perform tasks that computers are currently unable to do. It is one of the suites of Amazon Web Services. Amazon calls tasks that are difficult for computers but easy for humans as HITS (human intelligence tasks). Workers are referred to as Turkers and people designing the HITs are Requesters.

MTurk has been used in the realm of SMT for various applications. For ex. Ambati and Vogel [14] explored the effectiveness of Mechanical Turk for creating parallel corpora in the context of sentence translation. Ambati et al. [15] also explored the possibility of using active learning and MTurk in tandem for building low resource language pairs SMT system successfully. Zbib et al. [11] used MTurk crowd sourcing techniques to obtain machine translations for Arabic dialects.

3   Sentence Pivoting Approach

We leverage active learning to select most optimal English sentences from GALE corpus to translate to Hebrew using MTurk. We pivot English back to Arabic which results in the creation of a trilingual Arabic-English-Hebrew corpus. It is important to note that since the Arabic-Hebrew pivoted corpus has been created by manual translations performed by Turkers, so we minimize the error propagation that generally occurs while pivoting the outputs of two different SMT systems like Arabic-English and English-Hebrew. We create an intermediate English-Hebrew and Arabic-Hebrew SMT system. The English-Hebrew system and source (Arabic) side of Arabic-English bilingual corpus is then further used by the active learning approach to generate another batch of optimal English sentences, which is translated to Hebrew using MTurk. This process is repeated iteratively to create improving English-Hebrew and Arabic-Hebrew SMT systems until stopping criterion is reached. We have used the BLEU (Papineni et al., [5]) metric for evaluation purposes. Table 1 describes the tri-lingual corpus which is generated by our approach.

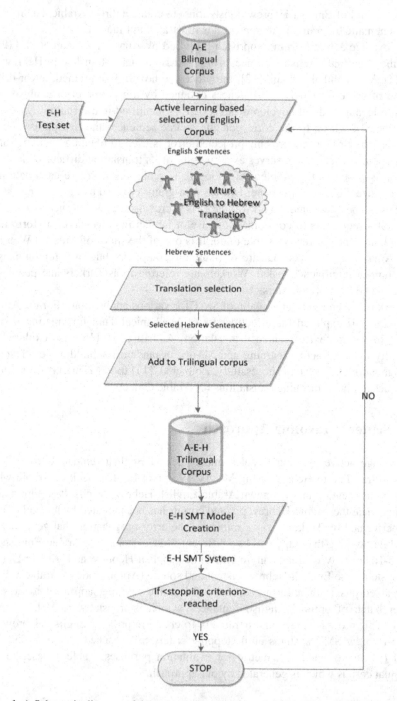

Fig. 1. A Schematic diagram of data annotation process and Arabic-Hebrew SMT systems

4   Cascade Approach

This approach considers the language pairs source-pivot and pivot-target independently. It consists of training and tuning two different SMT systems and combining them in a two-step process: first, we translate a source sentence using the source-pivot system. Then, we use the resulting sentence as input for the pivot-target translation. We have used 1-best results to avoid computational complexities. We have used English as a pivot language. We create a Hebrew to Arabic SMT system by cascading Hebrew-English and English-Arabic system. Similarly for Arabic to Hebrew SMT system, we cascade Arabic-English and English-Hebrew system.

Table 1. Description of Training/Test/Tune Corpus

Corpora	Arabic	English	Hebrew
Training sent.		13143	
Training words	0.28M	0.38M	0.27M
Training vocab.	40728	16796	44783
Dev. Sent.		816	
Dev. words	10774	14536	10861
Dev. vocab.	5003	3322	5384
Dev. OOV	917(*18%*)	305(*9%*)	1082(*20%*)
Test Sent.		732	
Test words	10498	14226	9875
Test vocab.	4857	3268	4891
Test OOV	829(*17%*)	370(*11%*)	1069(*22%*)

5   Phrase Generalization

This approach takes advantage of the fact that we have unused Arabic-English Bilingual corpora available as well as English-Hebrew Bilingual corpus. It is a similar approach to paraphrasing as proposed by Chris Callison-Burch et al. [1]. The Arabic-English corpus is part of the GALE corpus. We use the English-Hebrew corpus publically available through open source parallel corpus (OPUS). OPUS is a growing collection of translated texts from the web. As explained in the data annotation process, we already have Arabic-English-Hebrew trilingual corpus available to us. We extract Hebrew-English phrase pairs out of Hebrew-English corpora from the trilingual corpus. Similarly we extract all Arabic-English phrase pairs from the unused Bilingual Arabic-English corpus. We look for common English phrases in both groups of phrase pairs. Upon finding a common English phrase, we map corresponding Hebrew and Arabic phrase to create a Hebrew-Arabic generalized phrase pair. We subsequently extract all phrase pairs from Hebrew-Arabic trilingual corpus and then add both set of phrase pairs. In this way we can significantly enhance the number of resultant phrase pairs. We also conduct the same experiment by mapping out of domain Hebrew-English OPUS corpus with English-Arabic portion of the trilingual corpus.

6   MTurk Annotation Process

We have used MTurk to generate Hebrew translations of English sentences. During a pilot test, we found out that it's really important to set some qualification criteria's for Turkers, otherwise approx. 90% of work would be unusable. Here are some quality assurance steps taken by us in MTurk.

MTurk workers can be limited according to their country. Either they can be limited from being one particular country, or people from some particular countries can be banned from doing MTurk HITs. Turkers can also be limited based on what has been their approval rate on previous HITs. Anyone below a 90% approval rate was prevented from performing HITs. MTurk HITs were posted as an image rather than text to make cheating more difficult. It also prevents any automated systems to perform our HITs. Finally all translations were verified through an external English-Hebrew bilingual annotator to ensure the quality of Hebrew translations.

6.1   Batch Statistics

We submitted HITs in batches of approximately 1000 sentences so that we could closely monitor the MTurk annotation process. The batch was sub-divided into 4 smaller batches according to the median length of HITs in words. Our payment policy was 2 cents per source (English) word. The division of batch was necessary to ensure that there was no huge variation in median length of HITs of a particular batch, otherwise Turkers would be discouraged from doing larger HITs (since we pay equal amount of money for every HIT in a particular batch).

Table 2 presents the statistics on the size of batch, median length of HITs and amount paid for a single batch. We multiply the median length of batch by 2 to calculate payment per HIT in cents for a particular batch.

Table 2. MTurk HIT statistics for one batch

Batch	Number of HITs	Median Length of HITs	Payment per HIT
Subset 1	196	14	$0.28
Subset 2	264	21	$0.42
Subset 3	422	30	$0.60
Subset 4	118	41	$0.82

6.2   Country Statistics

We also categorized MTurk translation by Turker's Country. Our HITs were submitted by Turkers from more than 80 countries, but finally there were four major countries based on the IP Address analysis of each Turker. As expected, since Hebrew is the native language of Israel, so almost half of the accepted HITs were submitted by Israeli's Turkers. Georgia, USA and Germany were other three major countries. Table 3 shows the number of approved HITs performed by Turkers from each of these countries.

Table 3. MTurk translations by Turker's country of origin

Country	Number of Translations	Number of Turkers
Israel	6665	156
Georgia	3102	30
United States	1365	308
Germany	249	17

Table 4. English to Hebrew SMT Performance by Corpus type

Corpus	Translations	BLEU
Non Israel	4716	6.8
Non Israel + Bootstrap	6656	8.4
Israel	6665	8.9
Israel + Bootstrap	8605	10.2
Israel + Non Israel + bootstrap	13143	10.1

We also divided our HITs in two major categories: Israeli HITs and non-Israeli HITs. We measured English to Hebrew SMT BLEU statistics for both sub-groups, and also by adding bootstrap corpus to each one of them.

Table 4 presents the summary of this experiment. Comparison of just Israel and non-Israeli corpus (ignoring bootstrap) shows that SMT system based on Israel HITs performs 2.1 BLEU point better than non-Israeli SMT system, although Israeli corpus is also much larger than non-Israeli corpus. Interestingly Israel corpus supplemented with bootstrap corpus performs almost as well as all corpora combined. It would be reasonable to assume that limiting HITs to the native country in respect to target language would be a useful strategy in Mturk.

We used a country constraint criteria provided by MTurk while posting some of our batches to ensure submission of only native Israeli HITs. However this inevitably slows down the rate of HITs submission by Turkers (since only a fraction of existing Turkers can work on our HITs), so a better strategy could be to limit a portion of HITs (40%) to native countries and the rest to Turkers from all countries.

6.3   Iterative Analysis

Table 5 shows the per batch analysis of our MTurk setup. For example, first row explains that out of 1836 HITs submitted by Turkers, 1549 HITs were considered acceptable based on our Hebrew annotator's analysis. In total 98 Turkers submitted HITs out of which 62 performed acceptable HITs. We paid $1227 in total to Turkers and Amazon Mechanical Turk. The Hebrew translations obtained from MTurk were then supplemented with bootstrap set to create intermediate English to Hebrew SMT system. The active learning selection algorithm was applied on the intermediate SMT system to extract next batch of sentences for MTurk translation. We repeated this process through 9 iterations. Finally, we were able to create English to Hebrew parallel corpus of approximately 12.5k sentences while spending $8705. In total we spent approximately 70 cents for the translation of every English sentence to Hebrew.

Table 5. MTurk analysis per batch

Batch Number	Total HITs	HITs Accepted	Turkers Attempted	Turkers Accepted	Cost
1	1836	1549 (84%)	98	62	$1227
2	2000	1390 (69%)	126	52	$1050
3	2000	1761 (88%)	156	74	$1223
4	2000	1631 (81%)	85	51	$1134
5	1724	1186 (69%)	136	65	$850
6	1677	1414 (84%)	65	60	$968
7	2000	1420 (71%)	128	72	$911
8	2000	1128 (56%)	226	86	$687
9	2000	1086 (54%)	210	50	$652
Total	17285	12565(73%)			$8705

7   Experimental Paradigm

The tri-lingual corpus generated for our experiments has already been explained in Table 1. All experiments use the same methods for training, decoding and parameter tuning, and we only varied the corpora used for training, tuning and testing. The SMT system we have used is based on a phrase-based SMT model similar to that of Koehn et al. [9]. We used GIZA++ (Och and Ney, [3]) to align sentences. The decoder uses a log-linear model that combines the scores of multiple feature scores, including translation probabilities, smoothed lexical probabilities, in addition to a 4-gram language model.

8   Results and Discussion

8.1   Sentence Pivoting Results

We create English-Hebrew, Hebrew-Arabic and Arabic-Hebrew SMT models using the bootstrap system and by iteratively adding MTurk translations. The results are tabulated in Table 6. Our best performance for English-Hebrew SMT is 10.5 BLEU while for Hebrew-Arabic system, we achieve a BLEU of approximately 9. The Arabic-Hebrew SMT system performs a little worse at about 7.5. We have shown that by leveraging active learning to select English sentences to translate to Hebrew and then mapping Hebrew sentences back to Arabic, we create a Hebrew-Arabic SMT system, which improves in a manner consistent with English-Hebrew as evidenced in Figure 2. It shows that the trajectory of Hebrew-Arabic approx. mirrors to that of English-Arabic SMT system. It is important to note that Hebrew-Arabic and Arabic-Hebrew systems perform slightly differently. The difference can be mainly attributed to different Arabic and Hebrew LM.

Table 6. BLEU results across E-H, H-A and A-H

Batch Number	Corpus Size	E-H	H-A	A-H
Bootstrap	1940	6.47	3.18	3.04
+ Mturk batch 1	3199	8.17	5.12	5.08
+ Mturk batch 2	4331	9.23	6.20	5.94
+ Mturk batch 3	5606	9.14	5.82	5.68
+ Mturk batch 4	6952	9.62	6.75	6.29
+ Mturk batch 5	8169	9.81	7.57	6.88
+ Mturk batch 6	9557	10.48	8.11	6.76
+ Mturk batch 7	10962	10.65	8.34	7.01
+ Mturk batch 8	12071	10.41	8.99	7.67
+ Mturk batch 9	13143	10.05	8.96	7.49

8.1.1 Direct vs. Pivoted Test Set

It is important to note that we have used direct test set for Hebrew-Arabic and Arabic-Hebrew SMT system evaluation (results shown in Table 6), rather than using the indirect test set which uses English language as pivot. Figure 3 shows the BLEU trajectory of direct and pivoted Test sets for Hebrew-Arabic SMT system. It shows that direct test set performs slightly better than indirect test set, the fundamental reason being indirect test set consist of two different translations (obtained by fusing Arabic to English and English to Hebrew translations), which makes it more susceptible to errors.

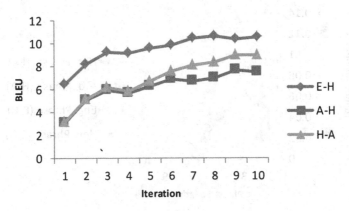

Fig. 2. Trajectory of BLEU (E-H, H-A, A-H)

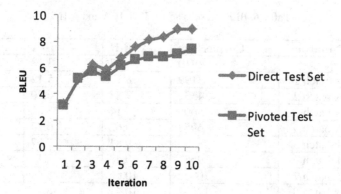

Fig. 3. Trajectory of Direct and Pivoted Test sets (Hebrew-Arabic SMT)

8.1.2 Phrase Length Analysis

We also computed statistics for unique phrases vs. phrase length for English-Hebrew and Hebrew-Arabic SMT system. Figure 4 shows the unique phrase statistics for both E-H and H-A systems. Since we have not performed any morphological tokenization for Hebrew and Arabic, so it can be seen that English phrases tend to be much longer compared to Hebrew phrases for E-H system. For H-A system, the area under the curve for Hebrew is larger compared to Arabic which signifies the greater number of unique phrase for Hebrew.

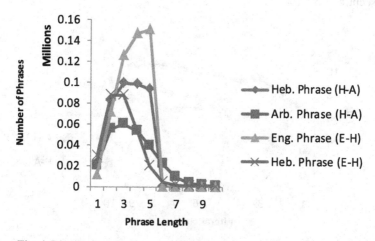

Fig. 4. Distribution of unique phrases by length for E-H and H-A systems

8.1.3 Effect of Corpus Size

Figure 5 shows the variation of the size of the phrase table over the various corpus sizes of English-Hebrew and Hebrew-Arabic systems. It can be seen that size of the phrase table increases uniformly with corpus size as expected.

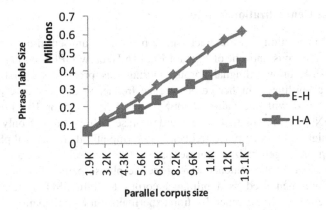

Fig. 5. Phrase Table variation with Corpus size

8.2   Cascade System Results

As mentioned in the Section 4, we create two cascaded SMT systems namely He-brew-Arabic (cascading Hebrew-English and English-Arabic) and Arabic-Hebrew (cascading Arabic-English and English-Hebrew). We have unused A-E bilingual cor-pora available to us. So, we create two different A-E SMT systems. Cascade-A uses the A-E corpus limited to the A-E-H trilingual corpus. Cascade-B uses all of A-E bilingual corpus available to us. The Arabic vocabulary size of full A-E corpora is 113k compared to 40k (Arabic vocabulary size in A-E-H trilingual corpus). Table 7 compares the performance of cascaded SMT systems with sentence pivoted SMT system. As Table 7 shows, Cascade-B system performs better than Cascade-A system in both H-A and A-H domain due to the larger A-E corpus used in Cascade-B com-pared to Cascade-A. It is important to note that sentence pivot systems only use the trilingual A-E-H corpus, so their results can only be compared to Cascade-A system which uses the same corpus. Cascade-A SMT system performs better than sentence pivot system in H-A domain by 0.8 Bleu points, while sentence pivot system outper-forms Cascade-A system by 1.8 Bleu in A-H domain. This can be explained by the fact that A-H cascaded SMT system is a combination of A-E and E-H system. A-E corpora is obtained from GALE corpus which is of significantly higher quality than E-H corpus obtained from Mturk users, coupled with the fact that the target language Hebrew (in case of E-H) is morphologically very complex language.

In the case of H-A, cascaded system is a combination of H-E and E-A SMT sys-tems. In this case, the target language of poorer quality corpus (H-E) is English which is morphologically much simpler than Hebrew.

Table 7. Comparison of Cascade and Sentence pivot systems

Technique	A-H	H-A
Cascade -A	5.7	9.8
Cascade -B	6.1	10.4
Sentence pivot	7.5	9.0

8.3   Phrase Generalization Results

As explained in Section 5, we extract phrase pairs from unused in-domain bilingual Arabic-English corpus and out of domain English-Hebrew OPUS corpus leveraging Arabic-English-Hebrew trilingual corpus. During this process, we first sorted the phrase table according to higher co-occurrence frequency of the source phrase. We then experimented with the value of topN (topN being 1, 5 or 10). For example, choosing topN as 1 implies that each source phrase is linked to the only one target phrase with highest co-occurrence frequency ignoring all other potential phrase pairs, while finding new generalized phrase pairs. The generalized phrase pairs thus obtained are added to the sentence pivot baseline phrase pairs. The new groups of phrase pairs are then used as a bilingual corpus to train SMT systems. Table 8 contains the result of phrase generalization experiments in H-A domain.

As seen in Table 8, we improve our Bleu performance by 3.9 points, when we add phrase pairs obtained from unused A-E corpus with topN being 5. The number of phrase pairs also increase approximately by 420k. It should be noted that A-E corpus is of the same domain as of the trilingual A-E-H corpus. Compared to the Cascade-B system which uses all of bilingual A-E corpus (please refer Table 7), the Bleu increases by 2.5. However in the case of out of domain E-H OPUS corpus, the increases in phrase pairs are only 200k with topN being 5. The Bleu increases only by 2.6 points. It demonstrates that an in-domain corpus (in case of A-E) is more helpful than out-of-domain corpora (in case of E-H). When we add phrase pairs obtained from both the E-H OPUS corpus and unused A-E corpus, the Bleu improves by 4.3 points.

Table 8. Table showing phrase generalization results, in H-A domain

System type	#phrase pairs	topN	Bleu
Sentence pivot Baseline	432k	N/A	9.0
+ A-E corpus	559k	1	12.2
+ A-E corpus	849k	5	12.9
+ A-E corpus	992k	10	12.8
+ H-E corpus	634k	5	11.6
+ H-E corpus	460k	1	10.7
+ A-E + H-E corpus	1051k	5	13.3
+A-E + H-E corpus	586k	1	12.7

9   Conclusion

In this paper we have explored a viable cost-effective method of creating a bilingual Arabic-Hebrew corpus. We also demonstrate and compare two different approaches for creating Arabic-Hebrew SMT system. Furthermore, we have shown the effectiveness of phrase generalization method in improving the SMT system both by using in-domain corpus and out of domain corpus. We believe that these approaches can be used to create SMT systems at a modest cost for a pair of foreign languages as long as a bilingual parallel corpus for one of those foreign languages and English exists.

This paper results in an annotated trilingual corpus of Arabic, English and Hebrew. At the same time, we have access to bilingual and monolingual corpus for Arabic, English and Hebrew languages. The next steps will involve leveraging all these data sources as well as statistics therein to improve SMT performance of the Arabic-Hebrew language pair.

References

1. Callison-Burch, C., Koehn, P., Os-borne, M.: Improved statistical ma-chine translation using paraphrases. In: Proceedings NAACL 2006 (2006)
2. Stefan, D., Munteanu, Marcu, D.: Improving machine translation perfor-mance by exploiting non-parallel corpora. Computational Linguistics 31(4), 477–504 (2005)
3. Och, F.J., Ney, H.: A systematic comparison of various statistical alignment models. Computational Linguistics 29(1), 19–51 (2003)
4. Wu, H., Wang, H.: Revisiting Pivot Language Approach for Machine Translation. In: Proceedings of the Joint Conference of the 47th Annual Meeting of the ACL and the 4th International Joint Conference on Natural Language Processing of the AFNLP, Suntec, Singapore, pp. 154–162. Association for Computational Linguistics (August 2009)
5. Papineni, K., Roukos, S., Ward, T., Zhu, W.-J.: BLEU: a method for automatic evaluation of machine translation. In: Proceedings of the 40th Annual Meeting of the Association of Computational Linguistics, ACL (2002)
6. Utiyama, M., Isahara, H.: Reliable measures for aligning Japanese-English news articles and sentences. In: ACL, pp. 72–79 (2003)
7. Utiyama, M., Isahara, H.: A Comparison of Pivot Methods for Phrase-Based Statistical Machine Translation. In: Human Language Technologies 2007: The Conference of the North American Chapter of the Association for Computational Linguistics; Proceedings of the Main Conference, Rochester, New York, pp. 484–491. Association for Computational Linguistics (April 2007)
8. Bertoldi, N., Barbaiani, M., Federi-co, M., Cattoni, R.: Phrase-Based Statistical Machine Translation with Pivot Languages. In: Proceedings of the International Workshop on Spoken Language Translation, Hawaii, USA, pp. 143–149 (2008)
9. Koehn, P., Och, F.J., Marcu, D.: Statistical phrase-based transla-tion. In: Proceedings of the 2003 Conference of the North American Chapter of the Association for Computational Linguistics on Human Language Technology, Morristown, NJ, USA, pp. 48–54. Association for Computational Linguistics (2003)
10. Resnik, P., Smith, N.A.: The web as a parallel corpus. Computational Linguistics 29(3), 349–380 (2003)
11. Zbib, R., Malchiodi, E., Devlin, J., Stallard, D., Matsoukas, S., Schwartz, R., Makhoul, J., Zaidany, O.F., Callison-Burch, C.: Machine Translation of Arabic Dialects. In: Proceedings of the 2012 Conference of the North American Chapter of the Association for Computational Linguistics: Human Language Technologies (2012)
12. Ananthakrishnan, S., Vitaladevuni, S., Prasad, R., Natarajan, P.: Source Error-Projection for Sample Selection in Phrase-Based SMT for Resource-Poor Languages. In: Proceedings of the IJCNLP 2011 (2011)

13. Cohn, T., Lapata, M.: Machine Translation by Triangulation: Making Ef-fective Use of Multi-Parallel Corpora. In: Proceedings of the 45th Annual Meeting of the Association for Computational Linguistics, Prague, Czech Republic, pp. 728–735 (2007)
14. Ambati, V., Vogel, S.: Can crowds build parallel corpora for machine translation systems? In: NAACL Workshop on Creating Speech and Language Data With Amazon's Mechanical Turk (2010)
15. Ambati, V., Vogel, S., Carbonell, J.: Active learning and crowd-sourcing for machine translation. In: Proceedings of Language Resources and Evaluation, LREC (2010)

Automatic and Human Evaluation on English-Croatian Legislative Test Set

Marija Brkić[1], Sanja Seljan[2], and Tomislav Vičić[3]

[1] University of Rijeka, Department of Informatics,
Radmile Matejčić 2, 51000 Rijeka, Croatia
mbrkic@uniri.hr
[2] Faculty of Humanities and Social Sciences,
Department of Information Sciences, Ivana Lučića 3, 10000 Zagreb, Croatia
sanja.seljan@ffzg.hr
[3] Freelance translator
ssimonsays@gmail.com

Abstract. This paper presents work on the manual and automatic evaluation of the online available machine translation (MT) service Google Translate, for the English-Croatian language pair in legislation and general domains. The experimental study is conducted on the test set of 200 sentences in total. Human evaluation is performed by native speakers, using the criteria of fluency and adequacy, and it is enriched by error analysis. Automatic evaluation is performed on a single reference set by using the following metrics: BLEU, NIST, F-measure and WER. The influence of lowercasing, tokenization and punctuation is discussed. Pearson's correlation between automatic metrics is given, as well as correlation between the two criteria, fluency and adequacy, and automatic metrics.

1 Introduction

Evaluation of machine translation (MT) is an extremely demanding task. Besides being time-consuming and subjective, there is no uniform opinion on "good quality" translation. However, the human translation, i.e. reference translation, is considered to be a "gold standard". There may be more than one reference translation set. Automatic evaluation metrics rely on different approaches, which all aim at performing evaluation as close as possible to human evaluation. The goal of evaluation can be comparing outputs of a single MT system through different phases, i.e. testing different parameter settings or system changes; comparing different systems based on different approaches; comparing similar systems, etc. Evaluation can be performed within a domain or across different domains. Automatic evaluation for morphologically rich under-resourced languages presents a domain of interest for researchers, educators and everyday users, especially when the language is to become one of official EU languages.

2 Related Work

A number of studies have explored correlation between human and automatic evaluation and conducted error analysis, especially for widely spoken languages.

A. Gelbukh (Ed.): CICLing 2013, Part II, LNCS 7817, pp. 311–317, 2013.

Qualitative analysis of MT output on a test set might point out some important general or domain-specific linguistic phenomena, especially when dealing with morphologically rich languages. In [12] the importance of qualitative view and the need for error analysis of MT output is pointed out. In [6] the complexity of MT evaluation is discussed and a framework for MT evaluation is defined, which relates the quality model to the purpose and the context, enabling evaluators to define usage context out of which a relevant quality model is generated. The main purpose is creating a coherent picture of various quality characteristics and metrics, providing a common descriptive framework and vocabulary, and unifying the evaluation process. [5] suggests a classification system of MT errors designed more for MT users than for MT developers. Error categories can be ranked according to the level of importance they have in the eyes of users, with regard to, for example, improvability and intelligibility. In [14] the relationship between automatic evaluation metrics (WER, PER, BLEU, and NIST) and errors found in translation is discussed. Errors are split into five classes: missing words, word order, incorrect words, unknown words and punctuation errors. The relationship between BLEU as an automatic evaluation measure and the expert human knowledge about the errors is discussed in [4]. Their results point to the fact that linguistic errors might have more influence on perceptual evaluation than other errors. Callison-Burch et al. in [1] evaluate MT output for 8 language pairs and conduct human evaluation in order to obtain different systems ranking and higher-level analysis of the evaluation process, and to calculate correlation of automatic metrics with human evaluation. Correlation between human evaluation of MT output and automatic evaluation metrics, i.e. BLEU and NIST, is explored in [2].

3   Evaluation Metrics

Four automatic metrics presented in subsequent sections are widely used in MT evaluation. However, there are not many researches on the evaluation of Croatian MT output, whereas Croatian is a highly inflected less widely spoken language that belongs to a group of Slavic languages. In Croatian, each lemma has many word forms, i.e. on average 10 different word forms for nouns, denoting case, number, gender and person. In this experimental study, GT-translated text has been evaluated by native speakers, errors have been analyzed, and, finally, correlation between automatic metrics separately, as well as between automatic metrics and human evaluation is given.

3.1   BLEU

Bilingual Evaluation Understudy (BLEU) is based on matching candidate n-grams with n-grams of the reference translation [11]. Scores are calculated for each sentence, and then aggregated over the whole test set. The algorithm calculates modified precisions in order to avoid MT over-generation of n-grams. For each candidate translation n-gram, BLEU takes into account the maximum number of times the n-gram appears in a single reference translation, i.e. the total count of each n-gram is clipped by its maximum reference count. The clipped counts are summed together and divided by the total number of n-grams in the candidate translation. Unigram precisions account for adequacy, while n-gram precisions account for fluency. In order to avoid too short candidates, the multiplicative brevity penalty factor is introduced. Some of the critiques directed towards BLEU are that it does not take into account the relative relevance of words, the overall grammatical coherence, it is quite unintuitive, and relies on the whole test set in order to correlate well with human judgments [8].

3.2 NIST

National Institute of Standards and Technology (NIST) is based on BLEU metric, but it introduces some changes. While BLEU gives the same weight to each n-gram in the candidate translation, NIST calculates how informative that n-gram is, namely the rarer the n-gram appears, the more informative it is, and more weight will be given to it. NIST also differs from BLEU in the calculation of brevity penalty factor, which does not influence result as much as the one in BLEU [3].

3.3 F-Measure

F-measure is widely used not only in MT, but also in information and document retrieval. This is the measure of accuracy which takes into account precision and recall, namely F-measure is a weighted average of both. It ranges from 0 to 1, 1 being the best value [10].

3.4 WER

Word Error Rate (WER) is a reference translation length-normalized Levenshtein distance [9]. Borrowed from speech recognition, it is one of the first metrics applied to statistical machine translation (SMT). Levenshtein distance can be defined as the minimum number of insertions, deletions and substitutions needed on a candidate or hypothesis translation so that it matches the reference translation [8]. WER is often criticized for being too harsh on word order. Namely, it does not allow any reordering [13]. If a candidate is exactly the same as its reference translation, WER equals to 0. Furthermore, it can be even bigger than 1 if a candidate is longer than its reference translation.

4 Experimental Study

4.1 Testset Descriptions

One part of the research has been conducted on English-Croatian parallel corpora of legislative documents, available at http://eur-lex.europa.eu/ and `http://ccvista.taiex.be/.` However, some additional editing has been deemed necessary for documents containing mostly tables and formulas, not usable for analysis, as well as typos and misspellings. For the purpose of analysis a total of 100 source sentences have been extracted, together with their reference translations. MT translation candidates have been obtained from Google Translate (GT) service, which has Croatian language support among others. Another part of the research has been conducted on the test set compiled from professional translations in different domains, i.e. religion, psychology, education, etc. 100 sentences have been extracted. The test set descriptions are given in Table 1.

Table 1. # of words in testset descriptions

	source	reference	translation
Testset 1	2.121	1.700	1.725
Testset 2	1.660	1.467	1.440

4.2 Human Evaluation

Human evaluation has been performed according to the criteria of fluency and adequacy, through an online survey. The survey has consisted of two polls for each criterion. Possible evaluation grades for fluency have been: Incomprehensible (1), Disfluent (2), Non-native (3), Good (4), Flawless (5). Adequacy evaluation grades having been: None (1), Little (2), Much (3), Most (4), All information preserved (5). The average obtained grade is 3.03 for fluency and 3.04 for adequacy on testset 1, and 3.30 for fluency and 3.67 for adequacy on testset 2.

4.3 Error Analysis

GT-translated sentences have been compared to the reference sentences. Although there have been many cases of several types of errors in a single sentence, the following errors have been distinguished: not translated/omitted words, surplus words in a translation, morphological errors/suffixes, lexical errors – wrong translation, syntactic errors – word order, and punctuation errors. The analysis has shown the highest number of morphological errors (on average 1.45 per sentence in testset 1 and 1.87 in testset 2), while other types of errors have been less represented. The next most represented error category has been that of lexical errors (on average 0.73 errors per sentence in testset 1 and 0.59 in testset 2), not translated words 0.41 errors per sentence in testset 1 and 0.4 in testset 2) and syntactic errors (0.48 errors per sentence in testset 1 and 0.47 in testset 2). The categories with the smallest number of errors detected have been surplus words (0.29 per sentence in testset 1 and 0.26 in testset 2) and punctuation errors (0.17 per sentence in testset 1 and 0.01 in testset 2).

4.4   Results

While in the first part of the experiment automatic scores have been configured to include case information, in the second part of the experiment case information has been omitted (Tables 2 and 3). The prefix l denotes case-insensitive part of the evaluation. The confidence intervals for BLEU and NIST have been calculated by bootstrapping and all the scores lie within the 95% interval [7].

Table 2. Automatic evaluation scores on testset 1 with respect to lowercasing, tokenization and punctuation removal

	no-preprocessing	tokenization	tok. and punct. removal
WER	76.12	57.20	58.78
lWER	75.76	56.50	57.62
F-measure	35.13	57.16	54.32
lF-measure	35.78	58.16	55.42
BLEU	33.70	33.64	31.61
lBLEU	34.32	34.25	32.19
NIST	6.2586	6.2539	6.0314
lNIST	6.3321	6.3271	6.1098

Table 3. Automatic evaluation scores on testset 2 with respect to lowercasing, tokenization and punctuation removal

	no-preprocessing	tokenization	tok. and punct. removal
WER	66.55	59.60	62.31
lWER	66.22	59.30	62.13
F-measure	47.74	55.82	51.89
lF-measure	48.89	56.83	53.11
BLEU	31.11	31.06	26.57
lBLEU	31.60	31.55	26.98
NIST	6.2628	6.2629	5.8507
lNIST	6.3432	6.3432	5.9309

5   Discussion

Before scoring with an automatic metric, the translated set and the reference set are usually preprocessed in order to improve the efficacy of the scoring algorithm [3]. Preprocessing usually implies lowercasing and tokenization. In addition to these two steps, we have added punctuation removal, and explored how these aspects affect the scores according to four automatic metrics. Lowercasing has systematically improved scores slightly. While tokenization has had enormous beneficial effect on WER and F-measure scores, especially for testset 1, i.e. the WER scores have dropped down for about 20 points, the F-measure scores have gone up for about 22 points, BLEU and NIST scores have slightly deteriorated.

This is due to the fact that the script used for calculating these scores performs internal tokenization which proved to be more beneficial than the one performed explicitly. Removing punctuation has had detrimental effect on all the scores, which has been expected because punctuation is translated more correctly. WER as an error measure has increased for more than 1 point compared to the tokenized testset 1 score, and for about 3 points on testset 2 score, irrespective of the case-sensitivity. The other three metrics scores have decreased, even more so on testset 2.

Pearson's correlation between WER and F-measure, as far as tokenization effects on true cased and lowercased test set are concerned, has proven statistically significant according to a two-tailed test at 0.05 significance level. As far as punctuation and tokenization is concerned, correlation between WER and F-measure, in addition to the correlation between BLEU and NIST, has proven statistically significant. Furthermore, WER and F-measure scores without punctuation have still beaten the baseline scores, i.e. the scores without tokenization and with punctuation included.

When all three aspects are taken into consideration, only WER and F-measure, as well as BLEU and NIST significantly correlate. WER and F-measure completely agree on the rankings of preprocessing techniques, while NIST seems to be less sensitive to tokenization when compared to BLEU.

The results indicate that when calculating WER and F-measure, an important pre-processing step should be tokenization, followed by lowercasing. As far as BLEU and NIST are concerned, lowercasing has proven to be of the biggest importance. However, all the above findings should be checked against correlation with human judgments.

For that purpose, we have divided our test sets into 5 different test sets, each containing 40 sentences, and calculated the correlation between human and automatic scores, with the above described aspects taken into consideration. None of the calculated correlations is statistically significant. We have also observed that NIST correlates much better with human adequacy, than human fluency scores, as in [3]. In our future work, we intend to explore correlations with human judgments in more detail.

References

[1] Callison-Burch, C., Fordyce, C., Koehn, P., Monz, C., Schroeder, J.: Meta-evaluation of machine translation. In: Proceedings of the Second Workshop on Statistical Machine Translation, pp. 136–158 (2007)
[2] Coughlin, D.: Correlating automated and human assessments of machine translation quality. In: Proceedings of MT Summit IX, pp. 63–70 (2003)
[3] Doddington, G.: Automatic evaluation of machine translation quality using n-gram co-occurrence statistics. In: Proceedings of the Second International Conference on Human Language Technology Research, pp. 138–145. Morgan Kaufmann Publishers Inc. (2002)
[4] Farrús Cabeceran, M., Ruiz Costa-Jussà, M., Mariño Acebal, J.B., Rodríguez Fonollosa, J.A., et al.: Linguistic-based evaluation criteria to identify statistical machine translation errors. In: Proceedings of EAMT, pp. 52–57 (2010)

[5] Flanagan, M.: Error classification for mt evaluation. In: Technology Partnerships for Crossing the Language Barrier: Proceedings of the First Conference of the Association for Machine Translation in the Americas, pp. 65–72 (1994)

[6] Hovy, E., King, M., Popescu-Belis, A.: Principles of context-based machine translation evaluation. Machine Translation 17(1), 43–75 (2002)

[7] Koehn, P.: Statistical significance tests for machine translation evaluation. In: Proceedings of EMNLP, vol. 4, pp. 388–395 (2004)

[8] Koehn, P.: Statistical Machine Translation, vol. 11. Cambridge University Press (2010)

[9] Leusch, G., Ueffing, N., Ney, H., et al.: A novel string-to-string distance measure with applications to machine translation evaluation. In: Proceedings of MT Summit IX, pp. 33–40 (2003)

[10] Melamed, I.D., Green, R., Turian, J.P.: Precision and recall of machine translation. In: Proceedings of the 2003 Conference of the North American Chapter of the Association for Computational Linguistics on Human Language Technology: Companion Volume of the Proceedings of HLT-NAACL 2003–Short Papers, vol. 2, pp. 61–63. Association for Computational Linguistics (2003)

[11] Papineni, K., Roukos, S., Ward, T., Zhu, W.J.: Bleu: a method for automatic evaluation of machine translation. In: Proceedings of the 40th Annual Meeting on Association for Computational Linguistics, pp. 311–318. Association for Computational Linguistics (2002)

[12] Stymne, S.: Blast: A tool for error analysis of machine translation output. In: Proc. of the 49th ACL, HLT, Systems Demonstrations, pp. 56–61 (2011)

[13] Tillmann, C., Vogel, S., Ney, H., Zubiaga, A., Sawaf, H.: Accelerated dp based search for statistical translation. In: European Conf. on Speech Communication and Technology, pp. 2667–2670 (1997)

[14] Vilar, D., Xu, J., d'Haro, L.F., Ney, H.: Error analysis of statistical machine translation output. In: Proceedings of LREC, pp. 697–702 (2006)

Enhancing Search: Events and Their Discourse Context

Sophia Ananiadou[*], Paul Thompson, and Raheel Nawaz

National Centre for Text Mining, Manchester Institute of Biotechnology,
University of Manchester, 131 Princess Street, Manchester, M1 7DN, UK
{sophia.ananiadou,paul.thompson}@manchester.ac.uk,
raheel.nawaz@cs.man.ac.uk

Abstract. Event-based search systems have become of increasing interest. This paper provides an overview of recent advances in event-based text mining, with an emphasis on biomedical text. We focus particularly on the enrichment of events with information relating to their interpretation according to surrounding textual and discourse contexts. We describe our annotation scheme used to capture this information at the event level, report on the corpora that have so far been enriched according to this scheme and provide details of our experiments to recognise this information automatically.

Keywords: event extraction, text mining, semantic search, discourse analysis.

1 Introduction

Data deluge makes finding relevant information increasingly difficult. Searching using keywords will usually return far more documents than are relevant to a query. A researcher interested in which proteins are *positively regulated* by IL-2 would typically expect the following sentence answering his query:

(1) p21ras proteins are activated by IL-2 in normal human T lymphocytes.

Using PubMed, a document containing (1) would be amongst the results retrieved using the search terms *IL-2* and *activate*. However, documents containing information directly relevant to the user's query may be hard to locate. Search engines view documents as "bags of words", omitting *relations* between search terms and do not incorporate variability in query terms e.g., acronyms, synonymous terms. Although users are interested in retrieving information about biological reactions that correspond to *positive regulations,* this can be expressed not only by the verb *activate* but other variations, e.g., *stimulate* or *affect*, or nominalisations such as *activation, activator, effect, stimulation.* Contextual interpretation is also important for a user, e.g., regulation may be negated: *p21ras proteins are <u>not</u> activated by IL-2 in normal human T lymphocytes.* Alternatively, there may be other types of information about the regulation specified in its textual context: *Our results <u>suggest</u> that p21ras proteins are <u>strongly</u> activated by IL-2 in normal human T lymphocytes.* The ability to specify

---

[*] Corresponding author.

A. Gelbukh (Ed.): CICLing 2013, Part II, LNCS 7817, pp. 318–334, 2013.

restrictions regarding interpretation or discourse function helps to further focus search results. For example, certain users may be interested specifically in negated interactions, whilst others may want to exclude them from their retrieved results. Other cases where interpretation can be important include matching hypotheses with experimental observations/evidence, or detecting contradictions that occur in the literature.

The above limitations of search engines can be alleviated through the integration of text mining methods [1–3] into customised search systems such as event-based search systems. Events are structured, semantic representations of pieces of knowledge contained within a text. In biomedicine, they include various biological processes, such as regulation, expression and transcription. Examples from newswire include terrorist attacks, company takeovers, personnel appointments, etc. In event-based search systems, searches take place over these structured events, not over unstructured text.

The sophistication of event-based search systems can be increased by automatic identification of contextual information, including information about discourse structure, such as causality relations [4], as well as level of certainty, negation, intensity of biological reaction, etc., and by allowing such types of information to be specified as restrictions on the types of events to be retrieved. We call these different types of information *meta-knowledge.*

In this paper, we firstly provide an overview of event-based text mining. We present our annotation scheme for enriched events with meta-knowledge information, corpora with event annotations, and describe how to train systems to recognise meta-knowledge information at the event level automatically.

2   Event-Based Text Mining

Recognising events in text usually involves separate identification and/or categorisation of several pieces of information in the text: triggers, the words around an event, and the event participants or arguments. Participants can include the instigator (or cause) of the event, the thing affected by the event (theme), etc. The information conveyed in (1) could thus be represented as a structured event as follows (based on [5]).

EVENT_TYPE: *positive_regulation*
TRIGGER: *activates*
CAUSE: *IL-2:PROTEIN*
THEME: *p21ras proteins:PROTEIN*
LOCATION: *normal human T lymphocytes:CELL*

The event representation above has been assigned an event type, drawn from an ontology of event types. Each participant is also assigned a named entity type. Participants can also be events themselves, i.e., events can be embedded within other events. Work on event extraction has not been limited to biomedical text, many earlier efforts were focused on newswire text.

The specific features of text, in terms of, e.g., the structure and language used, varies between domains. Event extraction systems thus must be adapted or reconfigured for different domains.

2.1   EventMine

EventMine [6] is a state-of the-art event extraction system, deployed for BioNLP Shared tasks on event extraction [7, 8]. It outperforms all systems in the BioNLP'09 ST subtask Task 2 and BioNLP'11 main tasks (GENIA, ID and EPI), achieving F-measures of 58.3%, 58.0%, 59.1%, 54.4%, other systems achieving F-measures of 57.4%, 56.0%, 57.6%, 53.3%, respectively. New features are constantly added, most recently by employing domain adaptation and coreference resolution [9]. It is a machine learning-based pipeline system with three detection modules for: (i) Event trigger/entity—assigns an appropriate trigger/entity category to each word that potentially constitutes the head word of an event participant; (ii) Event argument—finds semantic pair-wise relations among event participants; (iii) Multi-argument event—merges several pair-wise relations into complete event structures. It is designed to extract event structures from parser output. Any dependency parser could be substituted, but currently we use a combination of Enju [10] and GDep [11]. It extracts various token-related features (character n-grams, base form, parts-of-speech, etc). Contextual information is included in the feature set by taking into account dependency paths involving the focused word, n-grams of words surrounding the target word and its dependencies, and n-grams of words surrounding triggers and their identified arguments.

2.2   Semantic Event Searching: MEDIE

MEDIE[1] [12] facilitates event-based searching. A deep syntactic analyser tuned to the biomedical domain [13], an event expression recogniser and a named entity recogniser [14] provide its data. Queries take the form of <subject, verb, object> to specify an event, where *subject* and *object* refer to grammatical relations with the verb. In (1), the subject corresponds to the *Cause* participant, whilst the object corresponds to the *Theme*. One or more of the three "slots" in the query template can be left empty, in order to increase or decrease the specificity of the query: to find out which proteins are positively regulated by IL-2 we would specify: <IL-2, activate, ?>.

MEDIE addresses the issues of the simple keyword search engine, at least to a certain extent: (i) Only documents in which the specified grammatical relations hold between the search terms are retrieved, thus eliminating many of the spurious results retrieved by a traditional search engine; (ii) MEDIE detects named entities and event trigger terms, which are then linked with databases and ontologies. This allows automatic expansion of searches to include variants of search terms listed in these resources; (iii) Each sentence is automatically classified by title, objective, method, result or conclusion, and searches can specify which of these sentence types to consider when retrieving results [15]. For example, events in result sentences are likely to contain definite experimental results, whilst conclusion sentences will usually contain analyses or conclusions about experimental results.

Despite its advantages over a traditional search engine, MEDIE has limitations. It only allows two event participants (subject and object). Information on time, environmental conditions and manner is considered to be highly important to their correct interpretation [16]. MEDIE's search template is tied to the syntactic structure of the

---

[1] http://www.nactem.ac.uk/medie/

text. An approach in which users specify restrictions in terms of semantic rather than grammatical roles is more desirable. For instance, Cause and Theme semantic arguments do not consistently correspond to grammatical subject and object for all verbs. A semantic approach is even more desirable if additional participants (location, environmental conditions, etc.) may be specified as part of the search.

The meta-knowledge aspect of MEDIE involves classifying sentences as: *title, objective, method, result* or *conclusion*. Whilst useful, assignment of such information at the sentence level is often not sufficient when extracting information at the level of events. Sentence (2) helps to illustrate this:

(2) We conclude that the inhibition of the MAP kinase cascade with PD98059, a specific inhibitor of MAPK kinase 1, may prevent the inhibition of the alpha2 integrin subunit.

In (2), two "top-level" events can be identified:

a) A somewhat tentative conclusion: *the inhibition of the MAP kinase cascade with PD98059 may prevent the inhibition of the alpha2 integrin subunit.*
b) A general fact: *PD98059 is a specific inhibitor of MAPK kinase 1.*

Sentence (2) is likely to be classified by MEDIE as a *conclusion*. However, the two events identified here have different interpretations: a) is fairly typical of the type of event that would normally be expected to occur in a *conclusion*, i.e., an analytical conclusion based on experimental results. However, events with other types of interpretations can also occur in such sentences as in b). Thus, to support event-based searching, it is preferable for any information relating to discourse structure and interpretation to be assigned at the level of the event, rather than at the sentence level.

2.3 Semantic Event Searching

MEDIE's search strategy is largely based on syntactic analysis of text. By allowing specification of search criteria via an intuitive semantic template that abstracts from the way events are specified in text, users without linguistic expertise can easily specify their exact search criteria. An ideal template would allow specification of the following types of search options: (i) Specification of event types (chosen from a fixed set) as an alternative to specific event trigger words or phrases. Ontologies of event types provide the user with control over the level of generality of the results returned by the query; (ii) Use of semantic role types rather than grammatical relations when specifying restrictions on event participants; (iii) A flexible way of specifying restrictions on the values of particular participants, in the form of either particular terms (e.g., *NF-kappa B*), NE classes (e.g., *PROTEIN*), or a combination; (iv) Specification of meta-knowledge about the event, e.g., should only facts be retrieved or are experimental analyses also acceptable. If so, are highly speculative analyses of interest, or only more definite analyses? The main challenges of producing a system that can extract events to match such a template are the following: (i) How each ontological event type manifests itself in the text, i.e., which words and phrases can be used as triggers; (ii) How the syntactic structure of the text maps in various ways to the semantic arguments of different types of events; and (iii) How meta-knowledge information about the event can be derived from the textual context of the event.

Resources such as the GENIA event ontology [5] provide an inventory of relevant biomedical event types. It is linked to the Gene Ontology [17], and allows us to obtain potential trigger words for some event type. Also, the BioLexicon [18], a large-scale terminological resource, can help with syntax-semantics mapping in event extraction: it provides an account of the syntactic and semantic behaviour of biomedical verbs.

In general, whilst external resources can help to improve the accuracy of event extraction systems, they are usually not sufficient to facilitate the recognition of a customised set of event types. A well-established method of adapting text mining systems to new domains is through training using annotated corpora [19-21]. To facilitate training of event extraction systems, corpora containing semantically annotated events, e.g., BioInfer [22], GENIA event corpus [5], MLEE[2] [23] and GREC [24] have been used. These corpora vary in several ways, including the richness of the ontologies used to categorise events and named entities, the numbers and types of the semantic arguments identified, the types of meta-knowledge information included as part of the annotation and the overall corpus size. By far the largest is the GENIA event corpus, consisting of 1000 abstracts, containing a total of 36,858 events.

3   Event Interpretation, Discourse Structure and Meta-knowledge

The BioNLP shared tasks distinguish speculated and non-speculated events. However, speculation can be expressed to varying degrees, and the ability to distinguish between these is useful for certain tasks, e.g., slight hedging indicates the authors are quite confident about the results of their analyses, but they may include a hedging device as a safeguard. In contrast, larger amounts of speculation can indicate that the event should be taken as a hypothesis.

Events with no explicit specification of speculation may have different interpretations. An event may be presented as the subject of an investigation, a known fact, an experimental observation or as an outcome of analysing experimental results. We may also distinguish events that represent knowledge cited from a previously published paper and events that constitute part of the new knowledge contribution in the current paper. Depending on the nature and criticality of the task being undertaken, some or all of the above distinctions may be important when searching for events in text.

A more detailed distinction between events is needed, according to their intended interpretation, based upon their textual and discourse context. To facilitate the automatic recognition of such information at the event level, we have designed an annotation scheme, tailored to enrich event-annotated corpora with meta-knowledge [25]. Whilst the current scheme version is tailored to annotating biomedical events, it is possible to identify domain-specific and domain-independent aspects such that, by extending a core set of concepts, it can be tailored to other domains. Following our description of the current biomedical annotation model, we describe our preliminary

---

[2] http://www.nactem.ac.uk/MLEE/

efforts to adapt our model to the social history domain. Much of this work has concerned either speculation/certainty level detection [26, 27], or assignment of information relating to the general information content or discourse function of the sentences, which has been carried out on abstracts [28, 29] and full papers [30, 31].

A smaller number of annotation schemes and systems has considered annotation of either multiple aspects of meta-knowledge, e.g., assigning both a general information category and if the sentence refers to new or previous work [32, 33] or both negation and speculation [34]. Uniquely amongst the above cited corpora, [34] also annotates the clue expressions (i.e. the negative and speculative keywords) on which the annotations are based, as does [35], which annotates several types of information relating to the interpretation of information in newspaper articles.

Few schemes explicitly annotate meta-knowledge clue expressions, yet these have been shown to be highly important for the recognition of meta-knowledge. For example, corpus-based studies of hedging (i.e., speculative statements) in biological texts [36, 37] found that 85% of hedges are conveyed through lexical means. Specific lexical markers can also denote other types of information pertinent to meta-knowledge identification, e.g., markers of certainty [38], as well as deductions or sensory (i.e., visual) evidence [36]. We have also shown that different types of meta-knowledge may be expressed through different words in the same sentence [39]. Thus, although meta-knowledge is not always conveyed through lexical clues (and conversely, presence of a particular lexical clue in the sentence does not guarantee the "expected" meta-knowledge interpretation), we consider the identification of meta-knowledge clue expressions as one of the keys to accurate meta-knowledge identification.

Other annotation schemes consider e.g., clauses [40] or sentence segments [41], to account for the fact that several types of information can be specified in one sentence.

Our multi-dimensional scheme for enriching events with meta-knowledge takes inspiration from other schemes, but, given that event structures are different from continuous spans of text, it has been tailored to encode exactly the types of information that can be readily identified for events. Indeed, it has been shown that the information encoded by our scheme can provide complementary information to that encoded by sentence and clause-based schemes [42].

3.1   Meta-knowledge Scheme for Biomedical Events

Our multi-dimensional meta-knowledge scheme maximally captures useful information specified about events in their textual context. Each dimension consists of a set of complete, mutually-exclusive categories: an event belongs to just one category in each dimension. Moreover, the interplay between the different dimension values can be used to derive further information (hyper-dimensions) on event interpretation. To minimise annotation burden, the number of possible categories within each dimension has been kept as small as possible, whilst respecting important distinctions in meta-knowledge we have observed during our corpus study. A brief overview of the dimensions of our scheme and their possible values is provided below.

Knowledge Type (KT): Captures the event's general information content. Each event is classified as: Investigation (enquiries, examinations), *Observation* (direct experimental observations), *Analysis* (inferences, interpretations, conjectures), *Method* (experimental methods), *Fact* (general facts, well-established knowledge) or *Other* (default: expresses incomplete information, or the KT is contextually unclear).

Certainty Level (CL): Encodes the confidence or certainty level ascribed to the event according to three levels: *L3* (default: no expression of uncertainty), *L2* (high confidence or slight speculation) or *L1* (low confidence or considerable speculation).

Polarity: Identifies negated events. We define negation as the absence or non-existence of an entity or a process. Possible values are *Positive* (default) and *Negative*.

Manner: Captures information about the rate, level, strength or intensity of the event: *High* (event occurs at a high rate or level of intensity), *Low* (event occurs at a low rate or level of intensity) or *Neutral* (default: no indication of rate/intensity).

Source: Encodes the source of the knowledge being expressed by the event as *Current* (default: the current study) or *Other* (any other source).

Hyper-Dimensions: Correspond to additional information that can be inferred by considering combinations of some of the explicitly annotated dimensions. We have identified two such hyper-dimensions each with *Yes* or *No* values: *New Knowledge* (inferred from *KT, Source* and *CL*) and Hypothesis (inferred from *KT* and *CL*).

3.2   Meta-knowledge Annotation of Biomedical Corpora

The scheme was firstly applied to the GENIA event corpus of 1,000 abstracts (36,858 events) to create the GENIA-MK corpus [43]. Whilst the scheme was designed via examination of biomedical abstracts, it is also important that meta-knowledge should also be readily identifiable for events in full papers, especially given the recent trend of extending event extraction techniques to apply to full papers [44]. This means that a classifier trained on abstracts is unlikely to give optimal performance if applied to full papers.

We are currently creating a corpus of full papers with meta-knowledge annotation. Our preliminary set of 4 papers (1,710 events) has already been annotated via the GENIA event annotation scheme [45]. Future work will involve the meta-knowledge enrichment of full papers that have been annotated with different types of events, such as those made available following BioNLP 2011 and 2013 shared tasks.

3.3   Analysis of Meta-knowledge Annotations in Biomedical Corpora

We have analysed our two meta-knowledge enriched corpora (full papers and abstracts), to discover and compare their different types of characteristics of events. Table 1 reports our analysis, giving the relative frequencies (RF) of events assigned meta-knowledge values in abstracts (A) and full papers (FP). To make clearer differences in the distribution of meta-knowledge values between these two text types, difference in relative frequencies is also shown, together with percentage change.

Table 1. Comparison of meta-knowledge values in full papers and abstracts

Dim.	Cat.	RF (FP)	RF (A)	Diff. in RF (FP – A)	% Change in RF
KT	Ana.	22.2%	17.8%	4.4%	24.8%
	Inv.	3.8%	5.3%	-1.5%	-39.0%
	Obs.	36.3%	34.7%	1.4%	4.1%
	Fact	4.1%	8.1%	-4.0%	-98.7%
	Meth.	5.8%	2.6%	3.2%	120.8%
	Oth.	27.8%	31.3%	-3.5%	-12.7%
CL	L1	2.3%	2.1%	0.2%	9.7%
	L2	9.5%	6.0%	3.5%	57.6%
	L3	88.2%	91.9%	-3.7%	-4.2%
Polarity	Neg.	3.6%	6.1%	-2.5%	-66.7%
	Pos.	96.4%	93.9%	2.5%	2.6%
Manner	High	3.9%	3.8%	0.1%	2.2%
	Low	0.8%	0.8%	0.0%	0.0%
	Neut.	95.2%	95.3%	-0.1%	-0.1%
Source	Cur.	80.0%	98.5%	-18.5%	-23.1%
	Oth.	20.0%	1.5%	18.5%	1248.6%
Hyper-Dimensions	N.K	28.6%	43.4%	-14.8%	-51.7%
	Hypo.	15.2%	13.4%	1.8%	13.4%

In Table 1, in most cases, the "rankings" of each value within a particular dimension remain constant between full papers and abstracts and the absolute differences between the relative frequencies are small. However, the percentage change in relative frequencies between abstracts and full papers reveals significant differences.

Most notable is the difference between the relative frequencies of events that are assigned *Source=Other* in full papers and abstracts—full papers contain 12.5 times more such events than abstracts. Furthermore, citations, the most common way to denote previous work, are often not allowed in abstracts: full papers normally mention related work extensively, most notably in Background and Discussion sections.

There are differences in values of the KT dimension. For example, *Method* events are more than twice as abundant (in terms of relative frequency) in full papers. Since the average size of abstracts in the GENIA event corpus is 9 to 10 sentences [5], the relative frequency of *Fact* events in abstracts is high (over 8%). In full papers, factual events only appear half as frequently. The reason is that the only type of section in full papers in which *Fact* events occur with any significant frequency is *Background* (over 7% of all events in this section type), where the current state of knowledge is also discussed in detail. In contrast, other sections in full papers are more concerned with experimental details. A similar argument explains why *Investigation* events are more frequent in abstracts: most abstracts describe the purpose of the study, but a smaller proportion of full papers is devoted to describing investigations. In contrast to the previous two categories mentioned, the proportion of *Analysis* events is ~25% higher in full papers: in contrast to *Fact* and *Investigation* events, *Analysis* events are found with high frequency in several sections of full papers. There is much less variation in the *Observation* category, suggesting that the clear reporting of experimental observations is equally important throughout both full papers and abstracts.

Authors are more cautious in detailing their results in the body of papers, to maintain credibility in case these are later disproved. "Scientists gain credibility by stating the strongest claims they can for their evidence, but they also need to insure against overstatement." ([36] p. 257). Authors achieve this by using slight hedging (*L2*). Greater speculation (*L1*) is less common, as credibility is thus reduced. The fact that the proportion of slightly hedged *Analysis* events is particularly high in the *Results, Discussion* and *Conclusion* sections of full papers, rising as high as 51% in the *Discussion* sections, explains why *L2* events are over 57% more frequent in full papers.

Regarding *Polarity*, the relative frequency of negated events is significantly (67%) higher in abstracts. This is partly due to the fact that negative results are sometimes more significant than positive results [46], and are thus highlighted in the abstracts.

There is little difference in the relative frequencies of different values of *Manner* in both text types. For the hyper-dimensions, there is a higher proportion of hypotheses whilst for *New Knowledge*, there is a more significant difference. In abstracts, just under half of all events report new knowledge: unsurprising, given the previously specified main aims of abstracts. In contrast, there is much more room in full papers for describing and discussing previous work, and speculating about results.

3.4   Adapting the Meta-knowledge Scheme to a New Domain

We have investigated applying our scheme in the ISHER project[3] on social history, which aims to enhance search over digitised social history resources, through text mining-based rich semantic metadata extraction for collection indexing, clustering and classification, thus supporting semantic search. Semantic metadata include both named entities and events. As part of the training data, we use the Automatic Content Evaluation (ACE) 2005 corpus, which contains events, i.e., *Conflict* (Attack, Demonstrate) and *Justice* (e.g. Arrest-Jail, Sentence, Fine, etc.). We are enriching relevant events in the corpus with meta-knowledge annotation.

Three of the original meta-knowledge dimensions are useful for ACE, i.e., *Polarity, Source* and *CL*, as these dimensions and their values represent general characteristics of all text types. *Manner* is not relevant to the social history domain but *Knowledge Type (KT)* is, although a different set of values may be required for each different domain. The existing categories are very specific to academic papers, while an examination of events in the ACE corpus suggests that, although some categories may remain constant across different domains, other categories are domain specific. For example, although events describing facts and analyses of information can also be observed in the ACE corpus,

---

[3] http://www.nactem.ac.uk/DID-ISHER/

other types of information require different categories, e.g.: hypothetical events: *It could swell to as much as $500 billion if **we go to war in Iraq***, or opinions: *Dan Snyder of Baden, Pennsylvania writes, **"Bush should torture the al Qaeda chief operations officer."***

4   Automatic Detection of Meta-knowledge

We have extended EventMine to extract events and assign meta-knowledge to them. In EventMine-MK [47], meta-knowledge assignment is implemented as a separate module. We used two types of features: **event structure** concerned with the text surrounding both the event trigger and its arguments, both in terms of immediate context and dependency paths and **meta-knowledge clue features.** They include the position in the abstract of the sentence that contains the event, which is used since certain types of meta-knowledge (particularly events belonging to different values within the KT dimension) tend to appear in fixed places in abstracts (e.g., events with the KT type Fact or Observation often appear towards the beginning of an abstract). A citation feature refers to the presence of citations. Citations are extracted via a regular expression that matches parentheses or brackets surrounding numbers (e.g., [108]) or sequences ending in 4 digits (e.g., (..., 1998)). Clues for Other (Source dimension) often constitute citations, and thus are not covered by the clue dictionaries.

Through experimentation with different combinations of the above feature types, we found that the exact combinations of features that produce the best results vary according to the meta-knowledge dimension under consideration. However, since for each dimension, the difference between the performance of the best setting and the setting in which all features are enabled is less than 1%, we decided to enable all features for all dimensions, due to the extra computational and spatial costs that would be required to calculate and store a different set of features for each dimension.

Table 2. Results of applying EventMine-MK to the ST-MK corpus (**R**ecall, **P**recision, **F**-score)

Dimension	Average type	R/P/F	+GENIA (R/P/F)	Majority (R/P/F)
KT	Macro	56.5 / 60.9 / 57.3	56.2/ 59.7 / 57.3	16.7 / 6.8 / 9.6
	Micro	74.6 / 74.6 / 74.6	73.9/ 73.9 / 73.9	40.5 / 40.5 / 40.5
CL	Macro	57.0/49.3/52.3	66.8/87.1/69.2	33.3 / 32.2 / 32.8
	Micro	96.6 / 96.6 / 96.6	97.7 / 97.7 / 97.7	96.7 / 96.7 / 96.7
Polarity	Macro	84.5 / 77.2 / 80.3	82.5 / 79.8 / 81.0	50.0 / 47.9 / 48.9
	Micro	96.4 / 96.4 / 96.4	96.9 / 96.9 / 96.9	95.9 / 95.9 / 95.9
Manner	Macro	91.9 / 76.4 / 82.8	91.4 / 76.8 / 82.8	33.3 / 31.4 / 32.3
	Micro	96.2 / 96.2 / 96.2	96.3 / 96.3 / 96.3	94.1 / 94.1 / 94.1
Source	Macro	82.1 / 90.7 / 85.9	82.1/94.8/87.4	50.0 / 49.3 / 49.7
	Micro	99.3 / 99.3 / 99.3	99.4 / 99.4 / 99.4	98.6 / 98.6 / 98.6

EventMine-MK also makes use of a meta-knowledge prediction model, trained on the original GENIA-MK corpus, which has richer semantic information about events than the ST corpus (the corpus from the BioNLP'09 shared task, on which Event-Mine-MK was trained), and a much greater number of events. Thus, a model trained on the GENIA-MK corpus should predict meta-knowledge more accurately than one trained on the ST-MK corpus. Unfortunately, the differences in event types and distribution of meta-knowledge values in the ST corpus mean that direct application of the GENIA-MK trained model to predict meta-knowledge on the ST data will not produce optimal results. However, we did find that indirect use of the GENIA-MK model on the ST data (i.e., by adding additional features based on this model to the meta-knowledge model trained on the ST corpus), improved performance of meta-knowledge assignment. Table 2 reports results of applying EventMine-MK to the ST-MK corpus. For each dimension, micro and macro average scores are shown. For reference, a majority baseline is shown, i.e., the scores that would be achieved if each event was assigned the dimension value that appears most frequently in the corpus. Although performance is different for each dimension, it is in most cases superior to the baseline, in some cases by a significant margin. The +GENIA column shows the effects of adding features based on the model trained on the GENIA-MK corpus. For most dimensions, we see some improvement when these features are added. The effect is very noticeable when the macro average of the CL dimension is considered.

Table 3. Comparison of EventMine-MK with other systems on the task of negation and speculation detection

	Negation	Speculation	Total
EventMine-MK (+clues)	**29.96**/42.24 /**35.05**	21.63/36.59 / 27.19	**25.98**/39.79 / **31.43**
EventMine-MK	28.19/36.16 / 31.68	22.12/**41.82** / 28.93	25.29/38.33 / 30.47
[48]	22.03/49.02 / 30.40	19.23/38.46 / 25.64	20.69/43.69 / 28.08
[49]	18.06/46.59 / 26.03	**23.08**/40.00/ **29.27**	20.46/42.79 / 27.68
[50]	15.86/**50.74** / 24.17	14.98/50.75 / 23.13	16.83/**50.72** / 25.27

To evaluate and compare EventMine-MK, we applied it to the BioNLP'09 ST subtask (Task 3) of extracting events with associated negation and speculation information. Although this task does not deal with all aspects of meta-knowledge that can be predicted by our system, there are currently no other systems that can predict the values of other meta-knowledge dimensions at the event level, and so further comparisons cannot be undertaken. Two versions of EventMine-MK were trained, one on the ST-MK corpus, and one on the original ST corpus, which was annotated for negation and speculation, but not for negation and speculation clues. This latter corpus was the one used by the other systems compared for training, and so allows more direct comparison. Performance is reasonably low for all systems in Table 3, because the evaluation settings take into account event extraction performance as well as negation/speculation detection. We see that whether or not EventMine-MK is trained using meta-knowledge clues, it outperforms the top 3 systems that participated in the original task, in terms of both overall F-scores and F-scores for negation detection. We also see that meta-knowledge clue annotation helps improve performance, especially in detection of negated events. This provides strong evidence that our decision to annotate meta-knowledge clues was correct. For speculation, a small decrease in performance is observed when meta-knowledge clues are taken into account. However, this decrease reinforces the analysis by [44], that speculation annotations in the ST corpus do not conform to the standardised notion of speculation, i.e., in contrast to the events enriched with meta-knowledge annotation, events occurring with modal verbs (e.g., *may*) and epistemic adverbs (e.g., *probably*) are rarely annotated as speculative in the ST corpus. The model was also applied to the full-text subset of BioNLP-ST'11 GENIA corpus, to investigate differences in the distribution of meta-knowledge values in full papers and abstracts. Since the results obtained are somewhat different to our manual annotation efforts, this provides further evidence for our earlier hypothesis that different models need to be trained for abstracts and full papers.

4.1   Experiments with Individual Dimensions

EventMine-MK is designed to be robust and efficient, facilitating scalability to large scale event extraction. The overall efficiency of the framework used, together with spatial and computational costs, are all important considerations. The meta-knowledge module of EventMine-MK uses the same machine learning algorithm as other modules, i.e., SVMs, and also uses the same set of features for each meta-knowledge dimension. Although EventMine-MK produces very competitive results for negation and speculation detection, we decided to do some smaller-scale experiments to investigate whether the results for other meta-knowledge dimensions could be improved by using alternative feature sets and/or machine learning algorithms.

The meta-knowledge specific features of EventMine-MK take into account several more general observations about textual features that can affect meta-knowledge values. However, we decided to carry out a more in-depth analysis of individual dimensions to help suggest a customised set of features for each dimension, leading to improved prediction results. So far, we have carried out such analyses for two meta-knowledge dimensions, *Manner* and *Source*. Each study is characterised by detailed analysis of the contexts in which the different values of the given meta-knowledge dimension can occur, together with the different types of clues that can be used.

Both studies use a different set of core features, falling into six different categories, i.e., syntactic, semantic, lexical, lexicosemantic, dependency and constituency. The core features are more wide ranging than those used by the meta-knowledge prediction module of EventMine, particularly the use of semantic information about the bio-event (semantic types of events and participants, semantic roles assigned to participants, etc.), and the use of constituency features as well as dependency features.

Our detailed analysis of *Manner* cues [51] revealed that 8% of clues for *High* manner are of the form *n-fold*, in which *n* represents a number. Since *n* can vary, matching with clue lists is not the correct strategy here. In addition, the exact form of expression can vary, and in the GENIA-MK corpus, 13 different variants of this numerical expression have been annotated as *High* cues. Some examples include *2-fold*, *4-6 fold*, *5-to 7-fold*, etc. Accordingly, we use customised regular expressions to find such clues, which are subsequently included amongst the lexical clues extracted. In addition, the expression of negation inverts the polarity of a manner cue. For example, the word *significant* acts as a *High* cue, but its negated form (*no/not significant*) is a *Low* cue. Therefore, one of the lexical features used determines whether a negation cue is in the textual context of the event.

For the *Source* dimension [52] customised features include the tense of the main verb in the sentence (since events with *Source=Other* are often reported using the past tense). Also, positional features are included, as over 80% of *Other* events were found to occur in the first half of abstracts. A further interesting result of our analysis is that there is a correlation between event complexity and *Source* value. By event complexity, we mean whether an event is simple (i.e, if all of its participants are entities) or complex (i.e., if one or more of its participants is itself an event). Our analysis revealed that in abstracts, an arbitrary complex event is 2.6 times more likely than an arbitrary simple event to have knowledge source value of *Other*, whilst in full papers, an arbitrary complex event is 4.5 times more likely than an arbitrary simple event to have knowledge source value of *Other*.

For both the *Source* and *Manner* dimensions, training was done using the GENIA-MK corpus. The classifiers assume events have been pre-annotated: they do not attempt to recognise events as well as meta-knowledge. In both cases, we used the Random Forest algorithm to carry out the training, which develops an ensemble/forest of Decision Trees from randomly sampled subspaces of the input features. We used this algorithm since it has been successfully applied to various text mining and bioinformatics tasks. In particular, our recent experiments on detecting negated events [53] revealed that the Random Forest algorithm outperforms several other algorithms, including SVMs.

The experiments use 10-fold cross validation, so results can be compared to those produced when training the SVM classifier of the EventMine-MK meta-knowledge module on the GENIA-MK corpus, as we also report 10-fold cross validation results.

For *Source,* the best result achieved by the SVM classifier was micro-averaged F-Score of 98.4%, which is the same as majority baseline for this dimension (since the vast majority of events are assigned *Source=Current*). In comparison, the Random Forest classifier was able to achieve an improvement over the majority baseline, with a micro-averaged F-score of 99.4%. For *Manner,* the micro-averaged F-score for the SVM classifier was 95.4%, which again is virtually the same as the majority baseline for this dimension. In contrast, the micro-averaged F-Score for the Random Forest classifier is almost 3% higher, i.e., 98.3%. In terms of macro-averages, the gap is greater, with the SVM classifier achieving a macro-averaged F-Score of 59.2% for *Manner*, compared to 83.9% for the Random Forest classifier.

Results show that the Random Forest classifier is better suited to meta-knowledge value prediction than an SVM classifier, when used with customised feature sets. Future work involves a detailed analysis of combinations of features and learning algorithms to arrive at an optimal solution for automatic meta-knowledge prediction.

5   Conclusion

We have examined various aspects of event-based text mining. We have looked at how various text mining techniques can enhance users' search experience and help them to locate the information they require in a much more focussed and efficient way. In particular, we have discussed how analysing the structure of documents can help to restrict search results, how query expansion techniques can help to increase the number of relevant documents returned by a search; and how automatic recognition of various types of information (i.e., meta-knowledge) about the discourse and textual contexts of events provide even greater potential to refine event-based search queries according to the specific tasks being undertaken by individual users.

Using the MEDIE event-based search engine as an example, we showed that, whilst searching for events according to syntactic structures constitutes a robust method that offers improvements over traditional search engines, a more semantically based approach to extracting and searching for events provides greater power and flexibility to the user. The emergence of corpora with semantic annotation, coupled with the challenges posed by the BioNLP shared tasks, has helped to drive the development of more semantically-oriented event extraction systems. In particular, we provided details of EventMine, our own state-of-the-art event extraction system.

Since most event extraction systems to date have not attempted the detailed recognition of meta-knowledge information, the latter part of the paper has provided a summary of the various types of research that we have carried out within this area, including: the motivation for carrying out meta-knowledge recognition at the event level, rather than higher-level text units; the design of the original meta-knowledge scheme, tailored to enriching biomedical events; the application of the scheme to corpora of both abstracts and full papers; a comparison of the differences in the annotation results between abstracts and full papers; an investigation into how the meta-knowledge scheme could be adapted to other domains; and finally, an explanation of our most recent work, which has involved a number of efforts to train systems to recognise meta-knowledge automatically. This latest work consisted of two strands. The first of these extended EventMine with a module that is able to assign meta-knowledge information along the five dimensions of the scheme to automatically extracted events. The module follows the same structure as other modules in the EventMine pipeline, and is intended to strike a balance between accuracy, efficiency and robustness. Our second strand of work involved smaller scale experiments to investigate ways of improving on the prediction of meta-knowledge values, by considering different feature sets, coupled with a different machine-learning algorithm. The encouraging results achieved by these experiments suggest that there may be ways to improve the extraction of meta-knowledge in future versions of EventMine-MK.

As future work, we will analyse the effects that different machine learning algorithms and sets of features have on the accurate prediction of meta-knowledge information. We also aim to integrate more sophisticated event extraction technology in our search engines including MEDIE, Europe PubMed Central (a search engine over an archive of 25 million abstracts and 2 million full texts in the life sciences), as well as search engines operating on texts in other domains, such as ISHER.

Acknowledgements. The work described in this paper has been funded by the Meta-Net4U project (ICT PSP Programme, Grant Agreement: No 270893) and the JISC-funded Integrated Social History Environment for Research (ISHER)-Digging into Social Unrest project, which is part of the Digging into Data Challenge.

References

1. Zweigenbaum, P., Demner-Fushman, D., Yu, H., Cohen, K.B.: Frontiers of Biomedical Text Mining: Current Progress. Brief Bioinform. 8, 358–375 (2007)
2. Ananiadou, S., Kell, D.B., Tsujii, J.: Text Mining and its Potential Applications in Systems Biology. Trends Biotechnol. 24, 571–579 (2006)

3. Ananiadou, S., Nenadic, G.: Automatic Terminology Management in Biomedicine. In: Ananiadou, S., McNaught, J. (eds.) Text Mining for Biology and Biomedicine, pp. 67–98. Artech House, London (2006)

4. Mihăilă, C., Ohta, T., Pyysalo, S., Ananiadou, S.: BioCause: Annotating and Analysing Causality in the Biomedical Domain. BMC Bioinformatics 14, 2 (2013)

5. Kim, J., Ohta, T., Tsujii, J.: Corpus Annotation for Mining Biomedical Events from Literature. BMC Bioinformatics 9, 10 (2008)

6. Miwa, M., Saetre, R., Kim, J.D., Tsujii, J.: Event Extraction with Complex Event Classification using Rich Features. J. Bioinform. Comput. Biol. 8, 131–146 (2010)

7. Pyysalo, S., Ohta, T., Rak, R., Sullivan, D., Mao, C., Wang, C., Sobral, B., Tsujii, J., Ananiadou, S.: Overview of the ID, EPI and REL Tasks of BioNLP Shared Task 2011. BMC Bioinformatics 13 (suppl. 11), S2 (2012)

8. Pyysalo, S., Ohta, T., Rak, R., Sullivan, D., Mao, C., Wang, C., Sobral, B., Tsujii, J., Ananiadou, S.: Overview of the Infectious Diseases (ID) Task of BioNLP Shared Task 2011. In: BioNLP Shared Task 2011 Workshop, pp. 26–35. Association for Computational Linguistics (2011)

9. Miwa, M., Thompson, P., Ananiadou, S.: Boosting Automatic Event Extraction from the Literature using Domain Adaptation and Coreference Resolution. Bioinformatics 28(13), 1759–1765 (2012)

10. Miyao, Y., Sagae, K., Saetre, R., Matsuzaki, T., Tsujii, J.: Evaluating Contributions of Natural Language Parsers to Protein-Protein Interaction Extraction. Bioinformatics 25, 394–400 (2009)

11. Sagae, K., Tsujii, J.I.: Dependency Parsing and Domain Adaptation with LR Models and Parser Ensembles. In: Proceedings of the CoNLL 2007 Shared Task Session of EMNLP-CoNLL 2007, pp. 1044–1050. Association for Computational Linguistics (2007)

12. Miyao, Y., Ohta, T., Masuda, K., Tsuruoka, Y., Yoshida, K., Ninomiya, T., Tsujii, J.: Semantic Retrieval for the Accurate Identification of Relational Concepts in Massive Textbases. In: Proceedings of Coling/ACL, pp. 1017–1024. Association for Computational Linguistics (2006)

13. Hara, T., Miyao, Y., Tsujii, J.: Adapting a Probabilistic Disambiguation Model of an HPSG Parser to a New Domain. In: Dale, R., Wong, K.-F., Su, J., Kwong, O.Y. (eds.) IJCNLP 2005. LNCS (LNAI), vol. 3651, pp. 199–210. Springer, Heidelberg (2005)

14. Tsuruoka, Y., Tsujii, J.: Bidirectional Inference with the Easiest-First Strategy for Tagging Sequence Data. In: Proceedings of HLT/EMNLP 2005, pp. 467–474. Association for Computational Linguistics (2005)

15. Hirohata, K., Okazaki, N., Ananiadou, S., Ishizuka, M.: Identifying Sections in Scientific Abstracts using Conditional Random Fields. In: Proceedings of the 3rd International Joint Conference on Natural Language Processing, pp. 381–388. Association for Computational Linguistics (2008)

16. Tsai, R.T., Chou, W.C., Su, Y.S., Lin, Y.C., Sung, C.L., Dai, H.J., Yeh, I.T., Ku, W., Sung, T.Y., Hsu, W.L.: BIOSMILE: a Semantic Role Labeling System for Biomedical Verbs using a Maximum-Entropy Model with Automatically Generated Template Features. BMC Bioinformatics 8, 325 (2007)

17. Ashburner, M., Ball, C.A., Blake, J.A., Botstein, D., Butler, H., Cherry, J.M., Davis, A.P., Dolinski, K., Dwight, S.S., Eppig, J.T., Harris, M.A., Hill, D.P., Issel-Tarver, L., Kasarskis, A., Lewis, S., Matese, J.C., Richardson, J.E., Ringwald, M., Rubin, G.M., Sherlock, G.: Gene Ontology: Tool for the Unification of Biology. Nature Genetics 25, 25–29 (2000)

18. Thompson, P., McNaught, J., Montemagni, S., Calzolari, N., Del Gratta, R., Lee, V., Marchi, S., Monachini, M., Pezik, P., Quochi, V., Rupp, C.J., Sasaki, Y., Venturi, G., Rebholz-Schuhmann, D., Ananiadou, S.: The BioLexicon: a Large-Scale Terminological Resource for Biomedical Text Mining. BMC Bioinformatics 12, 397 (2011)
19. Kim, J.T., Moldovan, D.I.: Acquisition of Linguistic Patterns for Knowledge-Based Information Extraction. IEEE Transactions on Knowledge and Data Engineering 7, 713–724 (1995)
20. Soderland, S.: Learning Information Extraction Rules for Semi-structured and Free Text. Machine Learning 34, 233–272 (1999)
21. Califf, M.E., Mooney, R.J.: Bottom-Up Relational Learning of Pattern Matching Rules for Information Extraction. Journal of Machine Learning Research 4, 177–210 (2003)
22. Pyysalo, S., Ginter, F., Heimonen, J., Bjorne, J., Boberg, J., Jarvinen, J., Salakoski, T.: BioInfer: a Corpus for Information Extraction in the Biomedical Domain. BMC Bioinformatics 8, 50 (2007)
23. Pyysalo, S., Ohta, T., Miwa, M., Cho, H.-C., Tsujii, J.I., Ananiadou, S.: Event Extraction across Multiple Levels of Biological Organization. Bioinformatics 28, i575–i581 (2012)
24. Thompson, P., Iqbal, S.A., McNaught, J., Ananiadou, S.: Construction of an Annotated Corpus to Support Biomedical Information Extraction. BMC Bioinformatics 10, 349 (2009)
25. Nawaz, R., Thompson, P., McNaught, J., Ananiadou, S.: Meta-Knowledge Annotation of Bio-Events. In: Proceedings of LREC 2010, pp. 2498–2507. ELRA (2010)
26. Light, M., Qiu, X.Y., Srinivasan, P.: The Language of Bioscience: Facts, Speculations, and Statements in between. In: Proceedings of the BioLink 2004 Workshop at HLT/NAACL, pp. 17–24. Association for Computational Linguistics (2004)
27. Medlock, B., Briscoe, T.: Weakly Supervised Learning for Hedge Classification in Scientific Literature. In: Proceedings of ACL, pp. 992–999. Association for Computational Linguistics (2007)
28. Ruch, P., Boyer, C., Chichester, C., Tbahriti, I., Geissbühler, A., Fabry, P., Gobeill, J., Pillet, V., Rebholz-Schuhmann, D., Lovis, C.: Using Argumentation to Extract Key Sentences from Biomedical Abstracts. Int. J. Med. Informatics 76, 195–200 (2007)
29. McKnight, L., Srinivasan, P.: Categorization of Sentence Types in Medical Abstracts. In: Procedings of AMIA Annual Symposium, pp. 440–444. AMIA (2003)
30. Mizuta, Y., Korhonen, A., Mullen, T., Collier, N.: Zone Analysis in Biology Articles as a Basis for Information Extraction. Int. J. Med. Informatics 75, 468–487 (2006)
31. Teufel, S., Carletta, J., Moens, M.: An Annotation Scheme for Discourse-Level Argumentation in Research Articles. In: Proceedings of EACL, pp. 110–117. Association for Computational Linguistics (1999)
32. Liakata, M., Teufel, S., Siddharthan, A., Batchelor, C.: Corpora for the Conceptualisation and Zoning of Scientific Papers. In: Proceedings of LREC 2010, pp. 2054–2061. ELRA (2010)
33. Liakata, M., Saha, S., Dobnik, S., Batchelor, C., Rebholz-Schuhmann, D.: Automatic Recognition of Conceptualisation Zones in Scientific Articles and Two Life Science Applications. Bioinformatics 28(7), 991–1000 (2012)
34. Vincze, V., Szarvas, G., Farkas, R., Mora, G., Csirik, J.: The BioScope Corpus: Biomedical Texts Annotated for Uncertainty, Negation and their Scopes. BMC Bioinformatics 9, S9 (2008)
35. Rubin, V., Liddy, E., Kando, N.: Certainty Identification in Texts: Categorization Model and Manual Tagging Results. In: Shanahan, J.G., Qu, Y., Wiebe, J. (eds.) Computing Attitude and Affect in Text: Theory and Applications, pp. 61–76. Springer, Heidelberg (2006)

36. Hyland, K.: Talking to the Academy: Forms of Hedging in Science Research Articles. Written Communication 13, 251–281 (1996)
37. Hyland, K.: Writing without Conviction? Hedging in Science Research Articles. Applied Linguistics 17, 433–454 (1996)
38. Rizomilioti, V.: Exploring Epistemic Modality in Academic Discourse Using Corpora. In: Arnó Macià, E., Soler Cervera, A., Rueda Ramos, C. (eds.) Information Technology in Languages for Specific Purposes, pp. 53–71. Springer, New York (2006)
39. Thompson, P., Venturi, G., McNaught, J., Montemagni, S., Ananiadou, S.: Categorising Modality in Biomedical Texts. In: Proceedings of the LREC 2008 Workshop on Building and Evaluating Resources for Biomedical Text Mining, pp. 27–34. ELRA (2008)
40. de Waard, A., Pander Maat, H.: Categorizing Epistemic Segment Types in Biology Research Articles. In: Proceedings of the Workshop on Linguistic and Psycholinguistic Approaches to Text Structuring, LPTS 2009 (2009)
41. Wilbur, W.J., Rzhetsky, A., Shatkay, H.: New Directions in Biomedical Text Annotations: Definitions, Guidelines and Corpus Construction. BMC Bioinformatics 7, 356 (2006)
42. Liakata, M., Thompson, P., de Waard, A., Nawaz, R., Maat, H.P., Ananiadou, S.: A Three-Way Perspective on Scientific Discourse Annotation for Knowledge Extraction. In: Proceedings of the ACL Workshop on Detecting Structure in Scholarly Discourse (DSSD), pp. 37–46. Association for Computational Linguistics (2012)
43. Thompson, P., Nawaz, R., McNaught, J., Ananiadou, S.: Enriching a Biomedical Event Corpus with Meta-knowledge Annotation. BMC Bioinformatics 12, 393 (2011)
44. Cohen, K.B., Johnson, H.L., Verspoor, K., Roeder, C., Hunter, L.E.: The Structural and Content Aspects of Abstracts versus Bodies of Full Text Journal Articles are Different. BMC Bioinformatics 11, 492 (2010)
45. Nawaz, R., Thompson, P., Ananiadou, S.: Meta-Knowledge Annotation at the Event Level: Comparison between Abstracts and Full Papers. In: Proceedings of the Third LREC Workshop on Building and Evaluating Resources for Biomedical Text Mining (BioTxtM 2012), pp. 24–21. ELRA (2012)
46. Knight, J.: Negative Results: Null and void. Nature 422, 554–555 (2003)
47. Miwa, M., Thompson, P., McNaught, J., Kell, D.B., Ananiadou, S.: Extracting Semantically Enriched Events from Biomedical Literature. BMC Bioinformatics 13, 108 (2012)
48. Bjorne, J., Salakoski, T.: Generalizing Biomedical Event Extraction. In: Proceedings of the BioNLP Shared Task 2011 Workshop, pp. 183–191. Association for Computational Linguistics (2011)
49. Kilicoglu, H., Bergler, S.: Adapting a General Semantic Interpretation Approach to Biological Event Extraction. In: Proceedings of BioNLP Shared Task 2011 Workshop, pp. 173–182. Association for Computational Linguistics (2011)
50. Kilicoglu, H., Bergler, S.: Syntactic Dependency Based Heuristics for Biological Event Extraction. In: Proceedings of the BioNLP 2009 Workshop Companion Volume for Shared Task, pp. 119–127. Association for Computational Linguistics (2009)
51. Nawaz, R., Thompson, P., Ananiadou, S.: Identification of Manner in Bio-Events. In: Proceedings of the Eighth International Conference on Language Resources and Evaluation (LREC 2012), pp. 3505–3510. ELRA (2012)
52. Nawaz, R., Thompson, P., Ananiadou, S.: Something Old, Something New: Identifying Knowledge Source in Bio-Events. In: Proceedings of CICLing 2013 (2013)
53. Nawaz, R., Thompson, P., Ananiadou, S.: Negated Bio-events: Analysis and Identification. BMC Bioinformatics 14, 14 (2013)

Distributional Term Representations
for Short-Text Categorization

Juan Manuel Cabrera, Hugo Jair Escalante, and Manuel Montes-y-Gómez

Department of Computational Sciences,
Instituto Nacional de Astrofísica, Óptica y Electrónica,
Tonantzintla, 72840, Puebla, Mexico
{jmcabrera,hugojair,mmontesg}@inaoep.mx

Abstract. Everyday, millions of short-texts are generated for which effective tools for organization and retrieval are required. Because of the tiny length of these documents and of their extremely sparse representations, the direct application of standard text categorization methods is not effective. In this work we propose using distributional term representations (DTRs) for short-text categorization. DTRs represent terms by means of contextual information, given by document occurrence and term co-occurrence statistics. Therefore, they allow us to develop enriched document representations that help to overcome, to some extent, the small-length and high-sparsity issues. We report experimental results in three challenging collections, using a variety of classification methods. These results show that the use of DTRs is beneficial for improving the classification performance of classifiers in short-text categorization.

1 Introduction

During the last decade we have witnessed an exponential growth of the amount of textual information being generated every day. Therefore, efficient and effective tools for text organization and mining are required. Text classification (TC) is an essential task for the organization of textual information, it consists in associating documents with predefined thematic categories [24].

TC has been mainly faced as a supervised learning problem. Different classification methods have been used for TC, most notably naïve Bayes [7,12], K-nearest neighbor [27] and support vector machines [10], see [24] for a comprehensive review. Most TC approaches use the bag-of-words (BoW) representation for documents. Under BoW a document is represented by a vector indicating the weighted occurrence of words from a dictionary into the document. Since, only the words that appear in the document have non-zero entries in the corresponding representation vector, BoW can generate highly sparse representations; where the level of sparsity depends on both the length of documents and the narrowness of the vocabulary.

Albeit the sparsity issue, acceptable results have been obtained with the BoW representation in many TC applications dealing with regular-length documents [7,10,12,24]. However, the sparsity problem is much more critical in the categorization of short-texts, that is, documents composed of a few dozens of words at the most. Short-texts are rather common and abundant today as there has been an increasing spread of communication

A. Gelbukh (Ed.): CICLing 2013, Part II, LNCS 7817, pp. 335–346, 2013.

media that encourage the use of less words for sharing information. Examples of this type of media are social networks, micro-blogs, news summaries, FAQs, SMSs and scientific abstracts among others. The proliferation of these sources of information have posed a major challenge to researchers that must develop effective methods for the organization and access of such information.

Short-texts induce much more sparse representations than regular-length documents because only a few words occur in each short-text. In addition, in short-text domains the frequency of occurrence of words is rather low; that is, repeated occurrence of words in documents is rare and most words in the vocabulary occur a few times across the whole corpus. In consequence, vocabularies tend to be very large. For these reasons the usual approach to TC cannot be adopted directly.

This paper describes a new methodology for short-text categorization based on distributional term representations (DTRs) [11]. DTRs are a way to represent terms by means of contextual information, given by document-occurrence and term-co-occurrence statistics. Thus, the representation of a term is given by the documents in which it appears across the corpus or the other terms it co-occurs with. For short-text categorization we generate DTRs for terms in the vocabulary, and represent a document by combining the DTRs of terms that appear in it. In this way, a short-text is represented by the combination of the contexts of their terms, which reduces the sparseness of representations and alleviates, to some extent, the low frequency issue.

Since DTRs are based on occurrence and co-occurrence statistics, extracting them from short-text corpora may represent another challenge. Nevertheless, there are some domains in which one has available regular-length documents for developing the TC system, even when the ultimate goal of the system is the classification of short-texts [20]. For example, in databases of scientific articles we may have access to full texts (resp. abstracts) when developing the system and then we may want to categorize abstracts (resp. titles) of new documents. In this paper we focus on those domains for the application of DTRs. One should note that another option to generate useful DTRs is to rely on external resources, that is a research direction we may explore for future work.

In the following sections we present a review of related work on short-text categorization, describe the proposed methodology, and show results in three short-text corpora: the reduced Reuters R8 news corpus and two scientific abstracts collections: EasyAbstract and CICLing2002. Experimental results show that DTRs are more robust than the BoW representation for short-text categorization with different classification techniques and under different configurations. Results give evidence that DTRs capture better the semantic content of short-texts, even when direct word-occurrence information is scarce.

2   Related Work

In recent years different studies have recognized the relevance and complexity of short-text classification [22]. Many of these works have proposed document representations robust to sparsity and low term-frequency issues. In particular, most of them are based on document expansion [5,21,25,26,14]. The underlying hypothesis of these methods is to incorporate in a document representation a weight associated to terms that do not occur in the document, but that are associated to terms that actually occur. Thus, terms that do not occur in short-texts still can contribute to their representations.

Whereas the intuitive idea behind document expansion techniques is well sound, most approaches rely on external resources such as Wordnet for estimating the association between terms. This is an important limitation of this approach since the selection of an appropriate external resource to work with a particular collection is a problem itself, as we must guarantee the quality of the external resource, and most importantly, we must ensure that domains in the short-text collection and external resource are the compatible. In this paper we propose the use of document representations that expand a document by using contextual information. Opposed to previous works, we rely exclusively in information extracted from the same collection of short-texts, thus alleviating the need to obtain a reliable external resource.

Other type of methods modify the representation of documents with the goal of capturing discriminative information that may help to the classification of short-texts [28,23,19,16,15]. These kind of methods mainly use latent semantic analysis to project the document representations into another space in which documents that share semantically-related terms lie close to each other. Although all of these methods have reported acceptable results, they require of a large number of training samples to obtain satisfactory results. Therefore alternative techniques are required when dealing with small collections.

Finally, there are some techniques for short-text classification that use the BoW representation and aim to improve the classification method to obtain acceptable results in short-text classification. For example, Ramirez et al. propose a method that incorporates information from the similarity between test documents to improve the classification performance of the centroid-based classifier. Faguo et al. propose a classification method tailored for short-text domains in which adhoc statistics and rules are obtained [4]. This methods require a vectorial representation of documents, thus they are not restricted to the BoW representation. Therefore, the document representations proposed in this paper could be combined with the afore mentioned methods in order to further improve the classification performance.

On the other hand, DTRs have been mainly used in term classification and term clustering tasks [11], also they have been recently used in multimedia image retrieval [3]. DTRs, however, have not been used for short-text categorization, despite their potential benefits for document expansion. Actually, to the best of our knowledge, DTRs have not been used for TC at all.

3 Text Classification Using Distributional Term Representations

This section describes the proposed methodology for short-text classification. It is divided in two main phases: training and testing, see Figure 1. The training phase consists of calculating DTRs for terms, representing documents by combining DTRs from their terms and training a classifier by using the documents represented with DTRs. In this paper we considered two popular DTRs, namely, document-occurrence and term-co-occurrence representations [11]. For this stage, any learning algorithm can be used to build the classifier. In the second phase, test documents are represented by combining DTRs from terms as well; then, they are categorized by using the classifier trained in the previous stage. The rest of this section describes in detail the considered DTRs and the proposed document representation approach.

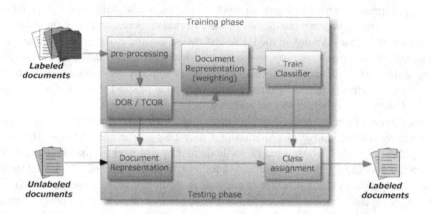

Fig. 1. General diagram of the proposed approach to short-text classification

3.1 Document Occurrence Representation (DOR)

The document occurrence representation (DOR) can be considered the dual of the *tf-idf* representation widely used in information retrieval [11]. DOR is based on the hypothesis that the semantics of a term can be revealed by a distribution of occurrence-statistics over the documents in the corpus. A term t_j is then represented as a vector of weights associated to documents $w_j = \langle w_{1,j}, \ldots, w_{N,j} \rangle$, where N is the number of documents in the collection and $0 \leq w_{k,j} \leq 1$ represents the contribution of document d_k to the specification of the semantics of t_j:

$$w_{k,j} = df(d_k, t_j) \cdot log \frac{|T|}{N_k} \tag{1}$$

where N_k is the number of different terms from the dictionary T that appear in document d_k, $|T|$ is the number of terms in the vocabulary, and

$$df(d_k, t_j) = \begin{cases} 1 + \log(\#(d_k, t_j)) & if(\#(d_k, t_j) > 0) \\ 0 & otherwise \end{cases} \tag{2}$$

where $\#(d_k, t_j)$ denotes the number of times term t_j occurs in document d_k. Intuitively, the more frequent the term t_j is in document d_k, the more important is d_k to characterize the semantics of t_j. Also, the more terms contains d_k, the less it contributes to characterize the semantics of t_j. The vector of weights is normalized so that $||w_j||_2 = 1$.

3.2   Term Co-Occurrence Representation (TCOR)

The term co-occurrence representation (TCOR) is based on co-occurrence statistics [11]. The underlying idea is that the semantics of a term t_j can be revealed by other terms it co-occur with across the document collection. Here, each term $t_j \in T$ is represented by a vector of weights $w_j = \langle w_{1,j}, \ldots, w_{|T|,j} \rangle$, where $0 \leq w_{k,j} \leq 1$ represents the contribution of term t_k to semantic description of t_j:

$$w_{k,t} = tff(t_k, t_j) \cdot log \frac{|T|}{T_k} \qquad (3)$$

where T_k is the number of different terms in the dictionary T that co-occur with t_j in at least one document and

$$tff(t_k, t_j) = \begin{cases} 1 + \log(\#(t_k, t_j)) & if(\#(t_k, t_j) > 0) \\ 0 & otherwise \end{cases} \qquad (4)$$

where $\#(t_k, t_j)$ denotes the number of documents in which term t_j co-occurs with the term t_k. The vector of weights is normalized to have unit 2-norm: $||w_j||_2 = 1$.

3.3   Representation of Documents Using DTRs

Previous sections have described how to obtain DTRs for terms. This section describes how to combine these DTRs to build document representations especially suited for short-text categorization. Let w_{t_j} denote the DTR of term t_j in the vocabulary, where w_{t_j} can be either the DOR or TCOR representations. The representation of a document d_i based on DTRs, d_i^{dtr}, is obtained as follows:

$$d_i^{dtr} = \sum_{t_j \in d_i} \alpha_{t_j} \cdot w_{t_j} \qquad (5)$$

where α_j is a scalar that weights the contribution of term $t_j \in d_i$ into the document representation. Thus, the representation of a document is given by the (weighted) aggregation of the contextual representations of terms appearing in the document. Scalar α_{t_j} aims to weight the importance that term t_j has for describing document d_i. Many options are available for defining α_{t_j}, in this work we considered three common weighting schemes from information retrieval: Boolean, term frequency and term frequency - inverse document frequency (tf-idf).

4   Experimental Evaluation

4.1   Setup and Datasets

For the evaluation of the proposed classification methodology we considered three data sets of varying complexities, namely, the reduced Reuters R8 news corpus and two scientific abstracts collections: EasyAbstract and CICLing2002. Documents in these collections are divided into two sections, titles and abstracts/bodies. We performed experiments using the whole documents for training and testing. We called this setting DD. With the aim of evaluating the benefits of DTRs for short-text categorization, we also assembled test collections consisting only of document titles. We refer to this setting as DT. In the following we describe the considered collections.

The R8 dataset is a subset of the Reuters-21578 collection that consists of documents labeled with the 8 most frequent categories, where each document belongs to a single class. The collection R8 was previously used, for example, by [21,20,17,2,9,18]. Table 1 describes the R8 data set.

Table 1. Main statistics of the R8 corpus. Column 3 shows the usual test partition (DD setting), while column 4 shows the reduced test partition (DT setting).

Feature	Train	Test (DD)	Test-Reduced (DT)
Vocabulary size	14,865	8,760	3,676
Number of Documents	4,559	2,179	2,179
Average terms per document	40.9	39.2	6.6

The EasyAbstracts data set was compiled by Rosas et al. [21], it has been widely used for the evaluation of methods for clustering of short-texts. The data set is composed of abstracts of papers published in proceedings of a conference. It comprises 4 classes (*machine learning, heuristic in optimization, autonomous intelligent agents,* and *automated reasoning*), and all of the abstracts are thematically related to the topic of *intelligent systems*. Table 2 shows some statistics for the EasyAbstracts corpus.

Table 2. Main characteristics of the EasyAbstracts corpus. We show informative statistics for the regular size (DD) and reduced (DT) versions of the data set.

Feature	Regular (DD)	Reduced (DT)
Vocabulary size	1136	206
Number of Documents	48	48
Average terms per doc.	60.3	5.85

The third corpus we considered for experimentation is CICLing2002. This corpus is composed of 48 scientific abstracts from the *computational linguistics* domain, the abstracts belong to one of the following classes: *linguistics, lexicon, ambiguity* and *text processing*. Thus, as with EasyAbstracts, the thematic content of documents is very close to each other. The CICLing2002 collection has been used by other researchers [13,21,8,9] mainly for the evaluation of clustering of short-texts. The corpus is described in Table 3.

One should note that there are no predefined training-test partitions for EasyAbstracts and CICLing2002 data sets. Thus, we adopted 10-fold cross validation for the evaluation of our method in these data sets. For the R8 collection we used the predefined partitions, which allows us to compare our results with previous works, e.g., [20]. In the following we will refer to a T-test at the 95% confidence level when mentioning statistically significant differences.

Table 3. Main characteristics of the CICLing2002 corpus

Feature	Regular (DD)	Reduced (DT)
Vocabulary size	813	180
Number of Documents	48	48
Average terms per doc.	45.06	4.8

4.2 Short-Text Classification with the Bag of Words Representation

In this section we report experimental results on the performance of the traditional bag of words (BoW) representation for short-text classification. The goal of the experiments is to verify the difficulties of the BoW for effectively representing the content of short-texts. We represented documents by using the BoW formulation under different weighting schemes[1] and evaluated the performance of five different learning algorithms representative of the wide diversity of methods available in the machine learning field[2]. Experimental results of this experiment are shown in Table 4. We report the obtained results when using the regular-length documents for training and testing (DD), and when using reduced documents for testing (DT) for each of the considered corpus. For R8 we report the performance obtained in the predefined testing partitions, while for the other corpora we report the average performance of 5 runs of 10-fold cross-validation.

From Table 4 we can see that acceptable classification performance was obtained by the different classifiers when regular-length documents were considered (DD columns). However, the performance of most classifiers dropped considerably when classifying short-texts (DT columns). Among the considered classifiers, SVM obtained the best results for most configurations of data sets and weighting schemes. On the contrary, it does not seem to be a *winning* weighting scheme for short-text classification (DT). The Boolean approach obtained the global best results for R8 and EasyAbstracts, but TF outperformed the other schemes in the CICLing2002 data set.

Macro F_1 dropped significantly when short-texts were classified (DT). The drop of accuracy was consistent for different weighting schemes and classifiers. The global average of decrements is of 38.66% and there are decrements of up to 72.74%.

Results presented in this section confirm those reported in previous works showing that the BoW representation is not well suited for short-text classification, not even when regular-length documents were available during training like in the DT setting. In the next section we report experimental results obtained with DTRs showing their usefulness for classifying short-texts.

4.3 Using DTRs for Short-Text Classification

In this section we evaluate the performance of DTRs for short-text classification (i.e., the DT setting). In particular, we are interested in assessing the added value offered

---

[1] One should not confuse the weighting schemes used in this section (for document representation under BoW) with those proposed in Section 3.3 (for document representation using DTRs.)

[2] We used the Weka implementation of the above described algorithms, where default parameters were considered for all of the classifiers [6].

Table 4. Classification results obtained with the BoW representation for regular-length documents (DD) and short-texts (DT); the best results for each data set and setting are shown in **bold**. Besides reporting the macro F_1 measure, we report the relative drop of accuracy (column *Decrease*) that occurs when classifying short-texts.

R8

	Boolean			TF			TFIDF		
	DD	DT	Decrease	DD	DT	Decrease	DD	DT	Decrease
AdaBoost	0.64	0.18	-72.74%	0.64	0.18	-72.74%	0.64	0.18	-72.74%
Knn1	0.69	0.39	-43.98%	0.47	0.34	-27.53%	0.47	0.34	-27.53%
Naive Bayes	0.87	0.66	-24.16%	0.82	0.34	-58.97%	0.82	0.34	-59.13%
RandomForest	0.80	0.54	-32.21%	0.80	0.57	-29.02%	0.82	0.74	-10.46%
SVMLineal	**0.91**	**0.83**	-7.85%	0.90	0.73	-19.29%	0.90	0.70	-22.59%

EasyAbstract

	Boolean			TF			TFIDF		
AdaBoost	0.41	0.27	-34.34%	0.40	0.25	-37.70%	0.40	0.25	-37.70%
Knn1	0.21	0.11	-46.14%	0.14	0.09	-38.74%	0.14	0.09	-38.74%
Naive Bayes	0.70	0.40	-42.89%	0.74	0.35	-53.09%	0.79	0.37	-52.93%
RandomForest	0.57	0.24	-57.82%	0.49	0.22	-54.34%	0.53	0.19	-64.01%
SVMLineal	**0.69**	**0.59**	-15.64%	0.90	0.16	-82.05%	0.85	0.30	-64.67%

CICLing

	Boolean			TF			TFIDF		
AdaBoost s	0.36	0.27	-22.76%	0.36	0.27	-22.76%	0.31	0.20	-35.32%
Knn1	0.29	0.10	-65.62%	0.14	0.16	10.62%	0.13	0.09	-31.31%
Naive Bayes	0.43	0.33	-23.50%	0.43	0.39	-10.50%	0.37	0.14	-61.30%
RandomForest	0.40	0.25	-38.01%	0.31	0.30	-1.10%	0.22	0.12	-46.91%
SVMLineal	0.45	0.35	-21.14%	**0.54**	**0.48**	-11.91%	0.21	0.14	-35.52%

by document representations based on DTRs over the BoW formulation. For this experiment, term representations based DOR and TCOR were obtained from the regular-length training documents. Next, training and test documents were represented as described in Section 3. Then the performance of the considered classifiers was evaluated. Table 5 shows the results obtained for this experiment under the proposed weighting schemes for document representation under DTRs, see Section 3.3.

From Table 5 we can see that results obtained with representations based on both DOR and TCOR, clearly outperformed those obtained with the BoW formulation for most configurations. In fact, in 62 out of the 90 results shown in Table 5 the improvements of DTRs over BoW were statistically significant, that is 68.88% of all of the results. DTRs did not outperform the BoW only in 7 results out of 90 (i.e., 7.8%).

Among the considered weighting schemes for DTRs, TFIDF was the most regular one (see the last 3 columns from Table 5), outperforming the BoW formulation for all of the classifiers and across all of the data sets. Regarding classification methods, it is clear that the combination of representations based on DTRs and SVM was the most effective. Since we considered a linear SVM classifier, DOR/TCOR based representations made short-texts more linearly separable, than when the BoW representation was used. Thus, we can say that DOR/TCOR based representations capture better the content of short-texts than BoW.

In average, results obtained with DOR and TCOR were very similar. Nevertheless, we claim that DOR is slightly better than TCOR. DOR based representations obtained the best results for R8 and CICLing2002 data sets, while the best result in the EasyAbstracts corpus was obtained with a TCOR based representation. One should

Table 5. Short-text classification results obtained with the proposed approach for the different classifiers and weighting schemes. Shaded cells indicate results where DTRs outperformed the BoW formulation; results in **bold** indicate a statistical significant difference between the results obtained with BoW and DOR/TCOR.

R8

Weigth	Boolean			TF			TFIDF		
Classifiers	BOW	DOR	TCOR	BOW	DOR	TCOR	BOW	DOR	TCOR
AB	0.175	0.645	0.668	0.175	0.632	0.651	0.175	0.591	0.667
KNN	0.386	**0.899**	**0.897**	0.337	**0.908**	**0.902**	0.337	**0.746**	**0.754**
NB	0.656	**0.881**	**0.893**	0.336	**0.874**	**0.886**	0.336	**0.785**	**0.854**
RF	0.543	**0.786**	**0.774**	0.565	**0.805**	**0.823**	0.736	**0.798**	**0.819**
SVM	0.834	**0.930**	**0.891**	0.728	**0.928**	**0.901**	0.699	**0.897**	**0.784**

EasyAbstract

	Boolean			TF			TFIDF		
AB	0.268	0.185	0.201	0.255	0.272	0.245	0.250	0.263	0.292
KNN	0.114	**0.600**	**0.482**	0.086	**0.666**	**0.712**	0.086	**0.571**	**0.541**
NB	0.402	**0.568**	**0.586**	0.345	**0.603**	**0.590**	0.370	**0.578**	**0.603**
RF	0.239	**0.495**	**0.332**	0.223	**0.507**	**0.582**	0.192	**0.588**	**0.550**
SVM	0.585	0.660	**0.639**	0.161	**0.728**	**0.733**	0.301	**0.622**	**0.589**

CICLIng2002

	Boolean			TF			TFIDF		
AB	0.274	0.188	0.244	0.274	0.129	0.224	0.199	0.201	0.232
KNN	0.099	**0.450**	**0.395**	0.156	**0.478**	**0.399**	0.089	**0.493**	0.44
NB	0.332	**0.473**	0.415	0.386	**0.426**	0.471	0.143	**0.506**	**0.399**
RF	0.249	0.184	**0.369**	0.304	0.279	0.374	0.119	**0.418**	**0.291**
SVM	0.354	**0.526**	0.414	0.48	0.504	0.502	0.135	**0.528**	**0.442**

note, however, that the difference of such result and the best one obtained with DOR based representations was of 0.005, which represents less than 0.7% of relative improvement. Thus, we can say that the use of DOR based representations is advantageous over TCOR based ones. Besides classification accuracy, DOR is advantageous over TCOR because it may result in document representations of much lower dimensionality.

4.4 Comparison of DTRs with Other Methods for Short Text Categorization

We have shown that DTRs outperformed BoW in short text categorization for reduced data sets (DT setting). Additionally, in preliminary work we showed evidence that DTRs also compare favorably against BoW when using the regular size data sets (DD setting) [1], although, as expected, improvements were lower for that setting because the BoW is less affected in a scenario when documents are large enough to capture its content, see Table 4.

In this section we further compare the performance of DTRs to alternative short text categorization techniques. In particular we consider three representative methods of the state of the art in short text categorization. We consider the method proposed by S. Zelikovitz, a representation-transformation approach implementing transductive latent semantic indexing (LSI) [28]. Also, we consider a representative method, based on word-sense-disambiguation (WSD), that expands the representation of documents using external resources [21]. Finally, we also considered a method that modifies a classifier to be suitable for short text categorization, the so called neighborhood consensus (NC) approach [20]. For the LSI and WSD methods we used the best configuration of parameters as suggested by their authors, while for NC we use the results reported in [20], as those authors used the same partitions we did for the R8 collection. Table 6 shows the results of this comparison for the DT setting.

Table 6. Comparison of the performance of the proposed approach to alternative methods for short text categorization

Setting / Method	BOW	NC [20]	LSI [28]	WDS [21]	DOR	TCOR
R8	74.23	78.5	60	78.78	92.97	89.72
EasyAbstracts	57.5	-	25	57.00	79.05	79.00
CICLing	34.31	-	30	51.00	60.52	60.51

We can see from Table 6 that the best results over the three collections were obtained with the proposed DTRs. DOR obtained slightly better results than TCOR, as reported in previous sections, although both DTRs achieve outstanding improvements over the other methods. Larger improvements were observed for the more complex data sets (i.e., EasyAbstracts and CICLing). Interestingly, the plain BoW representation outperformed the LSI approach in all collections and it achieved comparable performance to WSD in R8 and EasyAbstracts corpora. The NC method is based on the BoW representation, the improvement of NC over plain BoW was of $\approx 4\%$, thus we would expect that by applying the NC method with DTRs we could further improve the performance of our proposal.

5   Conclusions

We have introduced a way to take advantage of distributional term representations (DTRs) for short-text classification. Compared to regular-length text-categorization, the classification of short-texts poses additional challenges due to the low term-frequency occurrence, sparsity and term ambiguity. In this paper we aimed to overcome those issues by using DTRs. DTRs provide a natural way to expand the content of short-texts, which implicitly address the low-frequency, sparsity and term-ambiguity issues. We propose a new way to use the document occurrence representation (DOR) and the term-co-occurrence representation (TCOR) for this problem under three weighting schemes.

We reported experimental evidence that shows the proposed document representations significantly outperform the traditional bag of words (BoW) representation as well as the results by other state of the art approaches in short text classification, under different weighting schemes and using different classification methods. DOR obtained slightly better results than TCOR, besides, DOR induces lower dimensional representations. Therefore, we recommend the use of DOR for short-text classification.

Future work directions include using external resources for obtaining better DTRs. Exploring the use of information fusion techniques for combining information from multiple DTRs with the BoW formulation. Developing alternative weighting schemes for document representation.

Acknowledgements. This work was supported by CONACYT under project grant No. 134186 and scholarship 225734.

References

1. Cabrera, J.M.: Clasificación de textos cortos usando representaciones distribucionales de los términos. Master's thesis, Instituto Nacional de Astrofísica, Óptica y Electrónica (2012)
2. Cardoso-Cachopo, A., Oliveira, A.: Combining LSI with other classifiers to improve accuracy of single-label text categorization. In: First European Workshop on Latent Semantic Analysis in Technology Enhanced Learning, Netherlands (2007)
3. Escalante, H.J., Montes, M., Sucar, E.: Multimodal indexing based on semantic cohesion for image retrieval. Information Retrieval 15(1), 1–32 (2012)
4. Faguo, Z., Fan, Z., Bingru, Y.: Research on Short Text Classification Algorithm Based on Statistics and Rules. In: Third International Symposium on Electronic Commerce and Security, pp. 3–7 (July 2010)
5. Fan, X., Hu, H.: A New Model for Chinese Short-text Classification Considering Feature Extension. In: International Conference on Artificial Intelligence and Computational Intelligence, pp. 7–11. IEEE (October 2010)
6. Garner, S.R.: Weka: The Waikato environment for knowledge analysis. In: Proceedings of the New Zealand Computer Science Research Students Conference, pp. 57–64 (1995)
7. He, F., Ding, X.-q.: Improving naive bayes text classifier using smoothing methods. In: Amati, G., Carpineto, C., Romano, G. (eds.) ECIR 2007. LNCS, vol. 4425, pp. 703–707. Springer, Heidelberg (2007)
8. Ingaramo, D., Errecalde, M., Rosso, P.: A General Bio-inspired Method to Improve the Short-Text Clustering Task. In: Gelbukh, A. (ed.) CICLing 2010. LNCS, vol. 6008, pp. 661–672. Springer, Heidelberg (2010)
9. Ingaramo, D., Pinto, D., Rosso, P., Errecalde, M.: Evaluation of internal validity measures in short-text corpora. In: Gelbukh, A. (ed.) CICLing 2008. LNCS, vol. 4919, pp. 555–567. Springer, Heidelberg (2008)
10. Joachims, T.: Text categorization with support vector machines: learning with many relevant features. In: Nédellec, C., Rouveirol, C. (eds.) ECML 1998. LNCS, vol. 1398, pp. 137–142. Springer, Heidelberg (1998)
11. Lavelli, A., Sebastiani, F., Zanoli, R.: Distributional Term Representations: An Experimental Comparison. In: Italian Workshop on Advanced Database Systems (2004)
12. Lewis, D.D.: Naive Bayes at Forty: The independence assumption in information retrieval. In: Nédellec, C., Rouveirol, C. (eds.) ECML 1998. LNCS, vol. 1398, pp. 4–15. Springer, Heidelberg (1998)
13. Makagonov, P., Alexandrov, M., Gelbukh, A.F.: Clustering abstracts instead of full texts. In: Proceedings of the 10th International Conference on Text, Speech and Dialogue, pp. 129–136 (2004)
14. Nagarajan, M., Sheth, A., Aguilera, M., Keeton, K.: Altering Document Term Vectors for Classification - Ontologies as Expectations of Co-occurrence. In: ReCALL, pp. 1225–1226 (2007)
15. Phan, X.-H., Nguyen, C.-T., Le, D.-T., Nguyen, L.-M., Horiguchi, S., Ha, Q.-T.: A hidden topic-based framework towards building applications with short web documents. IEEE Transactions on Knowledge and Data Engineering 23(7), 961–976 (2011)
16. Phan, X.-H., Nguyen, L.-M., Horiguchi, S.: Learning to classify short and sparse text & web with hidden topics from large-scale data collections. In: Proceeding of the 17th International Conference on World Wide Web - WWW 2008, p. 91 (2008)
17. Pinto, D., Rosso, P.: On the Relative Hardness of Clustering Corpora. In: Proceedings of the 10th International Conference on Text, Speech and Dialogue, pp. 155–161 (2007)
18. Pinto, D., Rosso, P., Jimenez-Salazar, H.: A Self-enriching Methodology for Clustering Narrow Domain Short Texts. The Computer Journal, 1–18 (September 2010)

19. Pu, Q., Yang, G.-w.: Short-text classification based on ICA and LSA. In: Wang, J., Yi, Z., Żurada, J.M., Lu, B.-L., Yin, H. (eds.) ISNN 2006. LNCS, vol. 3972, pp. 265–270. Springer, Heidelberg (2006)
20. Ramírez-de-la-Rosa, G., Montes-y-Gómez, M., Solorio, T., Villaseñor-Pineda, L.: A document is known by the company it keeps: neighborhood consensus for short text categorization. Language Resources and Evaluation, 1–23 (to appear, 2013)
21. Rosas, V., Errecalde, M.L., Rosso, P.: Un Analisis Comparativo de Estrategias para la Categorización Semantica de Textos Cortos. Sociedad Española para el Procesamiento del Lenguaje Natural 44, 11–18 (2010)
22. Rosso, P., Errecalde, M., Pinto, D.: Language resources and evaluation journal: Special issue on analysis of short texts on the web (forthcoming, 2013)
23. Sahlgren, M., Cöster, R.: Using bag-of-concepts to improve the performance of support vector machines in text categorization. In: Proceedings of the 20th International Conference on Computational Linguistics, COLING 2004, pp. 1–7 (2004)
24. Sebastiani, F.: Machine learning in automated text categorization. ACM Computing Surveys 34(1), 1–47 (2002)
25. Wang, J., Zhou, Y., Li, L., Hu, B., Hu, X.: Improving Short Text Clustering Performance with Keyword Expansion. In: Wang, H., Shen, Y., Huang, T., Zeng, Z. (eds.) The Sixth International Symposium on Neural Networks (ISNN 2009). AISC, vol. 56, pp. 291–298. Springer, Heidelberg (2009)
26. Xi-Wei, Y.: Feature Extension for short text. In: Proceedings of the Third International Symposium on Computer Science and Computational Technology, pp. 338–341 (2010)
27. Yang, Y., Liu, X.: A re-examination of text categorization methods. In: Proceedings of the 22nd Annual International ACM SIGIR Conference on Research and Development in Information Retrieval, SIGIR 1999, pp. 42–49. ACM, New York (1999)
28. Zelikovitz, S.: Transductive LSI for Short Text Classification Problems. In: American Association for Artificial Intelligence (2004)

Learning Bayesian Network Using Parse Trees for Extraction of Protein-Protein Interaction

Pedro Nelson Shiguihara-Juárez and Alneu de Andrade Lopes

Institute of Mathematical Sciences and Computation
University of São Paulo
{pshiju,alneu}@icmc.usp.br

Abstract. Extraction of protein-protein interactions from scientific papers is a relevant task in the biomedical field. Machine learning-based methods such as kernel-based represent the state-of-the-art in this task. Many efforts have focused on obtaining new types of kernels in order to employ syntactic information, such as parse trees, to extract interactions from sentences. These methods have reached the best performances on this task. Nevertheless, parse trees were not exploited by other machine learning-based methods such as Bayesian networks. The advantage of using Bayesian networks is that we can exploit the structure of the parse trees to learn the Bayesian network structure, i.e., the parse trees provide the random variables and also possible relations among them. Here we use syntactic relation as a causal dependence between variables. Hence, our proposed method learns a Bayesian network from parse trees. The evaluation was carried out over five protein-protein interaction benchmark corpora. Results show that our method is competitive in comparison with state-of-the-art methods.

1   Introduction

The automation of protein-protein interaction (PPI) extraction from scientific papers is a critical and relevant task in the biomedical field. PPIs are important to understand the cell behavior and, consequently, to develop new drugs. Initially, manual extractions (curations) were used to perform this task. However, the PPI extraction has been adversely limited by the growing amount of papers and the time-consuming task involved [1]. Although this task has been addressed by various computational approaches, the extraction of PPI still challenges the Machine Learning community. For example, a sentence containing names of several proteins could involve multiple interactions. In addition, there is a large number of possibilities to express the same idea utilizing natural language in written form. Since these problems affect the performance of computational approaches, the task is commonly considered as a binary classification problem in order to reduce its complexity [2]. It consists in detecting whether an interaction involved between a pair of protein names exists or not.

Figure 1 illustrates the extraction of PPIs from a sentence[1] as a binary problem. Three pairs of protein names are candidates: *Actin−Iota toxin*, *Iota*

---

[1] This sentence belongs to the Bioinfer corpus. ID: "BioInfer.d27.s0".

A. Gelbukh (Ed.): CICLing 2013, Part II, LNCS 7817, pp. 347–358, 2013.

toxin−Profilin and *Actin−Iota toxin*, and the output for these pairs of protein names are "true", "false" and "true" respectively. Thus, only two pairs of protein names are considered interaction relationships.

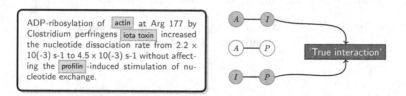

Fig. 1. *actin* (A), *iota toxin* (I) and *profilin* (P) proteins are evaluated in order to determine which pairs of these proteins describe an interaction. Two interactions were founded. The first one belongs to the *Actin−Iota toxin* (A,I) pair and, the last one belongs to the *Iota toxin−Profilin* (I,P) pair.

As stated earlier, the PPI task has been treated by different machine learning approaches, particularly, the kernel approaches using parse trees (PTs) [3] or dependency trees [2]. In contrast, there is a lack of recent research on Bayesian network (BN) models. Although BN models, using lexical features, have been proposed to solve the PPI problem before [1, 4], the PTs may play an important role to provide syntactic features for BN models. In BN, features and their relationships are modeled as a graph from unstructured data. However, this modeling may be enriched when relationships among features are also given from the training data set. In that sense, a PT represents syntactic information from a sentence in a tree form, where relationships are already given. Taking this into account, BNs can be learned from PTs. The rationale behind this hypothesis is that syntactic patterns among features may convey relevant information about the existence or not of PPIs.

To the best of our knowledge, this is the first work that proposes to learn a Bayesian network directly from parse trees to PPI extraction. The method combines the PTs, from sentences in training data, in order to create a (probably cyclic) graph. Then, the method removes edges and obtains a directed acyclic graph. Finally, we limit the maximum number of parents for each node in order to reduce the number of parameters and add a class node containing the values 'true interaction' or 'no interaction' in the BN. To complete the BN model (structure and parameters), the method uses the maximum likelihood estimation (MLE) [5] to calculate the parameters.

Models obtained by the proposed method were evaluated using five well-known PPI corpora. They are IEPA, AIMed, BioInfer, HPRD50 and LLL. Results demonstrated that the performance of our BN models are competitive compared to kernel-based methods applied on large corpora. However the performance of our models decrease when employed on small corpora. This limitation can be less latent since annotated corpora on PPI are becoming larger.

The reminder of the paper is organized as follows. In Section 2, we describe the related work on the PPI extraction problem. In Section 3, we explain the proposed method. In Section 4, we provide experimental results that demonstrate the competitive performance of the proposed method in comparison with the state-of-the-art methods. Finally, we conclude this work.

2   Related Work

According to [6], computational methods for PPI extraction can be organized in three groups. The first group consists of co-occurrence and rule-based approaches. The co-occurrence-based approaches evaluate the likelihood of two protein names co-occur in a same sentence. The rule-based approaches define a set of rules which is commonly applied on syntactic features, such as dependency trees [7]. These rules represent evidence of the existence of interactions. However, approaches based on co-occurrence lead to a low precision, whereas rule-based approaches lead to a low recall on the PPI extraction [1]. As a consequence, these methods have not succeeded on this task, needing further improvements. The second group is machine learning-based approaches which are commonly combined with natural language processing techniques. Recently, ensemble systems demonstrated a high performance in BioCreative II.5 [8]. Also, kernel-based machine learning approaches have been widely used in the PPI extraction task [2, 9–11, 6, 3, 12] with a high performance. They are considered state-of-the-art with the use of syntactic features. The third group is the combination from the approaches of the earlier two groups [7].

In machine learning-based methods, Dynamic Bayesian networks [4] were employed to solve the PPI extraction task in a multi-class context. In this case, words from sentences are used as features. However, the PPI corpora are getting larger and the use of all words as features in a Bayesian network increases computational cost. Bayesian networks learned by using Hill-climbing search algorithm [1] were also used in PPI extraction. These BN models use a fixed number of features (7 features). This small number of features could be little representative for all possible interactions existing in a sentence. A study [9] of the performance of the state-of-the-art methods in the PPI extraction indicates that these methods achieve between 19% and 30% of performance in terms of F-measure. In that sense, this study [9] proposes the use of an unified format of five benchmark corpora for more reliable evaluations. At present, these corpora are used by several works on PPI extraction problem and are also used in this work.

Currently, the kernel-based approaches are dominant on this task. Dependency trees were used to construct knowledge in a graph representation [2]. Afterwards, features are extracted from the graph. In the next step, these features are used on a regularized least squares kernel-based approach. In a similar way, kernel based on dependency paths [13] were used to cover different syntactic substructures and to obtain similarities (or dissimilarities) among sentences. Also, support vector machines using lexical and syntactic features [12] were used, obtaining little improvements. In an effort to achieve better results, multiple kernels combining lexical and syntactic features [10] obtained a high performance in comparison with the state-of-the-art approaches. Factors to improve the performance, such as pruning parse trees and tuning parameters of support vector machines, were employed in a simple but effective kernel-based method [3]. As suggested in [3], the pruning factor can be used independently of the machine learning-based approach employed. Thus, we also use prune method for parse trees to improve the performance and to reduce irrelevant information. In BioCreative II.5 challenge, the OntoGene text mining environment [14] obtained the best results. The OntoGene system employed dependency trees, demonstrating the importance of syntactic features on the PPI extraction task.

3   Extraction of PPI Using a Probabilistic Model

In PPI extraction, the relation extraction typically consider binary relations. This means that we have $\binom{n}{2}$ instances from a sentence, where n is the number of proteins. For example, we have $\binom{3}{2} = 3$ instances from the sentence described in Figure 1, since there are 3 protein names. These instances named (A, I), (A, P) and (I, P), must be evaluated to identify if there exist an interaction or not. Thus, currently, the relation extraction task can be seen as a binary classification problem [2]. The goal of this classification problem is to calculate the maximum a posteriori ($\hat{y}$) of the random class variable ($\mathcal{C}$), and a pair of proteins is extracted if there exist an interaction ($\mathcal{C} = 1$). In both cases ($\mathcal{C} = 1$ and $\mathcal{C} = 0$), we are given a classifier model $\mathcal{M}$, a training data set $\mathcal{D}$ and a test instance x denoted by (prot1, prot2), since it is assumed a binary relation. The maximum a posteriori is obtained among values of the class $\mathcal{C}$ in equation (1).

$$\hat{y} = \arg{}_\mathcal{C}\max \{P(\mathcal{C} = 0 \mid \mathcal{M}, \mathcal{D}, x), P(\mathcal{C} = 1 \mid \mathcal{M}, \mathcal{D}, x)\} \tag{1}$$

Thus, we use a function for relation extraction according to equation (2).

$$extractRelation\,(\hat{y}, x) = \begin{cases} \hat{y} = 1 & interactsWith(x.prot1, x.prot2) \\ \hat{y} = 0 & \emptyset \end{cases} \tag{2}$$

where the relation $interactsWith(x.prot1, x.prot2)$ is extracted whether a $\mathcal{M}$ classifier predicts an interaction ($\hat{y} = 1$) for the pair of protein names corresponding to the test instance x. For all test cases, the relation $interactsWith$ is defined as 'true interaction'. In this work, parse trees are considered as instances in the training data set $\mathcal{D}$ and, $\mathcal{M}$ is our Bayesian network model induced from $\mathcal{D}$.

3.1   Learning a Bayesian Network from Parse Trees

Given a Bayesian network model $\mathcal{M} = \langle G_\mathcal{M}, \theta \rangle$, where $G_\mathcal{M}$ is its structure (i.e., a directed acyclic graph) and θ is the set of parameters of the BN, the goal is to learn the model from parse trees in $\mathcal{D}$. The graph $G_\mathcal{M} = \langle Z_\mathcal{M}, E_\mathcal{M} \rangle$ contains a set of nodes ($Z_\mathcal{M}$) and edges ($E_\mathcal{M}$).

The basic assumption of the method is that non-terminal and pre-terminal nodes in parse trees can be considered as random variables $(Z_{\mathcal{M}})$ in order to learn a Bayesian network. In addition, an edge between a parent node and its child node, in the parse trees, can be regarded as a directed edge $(E_{\mathcal{M}})$ in the Bayesian network structure. This information is used to infer how these potential random variables are related to each other. Thus, we explain how our method allows to learn the structure and the parameters of a Bayesian network model.

Bayesian Network Structure. Our method follows 4 steps to learn the network structure: (1) create a (probably cyclic) graph structure using the non-terminal and pre-terminal nodes from parse trees, (2) remove all the cycles in the graph structure, (3) limit the maximum number of parents to d for each node z_k using mutual information to rank and keep only the d *best parents*, and (4) add a bi-valued class node to the graph in order to infer whether a sentence has or not an interaction.

In the first step, we consider a training data set containing M parse trees. Each parse tree PT_i $(1 \leq i \leq M)$ is defined as $PT_i = \langle Z_{PT_i}, E_{PT_i} \rangle$, where a non-terminal or a pre-terminal node $z_k \in Z_{PT_i}$ could also exist in other parse trees. In the i-th parse tree, PT_i, Z_{PT_i} and E_{PT_i} is the set of nodes and edges respectively. We aim to find the random variables of the Bayesian network $Z_{\mathcal{M}} = \{Z_{PT_1} \bigcup \ldots \bigcup Z_{PT_n}\}$. In this context, we considered the non-terminal and pre-terminal nodes from parse trees as random variables in the $Z_{\mathcal{M}}$ set. In a same way, the set of edges can be defined by $E_{\mathcal{M}} = \{E_{PT_1} \bigcup \ldots \bigcup E_{PT_n}\}$. An edge $e_x \in E_{\mathcal{M}}$ describes a relationship *parent* $\rightarrow$ *child* over the nodes in Z_{PT_i}. For example, assuming that we can learn a Bayesian network structure from only four parsed sentences (see Figure 2), the result $G'_{\mathcal{M}} = \langle Z'_{\mathcal{M}}, E'_{\mathcal{M}} \rangle$, after removing cycles, is shown in Figure 3.(a).

The second step involves the elimination of the cycles. We employ a depth-first search technique and put repeated nodes (with the corresponding parent) in a queue data structure. Since the preservation of the original relationships is desired, we only remove one edge in the queue. This edge contains the pair of nodes (a repeated node with its parent node) with the maximum distance from each other, when we apply the depth-first search. We do this second step again until no further repeated nodes are detected.

In the third step, each node $z_k \in Z_{\mathcal{M}}$ is restricted to have at most d number of parents. Such restriction minimizes the complexity of the Bayesian network structure in terms of the number of parameters. To select edges to be removed, we rank each edge containing a relationship between the node z_k and each of its parents $z_{\pi_{kj}} \in Z_{\mathcal{M}}$ by using the *mutual information* $(\mathcal{I})$ measure in equation (3). The node $z_{\pi_{kj}}$ is the j-th parent of the z_k node. The mutual information, between two variables z_k and $z_{\pi_{kj}}$, determines how similar is the joint distribution $p(z_k, z_{\pi_{kj}})$ in comparison to the factored distribution $p(z_k) p(z_{\pi_{kj}})$.

$$\mathcal{I} = \sum_{z_k} \sum_{z_{\pi_{kj}}} p(z_k, z_{\pi_{kj}}) \log \frac{p(z_k, z_{\pi_{kj}})}{p(z_k) p(z_{\pi_{kj}})} \tag{3}$$

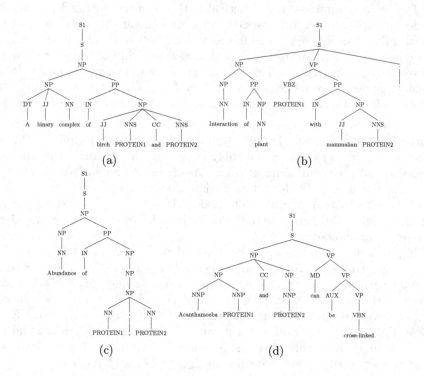

Fig. 2. This figure shows four parsed sentences

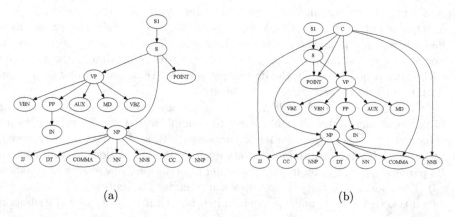

Fig. 3. (a) An acyclic graph containing non-terminal and pre-terminal nodes from the parse trees in Figure 2. (b) A Bayesian network structure with non-terminal and pre-terminal nodes and a class node $\mathcal{C}$.

For each node z_k, we rank the edges $e_{kj} = \left(z_k, z_{\pi_{kj}}\right)$, where $e_{kj} \in E_\mathcal{M}$, in a descendent order in terms of $\mathcal{I}$, such that, the first d edges in the rank are considered in the graph and remaining edges are removed.

Finally, in the last step, we add a class node $\mathcal{C}$ and use a hill-climbing local search algorithm and a Bayesian score [15] to connect this class node with the rest of nodes in the graph. This new graph structure is the final Bayesian network structure. For example, Figure 3(b) shows the graph of the Figure 3(a) but including the class node $\mathcal{C}$. The class node $\mathcal{C}$ is a binary variable, where $\mathcal{C} = 1$ means 'true interaction' and $\mathcal{C} = 0$, 'no interaction'.

Parameters of the Bayesian Network. To calculate the parameters θ for each variable z_k ($1 \leq k \leq n$) in a Bayesian network model, we use the *maximum likelihood estimation* (MLE) [5]. Given a training data set D with M samples, we use the MLE in equation (4).

$$L\left(\hat{\theta} : D\right) = max_{\theta \in \Theta} L\left(\theta : D\right) \tag{4}$$

$$L\left(\theta : D\right) = \prod_{m=1}^{M} P\left(z_1[m], \ldots, z_n[m] : \theta\right) \tag{5}$$

The node $z_k[m]$ corresponds to the z_k feature of the m-th sample. The likelihood estimation of the joint probability $L\left(\theta : D\right)$ given the parameters θ is specified in equation (5). We can decompose this joint probability into shorter and separate terms according to their conditional probabilities in equation (6) [5].

$$L\left(\theta : D\right) = \prod_{m=1}^{M} P\left(z_1[m] \mid z_{\pi_1}[m] : \theta\right) \ldots P\left(z_n[m] \mid z_{\pi_n}[m] : \theta\right) \tag{6}$$

Once we have the structure of the BN and its parameters, the BN model can be used to extract PPIs. In this case, for a test sentence x, we use the queries $P\left(\mathcal{C}[x] = 1 \mid z_1[x], \ldots, z_n[x]\right)$ and $P\left(\mathcal{C}[x] = 0 \mid z_1[x], \ldots, z_n[x]\right)$, in order to calculate the probabilities for the values $\mathcal{C}[x] = 1$ and $\mathcal{C}[x] = 0$. To infer if there exist an interaction or not, we calculate the maximum a posteriori in equation (1), of the probabilities $P\left(\mathcal{C}[x] = 1 \mid z_1[x], \ldots, z_n[x]\right)$ and $P\left(\mathcal{C}[x] = 0 \mid z_1[x], \ldots, z_n[x]\right)$. Finally, we extract a PPI interaction according to the equation (2).

3.2 Pruning Parse Trees

We employed the *path-enclosed tree method* to prune parse trees [16]. The generalization provided by the pruning has been demonstrated to improve the performance of extraction methods [3]. This method only considers the information surrounding the pair of protein names that are being analysed.

Figure 4 shows an example of pruning using the path-enclosed tree method. Note that after pruning, the leaves nodes "PROTEIN1" and "PROTEIN2" are the borders of the new parse tree.

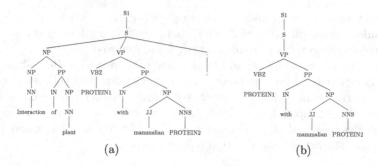

Fig. 4. (a) A parse tree from a sentence ID="BioInfer.d1.s2". (b) A pruned parse tree using the path-enclosed tree method.

4   Experimental Results

4.1   Evaluation Corpora

The proposed method was evaluated on five well-known PPI corpora. The corpora is formed by the following collections: AIMed, BioInfer, IEPA, HPRD50 and LLL; which are described elsewhere [9]. Table 1 summarize general information about the corpora, which are organized in an unified XML format [9]. BioInfer and AIMed corpora are greater than the rest of corpora in terms of instances and sentences. They have 9666 and 5834 instances respectively. In contrast, IEPA, HPRD50 and LLL have 817, 433 and 330 instances respectively. Also, sentences in PPI corpora are normalized, since only a pair of proteins can have an interaction and this interaction can be either positive or negative.

Table 1. General information of the Five PPI Corpora. "# inst. pos" and "# inst. neg" are the number of positive and negative sentences respectively.

	AIMed	BioInfer	HPRD50	IEPA	LLL
# documents	225	836	43	200	45
# sentences	1955	1100	145	486	77
# instances	5834	9666	433	817	330
# inst. pos	1000	2534	163	335	164
# inst. neg	4834	7132	270	482	166

The pair of protein names, forming an instance, is replaced by the following pair of alias: "PROTEIN1" and "PROTEIN2". The rest of protein names are replaced by "PROTEIN". This replacement can be observed in sentences of the Figure 2.

4.2   Experimental Settings

The Charniak parser[2] was used to obtain the parse trees from sentences. In the modelling of Bayesian networks, we extended the Weka 3.6 package [17] in order to employ the parse trees as training data. Furthermore, we use the inference algorithm of Weka over our BN models. In the evaluation of the models, we used the macro-averaged F-measure (ma-F) and employed a 10-fold cross-validation at document level [3].

4.3   Performance of the PPI Extraction

The results are shown in Table 2. BN models have a clear pattern, their performances vary according to the size of the training corpus and the test corpus used. Thus, our models overcome the previous approaches for the larger collections AIMed and BioInfer in comparison with the following best results of CM (2010) [3]. However, their performance is reduced when dealing with smaller corpora, such as HPRD50 and LLL.

Table 2. Results on PPI Corpora. PT-BN: our method. It uses the d parameter to limit the maximum number of parents for each node in the BN. CM (2010): [3]. Miwa (2009a): [10]. ma-F: Average of F-measure. σ_{ma-F}: standard deviation of F-measure.

	PT-BN $d = \infty$		PT-BN $d = 3$		CM (2010)		Miwa (2009a)	
	ma-F	σ_{ma-F}	ma-F	σ_{ma-F}	ma-F	σ_{ma-F}	ma-F	σ_{ma-F}
AImed	68.6	6.0	**76.4**	3.4	67.0	4.5	60.8	6.6
BioInfer	71.9	5.2	**72.8**	4.7	72.6	2.7	68.1	3.2
HPRD50	50.9	12.8	53.1	11.1	**73.1**	10.2	70.9	10.3
IEPA	68.2	4.6	67.4	0.5	**73.1**	6.0	71.7	7.8
LLL	55.43	1.7	59.0	19.3	**82.1**	10.4	80.1	14.1

We tested two types of models: *BN-PT (d = 3)* and *BN-PT (d = ∞)*. We selected an arbitrary number of parents, $d = 3$, since it proportionally increases the complexity of the search best BN structure. The complexity of the search procedure of the best BN is critical when $d \geq 2$ [5]. In addition, we consider all the parents of a node after our method generates the acyclic directed graph (i.e. without using the mutual information measure to remove edges). In this case, we denote $d = \infty$ to indicate that no edges were removed and all the parents were considered.

It turns out that the number of parents d influences the performance of the BN model. It is logical since the number of parents determines consequently the number of parameters in the BN. To calculate these parameters, the number of instances, in the training data set, must be large enough to learn the BN model. Thus, the results of our BN models in large corpora, such as BioInfer

---

[2] http://www.cs.brown.edu/people/ec/#software

Table 3. Results in cross-corpora. The comparison is performed in terms of F-measure. PT-BN: our method. d: maximum number of parents for each node in the BN. CM: [3]. Miwa (2009a): [10]. Airola (2008): [2]. Training corpora are distributed in rows and test corpora in columns.

	Methods	BioInfer	AIMed	IEPA	HPRD50	LLL
BioInfer	PT-BN $d = 3$	**72.8**	**69.3**	67.4	66.0	65.5
	CM (2010)	72.6	65.2	**72.9**	**71.9**	**78.4**
	Miwa (2009a)	68.1	49.6	71.4	68.3	76.9
	Airola (2008)	61.3	47.2	68.0	63.9	78.0
AIMed	PT-BN $d = 3$	**64.3**	**76.4**	47.4	54.4	39.5
	CM (2010)	**64.2**	67.0	59.0	**72.9**	62.7
	Miwa (2009a)	53.1	60.8	**68.1**	68.3	73.5
	Airola (2008)	47.1	56.4	67.4	69.0	**74.5**
IEPA	PT-BN $d = 3$	**67.3**	**64.2**	67.4	61.9	63.0
	CM (2010)	66.1	57.8	73.1	66.3	78.4
	Miwa (2009a)	55.8	40.4	71.7	66.5	**83.2**
	Airola (2008)	51.7	39.1	**75.1**	**67.5**	77.6
HPRD50	PT-BN $d = 3$	65.0	**67.0**	54.3	53.1	56.8
	CM (2010)	**65.5**	63.1	**69.3**	**73.1**	**73.7**
	Miwa (2009a)	48.6	43.9	67.8	70.9	72.2
	Airola (2008)	42.5	42.2	65.1	63.4	67.9
LLL	PT-BN $d = 3$	58.7	43.6	51.5	54.8	59.0
	CM (2010)	**64.4**	**55.9**	**71.4**	**69.4**	82.1
	Miwa (2009a)	48.90	38.60	65.6	64.0	**83.2**
	Airola (2008)	42.50	33.30	64.9	59.8	76.8

or AIMed, are competitive (and even better) compared to the state-of-the-art methods. Nevertheless, the performance of our BN models decrease with small corpora.

Results using cross-corpora experiments are shown in Table 3. It is worthy of note that the size of the corpus affects greatly the performance of the BN in terms of F-measure. Our method achieves a competitive performance, or even better, when it is used with IEPA, BioInfer or AIMed corpus as training data set and, BioInfer or AIMed corpus as test data set. Note that IEPA, BioInfer and AIMed have more than 480 sentences (486, 1,100 and 1,955 sentences respectively). On the other hand, the corpora HPRD50 and LLL have less than 150 sentences (145 and 77 sentences respectively). The diversity of sentences in large corpora could explain why our models performs better with them.

5   Conclusion

We conclude that the use of non-terminal nodes and their relationships from parse trees provide important information in order to learn BN models for the PPI extraction task.

Thus, the BN-PT method was proposed to learn Bayesian network models from parse trees. Like [3], we pruned the parse trees. Next, we use the parse trees as training data set to learn BN models. These BN models obtained good results in the PPI extraction task when they were used in large corpora such as AIMed and BioInfer. However, the performance considerably decreased in comparison with the state-of-the-art methods when they were employed in small corpora. To overcome this problem, one strategy would be tuning the number of parents d, obtaining a different value according to the corpus used. Nevertheless, taking into account that collections of biomedical texts tend to become larger, the proposed method can be useful in the task of extracting PPIs from these collections.

Acknowledgments. We acknowledge the Brazilian funding agencies CNPq and FAPESP for supporting this research.

References

1. Chowdhary, R., Zhang, J., Liu, J.S.: Bayesian inference of protein-protein interactions from biological literature. Bioinformatics 25(12), 1536–1542 (2009)
2. Airola, A., Pyysalo, S., Björne, J., Pahikkala, T., Ginter, F., Salakoski, T.: A graph kernel for protein-protein interaction extraction. In: Proceedings of the Workshop on Current Trends in Biomedical Natural Language Processing. BioNLP 2008, Stroudsburg, PA, USA, pp. 1–9. Association for Computational Linguistics (2008)
3. Choi, S.P., Myaeng, S.H.: Simplicity is better: revisiting single kernel ppi extraction. In: Proceedings of the 23rd International Conference on Computational Linguistics, COLING 2010, Stroudsburg, PA, USA, pp. 206–214. Association for Computational Linguistics (2010)
4. Rosario, B., Hearst, M.A.: Multi-way relation classification: Application to protein-protein interactions. In: Proceedings of HLTNAAC05 (2002)
5. Koller, D., Friedman, N.: Probabilistic Graphical Models: Principles and Techniques. MIT Press (2009)
6. Bui, Q.C.C., Katrenko, S., Sloot, P.M.: A hybrid approach to extract protein-protein interactions. Bioinformatics 27(2), 259–265 (2011)
7. Miwa, M., Saetre, R., Miyao, Y., Tsujii, J.: Entity-focused sentence simplification for relation extraction. In: Proceedings of the 23rd International Conference on Computational Linguistics, COLING 2010, Stroudsburg, PA, USA, pp. 788–796. Association for Computational Linguistics (2010)
8. Leitner, F., Mardis, S.A., Krallinger, M., Cesareni, G., Hirschman, L.A., Valencia, A.: An overview of biocreative ii.5. IEEE/ACM Trans. Comput. Biol. Bioinformatics 7(3), 385–399 (2010)
9. Pyysalo, S., Airola, A., Heimonen, J., Bjorne, J., Ginter, F., Salakoski, T.: Comparative analysis of five protein-protein interaction corpora. BMC Bioinformatics 9, S6 (2008)
10. Miwa, M., Saetre, R., Miyao, Y., Tsujii, J.: A rich feature vector for protein-protein interaction extraction from multiple corpora. In: Proceedings of the 2009 Conference on Empirical Methods in Natural Language Processing, EMNLP 2009, Stroudsburg, PA, USA, vol. 1, pp. 121–130. Association for Computational Linguistics (2009)

11. Miwa, M., Saetre, R., Miyao, Y., Tsujii, J.: Protein-protein interaction extraction by leveraging multiple kernels and parsers. International Journal of Medical Informatics 78(12), e39–e46 (2009)
12. Liu, B., Qian, L., Wang, H., Zhou, G.: Dependency-driven feature-based learning for extracting protein-protein interactions from biomedical text. In: Proceedings of the 23rd International Conference on Computational Linguistics: Posters. COLING 2010, Stroudsburg, PA, USA, pp. 757–765. Association for Computational Linguistics (2010)
13. Kim, S., Yoon, J., Yang, J.: Kernel approaches for genic interaction extraction. Bioinformatics 24(1), 118–126 (2008)
14. Rinaldi, F., Schneider, G., Kaljurand, K., Clematide, S., Vachon, T., Romacker, M.: Ontogene in biocreative ii.5. IEEE/ACM Trans. Comput. Biol. Bioinformatics 7(3), 472–480 (2010)
15. Cooper, G.F., Herskovits, E.: A bayesian method for the induction of probabilistic networks from data. Mach. Learn. 9, 309–347 (1992)
16. Zhang, M., Zhou, G., Aw, A.: Exploring syntactic structured features over parse trees for relation extraction using kernel methods. Inf. Process. Manage. 44(2), 687–701 (2008)
17. Witten, I.H., Frank, E.: Data Mining: Practical Machine Learning Tools and Techniques, 2nd edn. Morgan Kaufmann Series in Data Management Systems. Morgan Kaufmann Publishers Inc., San Francisco (2005)

A Model for Information Extraction in Portuguese Based on Text Patterns

Tiago Luis Bonamigo and Renata Vieira

Pontifícia Universidade Católica do Rio Grande do Sul (PUCRS)
Porto Alegre, RS, Brazil

Abstract. This paper proposes an information extraction model that identifies text patterns representing relations between two entities. It is proposed that, given a set of entity pairs representing a specific relation, it is possible to find text patterns representing such relation within sentences from documents containing those entites. After those text patterns are identified, it is possible to attempt the extraction of a complementary entity, considering the first entity of the relation and the related text patterns are provided. The pattern selection relies on regular expressions, frequency and identification of less relevant words. Modern search engines APIs and HTML parsers are used to retrieve and parse web pages in real time, eliminating the need of a pre-established corpus. The retrieval of document counts within a timeframe is also used to aid in the selection of the entities extracted.

1 Introduction

This paper proposes an information extraction model for the Portuguese language. As seed data, pairs of entities with semantic relations in common are used and text patterns that likely express that relation are extracted. After extracting such patterns, the complementary entity of a pair might be extracted when only one entity is provided. As an example, to identify who is the father of a certain person, one should provide pairs of entities with "is-son-of" relation as examples and who this person is. The model uses the examples to identify text patterns that likely express this relation and, by using them, it will attempt to identify who the father is. While similar methods are abundant for the English language, Portuguese is lacking in models for such task. Therefore the paper deals with the specific tools proposed for the Portuguese and its subsequent evaluation.

In this paper we provide a brief presentation of related works for both English and Portuguese (Section 2); a succint explanation of the model (Section 3); a description of the proof of concept (Section 4) and its respective results (Section 5); and in Section 6 some final observations regarding the model and the experiments.

A. Gelbukh (Ed.): CICLing 2013, Part II, LNCS 7817, pp. 359–368, 2013.

2   Related Work

The model proposed in this paper combines different relation extraction approaches applied to a large volume of data. Researches closer to the state of the art dealing with such volumes are mostly focused on extracting relations from English texts. There is also a substantial amount of language agnostic research with a more statistical approach. Our desire is to add to those models – language agnostic or adapted from English – resources and principles that can deal with Portuguese texts, therefore using lexical, syntactic and morphological information and using specific support tools to improve results. Works that were studied throughout this research include the systems DIPRE [1], StatSnowball [2], KnowItAll [3], TextRunner [4], WOE [5] and ReVerb [6]. For the Portuguese language, we studied the SeRELeP [7] and REMBRANDT [8] systems. Those systems were evaluated in the ReRelEM track, created in the second edition of HAREM [9], a joint evaluation conference for named entity recognition systems and related tasks.

Some of the relation extraction approaches studied are language agnostic and the one closest to the one proposed is DIPRE. DIPRE also identifies patterns between entities based on examples, but those are not necessarily text patterns. Given DIPRE had a goal to extract information from web pages within the World Wide Web, it also considers markup elements as relevant while identifying patterns that descriibe relations. Each of the other systems mentioned have at least one particular element that might be tested within a context of relation extraction for Portuguese texts. StatSnowball uses Markov networks to establish relations among entities; TextRunner employs bayesian classifiers, using morphological and syntactic information extract by parsers to identify those relations; WOE establishes parallels between relations extracted from texts and relations contained within Wikipedia infoboxes to enhance its results; ReVerb applies restrictions based on syntactic and lexical models to eliminate extracted relations that might not be coherent and KnowItAll started using generic extraction patterns to find a seed set of entities and relations and uses this information to extract more entities and relations.

The SeRELeP system identifies relations between entities in Portuguese texts. It mainly focuses on three relations: identity, included-in and located-in. It applies recognition of named entities through the usage of the PALAVRAS parser and classifies entities using annotations provided by the parser. After this, predefined heuristics are used for the relation extraction. Since it is based on annotations provided by the parser, SeRELeP only works with Portuguese texts. While the SeRELeP system focused on specific relations, the REMBRANDT system attempts to extract any given relation between two entities, using the Portuguese Wikipedia as an additional resource as well as generic heuristics. The Portuguese Wikipedia is used for disambiguation, to define the class of an entity so heuristics based on that information can identify such patterns.

Both systems present approaches relevant to this research. SeRELeP is a great example of how syntactic information can enrich the results, while REM-BRANDT proposes how pre-established relations provided by a repository such as Wikipedia can aid relation identification and extraction. Based on their experiments, we chose to use a search engine API, a part-of-speech tagger, a web scraper and a HTML parser as support tools to solve specific problems in the proposed model.

3   Our Proposed Model

The model proposed in this article needs seed information, in that case, a set of examples – pairs of entities – with a common relation between them. The main goal is to identify text patterns representing that relation. There is a risk related to the set selection, since there are other relations that might exist between two entities that might not be perceived during the selection. Take, for instance, the relation between lead singers and their respective bands. If the set of examples provided is only focused on a certain music style or only features lead singers of a specific gender, it will likely affect the results. In Portuguese article and substantive forms are often assigned to a specific gender, thus adding a new challenge that is far more frequent than in English. The patterns identified throughout the process can be inspected to verify whether the set of examples is being more restrictive or encompassing than the expected.

For each example within the set a search engine is used to find a list of 250 URLs of web pages containing both entities. The search engine API provides language filtering, so all documents analysed are in Portuguese. Web services interfaces to popular search engines such as Google and Bing as well as search engine frameworks such as Lucene provide those features. After retrieving the URL list, the text contained within each web page must be retrieved using a web scraper and a HTML parser. After the text is extracted, a part-of-speech tagger is used to identify adjectives and adverbs, and those are categorized into specific groups using a previously provided dictionary. Those groups are adverbs of affirmation and adjectives regarding quality, such as "melhor" ("best", in Portuguese) or "pior" ("worst", in Portuguese). Those words are then removed to improve the grouping of patterns. Take as an example the patterns "[PERSON] was certainly the best [CITY] mayor" and "[PERSON] was definitely the worst [CITY] mayor". If the example set featured mayors and their respective cities, without the removal of such words those would be considered to be distinct patterns. By removing the affirmation adverbs and adjectives, they become identical: "[PERSON] was the [CITY] mayor".

The pattern extraction begins shortly after the removal of adjectives and affirmation adverbs. Each retrieved web page is scanned to check whether it contains the two entities within a maximum distance of ninety characters from each other. If the web page contains the entities and fulfills the distance condition, an algorithm is used to extract the basic form of the relation pattern. It extracts the text between the two entities replacing the first and second entities with generic tags. Considering the sentence "Rafael Correa, presidente do Equador" (Rafael Correa, president of Ecuador), it would become "[ENTITY1], president of [ENTITY2]". The pattern is then stored in a database, indicating through which pair of entities this pattern was found. Afterwards it also retrieves three words preceding and succeeding the pattern and creates different combinations by concatenating those to the original pattern, storing more twelve patterns. After all patterns are extracted, a score of pattern relevance is calculated using the following formula:

$$Score_{pattern} = T_{oc} \cdot \frac{T_{dex}^{1.5}}{T_{ex}}$$

Where:

T_{oc}: Number of times the pattern was found in all examples
T_{dex}: Number of examples with which the pattern was found
T_{ex}: Total number of examples

After the training phase, that ends up with the calculation of the scores, it is possible to proceed to extraction. The first step is selecting which patterns will be used for search. So based on the score, ten patterns with the best scores are used. As an example we take regarding the "presidente de" ("president of") relation, "o presidente [ENTITY1] da [ENTITY2]" (the president [ENTITY1] of [ENTITY2]). The goal is discovering the first entity – the president – by providing the second entity – the country. The search string used to retrieve relevant web pages is formed by placing the known entity in the text pattern. For example, if the goal is to discover the president of Colombia (Colômbia in Portuguese), the second entity would be replaced by Colômbia and the first one would become a search phrase delimiter. The search string would become: "o presidente" "da Colômbia" and then it would be sent to the search API to retrieve 150 web pages. By using this search string the search engine will obviously return the URL of web pages in which the strings "o presidente" and "da Colômbia" are quite distant from each other, but this is compensated by the number of web pages captured.

For each web page retrieved, the part-of-speech tagger is used once again to remove adjectives and adverbs of affirmation, the original pattern is adapted into a search expression featuring the text pattern and the known entity. This is then adapted into a regular expression developed to capture proper nouns and is also used to check whether there are matches within the web page. If so, each match is then stored with the previously calculated score of the respective pattern. This information will be used to calculate the final score for each answer. As soon as all the patterns were used to retrieve the entities, the dataset featuring the matches and score is grouped by the string representing the probable entity and the sum of its pattern scores is calculated. To calculate the final score of the answer, the algorithm proceeds to retrieve document counts from the search API by searching ten patterns with the highest score. Those results are restricted to documents indexed within the last six months. It then proceeds to calculate the answer score using the following formula:

$$Score_{answer} = (\log S_p) * S_{docs}$$

Where:

S_p: Sum of the $Score_{pattern}$ for each extracted answer

S_{docs}: Sum of documents found for each of the ten patterns substituting the generic tags on the pattern by the two entities, the one provided and the one extracted.

The model considers that the match with the highest score is likely to be the correct one.

In the earlier versions of the software developed to test the model, an algorithm for pronominal anaphora resolution based on [10] was also applied, but there was no significant difference on the results and the speed was considerably compromised. So it was removed from the model.

4   Proof of Concept

To check whether this model provides relevant results, we chose to explore two different relations: the relation between companies and their respective CEOs and the relation between South American countries and their respective presidents. For each scenario, ten examples were provided. The selection of examples was based, respectively, on a report written by an specialized company featuring the ten most popular brands of 2011 [11] and the ten South American countries with the highest GDP estimate, provided by The World Factbook [12].

A software program was developed in C# to test the model. It automates the process of capturing web pages using multiple threads to speed up the process. On a computer using Windows 7 with 2.4 GHz Core 2 Duo processor, 4 Gb of RAM and a 10 MBit ADSL connection the training process and the extraction can be done in less than 40 minutes. This time could be reduced since some parallelism issues weren't fully addressed. As mentioned before, the model uses a search engine API, a web scraper, a HTML parser and a part-of-speech tagger. The Google Search API [13] was chosen to fulfill the search engine needs, the HTML parsing and web scraping capabilities were provided by the library HTMLAgilityPack [14] and the LX-Center part-of-speech tagger [15] Center was chosen due to its speed.

Our goal was not to provide an exact calculation of precision and recall, but to show that with available tools the model is prone to achieve relevant results. For that reason we chose a straightforward approach to evaluate the results. The software uses leave-one-out cross-validation over the example sets. For each example chosen to test the model in a particular scenario, the nine remaining examples were used in the training phrase, to identify the text patterns. After the proper identification of the patterns, it proceeds to extract information using the chosen example set, attempting to extract the respective entity to the first one of the pair being currently tested. The software returns a list of answers, each one featuring a different score. We considered as a correct result any answer in which the result with the highest score is manually tagged as a right answer. Variants of the names were also considered as right answers. Even though we are only considering the first result to calculate the precision, we will present the top 3 results to detail further some of the problems found.

Table 1. Entity pairs – companies and their respective CEOs – used for training

Company	CEO
Apple	Tim Cook
Coca-Cola	Muhtar Kent
Disney	Bob Iger
General Electric	Jeffrey Immelt
Google	Larry Page
HP	Meg Whitman
IBM	Ginni Rometty
Intel	Paul Otellini
McDonald's	James A. Skinner
Microsoft	Steve Ballmer

For companies and their respective CEOs (see Table 1), 80% of the entities retrieved were correct. For Latin American countries and their respective presidents (see Table 2), 100% of the respective entities retrieved were correct. The settings for the web page retrieval from the search engine API and HTML parser remained the same as proposed before. During the training phase 250 documents were retrieved for each example on the set. After the text patterns were extracted, 150 documents were retrieved for each of the top ten patterns with the highest score. Those documents were then used for the extraction of the respective entity.

A small test was conducted to verify on whether the results would be affected if the document count used in the $Score_{answer}$ was changed to retrieve the document count from the first semester of 2010 instead of the last six months. This test was solely applied to Brazil and Paraguay, since both countries have different presidents since then, and the results reflected that particular period. This suggests the model might be used to extract answers relative to a specific timeframe. Those results can be seen in Table 5.

5 Final Remarks

The results of the experiments conducted are promising. Even though popular relations and entities were chosen, the results of 80% and 100% of correct entities for the relations explored in the proof of concept are not to be dismissed.

This information extraction model is not too dissimilar to systems such as DIPRE or Open Information Extraction systems such as Espresso, nor the training part is too dissimilar from Kushmerick's approaches to wrapper induction. Still, the object of study and tools are different. This model considers solely

Table 2. Entity pairs used from training - Latin American coutries and their respective presidents

Country	President
Argentina	Cristina Kirchner
Bolvia	Evo Morales
Brasil	Dilma Rousseff
Chile	Sebastián Piñera
Colômbia	Juan Manuel Santos
Equador	Rafael Correa
Paraguai	Federico Franco
Peru	Ollanta Humala
Uruguai	José Mujica
Venezuela	Hugo Chávez

Table 3. Top three answers for each company

Company	CEO	Web pages	$Score_{answer}$
Apple	Tim Cook (correct)	137	10069,56
	Steve Jobs	52	1893,43
	Mike Scott	3	11,69
Disney	Bob Iger (correct)	18	306,9
	Robert Iger	75	134,57
	Jay Rasulo	18	10,1
General Electric	Jeffrey Immelt (correct)	14	4,85
	Adriana Machado	6	1,99
Google	Eric Schmidt (wrong)	82	655,44
	Larry Page	12	328,13
	Fábio José Slvia Coelho	9	29,62
HP	Meg Whitman (correct)	6	196,24
	Leo Apotheker	8	20,24
	Oscar Clarke	27	11,76
IBM	Rodrigo Kede (wrong)	6	24,07
	Ginni Rometty	5	22,78
	Lou Gerstner	6	17,32
Intel	Paul Otellini (correct)	33	642,11
	Gordon	8	48,26
	Fernando Martins	15	5,3
McDonald's	Don Thompson (correct)	3	4,05
	Jim Skinner	4	2,13
Microsoft	Steve Ballmer (correct)	182	4148,55
	Bill Gates	20	234,97
	Michel Levy	16	121,01
The Coca-Cola Company	Muhtar Kent (correct)	17	119,89
	Irial Finan	3	1,79

Table 4. Top three answers for each Latin American country

Country	President	Web pages	$Score_{answer}$
Argentina	Cristina Kirchner (correct)	24	24097,55
	Néstor Kirchner	6	2258,25
	Cristina Fernández de Kirchner	3	1503,24
Bolvia	Evo Morales (correct)	42	18342,57
	Alvaro Garca Linera	3	258,65
Brasil	Dilma Rousseff (correct)	2	12770,44
	Luiz Inácio Lula da Silva	3	3232,55
	Presidente Fernando Henrique Cardoso	3	2001,59
Chile	Sebastián Piñera (correct)	49	5739,77
	Michelle Bachelet	6	1614
	Michele Bachelet	3	180,82
Colômbia	Juan Manuel Santos (correct)	60	20133,35
	Angelino Garzn	13	605,87
	Manuel Santos	3	14
Equador	Rafael Correa (correct)	131	23437,13
	Rafael Correa Em	6	775,14
	Lenin Moreno	3	78,38
Paraguai	Federico Franco (correct)	66	18533,87
	Fernando Lugo	43	16932,81
	Nicanor Duarte Frutos	5	2,99
Peru	Ollanta Humala (correct)	75	11648,57
	Alan Garcia	6	490,94
Uruguai	José Pepe Mujica (correct)	6	15597,19
	José Mujica	95	12128,35
	Tabaré Vázquez	6	221,48
Venezuela	Hugo Chávez (correct)	36	41671,6
	Hugo Chaves	1	219,59

Table 5. Top three answers for Brazil and Paraguay restricting the document count to the first 2010 semester

Country	President	Web pages	$Score_{answer}$
Brasil	Luiz Inácio Lula da Silva (correct for 2010)	3	1896,07
	Dilma Rousseff	2	318,82
	Fernando Henrique Cardoso	3	186,65
Paraguai	Fernando Lugo (correct for 2010)	43	1770,07
	Federico Franco	66	492,05
	Nicanor Duarte Frutos	5	11,96

text patterns and not markups – unlike DIPRE, for example – and the goal was to enhance the model for the Portuguese language, given the particularities of the language, also akin to other Romance languages. Also some approaches as the ones developed by the University of Washington (TextRunner, WOE, Re-Verb, OLLIE) rely on English language-related information to improve pattern extraction. In our approach the part-of-speech tagger is solely used to clean up the text, the pattern scoring itself is purely statistical to increase speed.

Regarding the particularities of Romance language, at first only the removal of less relevant adverbs and adjectives was addressed, but other challenges exist: the inflection of verbs, which is far more complex than in English; the frequent use of suffixes to express diminutive or augmentative forms of a substantive or adjective ("little duck" is "patinho" in Portuguese, with "pato" meaning "duck" and the "-inho" suffix changing the formation to diminutive); different grammatical gender formations also happen more often, for example, "the worker" in English can be used to describe both female and male workers, while in Portuguese both article and substantive change regarding the gender – "a trabalhadora" for the female worker, "o trabalhador" for the male worker. Those aren't the only challenges, but the more apparent ones.

The model would also greatly benefit from cross-document coreference resolution. Taking for instance the entities retrieved by the software when provided the company entity Disney, both first and second results are correct, because Bob Iger and Robert Iger are coreferent. And while the coreference problems didn't affect this particular result considering our evaluation scheme, it affected greatly the extraction of the IBM CEO. The second result is related to the current CEO, Ginni Rometty, but she is also mentioned as Virginia Marie Rometty, Ginny Rometty, Virginia M. Rometty, Virginia Rometty and sometimes solely Rometty. With proper cross-document coreference the results would possibly be improved.

The results achieved in the proof of concept are quite positive, even though they do not prove on whether this model would be of any help given a more obscure set of examples, but still it is a positive starting point with a fair share of possibilities for improvement and even usage in other scenarios, such as open information extraction, automatic ontology construction, thesaurus building and other tasks.

Combining this model with another algorithm to provide seed data in the form of pairs of examples – be it from structured or unstructured data sources – it could become an open information extraction model for the Portuguese language. The idea of using Wikipedia, in systems such as WOE, to provide the seed examples is something we are bound to explore in the near future.

References

1. Brin, S.: Extracting patterns and relations from the world wide web. In: Atzeni, P., Mendelzon, A.O., Mecca, G. (eds.) WebDB 1998. LNCS, vol. 1590, pp. 172–183. Springer, Heidelberg (1999)
2. Zhu, J., Nie, Z., Liu, X., Zhang, B., Wen, J.R.: StatSnowball: a statistical approach to extracting entity relationships. In: Proceedings of the 18th International Conference on World Wide Web, WWW 2009, pp. 101–110. ACM, New York (2009)

3. Etzioni, O., Cafarella, M., Downey, D., Popescu, A.M., Shaked, T., Soderland, S., Weld, D.S., Yates, A.: Unsupervised named-entity extraction from the web: An experimental study. Artif. Intell. 165(1), 91–134 (2005)
4. Yates, A., Cafarella, M., Banko, M., Etzioni, O., Broadhead, M., Soderland, S.: TextRunner: open information extraction on the web. In: Proceedings of Human Language Technologies: The Annual Conference of the North American Chapter of the Association for Computational Linguistics: Demonstrations. NAACL-Demonstrations 2007, Stroudsburg, PA, USA, pp. 25–26. Association for Computational Linguistics (2007)
5. Wu, F., Weld, D.S.: Open information extraction using Wikipedia. In: ACL, pp. 118–127 (2010)
6. Etzioni, O., Fader, A., Christensen, J., Soderland, S.: Mausam: Open information extraction: The second generation. In: Walsh, T. (ed.) IJCAI, IJCAI/AAAI, pp. 3–10 (2011)
7. Brucksen, M., Souza, J.G.C., Vieira, R., Rigo, S.: Sistema SeRELeP para o reconhecimento de relações entre entidades mencionadas. In: Mota, C., Santos, D. (eds.) Segundo HAREM, ch. 14, pp. 247–260. Linguateca (2008)
8. Cardoso, N.: REMBRANDT – Reconhecimento de entidades mencionadas baseado em relações e análise detalhada do texto. In: Mota, C., Santos, D. (eds.) Segundo HAREM, ch. 11, pp. 195–211. Linguateca (2008)
9. Santos, D., Seco, N., Cardoso, N., Vilela, R.: Harem: An advanced ner evaluation contest for portuguese. In: Odjik, Tapias, D. (eds.) Proceedings of LREC 2006 (LREC 2006), Genoa, pp. 22–28 (2006)
10. Lappin, S., Leass, H.J.: An algorithm for pronominal anaphora resolution. Comput. Linguist. 20(4), 535–561 (1994)
11. Interbrand: 2011 ranking of the top 100 brands, http://www.interbrand.com/en/best-global-brands/best-global-brands-2008/best-global-brands-2011.aspx (January 2012) (acessado: June 12, 2012)
12. CIA: The world factbook 2009. Central Intelligence Agency, Washington D.C (2009)
13. Corporation, G.: Google Custom Search API (January 2012), https://developers.google.com/custom-search/v1/overview (acessado: November 19, 2012)
14. Mourier, S.: Htmlagilitypack (October 2012), http://htmlagilitypack.codeplex.com (acessado: November 14, 2012)
15. Branco, A., Silva, J.A.: Evaluating Solutions for the Rapid Development of State-of-the-Art POS Taggers for Portuguese. In: Lino, M.T., Xavier, M.F., Ferreira, F., Costa, R., Silva, R. (eds.) LREC 2004, pp. 507–510 (2004)

A Study on Query Expansion Based on Topic Distributions of Retrieved Documents

Midori Serizawa and Ichiro Kobayashi

Ochanomizu University
Advanced Sciences, Faculty of Sciences
2-1-1 Otsuka Bunkyo-ku, Tokyo, 112-8610
koba@is.ocha.ac.jp

Abstract. This paper describes a new relevance feedback (RF) method that uses latent topic information extracted from target documents.In the method, we extract latent topics of the target documents by means of latent Dirichlet allocation (LDA) and expand the initial query by providing the topic distributions of the documents retrieved at the first search. We conduct experiments for retrieving information by our proposed method and confirm that our proposed method is especially useful when the precision of the first search is low. Furthermore, we discuss the cases where RF based on latent topic information and RF based on surface information, i.e., word frequency, work well, respectively.

1 Introduction

In information retrieval, it is often that we cannot be satisfied with the result of information retrieval. This happens basically because user's query is usually short and does not contain enough information to achieve good performance in retrieval results. In this context, the method to expand initial query, called relevance feedback, was developed and has been actively studied. Relevance feedback (RF) is the method to retrieve information with expanded query reflected by the initial results for retrieving some relevant documents missed in the first retrieval round. To expand the initial query, users are basically required to judge the relevant documents among the retrieved documents. A basic RF techniques was proposed by Rocchio [1], and then many studies underlying Rocchio's algorithm have so far been proposed.

RF is a useful technique to raise retrieval performance, however, asking users to judge the retrieved result is usually expensive. Therefore, as an alternative solution for that, pseudo-relevance feedback (PRF), which expands user's second query based on the top-ranked retrieved documents in the first retrieval round, has been actively studied [2, 3]. The basic idea of PRF is to extract terms which serve to enhance user's requirement from the top-ranked documents in the first retrieval round and then expand a query used in the next retrieval round. PRF has been confirmed as it raises retrieval performance by several studies [2, 4, 5]. However, it is also pointed out that the result of PRF depends on the number and quality of feedback documents, since a user does not choose

A. Gelbukh (Ed.): CICLing 2013, Part II, LNCS 7817, pp. 369–379, 2013.

relevant documents by him/herself. In this context, many researches have been investigating proper feedback documents for PRF [6, 7] – Lee et al. [6] and He et al. [7] introduced a clustering technique to identify relevant documents from the retrieved documents. However, even though we could identify relevant documents for query expansion, those documents may contain irrelevant components in themselves, so we need another method to extract the essential information from the documents. In this context, another approach to extract latent topics from the documents and feed back the topics to expand query has been progressive. In this paper, we also propose a method to feed back the latent topics of relevant documents to expand query.

The reminder of this paper is organized as follows. In the second section, we introduce studies with topic-based relevance feedback related to our study. In the third section, we show our proposed method which feeds back topic distribution of top-ranked retrieved documents to the initial query. In the fourth section, we show an experiment with NTCIR data set; discuss the experiment result; and verify our proposed method. Finally in the fifth section, we conclude our study and show some future directions.

2   Related Studies

The methods dealing with latent semantics, such as probabilistic Latent Semantic Analysis (pLSA) [8] and Latent Dirichlet Allocation (LDA) [9], have been successfully applied in the field of IR. In particular, LDA has recently been widely applied to many IR studies [10–13]. Wei & Croft [10] proposed an LDA-based document model within the language modeling framework and confirmed that their proposed method improves the retrieval results over retrieval using cluster-based models. Yi & Allan showed several ways that topic models can be integrated into the retrieval process [11] and then explored the utility of different types of topic models[1] for retrieval purposes. Furthermore, they indicated that incorporating topics in the feedback documents for building relevance models can generally benefit the performance more for queries that have more relevant documents. Zhou & Wade [13] used latent topics of target documents for re-ranking initial retrieval results.

As for the recent studies about topic-based relevance feedback, Ye et al. [16] proposed a method to find a good query-related topic based on LDA, and confirmed through experiments that query expansion based on the derived topic achieved statistically significant improvements over a representative feedback model in the language modeling framework. Harashima et al. [17] proposed a method to re-rank retrieved results based on the surface and latent topic information in texts, so that documents with a similar topic distribution to the initial query will be re-ranked higher. They confirmed that their method is effective for both explicit and pseudo RF.

---

[1] They adopted the following three models, i.e., Mixture of Unigrams [14], LDA [9], and Pachinko Allocation Model [15].

On the other hand, we propose a method to feed back topic distributions for expanding the topic distributions of the initial query, unlike Ye's and Harashima's approaches – they feed back the topic distribution over words to the initial query.

3   Relevance Feedback Based on Latent Topics

We will explain a topic model we use in this study; the framework of language model; and an overview of our proposed method in the following.

3.1   Latent Dirichlet Allocation

Latent Dirichlet Allocation [9] is a probabilistic model for the process of generating a document. It can express that multiple topics are included in a document. A concrete process of the method is briefly explained in the following. A multinomial distribution over words for each topic is selected based on Dirichlet distribution. A multinomial distribution over topics in each document is also selected based on Dirichlet distribution. A topic for each word in a document is selected based on the multinomial distribution over topics, and then a word is finally selected based on the multinomial distribution over words for each topic. This process repeats until iteration reaches the number of the words in the target documents, and then it is regarded as the selected words compose a document.

3.2   Language Model for Query and Documents

In our study, a query model and a document model are represented based on maximum likelihood estimation (MLE, hereafter) and Dirichlet smoothed estimation (DIR, hereafter) .

The maximum likelihood estimation of word w_j is represented in equation (1)

$$P_{\mathbf{t}}^{MLE}(w_j) = \frac{tf(w_j, \mathbf{t})}{|\mathbf{t}|} \tag{1}$$

Here, $\mathbf{t}$ indicates a text, e.g., either a query or a document, and $tf(w_j, \mathbf{t})$ indicates the frequency of w_j in $\mathbf{t}$.

In order to adopt smoothing for the model, we use Dirichlet smoothed estimation (DIR)[2]. The DIR is represented in equation (2).

$$P_{\mathbf{t}}^{DIR}(w_j) = \frac{tf(w_j, \mathbf{t}) + \mu P_{\mathbf{D}_{all}}^{MLE}(w_j)}{|\mathbf{t}| + \mu} \tag{2}$$

Here, $\mathbf{D}_{all}$ indicates all target documents, and μ indicates the smoothing parameter that adjusts the degree of confidence in the frequency in $\mathbf{D}_{all}$ rather than in the frequency in $\mathbf{t}$.

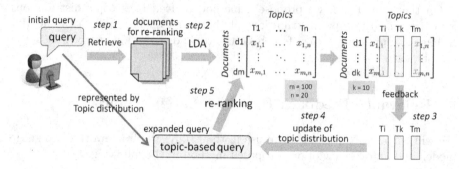

Fig. 1. Overview of our proposed method

3.3   Overview of Our Proposed Method

Figure 1 illustrates an overview of our proposed method. We will explain the process flow of our proposed method by following the arrows in Figure 1.

step 1.   The system retrieves documents with a query for acquisition of documents to be re-ranked. According to this documents ranking, top m ranked documents are used to be re-ranked. In this study, we set m to 100.

step 2.   The topic-based models of the initial query and the documents to be re-ranked are constructed. The latent topics of the retrieved documents and the query given at step 1 are extracted by means of LDA with the number of topics K, which we set to 50.

step 3.   Among the m documents at step 1, the top k ranked documents are regarded as documents which are relevant to user's need and used to construct the feedback model.

step 4.   The topic-based initial query model is updated using the feedback at step 3.

step 5.   The expanded query, which is represented in the form of topic distribution, is used for re-ranking the initial retrieved top m documents. The documents are re-ranked based on its topic distribution. Then, the user obtain the search results.

3.4   Initial Search for Acquisition of Documents to be Re-ranked (step 1)

In the initial search, we employ a KL-divergence retrieval model [18] to acquire documents to be re-ranked. First, the MLE-based query model $P_{\mathbf{q}}^{MLE}(\cdot)$ and the DIR-based document model $P_{\mathbf{d}}^{DIR}(\cdot)(\mathbf{d} \in D_{all})$ are constructed using equation (1) and (2), respectively. Then, the score of each document, which is defined as the KL-divergence function for the query model and the document model as follows, is computed.

$$KL\_score(\mathbf{d}, \mathbf{q}) = -KL(P_\mathbf{q}^{MLE}(\cdot)||P_\mathbf{d}^{DIR}(\cdot)) \tag{3}$$
$$KL(P_\mathbf{q}(\cdot)||P_\mathbf{d}(\cdot)) = -\sum_w P_\mathbf{q}(w) \log P_\mathbf{d}(w)$$

Finally, the target documents are ranked according to this score. We regard top m ranked documents in all ranked target documents as documents to be re-ranked $D_{initial}$.

3.5 Topic-Based Model and Feedback Model Construction (step 2 and 3)

First, we perform LDA on $D_{initial}$ and the initial query to infer the topic distribution. A topic distribution of each document is considered as the topic-based document model $P_\mathbf{d}^{TOPIC}(\cdot)$. Then, the topic-based initial query model $P_\mathbf{q}^{TOPIC}(\cdot)$ is constructed likewise.

Next, the topic-based feedback model $P_\mathbf{F}^{TOPIC}(\cdot)$ is constructed. In this study, the feedback model is the average of the topic distributions of the relevant documents. Therefore, the topics which dominate the topic distribution, i.e, topics with high probabilities, are reflected to the feedback.

3.6 Documents Re-ranking(step 4 and 5)

The documents is re-ranked using the KL-divergence model in the same way as the initial search. First, in this step, we update the initial query model. The new query model $P_\mathbf{q}^{NEW}(\cdot)$ is constructed by updating the initial query model $P_\mathbf{q}^{TOPIC}(\cdot)$ with the feedback model $P_\mathbf{F}^{TOPIC}(\cdot)$ using the interpolation parameter a as follows.

$$P_\mathbf{q}^{NEW}(z_j) = (1-a)P_\mathbf{q}^{TOPIC}(z_j) + aP_\mathbf{F}^{TOPIC}(z_j) \tag{4}$$

The interpolation parameter a controls the reliability of the feedback model. Then, the documents is re-ranked using the KL_score(3). Finally, we obtain the result of re-ranking.

4 Experiment

4.1 Experimental Settings

In the experiments, we used the test collection used in the Information Retrieval Task in the Second NTCIR Workshop [19]. The collection includes three materials: (i) Document data (Author abstracts of the Academic Conference Paper Database (1997-1999) and Grant Reports (1988-1997) = about 400,000 Japanese and 130,000 English documents), (ii) 49 Search topics (Japanese and English,) and (iii) Relevance Judgements. We used 1000 Japanese documents in the collection for each search topic. We experimented 30 Japanese search topics. In each task, there are three types of short texts available as queries: some words tagged with <TITLE>, a short sentence tagged with <DESCRIPTION>, and either long sentence or multiple sentences tagged with <NARRATIVE>. We used a short sentence as a query in our experiments. As shown in section 3.3, we used 1000 Japanese documents as initial documents, top 100 documents by the initial retrieval are processed by means of LDA. We employed the parameter settings for LDA as follows. The number of topics, K, is decided as 50 by a preliminary experiment; the hyperparameter α for topic distribution is K/50; the hyperparameter β for word distribution is 0.01; and the number of iteration is 200. We employed Gibbs sampling for estimation. The parameter μ for Dirichlet smoothing is 1000, following the prior studies [17, 20]. As for evaluation method, we employed Precision measured at rank 10 (P@10) and mean average precision (MAP). The methods to be compared are normal information retrieval without RF and the following three methods.

- topic-based RF
    - TOPIC: the method expressing a document with the topic distribution for the document.
    - TPCWORD: the method expressing a document with the topic distribution for the words in the document.
- DIR-based RF
    - WORD: information retrieval with RF

4.2   Experiment for Query Expansion Based on Topic Distribution

We have conducted two types of PRF experiments: one is the experiment which employs query expansion based on the topic distributions of the initial retrieved documents (topic-based RF, that is, our proposed method), and the other is the experiment which employs query expansion based on word frequency of the initial retrieved documents (DIR-based RF). These two types of experiments are conducted in order to compare the ability of our proposed PRF method, i.e, query expansion based on the topic distributions, with that of the conventional PRF method, i.e., query expansion based on DIR-based model. In terms of expanding the initial query, as for PRF via query expansion based on DIR-based model, the DIR of words in top k documents is processed as well as the probability of topics of our proposed method, that is, the importance of words are averaged and fed back to the importance of words in the initial query.

As for experimental conditions, since the objective of our study is to investigate how much feedback information provided as topic distributions works to improve the initial search result, therefore, we provide several conditions for the retrieved documents used for expanding the initial query and investigate the ability of our proposed method under the conditions. This is because the ability of RF especially depends on the feedback information, therefore, it is important to consider the conditions of feedback information.

We investigated the accuracy of the result of second retrieval round under various number of relevant documents included in top 10 retrieved documents at the first retrieval round used for expanding the initial query. Furthermore, we investigated the case where the proportion of the topic distributions between initial query and feedback are changed.

4.3   Experimental Result and Discussions

Figure 2 and Figure 3 show the results of TPCWORD and WORD, respectively. Each line in these Figures represents P@10 score with different ratios of relevant documents to all documents used in the construction of a feedback model. We also experimented on TOPIC, but we could not see any big difference from the result of TPCWORD.

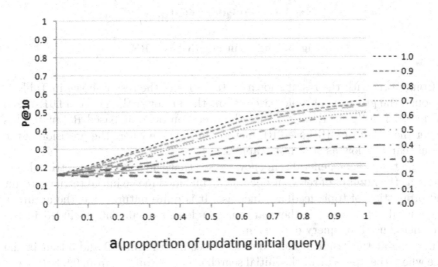

Fig. 2. The result of topic-based RF (proposed)

Table 1 shows the result of each method at the ratio a which provides the best accuracy in each method.

Table 1. Precision of all document sets

Methods	P@10	MAP
Initial search	0.331	0.239
TOPIC ($a = 0.9$)	0.283	0.225
TPCWORD($a = 1.0$)	0.286	0.215
WORD ($a = 0.9$)	0.348	0.239

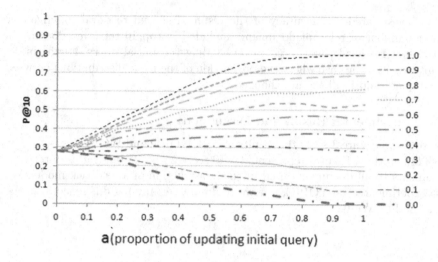

Fig. 3. The result of DIR-based RF

Comparing with the result shown in Table 1 and the result shown in Table 1, although the precision of topic-based RF method is lower than that of DIR-based RF method in Table 1, we see that the precision of topic-based RF method is higher than that of DIR-based RF method in the case where the precision of the initial search is low as shown in Table 2.

In the result, as a general tendency, we see that the accuracy of retrieval gets better as the number of relevant documents included in the initial retrieval result and proportion of topic feedback increase. It is quite natural that the accuracy of retrieval gets better, if there are many relevant documents included in the documents used for query expansion.

Looking at the result in more detail, we see from Figures 2 and 3 that in the case where the precision of the initial search is low as ranging from 0.0 to 0.3, the precision of DIR-based RF method decreases as the ratio a increases, moreover, in the case of only adopting feedback information for a new query, i.e., the case of $a = 1.0$, the precision gets lower than that of the initial search. On the other hand, as for topic-based RF, although the precision of second search does not get the precision as high as that of DIR-based as the proportion of feedback information increases, we see that the precision of second search keeps almost the same one as that of the initial search even in the case that the query for second search round is totally replaced by feedback information.

From these things, we see that the tendency of precision is defferent in the cases where the precision of the initial search is low and high, respectively. As for evaluation, looking at the document sets [2], the result is shown in Table 2.

Considering the reason for these results, we think that if the precision of the initial search is low, it is natural that the feedback information produced by the

---

[2] We selected the document sets whose precision of the initial search is less than 0.4. The number of the document sets we used in the experiment is 18.

Table 2. Precision of the document sets whose initial precision is low

Methods	P@10	MAP
Initial search	0.094	0.112
TOPIC($a = 0.2$)	0.141	0.118
TPCWORD($a = 0.0$)	0.124	0.106
WORD($a = 0.0$)	0.124	0.116

result of the initial search does not reflect user's request very much. As for DIR-based RF, the weight of the words which satisfy user's request will decrease as the proportion of feedback increases. Therefore, the precision by the second query is thought of as lower than that by the initial search. On the other hand, as for topic-based RF, a query itself is represented by topic distributions, therefore, even though there are not words which directly indicate user's request in the documents used for query expansion, the words which have similar meaning can be selected through topic distributions and therefore the recall by the method should be better, so we can guess that the second search can exceed the result of the initial search.

In the case that the precision of the initial search is high, for example, it is higher than 0.4, we see from Figures 2 and 3 that the precision gets higher as the proportion of feedback information is larger, and it is better than the precision of the initial search. Whereas, as for topic-based RF, the precision increases gradually as the proportion of feedback information increases, but it is not as the same high precision as DIR-based RF. We think the reason for this is because topic is estimated, reflecting the words which are not directly related to the words included in the initial query, in other words, it is probable that topic-based RF picks up noisy words in its estimation, therefore, when the precision of the initial search is high, DIR-based RF can be thought as useful.

5 Conclusions

We have proposed a method of PRF via query expansion based on topic distributions and conducted experiments to investigate the ability of the proposed method. In the experiments, in addition to the experiment of our proposed method, we have also conducted an experiment to compare our proposed method with the method based on surface information, i.e., word frequency. As a result, in the case that there is a little information in user's search request, topic-based RF is useful rather than DIR-based RF, on the contrary, in case that there is enough information in user's search request, DIR-based RF is useful rather than topic-based RF. Therefore, in order to make PRF work usefully, we think that feedback method should be switched based on the richness of the information of the initial query provided by a user. However, it is difficult for PRF to measure the degree of relevance between user's need and documents used as feedback information. If we would like to change feedback method based on user's initial query, we will further investigate the method of how we should measure the relevance. Besides, we have not yet investigated various parameters used in our method, i.e., interpolation parameter, etc. These things will be future issues in our study.

References

1. Rocchio, J.J.: Relevance feedback in information retrieval. In: The SMART Retrieval System: Experiments in Automatic Document Processing, pp. 313–323. Prentice-Hall Inc. (1971)
2. Lafferty, J., Zhai, C.: Document language models, query models, and risk minimization for information retrieval. In: Proceedings of SIGIR 2001, pp. 111–119 (2001)
3. Carpineto, C., de Mori, R., Romano, G., Bigi, B.: An informationtheoretic approach to automatic query expansion. ACM Transactions on Information Systems (TOIS) 19, 1–27 (2001)
4. Lavrenko, V., Croft, W.B.: Relevance based language models. In: Proceedings of the 24th Annual International ACM SIGIR Conference on Research and Development in Information Retrieval (SIGIR 2001), pp. 120–127 (2001)
5. Collins-Thompson, K.: Reducing the risk of query expansion via robust constrained optimization. In: Proceedings of the 18th ACM Conference on Information and Knowledge Management (CIKM 2009), pp. 837–846. ACM Press, New York (2009)
6. Lee, K.S., Croft, W.B., Allan, J.: A cluster-based resampling method for pseudo-relevance feedback. In: Proceedings of the 31st Annual International ACM SIGIR Conference on Research and Development in Information Retrieval (SIGIR 2008), pp. 235–242. ACM Press, New York (2008)
7. He, B., Ounis, I.: Studying query expansion effectiveness. In: Boughanem, M., Berrut, C., Mothe, J., Soule-Dupuy, C. (eds.) ECIR 2009. LNCS, vol. 5478, pp. 611–619. Springer, Heidelberg (2009)
8. Hofmann, T.: Probabilistic latent semantic indexing. In: Proceedings of the 22nd Annual International ACM SIGIR Conference on Research and Development in Information Retrieval (SIGIR 1999), pp. 50–57. ACM Press, NewYork (1999)
9. David, M., Blei, A.Y.N., Jordan, M.I.: Latent dirichlet allocation. Jounal of Machine Learning Research 3, 993–1022 (2003)
10. Wei, X., Croft, W.B.: Lda-based document models for ad-hoc retrieval. In: Proceedings of the 29th Annual International ACM SIGIR Conference on Research and Development in Information Retrieval (SIGIR 2006), pp. 178–185. ACM Press, NewYork (2006)
11. Yi, X., Allan, J.: Evaluating topic models for information retrieval. In: Proceedings of the 17th ACM Conference on Information and Knowledge Management (CIKM 2008), pp. 1431–1432. ACM Press, New York (2008)
12. Yi, X., Allan, J.: A comparative study of utilizing topic models for information retrieval. In: Boughanem, M., Berrut, C., Mothe, J., Soule-Dupuy, C. (eds.) ECIR 2009. LNCS, vol. 5478, pp. 29–41. Springer, Heidelberg (2009)
13. Zhou, D., Wade, V.: Latent document re-ranking. In: Proceedings of EMNLP 2009, pp. 1571–1580 (2009)
14. Xu, J., Croft, W.B.: Cluster-based language models for distributed retrieval. In: Proceedings of ACM SIGIR, pp. 254–261 (1999)
15. Li, W., McCallum, A.: Pachinko allocation: Dag-structured mixture models of topic correlations. In: Proceedings of ICML, Pittsburgh, PA, pp. 577–584 (2006)

16. Ye, Z., Huang, X., Lin, H.: Finding a good query-related topic for boosting pseudo-relevance feedback. Journal of the American Society for Information Science and Technology archive 62(4), 748–760 (2011)
17. Harashima, J., Kurohashi, S.: Relevance feedback using latent information. In: Proceedings of the 5th International Joint Conference on Natural Language Processing, Chiang Mai, Thailand, pp. 1037–1045 (2011)
18. Zhai, C., Lafferty, J.: Model-based feedback in the language modeling approach to information retrieval. In: Proceedings of CIKM 2001, pp. 403–410 (2001)
19. Kando, N.: Overview of the second ntcir workshop. In: Proceedings of the Second NTCIR Workshop on Research in Chinese & Japanese Text Retrieval and Text Summarization (2000), http://research.nii.ac.jpntcirworkshopOnline Proceedings2ovview-kando.pdf
20. Zhai, C., Lafferty, J.: A study of smoothing methods for language models applied to information retrieval. ACM Transactions on Information Systems 22(2), 170–214 (2004)

Link Analysis for Representing
and Retrieving Legal Information[*]

Alfredo López Monroy[1], Hiram Calvo[1],
Alexander Gelbukh[1] and Georgina García Pacheco[2]

[1] Centro de Investigación en Computación,
[2] ESIME-Zacatenco
Instituto Politécnico Nacional.
Av. Juan de Dios Bátiz e/M.O. de Mendizábal s/n, México, D.F. 07738
{alopezm301,ggpacheco}@ipn.mx,
hcalvo@cic.ipn.mx, gelbukh@gelbukh.com

Abstract. Legal texts consist of a great variety of texts, for example laws, rules, statutes, etc. This kind of documents has as an important feature, that they are strongly linked among them, since they include references from one part to another. This makes it difficult to consult them, because in order to satisfy an information request, it is necessary to gather several references and rulings from a single text, and even with other texts. The goal of this work is to help in the process of consulting legal rulings through their retrieval from a request expressed as a question in natural language. For this, a formal model is proposed; this model is based on a weighted, non-directed graph; nodes represent the articles that integrate each document, and its edges represent references between articles and their degree of similarity. Given a question, this is added to the graph, and by combining a shortest-path algorithm with edge weight analysis, a ranked list of articles is obtained. To evaluate the performance of the proposed model we gathered 8,987 rulings and evaluated the answer to 40 test-questions as correct, incorrect or partial. A lawyer validated the answer to these questions. We compared results with other systems such as Lucene and JIRS (Java Information Retrieval System)

1   Introduction

Legal documentation is made up of a wide variety of texts, laws, regulations, codes, statutes, guidelines, policies, etc. Such documents represent an important source of information because they contain rules or regulations governing the conduct of individuals up to organizations and operation of government and private institutions. For this reason, the legal documentation should be accessible to the public; however, its structure and the relationship between documents, among other things, hinder its wide understanding [1]. The rulings that make up the bulk of a legal document are

---
[*] Work done under partial support of the Mexican Government (CONACyT, SNI), PIFI and SIP-IPN, Mexico, and RITOS-2. We thank Conacyt for the grant 205897 and project funding 60557.

A. Gelbukh (Ed.): CICLing 2013, Part II, LNCS 7817, pp. 380–393, 2013.

often strongly interrelated, as reflected by the numerous textual references they usually contain. Also, it is possible to find (to a lesser extent) references between different standard documents. The example in Table 1 illustrates two cases:

Table 1. Article 14 of the Act refers to Article 13 of the same Act, and referfences an article of a different text as well

Organic Law of the Nat. Polytechnic Institute
Article 14. - The powers and duties of the General Director: ... XV. Naming the area secretaries, after consulting the Secretary of Education, who shall meet the qualifications set forth in **Article 13 of this Law itself**; ... XIX. Exercising the legal representation of the Institute with the broadest powers referred to the first two paragraphs of **Article 2554 of the Civil Code for the Federal District in common matters and for the Republic in federal matters**, and XX. Other priovided by this Act and other applicable ordinances.

It has been found that these intra-document and inter-document references difficult consulting legal texts, because satisfying a need for information may require complementing provisions belonging to one or more documents. This has been generically referred to as the *knowledge synthesis problem* [2]—The content of various documents may be irrelevant for a user, however, their combined use would allow to infer an answer. Table 2 and Table 3 illustrate this problem in the legal area from inquiries made as a natural language question; the answers are inferred from related provisions. The examples belong to a set of questions posed by the community of a government institution, which were answered by the legal office staff, from legislation and regulations of the same institution. In Table 2, the provisions answered are related by a reference, while in the example of Table 3, the provisions answered are not related explicitly, but with regard to the question posed. These examples illustrate that the problem involves locating and consulting not only one but several legal documents to find solutions to a given situation from its provisions.

Table 2. The answer can be inferred from both articles is that it is inappropriate for the deputy to be involved in the process again

Is it possible for a person who has been deputy twice, to participate in the election process for designating triads for the technical, administrative and academic deputies, and in its case, to exercise that position again?	
Article 182 of the Internal Rules. "Principals and assistant principals of schools, centers and units cease teaching and research in office ... II. At the end of the term of his office, as provided by **Article 21 of the Organic Law**, ... "	**Article 21 of the Organic Law**. "Principals of schools, centers and units of teaching and research, ... shall remain in office three years and may be appointed only once for another period."

Table 3. According to the articles-response, voting for their representatives is a right of every individual who is part of the academic staff of the institution

Are interim teachers assigned to a school entitled to vote or not?	
Article 208 of the Rules. The faculty of each academic program, or equivalent, will directly be elected by a majority of votes in their respective representatives, under the terms of **Article 27 of the Organic Law.**	**Article 6 of the Regulation of the Internal Working Conditions for the Academic Staff.** For the proper interpretation and application of this Regulation of the Internal Working Conditions for the Academic Staff, in the course of this instrument, shall be known as ... XV. Academic Staff: A person who assists the Institute ..., performing academic work in terms of these rules ... XXII. Interim Academic staff: One that covers a temporary license of the base academic staff, for up to six months ...

To facilitate understanding of the legal information is considered that this requires careful structuring, organization and systematization [1]. For this, it has been proposed to exploit the characteristics that the legal documents possess such as intra-document and inter-document references and their structure [2, 3, 4, 5]. Despite the diversity of legal documents, their content is organized in a specific way. As illustrated in Table 4, the documents usually begin with the name of the law, regulation, statute, etc.., Followed by a brief preamble which states the objective, legal basis, scope, etc. of the document. Subsequently, the rules designed to regulate the conduct of individuals are grouped in a section called general provisions, while the rules aimed to the authorities who must apply them are set out in the section called transitional provisions. In turn, these latter, as well as the general provisions are divided into one or more articles.

Table 4. Typical structure of a legal document

Title	Organic Law of the National Polytechnic Institute
preamble	Current text. New Law published ...
general provisions	Article 1. ... is the institution designed by the Sstate to consolidate through through the Education ... Article 34. ...
transitional provisions	TRANSITIONAL ARTICLES FIRST. - This Act shall take effect ...

According to the above, the objective of this work is to develop a model based on the characteristics that different types of legal documents have. This, for help in locating related items either explicitly or implicitly from a request for information expressed in the form of a question in natural language and reduce the problem with this synthesis of knowledge in the legal field. To investigate the ability of the proposed model we present the results of an experimental study. 8,987 items were used and a set of 40 questions with their corresponding answer. The results obtained were compared with those obtained using the free software Lucene and Passage Information Retrieval System JIRS (Java Information Retrieval System).

This paper is organized as follows: in Section 2 we provide a review of recent literature on methods applied to legal documentation. Section 3 describes in general terms the proposed model. The model is based on a non-directed weighted graph constructed from the articles of the general provisions of the documents in a predefined corpus. Specifically, the set of vertices represent articles, whereas, the set of edges represents similarity and/or references between them. Then, in Section 3.1, we describe our implementation. The experiments and the evaluation method are presented in Section 3.2. In Section 4 we provide and discuss the preliminary results. Finally, the conclusions of the study are given in Section 5.

2   Related Work

The generic task on which this paper focuses is textual information retrieval given an information request. One area that has focused on this task is known as Information Retrieval (IR). The most popular methods, the boolean model and the vector space model, have been applied to a wide variety of documents including legal [6]. Lucene is free software based on previous models, which has been used in legal information retrieval [7, 8].

Today there are various tools based on IR methods available through Web portals for both government and private users who help in the recovery of legal documentation. Due to the increase of legal information available on the Internet, search engines are also widely used for the recovery of legal documents [9]. However, legal information retrieval is still a difficult task. One of the drawbacks that has been reported is the number of documents than traditional information retrieval systems typically returned [10]. This implies that it is necessary to invest a considerable amount of time to locate the desired information. Hence, for a number of years, passage recovery has been researched [11]. Briefly, a system for recovering passages (or Passage Retrieval, PR) is an information retrieval system which, from a request, it returns text fragments (regions) instead of relevant documents.

In particular, JIRS is a passage retrieval system based on *n-grams* (an *n-gram* is a sequence of adjacent words in a text), which is based on the premise that a collection of documents must contain at least one *n-gram* associated with requests for information. By employing different *n-gram* models JIRS has the ability to locate structures present in the request for information in a document collection. The recovery is performed by looking for possible *n-grams* of the information request in the document collection. [12] provides a more detailed description. The information retrieval system JIRS has recently been used in the search for legal information [13].

While PR has helped to reduce the "information overload", generic information retrieval, and of course legal, is an unresolved task and even more when considering the problem of synthesis of knowledge. In order to improve access to legal information, it has been suggested to consider the structure and references to policy and legislative documents [14].

Recently, trying to exploit the structure of the legal documents has been explored using the markup languages of electronic texts. For example, [2] describes a representation of normative and legal texts based on XML (eXtended Markup

Language) for developing a system of verification of compliance with provisions. Moreover, [15] describes the first steps in the automatic transformation of legal textual information in formal models, illustrated through the text markup language XML and the Unified Modeling Language UML (Unified Modelling Language). [16] describes tools aimed at the creation and management of legal documents based on markup language XML for electronic texts.

Using document structure has been proposed too. Some works employ references to legal texts [3, 5]; however, the literature on the subject is still scarce. In [17], the authors describe a comparative analysis scheme developed to facilitate the retrieval of related provisions from different regulatory texts. The comparison is based on using a measure of similarity and the structure of the legal texts and references in the general provisions. In [18] the authors describe a method for retrieving legal information based on the structure of texts such as laws, regulations, etc. In addition, due to the evolution of the WWW (World Wide Web), there has been a lot of research related to academic citation analysis. Currently, research on different techniques and models, oriented citation analysis for application in various areas. For example, iterative algorithms on graphs such as Kleinberg's HITS algorithm [20] or PageRank [21], which had been mainly used in the analysis of Web page references, social networking, more recently have been successfully applied tool in the area of automatic text processing [22, 23].

In brief, a sorting algorithm on graph is a method for determining the significance of a vertex in a graph, which is used for global information recursively calculated on the complete graph instead of only local information specific to each node. The basic idea implemented by the management model is the voting or recommendation. When a vertex is linked to another, basically gives a vote to that other vertex. The greater the number of votes obtained a vertex greater importance.

These sorting algorithms based on graphs are based on the random walk model, which consists of traversing the graph node at random, with the route being modeled as a Markov process. Under certain conditions, this model converges to a stationary distribution of probabilities associated with the vertices. Based on the Ergodicity theorem for Markov chains [24], these algorithms guarantee convergence if the graph is aperiodic and irreducible. The first condition is fulfilled by any non-bipartite graph, while the second condition holds for any strongly connected graph. These two conditions are met in the graphs constructed for implementing retrieval measures considered in this paper.

A brief description of PageRank and HITS algorithms follows.

Let $G = (V, E)$ be a directed graph with the set of vertices V and the set of edges E, where E is a subset of $V \times V$. For a given vertex V_i, let $In (V_i)$ be the set of vertices that lead to the V_i node through an edge, and be $Out (V_i)$ the set of vertices that can be reached from the node V_i via an edge. PageRank [21] is probably the most popular sorting algorithm originally designed as a method for analyzing electronic page references.

$$PR(V_i) = (1 - d) + d * \sum_{V_j \in In(V_i)} \frac{PR(V_j)}{|Out(V_j)|} \qquad (1)$$

Where d is a parameter between 0 and 1.

HITS (Hiperlinked Induced Topic Search) [20] makes a distinction between "authorities" (pages heavily referenced by other pages) and "hubs" (pages with many references to other pages). For each vertex, HITS produces two values.

$$HITS_A(V_i) = \sum_{V_j \in In(V_i)} HITS_H(V_j) \qquad (2)$$

$$HITS_H(V_i) = \sum_{V_j \in Out(V_i)} HITS_A(V_j) \qquad (3)$$

For these algorithms, beginning with arbitrary values assigned to each vertex in the graph, iterate until the calculations converge on a predetermined threshold. After running the algorithm, each vertex will be associated with a value which represents the *importance* of each vertex in the graph. It is worth noting that the final values are not affected by the initial values assigned to the nodes, but only by the number of iterations required to achieve convergence to predefined threshold.

In this research the numerous references that represent significant legal documents, similar to electronic pages on the Web. However, in [17] the authors argue that despite the aforementioned argument, the legal domain differs from that of the WWW because the reference analysis assumes a single reference collection of documents with each other, while the collection of legal documents contain rather *islands* of information. Within an island, articles of the legal document collection are strongly related, while references between articles of other island are much lower. On the other hand, unlike the earlier view, in [19] Geist conducts a research on the use of technical analysis of references in search systems of legal information. The author concludes that, despite the difficulties of analysis algorithms references in the legal field, they could be used to improve the performance of existing information retrieval systems. This is because it shows the similarity between collections of legal documents and web pages on the WWW. However, even the analysis of references has been little explored in the development of algorithms for the recovery of legal information.

Because of this, in this paper we investigate a method for retrieving articles given an information request in natural language and citation analysis algorithms, instead of a traditional method, such as markup languages for Electronic Text in a formal model. The model we propose uses a non-directed weighted graph in which the provisions are represented (and their relationships) from a collection of legal and regulatory texts. In the following section we describe in detail our proposal.

3   Description of the Proposed Method

The proposed method consists of two stages. The first one consists of the representation of the general provisions of a set of legal documents, and the second one consists of retrieving the articles from a request for information as a natural language query. Both stages are based on a non-directed weighted graph $G = (A, S)$ as described below:

Stage 1. Representation: The items that make up the legal documents are represented by the set of vertices $A = \{a_1, ..., a_n\}$ of the graph G and through the edge set S their similarity and/or references containing intra-document and inter-document relations.
Stage 2. Recovery: The non-directed weighted graph G and the information collection is used to retrieve a set of items.

3.1   Implementation

To set the edges between vertices of the graph are compared with each other articles and intra-document and inter-document references were extracted.

For references between articles, first, we analyzed a set of references manually extracted from five documents and articles from these patterns were defined for their automatic extraction in the 37 remaining documents. In this way it was found that up to 22% of the articles used in the evaluation made reference to another article (Table 5).

Table 5. Number of references between items found in the document collection

Number of documents		42
Number of articles		8987
References	inter-document	170
between articles	intra-document	1804
Total number of references		1974

For the comparison between articles we used the cosine similarity function:

$$s(a_i', a_j') = \frac{a_i' \cdot a_j'}{\|a_i'\| * \|a_j'\|} \tag{4}$$

Where a_i', a_j' are vectors with the weights of each term in the articles corresponding to the vertices a_i and a_j.

In order to use the cosine similarity, the articles were previously processed. First, the contents of each article was converted to lowercase and lemmatized. Additionally, all punctuation, numbers, and a predefined list of words (known as stop-words) was eliminated. The weight value of the terms resulting from each article was obtained from the measured tf·idf (term frequency·inverse document frequency). In this way, the value of a term x in the k-th item of the collection of articles A, is defined as:

$$tfidf(x, a_k, A) = tf(x, a_k) * idf(x, A) \tag{5}$$

$$tf(x, a_k) = \text{Frecuency of } x \text{ in } a_k \tag{6}$$

$$idf(x, A) = \log\left(1 + \frac{\text{Total no. of articles } |A|}{\text{Articles with x}}\right) \tag{7}$$

From the value of similarity between articles and references between articles, an edge between the vertices a_i and a_j with associated value p_{ij} was established, given the following:

- If the items corresponding to the vertices a_i, a_j do not contain a reference to each other and $(a_i', a_j') > 0$, then $p_{ij} = s(a_i', a_j')$.
- In the case where the items represented by the vertices a_i, a_j contain a reference to each other, and $s(a_i', a_j') = 0$, then $p_{ij} = 1$. In case that $s(a_i', a_j') > 0$, then $p_{ij} = 2 * s(a_i', a_j')$. Figure 1 illustrates the proposed representation.

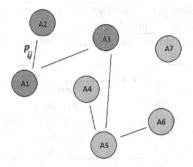

Fig. 1. Each vertex corresponds to an item of the general provisions of each document. The edges represent the likeness and/or references between articles.

For retrieving items from the graph, the process is described as follows: given a question q: First, the terms of q are divided into two sets, which are then represented in the non-directed weighted graph as if it were two new items. Once the graph vertices corresponding to the question are entered, a set of routes between them is obtained. Finally, the items corresponding to the vertices of the obtained paths are returned in the order in which they were found. Figure 2 illustrates the process for recovering articles from a given question.

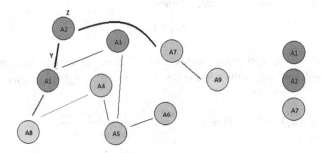

Fig. 2. Non-directed weighted graph. Vertices A8 and A9 represent the information query. A possible outcome consisting of the list of vertices A1, A2 and A7 is shown.

The following describes the process of splitting the question and recovering the articles.

Each question is first processed similarly to the articles, *i.e.*, converted to lowercase and lemmatized; punctuation, numbers, and the terms that do not appear in the set of articles are eliminated. Subsequently, the remaining terms are associated with a number, starting with 1 for the first term 2 for the second, etc. Then, the odd and the even terms are separated to form two sets of terms $q_a = \{2, 4, ...\}$ and $q_b = \{1, 3, ...\}$ as shown in Table 6. For comparison of q_a and q_b with articles and general provisions we used again the cosine similarity measure.

Table 6. Example of query splitting and the weights of its terms

How is the procedure performed by students electing representatives to the Technical Advisory Council School?			
procedure perform students election representative technical advisory council scholar			
q_a	a'_{n+1}	q_b	a'_{n+2}
procedure	0.31	perform	0.23
students	0.35	election	0.20
election	0.17	representative	0.23
technical	0.20	advisory	0.21
council	0.23	scholar	0.17

As described previously, in the graph both parts of the question q_a y q_b were represented by two new vertices a_{n+1} and a_{n+2}. An edge is generated, from the value of similarity between two vertices a_k and a_{n+1}, one corresponding to an article and the other to a part of the information request. If $s(a'_k, a'_{n+1}) > 0$, then $p_{k(n+1)} = s(a'_k, a'_{n+1})$, where a'_k y a'_{n+1} are vectors with the weights of the terms in a_k y q_a respectively. The paths between vertices a_{n+1} and a_{n+2} were obtained by a linear combination of the values associated with the edges and vertices associated values.

The route began operating at the vertex a_{n+1}, progress was made by choosing the node that will maximize the value of the expression 5, reaching the node a_{n+2}. Once a route is found, the vertices that composed it were eliminated. The process is repeated until all possible routes are obtained.

$$w_1 * p_{xy} + w_2 * v(a_y) \tag{8}$$

The value associated with each vertex was obtained from the reference analysis algorithm PageRank [20] in the version for non-directed weighted graphs.

$$v(a_i) = (1 - d) + d * \sum_{a_j \in E(a_i)} p_{ji} \frac{v(a_j)}{\sum_{a_k \in S(a_j)} p_{kj}} \tag{9}$$

Where a_i corresponds to a vertex, d to a fixed value preset d, a_j to a vertex adjacent to a_i, and p_{ji} to the value associated to the edge between vertices a_i and a_j.

w_1 and w_2 values were obtained experimentally by evaluating the implementation of the model with 21 questions. The best values were selected for the rest of the experiments ($w_1 = 0.99$ and $w_2 = 0.01$). The recommended value of $d = 0.70$ was used.

4   Experiment and Results

In total we used a set of 40 questions with answers. The system looked for the answer to these questions in a set of 8,987 articles from a total of 42 documents (two laws, one code, and other regulations). The answers are found in a subset of 11 documents which together have 1,117 articles. The system must discard the irrelevant documents.

The answer to each question specifically consisted of a set of articles (up to 5). The questions were classified into three types:

Questions with a yes/no answer. These are questions that relate to the ability to perform a certain action or not, usually expressed as: Is it possible ...?, Does it proceed that...?, Is it possible to ...?, As shown in Table 7.

Table 7. Sample yes/no question

P3. Is it appropriate for a student to request undergraduate honors if he chose their degree by the choice of schooling?
Answer: Article 13 and 43 of the Rules of Professional Qualifications.

Procedure questions. Require as an answer the requirements, or the official or accepted way to perform a certain task, fill a position or job, etc. (Table 8). Examples of this questions are: How is .... done?, What is the procedure for ...?, What are the requirements to ...?

Table 8. Example of *procedure question*

P1. How is the procedure performed by students electing representatives to the Technical Advisory Council School?
Answer: Article 28 of the Basic Law and, 206, 207, 209 and 213 of the Rules.

Definition question. Request information for understanding a concept. What is ...? What does it mean ...?, See Table 9.

Table 9. Example of *definition question*

P14. What is meant by having graduate?
Answer: Article 3 and 4 of the Rules of Graduate Studies.

The implementation of the model was evaluated based on its ability to retrieve items that answer the test questions. For each question, the articles (obtained by the process described previously) were enumerated from 1 to m, starting with the first article recovered. If this list of m elements contained articles that answered the question, that list scored as *correct*; if it contained fewer items than required, it was scored as *partially correct*; and finally, if it did not contain the required items it was scored as *incorrect*. For each question were up to 75 articles were used.

We used 3 collections of articles as follows:

- **Collection A**: 1117 articles. The articles of the 11 documents containing-response items.

- **Collection B**: 2162 articles. The items in the collection to more than 18 documents belonging to the same institution.
- **Collection C**: 8987 articles The articles in the two previous collections and those of 13 texts with more generally applicable standards, such as the law governing the relations between workers and employers across the country. The documents used were selected because of the collection items B contained a reference to articles of documents in the collection C.

We ran two experiments and comparisons which are detailed below:

Experiment 1: Collection A 1. Similarity: $w_1 = 1, w_2 = 0$ 2. PageRank (PR): $w_1 = 0, w_2 = 1$ 3. Similarity and PR: $\quad w_1 = 0.99, w_2 = 0.1$ Comparison with Lucene and JIRS.	**Experiment 2**: Collections B and C Similarity and PR: $w_1 = 0.99, w_2 = 0.1$ Comparison with Lucene and JIRS.

Table 10. Results of experiment 1

Answers	1. Sim	2. PR	3. Sim and PR
Correct	**10**	**5**	**36**
Incorrect	18	25	0
Partial	12	10	4

Table 11. Results of experiment 1

Position	1. Sim	2. PR	3. Sim and PR
<= 10	9	0	18
10 < P <= 20	0	0	3
20 < P <= 30	0	0	9
30 < P <= 40	0	1	3
40 < P <= 50	0	1	0
50 < P <= 60	1	1	1
60 < P <= 70	0	2	2
70 < P <= 80	0	0	0

Table 10 and Table 11, show that the best results are obtained by combining the similarity value between nodes and the value associated with each vertex given by citation analysis algorithm.

Table 12. Comparison 1 – answers

Answer	Lucene	JIRS
Correct	**31**	**32**
Incorrect	0	0
Partial	9	8

Table 13. Comparison 1 – positions

Position	Lucene	JIRS
<= 10	15	14
10 < P <= 20	8	9
20 < P <= 30	3	4
30 < P <= 40	4	3
40 < P <= 50	1	0
50 < P <= 60	0	2
60 < P <= 70	0	0
70 < P <= 80	0	0

Table 13 shows that the proposed method answers 36 questions; 5 more than Lucene, and 4 more than JIRS. Table 13 shows that the proposed method obtains answers for 30 questions on the top 30 places, while Lucene obtains answers for 26 questions, and JIRS for 27.

Table 14. Experiment 2 – answers

Answer	2162 articles	8987 articles
Complete	35	33
Incomplete	0	2
Partial	5	5

Table 15. Experiment 2 – positions

Position	Number of articles	
	2162	8987
<= 10	17	18
10 < P <= 20	4	3
20 < P <= 30	6	3
30 < P <= 40	3	1
40 < P <= 50	3	3
50 < P <= 60	0	1
60 < P <= 70	1	3
70 < P <= 80	1	1

Table 16. Experiment 2

Answers	2162		8987	
	Lucene	JIRS	Lucene	JIRS
Complete	31	31	30	30
Incomplete	0	0	1	0
Partial	9	9	9	10

5   Conclusions and Future Work

We have implemented a link analysis algorithm (PageRank) for legal document retrieval. As far as we know, this is the first system that takes advantage of the existing references in legal documents. The proposed model has better results for retrieving correct answers than the results that use only similarity. The fact that the graphs consider not only the similarity between the query and the documents but also between the latter improves the results with respect to the space vector model.

Despite that references (1,974) represent only 22% of the total items (8,987), there is appreciable impact on the retrieval of these items as shown by the results of the link analysis graph. Intra and Inter-document references improve the performance of the retrieval. The answer is not only correct, but the articles returned are better positioned in the results list. Note that the returned articles depend not only on the documents' structure, but also on the characteristics of the question itself, and its complexity as well. Finally, for the case for the case of simple and specific questions, the recovery using only similarity and similarity and link analysis is favored in a similar fashion, so that results are similar.

As future work, we plan to consider other properties of complex networks to improve our model, as well as implementing other natural language techniques to extend query understanding.

References

1. Ríos-Estavillo, J.J.: Derecho e informática en México. [en línea], México, Instituto de Investigaciones Jurídicas (Universidad Nacional Autónoma de México), Serie E: Varios (Núm. 83) (1997), Formato pdf, Disponible en Internet, http://www.bib liojuridica.org/libros/libro.htm?l=147 [citado 1-04-12]
2. Kerrigan, S., Law, K.H.: Logic-based regulation compliance-assistance. In: Proceedings of the 9th International Conference on Artificial Intelligence and Law, pp. 126–135. ACM, New York (2003)
3. Moens, M.F.: Improving Access to Legal Information: How Drafting Systems Help. In: Lodder, A.R., Oskamp, A. (eds.) Information Technology & Lawyers: Advanced Technology in the Legal Domain, from Challenges to Daily Routines, pp. 119–136. Springer, The Netherlands (2006)
4. Monroy, A., Calvo, H., Gelbukh, A.: NLP for Shallow Question Answering of Legal Documents Using Graphs. In: Gelbukh, A. (ed.) CICLing 2009. LNCS, vol. 5449, pp. 498–508. Springer, Heidelberg (2009)
5. Turtle, H.R.: Text Retrieval in the Legal World. Artificial Intelligence and Law 3(1), 5–54 (1995)
6. Alvite Díez, M.: Tendencias a la investigación sobre la recuperación de Información Jurídica. Revista española de Documentación Científica 26(2) (2003)
7. Peñas, A., Forner, P., Rodrigo, A., Sutcliffe, R., Forascu, C., Mota, C.: Overview of ResPubliQA 2010: Question Answering Evaluation over European Legislation. In: Notebook Paper of CLEF, Workshop on Question Answering in Multiple Languages (MLQA 2010), Padua, Italy (2010)

8. Peñas, A., Forner, P., Sutcliffe, R., Rodrigo, Á., Forăscu, C., Alegria, I., Giampiccolo, D., Moreau, N., Osenova, P.: Overview of ResPubliQA 2009: Question Answering Evaluation over European Legislation. In: Peters, C., Di Nunzio, G.M., Kurimo, M., Mandl, T., Mostefa, D., Peñas, A., Roda, G. (eds.) CLEF 2009. LNCS, vol. 6241, pp. 174–196. Springer, Heidelberg (2010)

9. Barmakian, D.: Better search engines for law. Law Library Journal 92(4), 399–438 (2000)

10. Daniels, J.J., Rissland, E.L.: Finding legally relevant passages in case opinions. In: Proceedings of the 6th International Conference on Artificial Intelligence and Law, pp. 39–46. ACM, New York (2000)

11. Salton, G., Allan, J., Buckley, C.: Approaches to passage retrieval in full text information systems. In: Korfhage, R., Rasmussen, E., Willett, P. (eds.) Proceedings of the 16th Annual International ACM SIGIR Conference on Research and Development in Information Retrieval, pp. 49–58. ACM, New York (1993)

12. Buscaldi, D., Rosso, P., Gómez, J.M., Sanchis, E.: Answering Questions with an n-gram based Passage Retrieval Engine. Journal of Intelligent Information Systems 34(2), 113–134 (2010)

13. Correa, S., Buscaldi, D., Rosso, P.: NLEL-MAAT at ResPubliQA. In: Peters, C., Di Nunzio, G.M., Kurimo, M., Mandl, T., Mostefa, D., Peñas, A., Roda, G. (eds.) CLEF 2009. LNCS, vol. 6241, pp. 223–228. Springer, Heidelberg (2010)

14. Stranieri, A., Zeleznikow, J.: Knowledge discovery from legal databases-using networks and data mining to build legal decision support systems. In: Lodder, A.R., Oskamp, A. (eds.) Information Technology & Lawyers: Advanced Technology in the Legal Domain, from Challenges to Daily Routines, pp. 81–117. Springer, The Netherlands (2006)

15. Mercatali, P., Romano, F., Boschi, L., Spinicci, E.: Automatic Translation from Textual Representations of Laws to Formal Models through UML. In: Moens, M.F., Spyns, P. (eds.) Proceedings of the, Conference on Legal Knowledge and Information Systems, pp. 71–80. IOS Press, Amsterdam (2005)

16. Francesconi, E.: The Norme in Rete Project: Standards and Tools for Italian Legislation. International Journal of Legal Information 34(2), 357–376 (2006)

17. Lau, G.T., Law, K.H., Wiederhold, G.: A relatedness analysis of government regulations using domain knowledge and structural organization. Information Retrieval 9(6), 657–680 (2006)

18. Monroy, A., Calvo, H., Gelbukh, A.: Using Graphs for Shallow Question Answering on Legal Documents. In: Gelbukh, A., Morales, E.F. (eds.) MICAI 2008. LNCS (LNAI), vol. 5317, pp. 165–173. Springer, Heidelberg (2008)

19. Geist, A.: Using Citation Analysis Techniques for Computer-Assisted Legal Research in Continental Jurisdictions (2009), Available at SSRN: http://ssrn.com/abstract=1397674 or http://dx.doi.org/10.2139/ssrn.1397674

20. Kleinberg, J.M.: Authoritative sources in a hyperlinked environment. Journal of the ACM 46(5), 604–632 (1999)

21. Brin, S., Page, L.: The anatomy of a large-scale hyper- textual Web search engine. Computer Networks and ISDN Systems 30(1-7), 107–117 (1998)

22. Mihalcea, R.: Graph-based Ranking Algorithms for Sentence Extraction, Applied to Text Summarization. In: Proceedings of the 42nd Annual Meeting of the Association for Computational Linguistics, Companion Volume (ACL 2004), Barcelona, Spain (2004)

23. Mihalcea, R., Tarau, P., Figa, E.: PageRank on Semantic Networks, with application to Word Sense Disambiguation. In: Proceedings of The 20st International Conference on Computational Linguistics (COLING 2004), Switzerland, Geneva (2004)

24. Grimmett, G., Stirzaker, D.: Probability and Random Processes. Oxford University Press (1989)

Discursive Sentence Compression

Alejandro Molina[1], Juan-Manuel Torres-Moreno[1], Eric SanJuan[1],
Iria da Cunha[2], and Gerardo Eugenio Sierra Martínez[3]

[1] LIA-Université d'Avignon
[2] IULA-Universitat Pompeu Fabra
[3] GIL-Instituto de Ingeniería UNAM

Abstract. This paper presents a method for automatic summarization
by deleting intra-sentence discourse segments. First, each sentence is divided into elementary discourse units and, then, less informative segments are deleted. To analyze the results, we have set up an annotation
campaign, thanks to which we have found interesting aspects regarding the elimination of discourse segments as an alternative to sentence
compression task. Results show that the degree of disagreement in determining the optimal compressed sentence is high and increases with the
complexity of the sentence. However, there is some agreement on the decision to delete discourse segments. The informativeness of each segment
is calculated using textual energy, a method that has shown good results
in automatic summarization.

1 Introduction

Previous studies in automatic summarization have proposed to generate summaries by extracting certain sentences of a given document; i.e., an *extraction
summary* [1]. Nevertheless, an *abstract*, as defined in [2], is by far the most
concrete and most recognized document summarization method.

Today, automatic summarization approaches have improved to the point that
they are able to identify, with remarkable precision, the sentences that contain
the most essential information for any given text. However, high-scored sentences
could contain a great amount of irrelevant information. Hence, a finer analysis
is needed to prune the superfluous information while retaining only the relevant
one [3]. Sentence compression shall produce grammatical condensed sentences
that preserve important content and it represents a valuable resource for automatic summarization systems. Indeed, some authors argue that this task could
be a first step towards abstract generation [4].

This work presents a sentence compression approach for summarization. First,
sentences are segmented using a discourse segmenter and, then, a compressed
text is generated erasing the less informative segments. Statistical methods allow
to determinate segment's informativeness and grammaticality.

The rest of the paper is organised as it follows. In section 2, the main concepts of sentence compression are covered. In section 3, the discourse segmentation is presented. Then the description of how to score the informativeness

A. Gelbukh (Ed.): CICLing 2013, Part II, LNCS 7817, pp. 394–407, 2013.
© Springer-Verlag Berlin Heidelberg 2013

of the segments is presented in section 4 and its grammaticaility in section 5. The experimental protocol and the results are shown in section 6 and section 7 respectively. Finally, section 8 presents our conclusions and future work.

2 Summarization by Compression

2.1 Classic Sentence Compression

In [4], the sentence compression task is defined as: "Consider an input sentence as a sequence of n words $W = (w_1, w_2, ..., w_n)$. An algorithm may drop any subset of these words. The words that remain (order unchanged) form a compression". The authors included a standard corpus for sentence compression. Later, [5] confirmed that results could be interesting for text summarization and [6] used a similar approach for speech summarization. In [7], a sentence compression corpus, was annotated by humans considering the context. Nonetheless, the criteria used to elicit the compressions remain quite artificial for summarization. The autors asked the annotators to delete individual words from each sentence, but humans also tend to delete long phrases in an abstract. In all of these works it should be noticed a major drawback associated to individual words deletion: deleting individual words could be very risky in terms of grammar, and too poor in terms of compression rate. One single word deletion can seriously affect the sentence, for instance, erasing a verb or a negation.

2.2 More Recent Approaches in Sentence Compression

Recent studies have found outstanding results using clauses or discourse structures, instead of isolated words. An algorithm, proposed in [8], divides sentences into clauses prior to any elimination. Although the results of this last work are good in general, in some cases the main subject of the sentence is removed. The authors attempted to solve this issue by including features in a machine learning approach [9].

Discourse chunking [10] is an alternative to discourse parsing, thereby, showing a direct application to sentence compression as shown in [11]. The authors of these last two works argued that, while discourse parsing at document level is a significant challenge, discourse chunking at sentence level could present an alternative in human languages with limited language processing tools. In addition, some sentence-level discourse models have shown accuracies comparable to human performance [12].

3 Discourse Segmentation

3.1 Sentence Level Discourse Segmentation

In this work, we use a sentence-level discourse segmentation approach. Formally, "Discourse segmentation is the process of decomposing discourse into Elementary Discourse Units (EDUs), which may be simple sentences or clauses in a complex sentence, and from which discourse trees are constructed" [13]. The first step of discourse parsing is discourse segmentation (the next steps are detection of rhetorical relations and building of the discourse tree). However, we can consider segmentation at the sentence level in order to identify segments to be eliminated in the sentence compression task. The decomposition of a sentence into EDUs using only local information is called *intra-sentence discourse segmentation*. Today, automatic discourse segmentation systems exist for several languages such as English [13], Brazilian Portuguese [14], Spanish [15] and French [16].

3.2 Compression Candidates Generation

In this work we propose to generate compression candidates by deletion of some discourse segments from the original sentence. Let be a sentence S the sequence of its k discourse segments: $S = (s_1, s_2, ..., s_k)$. A candidate, CC_i, is a subsequence of S that preserves the original order of the segments. The original sentence always forms a candidate, i.e., $CC_0 = S$, this is convenient because sometimes there is no shorter grammatical version of the sentence, especially in short sentences that conform one single EDU. Since we do not consider the empty subsequence as a candidate, there are $2^k - 1$ candidates.

3.3 The DiSeg Discourse Segmenter

The discourse segmenter used in our experiments, DiSeg, is described in [15] and is based on the Rhetorical Structure Theory [17]. This system detects discourse boundaries in sentences. First, a text is pre-processed with sentence segmentation, POS tagging and shallow parsing modules using the Freeling toolkit[18]. Then, an XML file is generated with discourse marker annotations. Finally, several rules are applied to this file. The rules are based on: discourse markers, as "while" (*mientras que*), "although" (*aunque*) or "that is" (*es decir*), which usually mark the relations of CONTRAST, CONCESSION and REFORMULATION, respectively; conjunctions, such as "and" (*y*) or "but" (*pero*); adverbs, as "anyway" (*de todas maneras*); verbal forms, as gerunds, finite verbs, etc.

3.4 Adapting DiSeg to Sentence Compression: The CoSeg Segmenter

We have adapted the discourse segmenter DiSeg for the sentence compression task simply by modifying its original rules in order to ease the definition of EDUs. While in DiSeg it is mandatory that every EDU contains a principal verb, in CoSeg, a segment could not have any verb. In CoSeg, if a fragment contains a discourse marker it must be segmented. We also consider that punctuation marks, as parenthesis, comas or dashes, are natural boundaries in sentences. Afterall, the final goal is to create a sentence compression system based on this adapted version of DiSeg.

4 Informativeness of Discourse Segments

4.1 The Textual Energy

$$S = \begin{pmatrix} s_1^1 & s_1^2 & \cdots & s_1^N \\ s_2^1 & s_2^2 & \cdots & s_2^N \\ \vdots & \vdots & \ddots & \vdots \\ s_P^1 & s_P^2 & \cdots & s_P^N \end{pmatrix} ; \quad s_\mu^i = \begin{cases} TF_i \text{ if word } i \text{ is present in segment } \mu \\ 0 \quad \text{elsewhere} \end{cases} \quad (1)$$

Textual energy is a similarity measure used in several NLP tasks: automatic summarization [19], topic segmentation [20] and text clustering [21]. In this method, words and sentences are taken as a magnetic system composed of spins (words coded as 1's and 0's). In its original description [22], the minimal unit of processing is the sentence and the main idea is to rank all of the sentences in a text. In this work, we use the textual energy for the evaluation of discourse segments. First, documents are pre-treated with classical algorithms like filtering and lemmatisation to reduce the dimensionality (see [23] for details). Then, a *bag of words*, representing the segments, produces the matrix $S_{[P \times N]}$ (1) of word frequencies/absences consisting of $\mu = 1, \cdots, P$ segments (rows) and the vocabulary of $i = 1, \cdots, N$ terms (columns).

Let us consider the segments as sets σ of words. These sets constitute the vertices of a graph like that of the Figure 1. We can draw an edge between two of the vertices σ_μ, σ_ν every time they share at least a word in common $\sigma_\mu \cap \sigma_\nu \neq \emptyset$. We obtain the graph $I(S)$ from intersection of the segments. We evaluate these pairs $\{\sigma_1, \sigma_2\}$, which we call edges, by the exact number $|\sigma_1 \cap \sigma_2|$ of words that share the two connected vertices. Finally, we add to each vertex σ an edge of reflexivity $\{\sigma\}$ valued by the cardinal $|\sigma|$ de σ. This valued intersection graph is isomorphic with the adjacency graph $G(S \times S^T)$ of the square matrix $S \times S^T$. In fact, $G(S \times S^T)$ contains P vertices. There is an edge between two vertices μ, ν if and only if $[S \times S^T]_{\mu,\nu} > 0$. If it is the case, this edge is valued by $[S \times S^T]_{\mu,\nu}$ and this value corresponds to the number of words in common between the segments μ and ν. Each vertex μ is balanced by $[S \times S^T]_{\mu,\mu}$, which corresponds to the addition of an edge of reflexivity. It results that the matrix of Textual Energy E is the adjacency matrix of the graph $G(S \times S^T)^2$. The textual energy of segments interaction can be expressed by (2).

$$E = -\frac{1}{2} S \times (S^T \times S) \times S^T = -\frac{1}{2}(S \times S^T)^2 \quad (2)$$

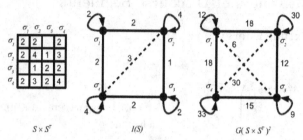

Fig. 1. Graph from the matrix of energy

4.2   Segment Energy

Given that, the sum of the row j in the energy matrix (2) gives the lexical link strengths of segment j, therefore we are able to determine how relevant segment j is in the text. Textual energy matrix connects segments having common words, as well as segments sharing the same neighbourhood but not necessarily identical vocabulary.

Figures 2 and 3 show a text extracted from our corpus. Each row corresponds to a segment while first and second columns correspond to the energy values considering the original sentence energy and individual segment energy. The gray tonality exhibits the degree of informativeness of the segments considering the whole text context: darker segments are the less informative. Bottoms of the figures show the density plot of energy values. Table 1 shows the approximate translations for both segmenters with the original sentences numbered.

5   Grammaticality of Discourse Segments

5.1   Scoring Discourse Segments with Language Models

Statistical language modeling [23] is a technique widely used to assign a probability to a sequence of words. The probabilities in a Language Model (LM) are estimated counting sequences from a corpus. Even though we will never be able to obtain enough data to compute the statistics for all possible sentences, we can base our estimations using large corpora and interpolation methods. In the experiments we use a big corpus with 1T words (LDC Catalog No.: LDC2009T25) to obtain the sequence counts and a LM interpolation based on Jelinek-Mercer smoothing [24]. In a LM, the maximum likelihood estimate of a sequence is interpolated with the smoothed lower-order distribution. We use the Language Modeling Toolkit SRILM [25] to score the segment likelihood probability. We assume that good compression candidates must have a high probability as sequences in a LM.

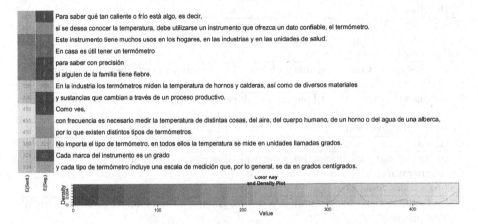

Fig. 2. Textual energy values for DiSeg segments in a text

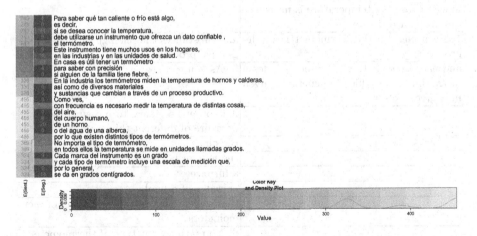

Fig. 3. Textual energy values for Coseg segments in a text

Table 1. Approximate segmentations translated for both segmenters

DiSeg	CoSeg
1. To find out how hot or cold something is, that means,	**1.** To find out how hot or cold something is,
if you want to know the temperature, use an instrument that provides reliable data, the thermometer.	that means,
2. This instrument has many uses at home, industries and care units.	if you want to know the temperature,
3. Having a thermometer at home is useful	use an instrument that provides reliable data,
to know with precision	the thermometer.
if a family member has fever.	**2.** This instrument has many uses at home,
4. In the industry thermometers measure the temperature of furnaces and boilers, as well as various materials	industries and care units.
and substances that change through a productive process.	**3.** Having a thermometer at home is useful
5. As you can see,	to know with precision
it is often necessary to measure the temperature of different things, the air, the human body, a furnace or water of a pool,	if a family member has fever.
that's why there are different types of thermometers.	**4.** In the industry thermometers measure the temperature of furnaces and boilers,
6. No matter what type of thermometer, in every case the temperature is measured in units called degrees.	as well as various materials
7. Each mark in the thermometer is a degree	and substances that change through a productive process.
and each type of thermometer includes a measuring scale, which in general, is given in degrees Celsius.	**5.** As you can see,
	it is often necessary to measure the temperature of different things,
	the air,
	the human body,
	a furnace
	or water of a pool,
	that's why there are different types of thermometers.
	6. No matter what type of thermometer,
	in every case the temperature is measured in units called degrees.
	7. Each mark in the thermometer is a degree
	and each type of thermometer includes a measuring scale, which
	in general,
	is given in degrees Celsius.

6 Experiments

6.1 To Delete or Not to Delete, That Is the Task

We have set up a campaign of text annotation with non-expert volunteers (a citizen science project). First, we have chosen 30 short texts. Then, each text was segmented twice: one using DiSeg and the other one with CoSeg. We asked human annotators to chose which segments must be preserved to form an abstract, following the criteria in section 6.2 hereafter. Figure 4 shows the interface used during the annotation campaign and its main components:

1. **Segments** are text fragments that can be activated or eliminated by clicking on them.
2. **Original text** contains the initial text. Segments can be read even after being deleted.
3. **Compressed text** displays the resulting text after eliminating selected segments.
4. **Button "Restart text"** restores the initial text before removing any segment.
5. **Button "End text"** sends the compressed text to the database and displays the next text to analyze.

We have recruited 66 volunteers, all native Spanish speakers, most of them undergraduate students and in 10 weeks we have collected 2 877 summaries (48 user summaries for each text in average). The system demo can be tested on the Web[1].

6.2 Criteria for Compression

What follows sets the criteria that had to be considered by annotators. These criteria have been used to analyze each sentence individually. Moreover, it was required that the resulting compressed text had to be entirely coherent.

Conservation. At least one segment must be kept for each sentence.
Importance. The main idea of the original text must be retained.
Grammaticality. The compressed sentences should be understandable and should not have problems of coherence (e.g., sentences must have a main verb).
Brevity. It should be compressed as much as possible. This means, deleting words as long as it keeps the same meaning, but with fewer words.

All criteria are equally important.

---

[1] http://dev.termwatch.es/~molina/compress4/man/.

Fig. 4. Annotation system interface for the annotation campaign

7   Results

7.1   Search Space and Solution Space

The search space associated with a sentence is the number of its possible compression candidates. According to section 3.2, for a sentence with k segments, the size of its search space is $2k - 1$. In order to define which is the optimal compression of a sentence, we deal with the dilemma of deciding if one person compressed a given sentence better than another. At the moment we can only consider that if a solution was given by someone during the annotation campaign it must be considered as a candidate solution. Table 2 shows the search space size and the average solutions space size for each segmenter. In general, the search space is larger than the solution space proposed by annotators. This is a fundamental fact because it points out some regularities in the compressions proposed by the annotators.

7.2   Annotators Agreement

As seen in section 7.1, the solution space proposed is shorter than the theoretical search space. However, in most cases, there is a high degree of ambiguity in the designation of the optimal compression. As the sentence complexity increases, the variability in the number of proposed solutions also increases considerably. In many cases, the most voted compression reached only 25% of the votes. Moreover, in some cases there is no clear trend in the optimal solution. Figure 5 shows the proportion of ambiguous cases for five classes defined by the number of words

Table 2. Theoretical search space and average number of human solutions using two discourse segmenters

k	Theoretical value $(2^k) - 1$	Avg. solutions using DiSeg	Avg. solutions CoSeg
1	1	1	1
2	3	2.6	2.5
3	7	4.6	4.3
4	15	7	6.1
5	31	10.8	8.8
6	63	16	11.6
7	127	-	12.3
8	255	16	14.3
9	511	26	21
10	1023	-	18
11	2047	-	16
12	4095	-	28
13	8191	-	39
21	2097151	-	40

using different votes thresholds. In this figure only multi-segment sentences were considered (sentences with a single segment are not relevant in our analysis). In order to ensure distribution's uniformity for both segmenters, classes were defined as it follows: C1 (20 words or less), C2 (21 to 27 words), C3 (28 to 34 words), C4 (35 to 45 words) and C5 (46 words or more).

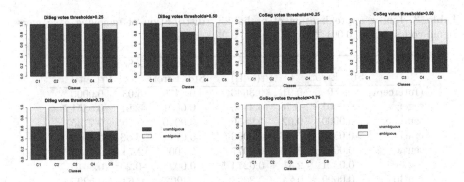

Fig. 5. Ambiguity proportion for different votes thresholds

7.3 Linear Models of Segment Removal Probability

We studied the possibility to predict the probability of whether a segment will be removed or not by annotators. We considered generalized linear models. Probability distributions are defined for each segmenter ($P_D(x)$ for DiSeg and $P_C(x)$

for CoSeg) and every segment x as the number of evaluators that chose to re-move the segment over the total number of annotators. In our data, the mean probability to remove a DiSeg segment is 25% and the median probability is only 10%. These probabilities increase for the CoSeg segments reaching a mean of 32% and a median of 25%. T-test and Wilcoxon tests show that these differences are highly significant (p-value $< 10^{-6}$).

Now, if we consider only non-ambiguous deleted segments, which were re-moved by at least 75% of the annotators, these are 33 over 412 (7%) segments for Diseg, and 81 over 740 (10%) for CoSeg. This shows that CoSeg gives the reader more opportunity to delete single segments for compression than DiSeg. However, can the probability of deletion be estimated using segment and sen-tence properties? For that, we consider the following indicators:

segener: Textual energy of the segment.
sentener: Textual energy of the sentence.
eratio: segener/sentener ratio.
segw: Segment length in number of words.
sentw: Sentence length in number of words.
wratio: segw/sentw ratio.
sentlp: Sentence likelihood probability.
seglp: Segment likelihood probability.
lpratio: seglp/sentlp ratio.
segpos: Segment position in the sentence.
relpos: Segment position relative to the number of segments.
nsegs: Number of segments in the sentence.

Table 3. Linear approximation of probability distribution of removing a segment. Signif. codes: *** < 0.001, ** < 0.01, * < 0.05, . < 0.1

	DiSeg			CoSeg		
	Std. Error	t	Pr(>|t|)	Std. Error	t	Pr(>|t|)
(Intercept)	0.0607	0.91	0.3646	0.0476	6.03	0.0000 ***
segener	0.0001	-0.76	0.4490	0.0001	-2.01	0.0451 *
sentener	0.0000	0.02	0.9859	0.0000	0.29	0.7735
eratio	0.0558	-3.47	0.0006 ***	0.0714	-4.86	0.0000 ***
sentw	0.0012	-1.04	0.2977	0.0009	2.74	0.0063 **
segw	0.0021	2.31	0.0214 *	0.0029	-0.20	0.8410
sentlp	0.0000	0.24	0.8083	24902	0.63	0.5297
seglp	6.3080	-0.87	0.3841	0.9858	-1.78	0.0759
lpratio	0.0942	-6.80	0.0000 ***	0.1049	-3.94	0.0001 ***
segpos	0.0224	-1.59	0.1130	0.0083	0.11	0.9158
nsegs	0.0188	1.65	0.0998 .	0.0075	-3.05	0.0024 **
relpos	0.0819	8.51	0.0000 ***	0.0604	5.70	0.0000 ***

Table 3 shows the output of linear model fitting function, in R software, to predict segment deletion probabilities for DiSeg and CoSeg. It reveals that deletions of CoSeg segments are correlated to a larger subset of segment and sentence descriptors. Deducing the two linear models by restricting previous linear fittings to significantly correlated indicators gives these models:

$$P_D(x) \sim 0.581\mathbf{segpos} - 0.523\mathbf{lpratio} - 0.214\mathbf{eratio} + 0.002\mathbf{segw}$$

$$P_C(x) \sim -0.342\mathbf{eratio} + 0.003\mathbf{sentw} - 1.574\mathbf{seglp} - 0.416\mathbf{lpratio}$$
$$-0.022\mathbf{nsegs} + 0.35\mathbf{relpos}$$

Both models are significantly correlated to the targeted probability distributions (Pearson's product-moment correlation p-value < 2.2e-16). Pearson's estimate is above 0.73 for $P_D(x)$ and only above 0.57 for $P_C(x)$. This would mean that DiSeg points out segments that are easier to characterize in terms of compression based on textual energy and likelihood than CoSeg.

7.4 Evaluation

A 20% folded cross-check experiment using $\frac{7}{8}$ fraction of the corpora to predict the probability distribution over the remaining $\frac{1}{8}$ fraction of segments shows that the above linear model is an efficient inference model for $P_D(x)$ but not for $P_C(x)$. Indeed, on the 20% test sets, Pearson's estimate is above 0.69% and below 76% with a median of 0.73 for Diseg p_S meanwhile Pearson's estimate can be very low on some 20% test sets (below 10%) with a maximum of 72%.

To evaluate qualitatively the results we have designed a Turing like test using 39 abstracts. Two summaries were presented to each judge (other than the annotators): one prepared by (A) a human and the other developed by (B) the computer. The judge had to decide which of the abstracts had been accomplished by (A) and which one by (B). The final result was that the judges properly allocated 13 of the abstracts; 15 of the abstracts prepared by the computer were mistakenly taken as summaries made by humans and 14 of the abstracts prepared by humans were mistakenly taken by the computer. In brief, $\frac{2}{3}$ of the judges were confused in this game of imitation.

8 Conclusions and Future Work

In this article we have described a new method for automatic summarization by compression of discourse segments into each sentence, and using the textual energy to weight the informativeness of these segments. Thanks to our annotation campaign, we tested various interesting aspects regarding the elimination of discourse segments for the automatic summarization. Our study revealed that, in general, there is disagreement to determine the optimum compression and the degree of disagreement increases as the sentence complexity increases. However, there is a general agreement to preserve segments with high energy values. We have proposed a generalized linear model to predict the probability of deleting a segment based on simple features. Another interesting result is that there is a human tolerance to non grammatical compressions if they allow to keep pertinent information in a short dense summary, which has led us to consider cooperative human-machine systems as an alternative to fully automatic summarization. Finally, we have performed a Turing test using the imitation game for evaluation. We believe that this kind of evaluation is more convincing than other automatic methods based on frequencies of n-grams because it implies human judging.

Acknowledgments. We would like to thank our contributors for their help with the corpus annotation. This work was partially supported by a CONACyT grant 211963 and Imagweb project.

References

1. Edmundson, H.P.: New Methods in Automatic Extraction. Journal of the Association for Computing Machinery 16, 264–285 (1969)
2. American National Standards Institute Inc.: American National Standard for Writing Abstracts. Technical Report ANSI Z39.14 – 1979, American National Standards Institute, New York (1979)
3. Witbrock, M.J., Mittal, V.O.: Ultra-Summarization: A Statistical Approach to Generating Highly Condensed Non-Extractive Summaries. In: Proceedings of the 22nd Conference SIGIR 1999, Berkeley, CA, Etats-Unis, pp. 315–316. ACM (1999)
4. Knight, K., Marcu, D.: Statistics-based summarization – step one: Sentence compression. In: Proceedings of the 17th National Conference on Artificial Intelligence and 12th Conference on Innovative Applications of Artificial Intelligence, Austin, TX, Etats-Unis, pp. 703–710 (2000)
5. Lin, C.Y.: Improving summarization performance by sentence compression-a pilot study. In: Proceedings of the 6th International Workshop on Information Retrieval with Asian Languages, Sapporo, Japon, pp. 1–8 (2003)
6. Hori, C., Furui, S.: Speech summarization: an approach through word extraction and a method for evaluation. IEICE Transactions on Information and Systems (Institute of Electronics, Informatics and Communication Engineering) 87, 15–25 (2004)
7. Clarke, J., Lapata, M.: Modelling compression with discourse constraints. In: Proceedings of the 2007 Joint Conference on Empirical Methods in Natural Language Processing and Computational Natural Language Learning (EMNLP-CoNLL), pp. 1–11 (2007)
8. Steinberger, J., Jezek, K.: Sentence compression for the lsa-based summarizer. In: Proceedings of the 7th International Conference on Information Systems Implementation and Modelling, pp. 141–148 (2006)
9. Steinberger, J., Tesar, R.: Knowledge-poor multilingual sentence compression. In: 7th Conference on Language Engineering (SOLE 2007), Cairo, Egypt, pp. 369–379 (2007)
10. Sporleder, C., Lapata, M.: Discourse chunking and its application to sentence compression. In: Proceedings of the Conference on Human Language Technology and Empirical Methods in Natural Language Processing, Association for Computational Linguistics, pp. 257–264 (2005)

11. Molina, A., Torres-Moreno, J.-M., SanJuan, E., da Cunha, I., Sierra, G., Velázquez-Morales, P.: Discourse segmentation for sentence compression. In: Batyrshin, I., Sidorov, G. (eds.) MICAI 2011, Part I. LNCS (LNAI), vol. 7094, pp. 316–327. Springer, Heidelberg (2011)
12. Soricut, R., Marcu, D.: Sentence level discourse parsing using syntactic and lexical information. In: HLT-NAACL (2003)
13. Tofiloski, M., Brooke, J., Taboada, M.: A syntactic and lexical-based discourse segmenter. In: Proceedings of the ACL-IJCNLP 2009 Conference Short Papers. ACLShort 2009, Stroudsburg, PA, USA, pp. 77–80. Association for Computational Linguistics (2009)
14. Maziero, E., Pardo, T., Nunes, M.: Identificaç ão automática de segmentos discursivos: o uso do parser palavras. Série de relatórios do núcleo interinstitucional de lingüística computacional, Universidade de Sao Paulo, São Carlos, Brésil (2007)
15. da Cunha, I., SanJuan, E., Torres-Moreno, J.-M., Lloberes, M., Castellón, I.: Discourse segmentation for spanish based on shallow parsing. In: Sidorov, G., Hernández Aguirre, A., Reyes García, C.A. (eds.) MICAI 2010, Part I. LNCS, vol. 6437, pp. 13–23. Springer, Heidelberg (2010)
16. Afantenos, S.D., Denis, P., Muller, P., Danlos, L.: Learning recursive segments for discourse parsing. CoRR abs/1003.5372 (2010)
17. Mann, W.C., Thompson, S.A.: Rhetorical Structure Theory: A Theory of Text Organization. University of Southern California, Information Sciences Institute, California, Marina del Rey (1987)
18. Atserias, J., Casas, B., Comelles, E., González, M., Padró, L., Padró, M.: FreeLing 1.3: Syntactic and semantic services in an open-source NLP library. In: Proceedings of the 5th International Conference on Language Resources and Evaluation (LREC 2006), pp. 48–55 (2006)
19. Torres-Moreno, J.M.: Résumé automatique de documents: une approche statistique. Hermés-Lavoisier, Paris (2011)
20. da Cunha, I., Fernández, S., Velázquez Morales, P., Vivaldi, J., SanJuan, E., Torres-Moreno, J.-M.: A new hybrid summarizer based on vector space model, statistical physics and linguistics. In: Gelbukh, A., Kuri Morales, Á.F. (eds.) MICAI 2007. LNCS (LNAI), vol. 4827, pp. 872–882. Springer, Heidelberg (2007)
21. Sierra, G., Torres-Moreno, J.M., Molina, A.: Regroupement sémantique de définitions en espagnol. In: Proceedings of Evaluation des Méthodes d'extraction de Connaissances Dans les Données (EGC/EvalECD 2010), Hammamet, Tunisie, pp. 41–50 (2010)
22. Fernández, S., SanJuan, E., Torres-Moreno, J.-M.: Textual energy of associative memories: performant applications of enertex algorithm in text summarization and topic segmentation. In: Gelbukh, A., Kuri Morales, Á.F. (eds.) MICAI 2007. LNCS (LNAI), vol. 4827, pp. 861–871. Springer, Heidelberg (2007)
23. Manning, C.D., Schütze, H.: Foundations of Statistical Natural Language Processing. MIT Press, Cambridge (1999)
24. Chen, S., Goodman, J.: An empirical study of smoothing techniques for language modeling. Computer Speech & Language 13, 359–393 (1999)
25. Stolcke, A.: Srilm – an extensible language modeling toolkit. In: Intl. Conf. on Spoken Language Processing, Denver, vol. 2, pp. 901–904 (2002)

A Knowledge Induced Graph-Theoretical Model for Extract and Abstract Single Document Summarization

Niraj Kumar, Kannan Srinathan, and Vasudeva Varma

IIIT-Hyderabad, Hyderabad-500032, India
niraj_kumar@research.iiit.ac.in,
{srinathan,vv}@iiit.ac.in

Abstract. Summarization mainly provides the major topics or theme of document in limited number of words. However, in extract summary we depend upon extracted sentences, while in abstract summary, each summary sentence may contain concise information from multiple sentences. The major facts which affect the quality of summary are: (1) the way of handling noisy or less important terms in document, (2) utilizing information content of terms in document (as, each term may have different levels of importance in document) and (3) finally, the way to identify the appropriate thematic facts in the form of summary. To reduce the effect of noisy terms and to utilize the information content of terms in the document, we introduce the graph theoretical model populated with semantic and statistical importance of terms. Next, we introduce the concept of weighted minimum vertex cover which helps us in identifying the most representative and thematic facts in the document. Additionally, to generate abstract summary, we introduce the use of vertex constrained shortest path based technique, which uses minimum vertex cover related information as valuable resource. Our experimental results on DUC-2001 and DUC-2002 dataset show that our devised system performs better than baseline systems.

Keywords: Single document summarization, Extract summary, Abstract summary, Minimum vertex cover, Semantic relatedness, Weighted minimum vertex cover and Vertex constraint shortest path.

1   Introduction

These days World Wide Web (WWW) and digital libraries contain huge amount of text resources, like: web pages, news documents, educational materials etc. These all again contain huge amount of text information and we have less time to go through. It is remarkable to note that all such documents do not always contain human supplied summaries.

We believe that an unsupervised approach to generate extract and abstract summary for single document by using limited linguistic resources can solve this problem. Facts like: generic summaries outperform over (1) query-based and (2) hybrid summaries in the browsing tasks and thus help users in browsing [11], also support our view. But, there are a lot of issues which affect the quality of summary (other than issues discussed in abstract). For example: (1) if any document contains

A. Gelbukh (Ed.): CICLing 2013, Part II, LNCS 7817, pp. 408–423, 2013.
© Springer-Verlag Berlin Heidelberg 2013

sentences which are rich in information, extract summary can be an alternative (useful but not the best alternative) next, (2) if any document contains sentences with majority of talkative terms (e.g., blogs and somewhat in news articles etc.); abstract summary can be a better option. This is due to the compactness of useful information in abstract summary sentences. But, generating abstract summary is a tough task compare to the generation of extract summary. Very few research works have been done in the area of combining the scattered information among multiple sentences. Linguistic complexity is also an issue. Next, it is important to note that single document summarization is slightly different from multi-document summarization [9], as earlier contains less information (i.e. significantly less amount of words) with respect to later. So, it requires more efficient strategy.

In view of these facts, we focus our attention towards the development of graph theoretical model for extract and abstract single document summarization.

Our Contribution: Our contributions toward the development of entire system can be summarized as:

➢ We introduce the use of bigram based model and word graph populated with semantic and statistical information of words of bigrams. This arrangement is helpful in reducing the impact of noisy or less important words and effectively handles the words having different levels of importance in document.
➢ To identify the most representative and thematic facts from each of the identified topic, we introduce the weighted minimum vertex cover based scheme.
➢ To generate the abstract summary, we introduce the use of vertex constrained shortest path based scheme, which uses identified set of weighted minimum vertex cover as prior information.

Paper Organization: In Section 2, we briefly describe the related works of this area. In Section 3, we explore the performance related issues and motivation behind our techniques to solve it, under the heading of "problem statements and motivation". In Section 4, we explore the bigram based model and calculate the semantic and statistical (local) importance of bigrams. In Section 5, we identify sentence clusters for given document and rank them. In Section 6 we discuss our method to calculate weighted minimum vertex cover in every identified topic. In Section 7, we discuss the method to generate extract summary. In Section 8, we discuss our method to generate abstract summary by using vertex constrained shortest path. In Section 9, we present pseudo code for entire system. In Section 10 we present experimental evaluation and finally in Section 11, we concluded the work.

2   Literature Survey

Brief literature survey of some latest related works of this area is given below:

Most of the graph-based unsupervised methods use either sentence or word as node of graph. For example: [12], uses every sentence as node of graph and apply Page Rank's "random surfer model" to rank the sentences. [14], incorporates the cross-document relationships between sentences in a cluster and finally apply graph ranking based summarization algorithm.

For given cluster of documents (CollabSum [14]), designed three summarization methods based on the use of cross-document relationship between sentences in the cluster of document.

Some latest advancement in single document summarization is proposed by [1], [2] and [13]. [1], describes a fusion of syntactic and semantic techniques used in single document sentence extraction, to beat the baselines. [2], combines syntactic, semantic, and statistical methodologies, and reflect psychological findings that pinpoint specific selection patterns as humans construct summaries. [13], proposes a unified approach to simultaneous single-document and multi-document summarization by making use of the mutual influences between the two tasks.

Some recent approach on abstract summarization methods like: [7], introduces the sentence compression scheme, by using word graph of text. The algorithm uses k-shortest path algorithm and filter all those paths, which are shorter than 8 words or do not contain a verb. It uses offset position of words and occurrence frequency in calculation of edge weight and includes salient words (selected based on occurrence frequency) in shortest path. [6], proposes a method which uses a compendium to decide, which sentences are suitable to include in abstract sentence(s). This system also incorporates frequency and page rank score of words.

3   Problem Statements and Motivation

In this section we discuss, some basic issues which affects the summarization task and motivation behind our techniques to solve it.

3.1   Handling Noisy Terms and Terms Having Different Levels of Importance

Problem Statements: Actually, due to high occurrence frequency and other supporting statistical features, majority of the times, noisy terms also get high importance score. Thus it reduces the performance of system. Next, less important terms, Like: (1) terms which are semantically important but, not very much useful for the document or (2) terms which are locally/statistically important in the document, but have less semantic importance (e.g., terms which exist in very few knowledgebase documents of sufficiently large collection), also degrade the performance of system.

Motivation: To solve these problems, we use bigram based model (instead of using Bag of words 'BoW' based model) and word graph, populated with semantic and statistical importance of words of bigrams.

Bigram Based Model: According to [3], the main drawback of the BoW representation is in destruction of semantic relations between words. Indeed, stable phrases, such as "White House" or "Bill Gates", are represented in the BoW as separated words so their meaning is lost. Given a BoW of a document in which words "bill" and "gates" occur, one can suggest that the document is about accounting or gardening, but not about computer software. Whereas given a document representation that contains a phrase "bill gates", the reader will hardly be mistaken about the topic of discussion. Thus to preserve the semantic relation at some extent, we use bigram based model.

Word Graph Populated with Local and Semantic Importance of Bigrams: We use the word graph populated with semantic and statistical (local) information of words of bigrams. The local importance of any bigram reflects its importance in given text document due to (1) occurrence frequency, (2) position of sentence in which given bigram occurs and (3) position of bigram in document etc. Some of these features have been effectively utilized by several keyphrase extraction algorithms, e.g., [8]. However, due to statistical nature of features like occurrence frequency, position of sentence, etc., some frequent noisy or less important bigrams may also get high importance.

To effectively reduce the chances of getting high importance by such frequent noisy or less important bigrams, we introduce the use of product of semantic relatedness score and local importance of bigrams (different from [15], which is based upon only semantic information). Thus the product of both scores automatically boosts the strength of useful bigrams in document (i.e. bigrams, which are both, locally and semantically important). The combined weight of bigrams is further used in calculation of importance of topics covered in document and in calculation of weighted minimum vertex cover.

3.2   Identifying Appropriate Thematic Facts

Motivation: We introduce the concept of weighted minimum vertex cover to identify the thematic facts from every identified topic or sentence cluster.

Why Weighted Minimum Vertex Cover?   Actually finding the minimum vertex cover is a classical optimization problem [4] and quality of identified vertex cover depends upon the information and resources used to calculate it [5]. As every word in document may have different levels of importance and it can be calculated by using its statistical and semantic information. We use this information to get optimal solution for minimum vertex cover. For this, we use the word graph of sentences populated with semantic and statistical information. We named it as "weighted minimum vertex cover" and further use it in identifying the most representative terms in every identified topic. But, before going into detail, we discuss about minimum vertex cover and requirement of weighted minimum vertex cover.

Minimum Vertex Cover: For a graph $G = (V, E)$, where, $V = \{V_1, V_2, ..., V_n\}$ denotes the vertex set and link set $(V_j, V_i) \in E$ if there is a link between V_j and V_i. A set $S \subseteq V$ s.t. for every $(u,v) \in E$, $u \in S$ or $v \in S$, where $|S|$ is minimized. In other words, a minimum vertex cover in graph is a vertex cover that has smallest number of vertices among all possible vertex covers. It depends upon the search strategy applied. For example in Figure 1, the minimum vertex cover may be {2, 3, 4} or {2, 4, 5}. Thus to get optimal set of minimum vertex cover, we need some additional information and it creates the requirement of "weighted minimum vertex cover" as discussed earlier.

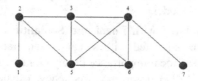

Fig. 1. Undirected graph

3.3   Sentence Abstraction

Problem Statement: Combining important information spread in different sentences to generate the compact summary sentence for abstract summarization is a challenging problem. Techniques like: [6], [7] use some fixed set of thresholds e.g., length of sentences, topical information and/or presence of some predefined specific words etc., for abstraction of documents of different nature. Due to the variable nature of documents fixing such information may not be a good idea and may even downgrade the performance of system.

Motivation: Different from above discussed methods, we incorporate the information content in extract summary (specially the presence of weighted minimum vertex cover in highest scoring sentences), as a rough idea about the nature and information content of the abstract summary sentences, which may vary from document to document. Next, to achieve the goal of abstract summarization, we introduce the use of vertex constrained shortest path based scheme, which use weighted minimum vertex cover related information as valuable resources. Thus due to the use of prior knowledge and word graph populated with semantic and statistical information, vertex constrained shortest path based scheme shows improvement in the quality of performance.

4   Calculating Importance of Bigrams

4.1   Input Cleaning and Pre-processing

Our input cleaning step includes: (a) removal of noisy symbols and stopwords and (b) stemming of text (by using Porter stemming algorithm[1]).
    In the pre-processing step, we filter the sentences.

4.2   Calculating Local Importance of Bigrams

To calculate the local importance/weight of every bigram in the given document, we use simple statistical, linguistic features and heuristics. The description about selection of these features and calculation of final weight is given below:

---

[1] http://tartarus.org/martin/PorterStemmer/

Position of Sentence in Document: It is already known that, the important terms come in earlier sentences [8]. To capture this we use the index of sentence in which given bigram occurs first. By using this feature, we can calculate the weight of bigram $W_1(i, j)$, which contains word pair w_i and w_j:

$$W_1(i, j) = \frac{(\#S - S_{index}(i, j))}{\#S} \tag{1}$$

Where, $\#S$ = Total number of sentences in document, $S_{index}(i, j)$ = Index of sentence in which given bigram occurs first.

Coverage: Here we consider the coverage strength of given bigram. i.e., count of sentences in which the given bigram exist. Next, we normalize this feature by dividing it with total number of sentences in document.

$$W_2(i, j) = \left(\frac{S_{count}(i, j)}{\#S} \right) \tag{2}$$

Where, $W_2(i, j)$ = Weight due to coverage strength of given bigram, $S_{count}(i, j)$ = Count of sentences which contains the given bigram.

Position of Bigram in Sentence: Position of bigrams in sentence is also important. The bigram which comes earlier in sentence plays more important role with respect to other bigrams which come either, at the end of the sentence or at any other part of the sentence. We use this feature in calculation of importance of bigrams.

$$W_3(i, j) = \frac{\sum \left((2 \times S_{Word\_Count}(i, j)) - S_{index}(i, j) \right)}{\sum \left(S_{Word\_Count}(i, j) \right)} \times 2 \tag{3}$$

Where, $W_3(i, j)$ = Weight due to position of given bigram in sentence, $S_{Word\_Count}(i, j)$ = Count of bigrams in sentence 'S', which contains the given bigram, $S_{index}(i, j)$ = Index of given bigram in sentence 'S', $\sum \left(S_{Word\_Count}(i, j) \right)$ = Count of total number of bigrams in all sentences, which contains the given bigram.

Description: The value of Equation 3 varies from 2.00 to 1.00. It gives highest score, if bigram comes at starting position and gives score 1.00 if the given bigram comes at the end of sentence. We calculate this score for all sentences, in which the given bigram exist. Finally we normalize the score, by dividing it with sum of number of bigrams in all sentences in which the given bigram exist.

Calculating Final Local Importance of Words: To calculate the final weight, we combine all three weighting schemes i.e. (a) Weight due to position of sentence in document, (b) Weight due to coverage strength and (c) Weight due to position of given bigram in the sentence. The final weight of given bigram can be calculated as:

$$W(i, j) = W_1(i, j) \times W_2(i, j) \times W_3(i, j) \tag{4}$$

Where, $W(i, j)=$ Final weight/statistical/local importance of given bigram which contains word pair w_i and w_j.

4.3   Calculating Semantic Importance of Bigrams in Document

To avoid the use of noisy or less important terms in calculation of semantic relatedness score of bigrams, we use frequent occurrences of bigrams in Wikipedia extended abstracts[2] (which contains extended summary of all Wikipedia articles). Additionally, we believe that if words of bigram co-occur (may or may not in adjacency) in more than one sentences of same document then semantic and topical relatedness strength of that bigrams will increase with the presence of more number of such documents. To calculate this we use Point Wise Mutual Information of bigrams.

$$PMI(i, j) = \log_2 \left(\frac{N \times Pt(i, j)}{P(i) \times P(j)} \right)$$ (5)

Where, $PMI(i, j)=$ Pointwise Mutual Information of bigram, containing word w_i and word w_j, $Pt(i, j)=$ Number of Wikipedia extended abstracts which contain at least two sentences, where each of them contain both words i.e., w_i and w_j (may or may not in adjacency), N = Total number of Wikipedia extended abstracts, $P(i)=$ Number of Wikipedia extended abstracts which shows twice occurrences of word w_i, $P(j)=$ Number of Wikipedia extended abstracts which shows twice occurrences of word w_j.

4.4   Calculating Combined Importance of Bigrams

To calculate the final/combined importance score of all distinct bigrams of given document, we take the product of (1) local importance and (2) semantic importance of bigrams in document.

$$CW(i, j) = W(i, j) \times PMI(i, j)$$ (6)

Where, $CW(i, j)$ represents the combined weight of bigram which contains word pair w_i and w_j. We use it in calculation of link weight and ranking topics at later stage.

5   Identifying Topics Covered in Document and Ranking Them

Identifying Topics: To identify the topics covered in document we use Group Average Agglomerative Clustering scheme (GAAC) (i.e. same as used in [16]). In this case, the topic is considered as set of sentences related to the same concept. To apply the GAAC on sentences we use a sentence vector representation of document.

---

[2] It contains full abstracts of Wikipedia articles, usually the first section and can be downloaded from: http://wiki.dbpedia.org/Downloads38#extended-abstracts

GAAC, uses average similarity across all pairs within the merged cluster to measure the similarity of two clusters. In this scheme, the average similarity between two clusters (say, c_i and c_j) can be computed as:

$$sim(c_i, c_j) = \frac{1}{|c_i \cup c_j|(|c_i \cup c_j| - 1)} \sum_{\bar{x} \in (c_i \cup c_j)} \sum_{\bar{y} \in (c_i \cup c_j): \bar{y} \neq \bar{x}} sim(\bar{x}, \bar{y}) \tag{7}$$

Where, $sim(\bar{x}, \bar{y})$ = count of co-occurring words in $\bar{x}$ and $\bar{y}$.

In the entire evaluation, we use the threshold 0.4.

Ranking Identified Sentence Clusters: At this step we rank the identified sentence clusters in descending order of their importance. To calculate the importance (weight) of every identified sentence cluster, we calculate the sum of weighted importance of all bigrams in the given sentence cluster. The calculation of weighted importance of any sentence cluster can be given as:

$$W(C) = \sum W(B_i) \tag{8}$$

Where, $W(C)$ = weight of given sentence cluster 'C', $\sum W(B_i)$ =weight of all bigrams in the given sentence cluster. (See Eq-6, for calculation of weight of bigrams). Finally, we rank all sentence clusters according to descending order of their importance (weight).

6   Calculating Weighted Minimum Vertex Cover

To calculate the weighted minimum vertex cover, we use (1) undirected word graph of sentences and (2) combined importance score of bigrams (see Eq-6). Before going into detail we, first go through the construction of word graph of sentences.

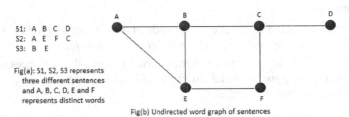

S1: A B C D
S2: A E F C
S3: B E

Fig(a): S1, S2, S3 represents three different sentences and A, B, C, D, E and F represents distinct words

Fig(b) Undirected word graph of sentences

Fig. 2. Undirected word graph of sentences

Constructing Word Graph of Sentences: For this, we treat every word of given sentence cluster as node of graph. We add link between any two words, if they co-occur together (i.e. bigram) in any sentence of given sentence cluster. Formally, we can define undirected graph as, $G = (V, E)$, where, $V = \{V_1, V_2, V_3, ... V_N\}$ represents set of vertices and $E = (V_i, V_j)$ represents edge, if there exist link(s) between V_i and V_j. To calculate edge weight between two adjacent vertices/words w_i and w_j (i.e., bigrams, also represented as V_i and V_j), we multiply the combined importance score of bigrams (Sub Section 4.4) with their occurrence frequency in given sentence cluster.

$$Edge\_Wt(V_i, V_j) = CW(V_i, V_j) \times Occr\_Freq(V_i, V_j) \tag{9}$$

Where, $Edge\_Wt(V_i, V_j)$ represents the edge weight between two vertices V_i and V_j; $Occr\_Freq(V_i, V_j)$ represents co-occurrence frequency of V_i and V_j in given sentence cluster.

Calculating Weighted Minimum Vertex Cover: for this we apply the following greedy algorithm.

Input: Graph $G = (V, E)$

Assumption: let 'S' be the set containing vertices of maximum vertex cover, E' be a temporary list to represent the edges of graph 'G'

Output: a minimum vertex cover set 'S' which contains vertexes and their weight (i.e. selected as minimum vertex cover).

Algorithm:

1. Initialize: $S \leftarrow \phi$, $E' \leftarrow E$
2. While $E' \neq \phi$ do
   - Let v be the vertex, for which the sum of weight of all incident edges are maximum in $G' = (V, E')$, where ties are broken arbitrarily.
   - $S \leftarrow S \cup \{v\}$ (add v to cover).
   - $E' \leftarrow E' \setminus \{(u, v)\}$ (remove from E' all edges incident to v)
   - Calculate the sum of weight of all removed edges and assign it as the weight of vertex 'v'.
3. Return (S)
4. For the rest of the words (treated as node in the graph), which do not exist in {S}, we put weighted minimum vertex cover score to zero.

Note: we use the weighted minimum vertex cover to rank the sentences in given sentence cluster. We apply the weighted minimum vertex cover scheme to sentence clusters having more than one sentence.

7   Generating Extract Summary

Before getting the extract summary, we first rank the sentences in every identified sentence cluster. For this we add the "weighted minimum vertex cover" score of all words present in that sentence and then rank the sentences in descending order of their weight. The calculation of weight/score of any sentence can be given as:

$$W(s) = \sum WVC(Wd) \tag{10}$$

Where, $W(s)=$ Weight of given sentence 's'; $\sum WVC(Wd)=$ Sum of weighted minimum vertex cover score of words in given sentence. (See Section 6, for calculation of weighted minimum vertex cover score).

Extracting Sentences:

1. Rank all identified sentence clusters in descending order of their weight (See Section 5). Next, select top ranked sentence from each sentence cluster and put them according to the order of ranked sentence clusters. In extraction of top sentence from every identified sentence cluster, we check, if the length of sentence is more than 40 words then we discard that sentence and select the next top ranked sentence having length less than 40 words.

2. Finally to get the summary of 'n' words, we select first 'n' words from collection of extracted sentences.

8   Generating Abstract Summary

For this, we use vertex constrained shortest path based scheme, which use the minimum weighted vertex cover related information from top scored sentence(s) of each identified topic as prior knowledge.

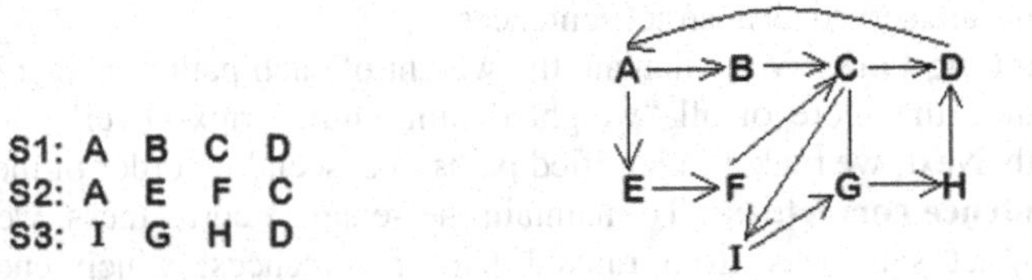

S1: A B C D
S2: A E F C
S3: I G H D

Fig: Sentences from any given identified topic   Fig: Directed word-graph of text
(after input cleaning and pre-processing)

Fig. 3. Directed word graph of text

Vertex Constrained Shortest Path: Consider a directed graph $G=(V,E)$ with vertices $V=\{v_1,v_2,.....,v_n\}$ and edge set $E=\{e_1,e_2,....,e_k\}$ (see Figure 3) with associated costs $c_1,c_2,.....,c_k$. The problem is to find the shortest paths from initial vertex 'S' to multiple targets t_1, t_2, ..., t_k by taking into account these costs, such that it pass through a set of specified vertices (or a number of vertices chosen from specified subset(s)).

Generating Abstract Summary: We use vertex constrained shortest path based technique to extract the top abstract sentence(s) from every identified sentence cluster. The detailed steps are given below:

1. **FOR** every identified sentence cluster (See Section 5), we apply the following procedures:

    1.1 **Preparing Directed Word Graph of Text**: Different from word graph of sentences (as used in Section 6 and 7), we consider entire text of given sentence cluster as sequence of words, (where sentences are ordered according to their occurrences in parent document) [7]. Next, we add directed link for every adjacent word pairs. (See Figure 3). The main aim behind such graph is to maintain the flow of information similar to that in main (parent) document.

1.2 **Calculating Path Length:** we believe that more similar nodes (tightly joined or node pairs having higher importance score), will have less path length. Thus, to calculate the path length we take inverse of edge weight. To calculate the edge weight i.e., $Edge\_Wt(V_i, V_j)$, we apply same procedure as used in Eq-9. We use this path length in calculation of vertex constrained shortest paths. The path length of path from V_i to V_j i.e., $Path\_len(V_i, V_j)$ can be given as:

$$Path\_len(V_i, V_j) = 1/Egde\_Wt(V_i, V_j)$$
(11)

1.3 **Applying Vertex Constrained Shortest Path Based Scheme:** We store all "weighted minimum vertex cover" nodes/vertices of top scored sentence of given sentence cluster as "set of specified vertices". Next, we sort all vertices present in "set of specified vertices" in descending order of the weighted minimum vertex cover score. Finally, we use all pairs of nodes to calculate the vertex constrained shortest paths, which pass through at least 70% of top ranked vertices present in the "set of specified vertices". These identified paths are served as abstract sentences.

1.4 **Ranking Paths:** We calculate the weight of each path (abstract sentence) by adding the score of all "weighted minimum vertex cover", present in the path. Next, we rank all identified paths in descending order of their scores.

1.5 **Sentence correctness**: To maintain the sentence correctness, we remove the abstract sentences from ranked list of sentences, which ends with any stopwords. We also remove abstract sentences whose size is more than 40 words.

2. **End FOR.**
3. Rank all identified sentence clusters in descending order of their weight (See Section 5). Next, select top ranked abstract sentence from each sentence cluster and put them according to the order of related ranked sentence clusters.
4. Finally to get the summary of 'n' words, we select first 'n' words from collection of sentences.

9   Pseudo Code

The pseudo-code for entire scheme is given below:

Input: (1) ASCII text document, (2) Wikipedia extended abstracts

Output: (1) Extract and (2) abstract summary.

Algorithm:

St1. Apply input cleaning and preprocessing for given document. (See Sub-section 4.1).

St2. Identify all bigrams in the given document and calculate their importance (which is actually the product of local and semantic importance of bigrams).)See Sub sections, 4.2, 4.3 and 4.4).

St3. Use group average agglomerative clustering algorithm to identify the sentence clusters in the given document. Next, use the score of bigrams (calculated above) and calculate the weight of all identified sentence clusters. Finally, rank all sentence clusters in descending order of their weight (Section 5).

St4. **Generating extract summary:**
   a.   **For** every Identified sentence cluster do the following:
   b.   Prepare undirected word graph of sentences and identify the weighted minimum vertex cover with their score (Section 6). Next, calculate the weight of every sentence by adding the weighted minimum vertex cover score of words in that sentence. Next, rank the sentences in descending order of their scores.
   c.   **End For**
   d.   Finally, extract top sentence from every identified sentence cluster and order them according to the order of identified sentence cluster. Select top 'n' words and present it as extract summary. (See Section 7 for detail).

St5. **Generate abstract summary:**
   a.   **For** every Identified sentence cluster do the following:
   b.   Use directed word graph of text and vertex constrained shortest path based technique to generate abstract sentences in every identified topic. Use weighted minimum vertex cover related information as vertex constraints. Remove incorrect sentences. Next, calculate the weight of every abstract sentence by adding the weighted minimum vertex cover score of words in that sentence. Next, rank the sentences in descending order of their scores. (See Section 8).
   c.   **End For**
   d.   Finally extract top abstract sentence from every identified sentence cluster and order them according to the order of identified sentence cluster. Select top 'n' words and present it as extract summary. (See Section 8 for detail).

10   Evaluation

We use DUC-2001 and DUC-2002 dataset for evaluation. The Task 1 (i.e. generating 100 words summary) of DUC-2001 and DUC-2002 is considered in entire evaluation process. The details of both dataset are given below:

DUC-2001 Dataset: It contains 309 news articles collected from TREC-9, divided into 30 sets.

DUC-2002 Dataset: It contains 59 document sets with total 567 articles (D008 is excluded from the original 60 document sets by NIST). TREC-9 is the main data source for all articles.

10.1   Systems Used in Evaluation and Evaluation Metrics

Baseline Systems: We compare our devised system's result with best published results of (1) CollabSum [14], (2) DUC-2002's best performing System (As, it performs better than some current baseline systems), (3) Text Rank [12] and (4) "Unified approach to Simultaneous Single-Document and Multi-Document Summarization" [13]. The details are given below:

(1) *CollabSum [14]*: For given cluster of documents, it designed three summarization methods based on the use of cross-document relationships between sentences in the cluster of document.

- *Uniform Link*: The method computes the informativeness score of a sentence based only on the cross-document relationships between sentences.

- *InterLink*: The method computes the informativeness score of a sentence based only on the cross-document relationships between sentences, i.e.

- *UnionLink*: The method computes the informativeness scores of sentence based on: (1) cross-document relationships and (2) within-document relationships between sentences respectively, and then combine them to get the final informative score.

- *IntraLink*: The method computes the informativeness score of a sentence based only within-document relationships between sentences.

  NOTE: among all four implementations UnionLink with Gold Clustering scheme shows the best performance with DUC-2001 dataset and UniformLink with gold clustering scheme shows the best performance with DUC-2002 dataset. We have taken top two scores related to both i.e. DUC-2001 and DUC-2002 in comparison. Here Gold clustering means manual categorization of documents and Average link is similar to group average agglomerative clustering algorithm.

(2) *DUC-2002's best performing System*: We also used DUC-2002's best performing system (see Table 1), represented as System-28 and System-21 (See Table 2).

(3) *Text Rank [12]*: We re-implemented the text rank algorithm (based on the description given in paper) and generated 100 words summary on DUC-01 and 02 dataset.

(4) We use the published results of *"Unified approach" [13]*, on DUC-2001 and DUC-2002 single document summarization.

Description of Our System: to properly evaluate the techniques applied in our devised system, we used three different systems.

1) *Our System (Extract summary)*: This system uses top extracted sentences, trimmed to achieve the required number of words as summary (Section 7, 9).

2) *Our System (Abstract summary)*: This system uses vertex constrained shortest path based scheme to generate abstract summary (see Section 8 and pseudo-code at Section 9). The resultant summary is trimmed to achieve the required number of words.

3) *Our System (MVC)*: This system uses "minimum vertex cover based scheme" (See Sub-Section-1.1), instead of using weighted minimum vertex cover (see Section

6). The main aim of this setup is to test the impact of weighted minimum vertex cover in entire result. We use node-degree to calculate the score of nodes selected as minimum vertex cover. Due to existence of more than one set of nodes, we use average score of three different readings in evaluation.

Evaluation Metric: We use ROUGE toolkit (version 1.5.5) to measure the summarization performance. To properly evaluate the summary we use ROUGE-1, ROUGE-2 and ROUGE-W based measures. The rest of the details and package is available at [10].

10.2   Analysis of Result

In this experiment we compare our devised system's result with results of baseline systems discussed in previous Sub-section. Results are given in Table 1 and in Table 2. Bold font is used to represent the highest scores in each table. From results it is clear that our system performs better than baseline methods. The results, related to baseline systems in both tables are presented in sorted form (decreasing order).

Table 1. Comparison results for single document summarization on DUC-2001

SYSTEM	ROUGE-1	ROUGE-2	ROUGE-W
Our System (Extract summary)	0.46331	0.17891	0.14994
Our System (Abstract summary)	**0.49152**	**0.18461**	**0.15445**
Our System (MVC)	0.41698	0.16102	0.13495
Unified approach	0.45377	0.17649	0.14328
CollabSum (Union Link, Gold)	0.44038	0.16229	0.13678
CollabSum (Union Link, AverageLink)	0.43950	0.16108	0.13679
Text Rank	0.43407	0.15696	0.13629

Table 2. Comparison results for single document summarization on DUC-2002

SYSTEM	ROUGE-1	ROUGE-2	ROUGE-W
Our System (Extract summary)	0.49037	0.22764	0.17238
Our System (Abstract summary)	**0.52002**	**0.23402**	**0.17756**
Our system (MVC)	0.44378	0.20601	0.15600
Unified approach	0.48478	0.21462	0.16877
System 28	0.48049	0.22832	0.17073
System 21	0.47754	0.22273	0.16814
CollabSum (UniformLink, Gold)	0.47187	0.20102	0.16318
CollabSum (Union Link, Gold)	0.47028	0.20046	0.16260
Text Rank	0.46261	0.19457	0.16018

Our devised systems i.e., (1) Our System (Extract summary) and (2) Our System (Abstract summary) show better performance with DUC-2001 dataset, i.e. it shows comparatively higher score with respect to the baseline systems.

With DUC-2002 dataset, our devised systems i.e., (1) Our System (Extract summary) and (2) Our System (Abstract summary) again performs better then baseline systems. However ROUGE-2 score for "Our System (Extract summary)" is slightly lower than System 28 of DUC-2002. Finally results, show that our devised system performs better then baseline systems of this area (except one case).

Correctness and Quality of Abstract Summary: To check the grammaticality of sentences, we use numeric scoring scheme which ranges from 0 to 5 (i.e., 0, 1, 2, 3, 4, 5). Here, score zero (0) means grammatically wrong (no hope at all) and five (5) means grammatically perfect. We use the average score of three different observations (by three different people) as correctness measure. The final grammaticality score was 3.68 on the scale of 0-to-5. This score is empirically comparable/"slightly better" than the grammaticality score produced by [7], on Google news dataset (which was 1.44 on scoring scheme ranges from 0 to 2 with English language). The main reason may be the use of information content of extract summary as prior knowledge in addition to the use of directed word graph of text and some simple correctness schemes.

As, from score given in Table 1 and Table 2, it is clear that "Our System (Abstract summary)" performs better than extract summary i.e., Our System (Extract summary). This also serves our purpose of development of abstract summarization technique.

Effect of Weighted Minimum Vertex Cover in Result: To test the effect of weighted minimum vertex cover, we prepare the same system, by using traditional minimum vertex cover (i.e., Our System (MVC)). The comparison results are given in Table 1 and Table 2. From differences in results w.r.t. "Our System (Extract summary)", it is clear that system get remarkable improvements after applying weighted minimum vertex cover.

11   Conclusion and Future Scope

This paper presents an unsupervised, extract and abstract single document summarization technique, which uses limited linguistic support (i.e. limited to include only stopwords, stemmers and punctuation marks). To achieve this goal, we introduce three different techniques:

➤ We introduce the word graph of text populated with semantic and statistical importance of words in document. The main aim of this scheme is to reduce the effect of noisy and less important terms in document and utilizing the importance or terms in calculation process. This scheme can be extended in other text graph framework, which suffers from the effect of noisy or less important terms.

➤ Calculation of minimum vertex cover is actually an optimization problem. Here we use local and semantic information (as discussed above) in calculation of weighted minimum vertex cover, which represents the most representative terms in document. This technique can be extended in topic detection etc.

➢ To calculate abstract summary, we introduce the use of vertex constrained shortest path based technique. This technique can be utilized in question answering task to combine the scattered answer fragments into a single sentence etc.

Thus techniques used in our devised system, not only useful for extract and abstract single document summarization, but also have some additional future scopes.

References

1. Barrera, A., Verma, R.: Automated Extractive Single-document Summarization: Beating the Baselines with a New Approach. In: SAC 2011, pp. 268–269 (2011)
2. Barrera, A., Verma, R.: Combining syntax and semantics for automatic extractive single-document summarization. In: Gelbukh, A. (ed.) CICLing 2012, Part II. LNCS, vol. 7182, pp. 366–377. Springer, Heidelberg (2012)
3. Bekkerman, R., Allan, J.: Using Bigrams in Text Categorization. CIIR Technical Report IR-408 (2004)
4. Cai, S., Su, K., Sattar, A.: Local search with edge weighting and configuration checking heuristics for minimum vertex cover. Artif. Intell. 175(9-10), 1672–1696 (2011)
5. Cai, S., Su, K., Sattar, A.: Two New Local Search Strategies for Minimum Vertex Cover. In: AAAI 2012 (2012)
6. Lloret, E., Palomar, M.: Analyzing the Use ofWord Graphs for Abstractive Text Summarization. In: IMMM 2011 (2011)
7. Filippova, K.: Multi-Sentence Compression: Finding Shortest Paths in Word Graphs. In: COLING 2010, pp. 322–330 (2010)
8. Kumar, N., Srinathan, K.: Automatic keyphrase extraction from scientific documents using N-gram filtration technique. In: ACM DocEng 2008, pp. 199–208 (2008)
9. Kumar, N., Srinathan, K., Varma, V.: Using wikipedia anchor text and weighted clustering coefficient to enhance the traditional multi-document summarization. In: Gelbukh, A. (ed.) CICLing 2012, Part II. LNCS, vol. 7182, pp. 390–401. Springer, Heidelberg (2012)
10. Lin, C.: Rouge: a package for automatic evaluation of summaries. In: Proceedings of Workshop on Text Summarization Branches Out, Post-conference Workshop of ACL 2004, Barcelona, Spain (2004)
11. Mcdonald, D.M., Chen, H.: Summary in context: searching versus browsing. ACM Transactions on Information Systems 24(1), 111–141 (2006)
12. Mihalcea, R., Tarau, P.: Textrank: bringing order into texts. In: Proceedings of the 2004 Conference on Empirical Methods in Natural Language Processing (2004)
13. Wan, X.: Towards a Unified Approach to Simultaneous Single-document and Multi-document Summarizations. In: COLING 2010, pp. 1137–1145 (2010)
14. Wan, X., Yang, J.: Collabsum: exploiting multiple documents clustering for collaborative single document summarizations. In: Proc. of SIGIR 2007, Amsterdam, The Netherlands, pp. 143–150 (2007)
15. Tsatsaronis, G., Varlamis, I., Nørvåg, K.: SemanticRank: Ranking Keywords and Sentences Using Semantic Graphs. In: COLING 2010, pp. 1074–1082 (2010)
16. Kumar, N., Srinathan, K., Varma, V.: Using graph based mapping of co-occurring words and closeness centrality score for summarization evaluation. In: Gelbukh, A. (ed.) CICLing 2012, Part II. LNCS, vol. 7182, pp. 353–365. Springer, Heidelberg (2012)

Hierarchical Clustering in Improving Microblog Stream Summarization

Andrei Olariu

University of Bucharest, Faculty of Mathematics and Computer Science
14 Academiei, RO-010014, Bucharest, Romania
andrei@olariu.org

Abstract. Microblogging has shown a massive increase in use over the past couple of years. According to recent statistics, Twitter (the most popular microblogging platform) has over 500 million posts per day. In order to help users manage this information overload or to assess the full information potential of microblogging streams, a few summarization algorithms have been proposed. However, they are designed to work on a stream of posts filtered on a particular keyword, whereas most streams suffer from noise or have posts referring to more than one topic. Because of this, the generated summary is incomplete and even meaningless. We approach the problem of summarizing a stream and propose adding a layer of text clustering before the summarizing step. We first identify the events users are talking about in the stream, we group posts by event and then we continue by clustering each group hierarchically. We show how, by generating an agglomerative hierarchical cluster tree based on the posts and applying a summarization algorithm, the quality of the summary improves.

Keywords: Microblog, Summarization, Text Clustering, Event Detection.

1 Introduction

Microblogging is a form of blogging characterized by very short posts, sometimes followed by a link to an article, photo or video. The term became popular in 2006-2007 to describe social networking platforms like Twitter and Tumblr.

During the past couple of years, microblogging experienced a sharp growth. In October 2012, Twitter had over 500 million posts per day (compared to 90 million posts per day in September 2010). Facebook has over one billion active users, while Tumblr has 70 million posts per day.

Among microblogging platforms, Twitter received most of the scientific research focus because of its publicly accessible posts (whereas on Facebook, for example, most posts are private). Tweets (Twitter posts) are limited to 140 characters, greatly influencing the writing style of the users. Abbreviations, internet slang and misspelled words are common. The writing style is mostly colloquial. Since most NLP techniques were developed for long, structured text, tweets are difficult to process and care must be taken to ensure proper normalization [8,10].

A. Gelbukh (Ed.): CICLing 2013, Part II, LNCS 7817, pp. 424–435, 2013.

Summarizing microblogging streams (a stream being a sequence of posts) has been a popular research topic in the past couple of years. For example, algorithms were proposed for detecting the highlights of a sporting event based on tweets [11,4,16]. There are also web applications that help users fight information overload by summarizing streams and extracting the most important news. For example, services like Summify[1] (now integrated into Twitter) and Prismatic[2] find the top most shared links, relative to a user's social network or interests.

In this paper, we tackle the problem of summarizing any type of streams, not just the ones filtered on certain criteria. The contribution of this paper is combining event detection, text clustering and text summarization in order to develop an algorithm for summarizing microblogging streams. This is the first summarization algorithm, to the best of our knowledge, that does not require any restrictions or prior information regarding the analyzed stream. Summarizing structured events has been approached before [4], but only for events matching a predefined format. Our approach can handle any type of event, being completely unsupervised. Work in progress has been published in [13].

We evaluate our solution on a corpus of Twitter posts. We show that the output is grammatically reasonable and informationally comprehensive. For comparison, we also apply summarization algorithms without event detection or clustering.

With the help of our system, users can easily skim over large volumes of messages and understand the important events as relative to the input stream, while also having the possibility of focusing on a certain event and understanding its different sides. The generated summaries improve as the user navigates deeper down the cluster tree, as can be seen in the results section.

The paper is organized as follows. Section 2 outlines previous research related to this paper. Section 3 presents the problem formulation and an overview of the proposed solution. Sections 4 and 5 describe in detail the components for clustering and summarizing streams. Experiments and results are presented in Section 6. Section 7 formulates a conclusion based on our results and proposes a series of ideas to improve and develop this technique.

2   Related Work

2.1   Microblog Event Detection

Event detection based on microblogging streams is the problem of analyzing a stream of posts and detecting the most important events mentioned in the stream. The extracted events are usually represented as groups of relevant keywords.

O'Connor et al. [12] are the first to propose an algorithm for detecting the main topics in a given stream. Their exploratory search application provides users with a set of subtopics for the given query, along with some example tweets.

---

[1] http://summify.com/
[2] http://getprismatic.com/

In [7], the authors propose an algorithm (called ETree) for hierarchically modeling events. ETree first clusters similar messages into information blocks. The event structure tree is then built by applying a hierarchical incremental modeling technique. Finally, causal relationships between information blocks are discovered by analyzing their temporal characteristics.

Nichols et al. [11] detect sharp increases in word frequencies (or spikes) in order to determine the highlights of football matches. Weng et al. [17] propose an algorithm that applies Fourier theory in analyzing spikes. They handle words as signals, thus enabling the clustering module to use a signal cross-correlation measure.

A very good framework for event detection and analysis, with emphasis on scalability, is presented in [18].

A similar problem to event detection is trending topic detection, approached in [2,3,9,1].

2.2 Multi-sentence Compression

Multi-sentence compression is the task of summarizing the most salient themes from a group of similar sentences into a single sentence. We highlight the approach by Filippova [5]. This algorithm does not require building a syntactic parse tree or applying a predefined set of language rules. It only needs a tokenizer and a POS tagger for the language the posts are in. We will present it in Section 5.1.

A similar approach is used in generating abstractive summaries of highly redundant opinions [6].

2.3 Summarization Applied to Twitter Streams

There are two main approaches to Twitter summarization: generating a short sentence based on the stream or selecting the post that seems to best describe the input stream.

The first approach is represented by the „Phrase Reinforcement" algorithm [14], for summarizing a stream filtered by a specific keyword. We will present it in Section 5.2.

The second approach is based on finding the post that best describes the input stream. This has been interpreted as finding the post that minimizes the average or maximal distance to all other posts in the corpus. Takamura et al. [16] reduce this to an optimization problem (facility location problem) and propose an exact algorithm and a faster, approximate algorithm. Chakrabarti et al. [4] build a modified Hidden Markov Model capable of splitting the corpus into events, each event being represented by a set of relevant messages. Both techniques were applied in summarizing tweets related to sporting events.

3   Problem Formulation and Approach Outline

The problem tackled by this paper is generating a set of sentences that best summarizes a stream. Unlike previous research, the input stream is not restricted to a given topic, nor filtered based on a given keyword.

The algorithm we propose is composed out of two modules. The first one (described in detail in Section 4) detects the important events in the corpus and tags posts as referring to one of the events or not. This module employs hierarchical clustering for decomposing events, helping in generating more detailed summaries.

The second module (Section 5) receives all of the posts referring to an event and attempts to generate a summary. For this module, we tested two different approaches in summary generation in order to have a better, less biased understanding on the improvement of our solution over the original summarization algorithms.

4   Hierarchical Event Analysis

4.1   Event Detection

As for other algorithms working on external data, the first step is preprocessing. In our case, considering the input data is a stream of Twitter posts, we replace all URLs and Twitter user names with the placeholder keywords URL and TWID. In the case of hashtags, we remove this special character and treat the hashtag as a normal word. We also remove stopwords and apply a stemming function to the remaining words.

Our approach on event detection is based on discovering words that show an unusual increase in frequency in the current window, in comparison to a background, default frequency. The goal is to help maximize the results of the summarization module and, at the same time, maintain the time and memory efficiency.

We compute the background frequency by using tweets from previous windows. For example, we may want to discover events in the last 24 hours. For this, we use all tweets in this 24 hour window and compare the word frequencies to those from one or more previous windows, keeping only the words that experience a noticeable increase (above a certain threshold). By using previous posts (instead of a fixed dataset) in finding spiking words, we reduce region-specific or topic-specific noise. The disadvantage of this approach may be a failure to detect long-term events that are mentioned in both the current and the background windows.

After applying this filter, the size of the dictionary of words to process shrinks to just a few tens, speeding up the following steps. Iterating through the window again, we compute the correlations between the remaining words, defined as:

$$corr(w_1, w_2) = \frac{\sum_{s \in S} 1_s(w_1) \times 1_s(w_2)}{\sqrt{tf(w_1) \times tf(w_2)}},$$

where w_1 and w_2 are words, S is the set of sentences in the current window, $tf(w)$ is the term frequency for a word w over the current window and $1_s(w)$ is the indicator function for word w on the set of words defined by the sentence s. The numerator computes the number of sentences that mention both of the given words. The denominator is used to normalize the word frequency.

The correlations are used in clustering keywords, such that the average in-cluster correlation between words is higher that a given threshold. A high threshold has the risk of generating two or more clusters referring to the same event. A small threshold might cluster together words that are just loosely correlated, thus increasing the amount of noise. Each group of keywords represents an event mentioned by a significant number of tweets from the input stream.

4.2   Message Clustering

Having determined events represented by groups of keywords, we assign tweets to one of them using the following similarity function:

$$simm(i,j) = \frac{\sum_{w \in (s_i \cap K_j)} \frac{1}{ln(tf(w))}}{\sqrt{1 + |K_j \setminus s_i|}},$$

where i is the index for a sentence, j is the index for a cluster of keywords, s_i is the sentence (represented as a set of words) having index i and K_j is the set of keywords having index j. This score gives a higher importance to less frequent keywords (in the nominator) and it also penalizes sentences that don't mention all of a cluster's keywords (in the denominator). The tweet is assigned to the cluster given by:

$$C(i) = \begin{cases} argmax_{j=1,|K|} \, simm(i,j), & if \, simm(i,j) > th \\ 0, & otherwise \end{cases},$$

where th is a threshold and 0 (in the "otherwise" branch of the function) is the index for noise. By clustering the messages based on events and removing noise, we provide the summarization component with a higher quality input, which is essential in generating relevant summaries.

4.3   Hierarchical Event Analysis

Having determined clusters of messages, each cluster relevant to one event, we would like to further improve our analysis of an event by discovering its different facets, as expressed by the messages assigned to it.

The hierarchical event analysis module receives as input a group of messages and outputs a tree structure, the result of an agglomerative clustering applied to information blocks (name introduced by [7]). An information block is a group of very similar messages, and by their high similarity we expect each of them to roughly express the same information. We use the cosine similarity based on n-grams over a sentence's words, with n going from 1 to 4.

We discard information blocks that consist of a small amount of messages. This helps in removing noise and speeding up the rest of the algorithm. The remaining information blocks highlight the important aspects of an event, relative to the microblogging community.

Finally, we apply an agglomerative clustering algorithm to the information blocks. An information block is represented as a single document (by merging its corresponding messages). At each step, the two information blocks with the highest cosine similarity are combined.

5   Summarization

5.1   Multi-sentence Compression

For summarizing clusters of tweets, we use two different approaches. The first one is a multi-sentence compression algorithm [5], referred to as MSC in the rest of this paper. We use this algorithm because it is unsupervised and it does not require syntactic parsing or a predefined set of language rules. Instead, it requires a tokenizer, a part of speech tagger and a stopwords list. Therefore, it is flexible regarding the input data, being easily extended to other languages. The disadvantage is that it is not able to work with large datasets or with datasets having a lot of unrelated sentences (as can be seen in the results section).

The algorithm builds a word graph from the input sentences. A word graph is a directed graph having words as nodes. An edge from word A to word B represents an adjacency relation. Two additional nodes, marking the start and end of sentences, are also used. The graph is constructed by iteratively adding sentences to it. A word from the current sentence is mapped onto a node in the graph if they have the same word form and the same part of speech tag and if no word from this sentence has already been mapped to this node.

After constructing the word graph, the algorithm searches for paths in the graph that maximize a weighting function. The weight for an edge is given by:

$$weight(e_{w_1,w_2}) = \frac{tf(w_1) + tf(w_2)}{tf(w_1) \times tf(w_2) \times \sum_{s \in S} d(s, w_1, w_2)^{-1}},$$

where e_{w_1,w_2} is the edge between the (already mapped) words w_1 and w_2 and $d(s, w_1, w_2)$ refers to the distance between the offset positions $(p(s, w))$ of the two words in the sentence s:

$$d(s, w_1, w_2) = \begin{cases} p(s, w_2) - p(s, w_1), & if \ p(s, w_1) < p(s, w_2) \\ 0, & otherwise \end{cases}$$

The phrase chosen as summary is given by the path between the start and end nodes having the smallest average edge weight and also satisfying a minimum length requirement. Other restrictions can also be applied in order to filter out ungrammatical phrases, for example checking if the phrase contains a verb.

5.2   Frequent Phrase Summarization

The second summarization approach is based on the Phrase Reinforcement algorithm [14]. Phrase Reinforcement was proposed for summarizing a collection of microblogging posts starting from one or several words. The phrase that maximizes a scoring function in selected as summary.

The algorithm builds a graph of words, having the topic as the root node. Each input sentence is converted to a word chain and added to the graph, updating counts for phrases that appear more than once.

The fact that the original Phrase Reinforcement algorithm starts from a specific phrase is limiting its applicability. For our current problem, we generalize this approach in order to find the most frequent phrase, regardless of a given topic. We use the sentences to build a prefix tree, where each edge is a word and each node is the sequence of words from the root to this node. We then search the tree for the phrase that maximizes a score computed using the length of the phrase and the log of the phrase's frequency. This algorithm will be referred to as FPS (Frequent Phrase Summarization).

6   Experiments and Results

6.1   Data Set

We used the Twitter Search API to build our corpus. We filtered the search on English posts, in order to reduce the amount of noise. The algorithm can be applied, however, to posts in any language (as long as a tokenizer, a POS tagger and a stemmer are available). We also restricted the search to tweets geolocated in the New York City area. We added this restriction in order to simplify interpreting the results and we picked New York City because of the popularity Twitter has in that area.

Our experiments are performed on 1.6 million tweets collected between the 4th and the 8th of July 2012. We also used 1.7 million tweets, collected during the previous week, as background corpus for the event detection module.

6.2   Events Detected in the Data Set

The algorithm detected, on average, 20 events per day. We divided the events into two categories: real (33% of the events were marked as real) and virtual (67%). Real events are those that take place outside of the virtual space. Most of the events detected in this category were sporting events, related to wrestling, basketball (Steve Nash's transfer to the Lakers), football (rumors of Robin van Persie leaving Arsenal London), baseball or tennis (Serena Williams winning Wimbledon).

The event that generated the highest number of messages was Independence Day. Since this holiday has several popular names, our algorithm generated four different clusters related to it on the 4th of July and another one on the 5th. Other detected events were about celebrities (Ringo Star's birthday, Andy Griffith's death), finding the Higgs boson, the European debt crisis, the commemoration of the 7 July 2005 London bombings, a new movie, a concert and even an eating contest.

We created a category for internet-specific (or virtual) events. Here we included memes, words or short phrases that are very popular, especially on Twitter. They are difficult to summarize, due to the diversity of ideas and variations. For example, a popular meme is "thingsidislike", which invites users to share something they dislike.

Another type of virtual event is the one generated by retweets. For example, the pop singer Lady Gaga has 27 million followers on Twitter. Each one of her messages gets retweeted thousands of times. In the future, retweets might be skipped when detecting events.

After determining the events, the tweets were assigned to one of them or were marked as noise, if no event was a fit. For our data, 2.9% of the 1.6 million tweets were assigned to events as detected by the event detection module. The other 97% were marked as noise. This is consistent with a 2009 study[3] on a sample of 2000 tweets, out of which 40% were classified as pointless, 37% as conversational and only 3.6% as news.

The thresholds mentioned in Sections 4.2 and 4.3 were determined empirically. We experimented with different settings until we were satisfied by the amount of noise and the size and quality of the event clusters.

6.3   Evaluation Metrics for Summaries

The final step of the algorithm is generating the summary. We generated summaries using both algorithms presented in Section 5 (MSC and FPS) and we evaluated each summary using two metrics: completeness and grammaticality, both graded on a scale of 1 to 5. A rating above 3 is considered good, while a rating below 3 is poor. Completeness measures how much information the summary expresses relative to the detected event and to the information available in the messages, where 1 means it provides no information, while 5 means the summary expresses the most important ideas for that group of messages. Regarding grammaticality, a rating of 1 means it lacks a grammatical structure or it has a lot of errors, while 5 means it is grammatically correct, with no errors.

For evaluating the improvements made by hierarchical summarization over simply summarizing the whole cluster, we cut the cluster tree to the level where it has four clusters and we provide the volunteers with the four sentences generated by summarizing the clusters. If the cluster tree was generated with less than four information blocks, then we didn't consider it in computing the results. This is the case with clusters generated by retweets, where the summary matches the popular post. Being of very low complexity, such clusters are not relevant in highlighting the characteristics of our approach and are likely to bias the results. After removing these clusters, we were left with 50 events.

Besides completeness and grammaticality, we also asked our volunteers to rate the level of non-redundancy. A rating of 1 means the four summaries repeat the

---
[3] http://www.pearanalytics.com/blog/wp-content/uploads/2010/05/
Twitter-Study-August-2009.pdf

same information, whereas a rating of 5 means the summaries present different aspects of the event in question.

We had a total of 4 volunteers that rated all of the 50 events. They were given access to a web interface. Each event was represented on a separate web page as a set of four summaries: two single-sentence summaries (one for each summarizing approach - MSC and FPS) and two four-sentence summaries (again, one generated by MSC and the other by FPS). The volunteers had to rate the summaries based on completeness, grammaticality and, for the four-sentence summaries, non-redundancy. The web page also showed the tweets associated with the current event, so that the volunteers would be able to better understand the sometimes cryptic Twitter hashtags and also assess the completeness rating of a summary.

6.4   Evaluation Results

The results of the evaluation process are presented in Table 1. We checked inter-rater reliability using the intra-class correlation coefficient ($ICC(3, k)$ as presented in [15]). We are interested in the extent to which the volunteers behave similarly on rating each summary, in order to assess their consistency. An ICC value of 1 represents perfect agreement, while 0 means there is no agreement at all. The ICC value computed using our ratings was 0.9788, indicating very good agreement.

The average completeness ratings for single-sentence summaries were 3.05 (for MSC) and 3.28 (for FPS). When considering hierarchical clustering and generating four-sentence summaries, the completeness ratings increased to 4.28 (MSC) and 4.11 (FPS). We can notice substantial improvements in both approaches.

Regarding grammaticality, the scores for single-sentence summaries are good (4.05 for MSC and 4.00 for FPS). When increasing the summary size to four sentences, the MSC's ratings show a negligible drop (to 4.00), while the FPS ratings drop by 15%, to 3.61. This decrease in grammaticality was noticed mostly on events that contain a popular message. FPS selects that message as a single-sentence summary. When generating four-sentence summaries, the popular message dominates one of the clusters. The other three clusters show high variety, leading to low grammaticality ratings.

Non-redundancy ratings are high: 4.01 for MSC and 3.82 for FPS. The volunteers considered that the amount of duplicate information in the four-sentence summaries was low, which is to be expected, considering the hierarchical clustering step performed before applying the summarizing algorithms.

6.5   Analysis of the Summarizing Algorithms

We noticed a few types of common summarization errors. Regarding MSC, we noticed it behaves poorly when confronted with very different sentences. Common bigrams (event independent) are selected over the event-specific bigrams, thus generating a meaningless summary. This effect is most obvious when

Table 1. The results of the evaluation process

Rated feature	Summary size	Average rating (standard deviation)	Improvement
MSC completeness	One sentence	3.05 (1.03)	40.3%
	Four sentences	4.28 (0.85)	
FPS completeness	One sentence	3.28 (0.99)	25.3%
	Four sentences	4.11 (0.86)	
MSC grammaticality	One sentence	4.05 (1.21)	-1.2%
	Four sentences	4.00 (1.00)	
FPS grammaticality	One sentence	4.25 (1.10)	-15.0%
	Four sentences	3.61 (1.10)	
MSC non-redundancy	Four sentences	4.01 (1.14)	-
FPS non-redundancy	Four sentences	3.82 (1.16)	-

summarizing all the posts for a day. For example, given all the posts on the 4th of July, the output of MSC is "rt TWID you to the TWID URL".

Meanwhile, when confronted with a set of different sentences, FPS will pick a long and frequent phrase. We tried summarizing all the messages from a day. Since retweeting is one of the core actions on Twitter, it is not surprising that the summaries generated for the 4 days were all popular posts.

When it comes to hierarchical message clustering, followed by summarization, we noticed that, the deeper we move down the hierarchical cluster tree, the better the summary becomes. This is due to the increasing similarity of the sentences that are being summarized. For example, the cluster of messages regarding rumors of Robin van Persie (football player) leaving Arsenal London was summarized by MSC as "rt TWID rvp is not acceptable arsenal". This sentence does not make sense. Meanwhile, FPS selects a popular post as the summary: "rt TWID wenger/gazidis have to speak out today i want answers losing our best player & captain 2 seasons running is not acceptable arsenal". This summary captures key facts regarding the event, but it is still far from complete.

By agglomeratively clustering the posts, cutting the tree and then summarizing, we get the following set (both FPS and MSC had the same output):

- rt TWID "unfortunately in this meeting it has again become clear to me that we in many aspects disagree on the way arsenal should move forward." rvp
- rt TWID look at this TWID statement URL what the hell did wenger/gazidis say to drive him away? arsenal
- rt TWID wenger/gazidis have to speak out today i want answers losing our best player & captain 2 seasons running is not acceptable arsenal
- rt TWID gazidis is "on a 2-week holiday in america" really? get your sorry a** back to london and sort this fiasco out arsenal TWID

This summary manages to express the key aspects of the current event, having a higher completeness rating than the single-sentence summary, while showing very low redundancy.

7   Conclusion and Future Work

The task of summarizing a stream of microblogging messages referring to a certain event has received significant interest in the past couple of years. Yet generalizing summarization algorithms to nonspecific streams has not been tackled before. In this paper, we presented an approach for hierarchically summarizing streams, without requiring any previous information or any specific properties for the streams.

We applied two summarizing algorithms (introduced by [5,14]) to a generic Twitter stream and we showed that the results are incomplete and even meaningless. After applying a hierarchical event detection module to the input stream, the quality of the summaries improved. The system split the stream into events and produced single-sentence summaries for each event, being assessed by a group of volunteers as having reasonable completeness and good grammaticality. When generating four-sentence summaries, the completeness was rated as good, with low redundancy and a small decrease in grammaticality.

We believe our system can be applied in processing microblogging streams, helping users fight information overload, understand events and how events are perceived by microblogging communities. The ability to summarize streams hierarchically, delving deeper into complex events or skimming over simple ones, is not employed to its full potential in this paper. We plan to develop a visual interface capable of rendering hierarchical summaries and investigating how large streams can be analyzed by users.

We have not put emphasis on dealing with big data. This is an important aspect, considering the popularity of microblogging platforms. We hope to develop, in the following months, an updated system capable of online summarization. The current system uses batch processing, which is appropriate for historical data. Online processing is essential in real-time analysis and would be recommended in integrating with a visual interface.

References

1. Alvanaki, F., Michel, S., Ramamritham, K., Weikum, G.: See what's enblogue: real-time emergent topic identification in social media. In: Proceedings of the 15th International Conference on Extending Database Technology, EDBT 2012, pp. 336–347. ACM, New York (2012)
2. Benhardus, J.: Streaming trend detection in twitter. Information Retrieval, 1–7 (2010)
3. Cataldi, M., Di Caro, L., Schifanella, C.: Emerging topic detection on twitter based on temporal and social terms evaluation. In: Proceedings of the Tenth International Workshop on Multimedia Data Mining, MDMKDD 2010, pp. 4:1–4:10. ACM, New York (2010)
4. Chakrabarti, D., Punera, K.: Event summarization using tweets. In: Proceedings of the 5th Int'l AAAI Conference on Weblogs and Social Media, ICWSM (2011)
5. Filippova, K.: Multi-sentence compression: finding shortest paths in word graphs. In: Proceedings of the 23rd International Conference on Computational Linguistics, COLING 2010, Stroudsburg, PA, USA, pp. 322–330. Association for Computational Linguistics (2010)

6. Ganesan, K., Zhai, C., Han, J.: Opinosis: a graph-based approach to abstractive summarization of highly redundant opinions. In: Proceedings of the 23rd International Conference on Computational Linguistics, COLING 2010, Stroudsburg, PA, USA, pp. 340–348. Association for Computational Linguistics (2010)
7. Gu, H., Xie, X., Lv, Q., Ruan, Y., Shang, L.: Etree: Effective and efficient event modeling for real-time online social media networks. In: Proceedings of the, IEEE/WIC/ACM International Conferences on Web Intelligence and Intelligent Agent Technology, WI-IAT 2011, vol. 01, pp. 300–307. IEEE Computer Society, Washington, DC (2011)
8. Kaufmann, M., Kalita, J.: Syntactic normalization of Twitter messages. In: Proceedings of the 8th International Conference on Natural Language Processing, ICON 2010. Macmillan India, Chennai (2010)
9. Mathioudakis, M., Koudas, N.: Twittermonitor: trend detection over the twitter stream. In: Proceedings of the International Conference on Management of Data, SIGMOD 2010, pp. 1155–1158. ACM, New York (2010)
10. Mosquera, A., Lloret, E., Moreda, P.: Towards facilitating the accessibility of web 2.0 texts through text normalisation. In: Proceedings of the LREC Workshop: Natural Language Processing for Improving Textual Accessibility (NLP4ITA), Istanbul, Turkey, pp. 9–14 (2012)
11. Nichols, J., Mahmud, J., Drews, C.: Summarizing sporting events using twitter. In: Proceedings of the ACM International Conference on Intelligent User Interfaces, IUI 2012, pp. 189–198. ACM, New York (2012)
12. O'Connor, B., Krieger, M., Ahn, D.: TweetMotif: Exploratory Search and Topic Summarization for Twitter. In: Cohen, W.W., Gosling, S., Cohen, W.W., Gosling, S. (eds.) ICWSM, The AAAI Press (2010)
13. Olariu, A.: Clustering to improve microblog stream summarization. In: Proceedings of the 14th International Symposium on Symbolic and Numeric Algorithms for Scientific Computing (SYNASC) (2012) (to appear)
14. Sharifi, B., Hutton, M.-A., Kalita, J.: Summarizing microblogs automatically. In: Human Language Technologies: The 2010 Annual Conference of the North American Chapter of the Association for Computational Linguistics, HLT 2010, Stroudsburg, PA, USA, pp. 685–688. Association for Computational Linguistics (2010)
15. Shorut, P.E., Fleiss, J.L.: Intraclass correlations: Uses in assessing rater reliability. Psychological Bulletin 86(2), 420–428 (1979)
16. Takamura, H., Yokono, H., Okumura, M.: Summarizing a document stream. In: Clough, P., Foley, C., Gurrin, C., Jones, G.J.F., Kraaij, W., Lee, H., Mudoch, V. (eds.) ECIR 2011. LNCS, vol. 6611, pp. 177–188. Springer, Heidelberg (2011)
17. Weng, J., Yao, Y., Leonardi, E., and Lee, F. Event Detection in Twitter. Tech. rep., HP Labs (2011)
18. Yang, X., Ghoting, A., Ruan, Y., Parthasarathy, S.: A framework for summarizing and analyzing twitter feeds. In: Proceedings of the 18th ACM SIGKDD International Conference on Knowledge Discovery and Data Mining, KDD 2012, pp. 370–378. ACM, New York (2012)

Summary Evaluation:
Together We Stand NPowER-ed

George Giannakopoulos and Vangelis Karkaletsis

NCSR Demokritos
Institute of Informatics and Telecommunications
GR-15310, Aghia Paraskevi, Attiki, Greece
{ggianna,vangelis}@iit.demokritos.gr

Abstract. Summary evaluation has been a distinct domain of research for several years. Human summary evaluation appears to be a high-level cognitive process and, thus, difficult to reproduce. Even though several automatic evaluation methods correlate well to human evaluations over systems, we fail to get equivalent results when judging individual summaries. In this work, we propose the NPowER evaluation method based on machine learning and a set of methods from the family of "n-gram graph"-based summary evaluation methods. First, we show that the combined, optimized use of the evaluation methods outperforms the individual ones. Second, we compare the proposed method to a combination of ROUGE metrics. Third, we study and discuss what can make future evaluation measures better, based on the results of feature selection. We show that we can easily provide per summary evaluations that are far superior to existing performance of evaluation systems and face different measures under a unified view.

1 Introduction

Summarization research becomes a necessity in the overwhelming amount of information of our age. The effort to achieve good summaries through automated Natural Language Processing (NLP) can be significantly boosted if one can automatically determine whether a generated summary is good or not. Based on this need, summarization system evaluation research has progressed as a new domain of focus for researchers.

For several years the evaluation community has relied on evaluation measures born in or derived from related NLP tasks (e.g., ROUGE [1] and the related BLUE measures [2]). However, several studies, as well as the experience on new summarization tasks, have shown the need for better evaluation measures [3, 4, 5]. This requirement for new measures is related to a variety of needs, ranging from better discrimination between acceptable and good (human-performance) systems [6] to multi-lingual summarization evaluation [7]. Furthermore, even though existing methods of automatic evaluation do well when judging whole systems, they perform average when judging individual summaries [8].

In this work, we try to "stand upon the shoulders" of existing metrics, which have been proposed over the years. We study whether it makes sense to combine

A. Gelbukh (Ed.): CICLing 2013, Part II, LNCS 7817, pp. 436–450, 2013.

existing, language-agnostic evaluation measures into a single, combined evaluation via optimization. If such a combination is effective, we examine if there exists a subset of features that are adequate for the task at hand. We also perform experiments trying to emulate different aspects of summary evaluation (responsiveness, Pyramid score) under the same, unified perspective.

The paper is structured as follows. We present an overview of summary evaluation literature (Section 2). We desribe the NPowER evaluation method (Section 3) and perform various analyses to determine good strategies for evaluation methods combinations (Section 4). We then conclude, summarizing the findings of this work (Section 5).

2   Summary Evaluation Overview

Summary evaluation allows us to identify errors and reiterate or reformulate certain aspects of the process to optimality. While this is common ground, the notion of automatic evaluation is not. For some time now, the domain of automatic evaluation of summaries was only superficially addressed, because many of the required summary qualities could not be automatically measured. Therefore, human judges have been widely used to evaluate or cross-check the summarization processes [9, 10, 3]. Below, we overview different evaluation types and methods.

An evaluation process can be either intrinsic or extrinsic (e.g., [9, 11]). Intrinsic evaluation operates on the characteristics of the summary itself, trying for example to capture how many of the ideas expressed in the original sources appear in the output. On the other hand, extrinsic evaluation decides upon the quality of a summary depending of the effectiveness of using the summary in a specific task. An extrinsic evaluation case is when we use summaries, instead of source texts, to answer a query. The evaluation is then based on whether the answer is equivalent to the answer derived from source texts. On the contrary, using a *gold standard* summary, i.e., a human-generated summary viewed as the perfect output, and estimating the similarity of the summary to the gold standard, is an intrinsic evaluation case (e.g., [12]).

Sparck-Jones argues [13] that the classification of evaluation methods as intrinsic and extrinsic is not enough and proposes an alternative schema of evaluation methods' classification. This schema is based on the degree to which the evaluation method measures performance, according to the intended purpose of the summary. Therefore, defining new classes that elaborate on the definitions of extrinsic and intrinsic, Sparck Jones classifies evaluation methodologies as: semi-purpose, e.g., inspection of proper English; quasi-purpose, based on comparison with models, e.g., n-gram or information nuggets; pseudo-purpose, based on the simulation of task contexts, e.g., action scenarios; full-purpose, based on summary operation in actual context, e.g., report writing.

In [14] we find a comment (part 3.4) referring to intrinsic evaluation, where the authors suggest that 'only humans can reliably assess the readability and coherence of texts'. This statement indicates the difficulty of that kind of evaluation. But do humans perform perfect in the evaluation of summaries? And what does *perfect* account for?

Humans tend to be able to identify good texts, in a qualitative manner. There is an issue of how to make human assessors grade the quality of a text in uniform and objective ways (see for instance [11, 12] for indications of the problem). At this point numerous efforts have pointed out the inter-judge agreement problem [15, 16, 17, 18]. People tend to have similar, but surely not too similar opinions. This led to looking for subjective measures correlated to human subjectivity. In other words, if our measures behave similarly to human evaluation, we will have reached an adequate level of acceptance for our (automatic) quality measures. In [12] partial inter-judge agreement is illustrated among humans, but it is also supported that, despite the above, human judgements generally tend to bring similar results. Thus, perfection is subjective in the summarization domain: we can only identify good enough summaries for a significant percentage of human assessors.

Pyramid evaluation [15] uses humans to evaluate summaries in a controlled process. The humans are called to identify the segments of the original text, from which pieces of the judged summary are semantically derived. In other words, the method makes use of a supposed (and argued) mapping between summary sentences and source documents, where summarization content units (SCUs) are identified. SCUs are minimal units of informative ability that also appear in the summary output. According to the number of human judges agreeing on the origin of an SCU (i.e., the text span that corresponds to the SCU), the SCUs are assigned weights, corresponding to pyramid layers. Thus, the SCUs higher in the pyramid are supposed to be the most salient pieces of information in the original sources. A summary is then evaluated by locating the SCUs present in the summary output and using a summing function to account for the weights. Doing so, two measures are defined: the pyramid score, which corresponds to precision, and the modified pyramid score, which corresponds to recall. Nenkova argues that the above evaluation process can suppress human disagreement and render useful results. Pyramid evaluation was also applied in DUC and TAC, and the use of a new set of directives for evaluators in DUC 2006 provided better results than DUC 2005 [19], though not reaching the effectiveness of automatic methods. This indicates that manual evaluation methods can be highly dependent on the instructions given to the evaluators.

A number of different intermediate representations of summaries' information have been introduced in existing summarization evaluation literature, ranging from automatically extracted s nippets to human-decided sub-sentential portions of text. These representations form the basis for the comparison between summaries. More specifically, the "family" of BE/ROUGE[1] [21, 1] evaluation frameworks, uses statistical measures of similarity, based on n-grams (of words), although it supports different kinds of analysis, ranging from n-gram to semantic [21]. The intuition behind the BE/ROUGE family is that, in order two texts to have similar meaning, they must also share similar words or phrases. One can take into account simple unigrams (single words) in the similarity comparison, or may require larger sets of words to be shared between compared texts. *Basic Elements* (BE) are

---

[1] See also [20] for the BLEU method on machine translation.

considered to be 'the head of a major syntactic constituent' and its relation to a single dependent. BEs are decided upon in many ways, including syntactic parsing and the use of cutting rules [21]. BEs can be matched by simple string matching, or by semantic generalization and matching, according to the proposed framework [3, 22]. A more recent work [23] uses variations on dependecies and external information (e.g., WordNet) to overcome the problems that arise from different formulations of model summaries.

An alternative to the aforementioned representations is that of the n-gram graphs [24], where mostly n-grams of characters are used to represent documents. Given a set of "gold standard" texts and their n-gram graphs, the similarity (Value Similarity [24]) between the graph of a judged summary ("peer" summary) and the "gold standard" graphs is used as a grade. This approach has offered two main variations:

- the AutoSummENG [24] original approach. This approach calculates the average of the similarities between the peer summary graph and the gold standard graphs. This average is the grade assigned as a score to the peer summary.
- the MeMoG (Merged Model Graph) variation [25]. In this case, the gold standard graphs are merged into a representative graph. Then, the score assigned to the peer is the similarity between the peer summary graph and the representative graph.

Other variations, based on the notion of the n-gram graph, are the Hierarchical Proximity Graphs (HPG) [25] (using a hierarchy of recursive n-gram graphs) and context chains [26] (n-gram graphs based on co-reference chains).

The most recently faced problems of automatic evaluation relate to:

- the ability of evaluation measures to take into account redundancy over subsequent summaries on the same topic. This task ("update" task in TAC) gave birth to measures like Nouveau-ROUGE [27], that take into account previous summaries to measure redundancy.
- the power of evaluation measures to distinguish consistently between "good" summarizers (usually human) and "bad" or "mediocre" summarizers (usually automatic methods), even across corpora [28]. Rankel et al. [6] use a variety of statistical features from the texts to create a regression-based prediction model that can assign a grade to a given summarization system.
- the lack of completely unsupervised methods (without gold standard summaries) for the evaluation of peer summaries. These methods solving this problem [29, 30] rely on statistical analysis of the content of summaries (term distribution), as well as the source documents to determine the quality of a summary.
- the lack of complementary evaluation measures, that can provide information about different aspects of summary quality (e.g., see [8, 31]).

For an overview of recent summarization evaluation efforts, please also consult [32, Section 5].

In this work, given the numerous efforts on summary evaluation, we study and provide an answer to the following questions:

- Can the combination of existing evaluation measures allow the creation of improved ones, with minor changes? Previous work has shown that some improvement can be achieved by adding liguistic quality information [5] or redundancy checking [27]. Can we achieve a good combination with simply surface-based measures (i.e., minimal preprocessing and no linguistic features)?
- If so, how should we combine these measures?
- Can we use different combinations of methods to grade different aspects of summaries?

In order to answer these questions we combine the well-established n-gram graph methods, under a machine learning perspective. In doing so, we use the individual evaluations as features that describe a single summary and apply regression to model how n-gram graph evaluations can be combined to form the final grade of a summary. We, thus, create a second-level grade estimator (in contrast e.g., to [33]) built as a regression problem, estimating a target grade (e.g., responsiveness or Pyramid score) based on the primary evaluation scores of different methods. We specifically focus on the n-gram graph based approaches (AutoSummENG and MeMoG) due to their purely statistical and language agnostic nature.

3   NPowER: N-Gram Graph Powered Evaluation via Regression

Our method is based on the following simple idea: if there exist a number of rather good grading systems for summaries and these grading systems are not always in agreement, it makes sense to supply an independent judge that can combine the graders' individual estimates to provide a better estimate on the final grade (see also [5]).

In our case we want to determine whether only using surface methods, based on n-gram graphs, we can estimate well-enough (i.e., with a strong correlation to humans) the grades of individual summaries. This is essentially a stronger requirement than that of correlating over whole systems. This is due to the fact that we judge a system based on the average of all its summaries. In fact, we can judge a system well even by taking turns in underestimating it and overestimating it in different summaries, due to the averaging effect. In the case where we judge single summaries this cannot happen.

To answer the questions posed in the previous section we build upon the notion of regression from the domain of statistics and machine learning.

Given a vector of descriptive (independent) features $\overline{x} \in \overline{X}$ and a target (dependent) numeric feature $y \in \mathbb{R}, y = f(\overline{x})$, with f unknown, we want to estimate a (combination) function

$$\tilde{f} : \overline{X} \to \mathbb{R} : \sum (\tilde{f}(\overline{x}) - f(\overline{x}))^2 \to 0, \forall \overline{x} \in X$$

of the descriptive features to best estimate the target feature. In the machine learning literature we find a variety of methods for regression ranging from simple linear regression, to logistic regression (see e.g., [34]) to Support Vector Regression (e.g., [35]). We use a linear regression, where features included in the regression model are selected based on the Akaike Informaion Criterion (AIC) [36], which selects features that best help the estimation without adding too much complexity. The implementation of the linear regression was provided as "Linear Regression" in the WEKA machine learning package [37, Version 3.7].

In the summary evaluation case we consider that the automatic evaluation methods of summaries consist good, descriptive features $\overline{x}$. The target feature y is the manual, human assigned grade. Since in summarization evaluation there exists a variety of human assigned grades, such as responsiveness or Pyramid score, we will need different applications of regression per case.

We examine three different approaches to see whether it makes sense to combine lots of evaluation methods or few, carefully selected ones:

- All: In this case a big set of automatic evaluations (submitted in the AESOP task of Text Analysis Conference) are used as $\overline{x}$ features.
- Only baselines: Only baseline systems are used as $\overline{x}$ features. We consider baselines systems which have been widely used for summary evaluation (i.e., ROUGE-based and BE evaluation).
- Only n-gram graph based: Only the proposed combination of methods is used, namely AutoSummENG and MeMoG, keeping the language-neutral approach of analysis.

We name the application of (linear) regression on the output scores of n-gram graphs methods the *NPowER* method: N-gram graph Powered Evaluation via Regression. We show in following sections that it constitutes a robust, high performing method for summary evaluation, even at the summary level.

Furthermore, we study which $\overline{x}$ features are the most informative, using feature selection, by viewing the evaluation problem as a classification problem. In the following section we report on the experiments and corresponding findings.

4 Experimental Setting and Results

In this section, we describe the data used for the evaluation of the NPowER method and we study how different features contribute to the performance of the system.

4.1 Data

We use the data generated within the AESOP task of the Text Analysis Conferences of 2009 and 2010 (TAC 2009 and TAC 2010[2]). The summaries in the AESOP test data of TAC 2009 consist of all the model summaries and "peer"

---

[2] See http://www.nist.gov/tac (Last visit: Dec 20, 2012) for more information.

(automated, non-model) summaries produced within the TAC 2009 Update Summarization task. 8 human summarizers produced a total of 352 model summaries, and 55 automated summarizers produced a total of 4840 peer summaries. The set of systems included three baseline summarizers. The summaries are split into Initial Summaries (Set A) and Update Summaries (Set B). Update summaries are supposed to take into account the corresponding initial summary and not repeat information on a given topic. In 2009 a total of 12 participants submitted 35 different AESOP metrics, in addition to 2 baselines.

In the 2010 Guided Summarization task, 8 human summarizers produced a total of 368 model summaries, and 43 automatic summarizers produced a total of 3956 automatic summaries. The summaries are split into Main (or Initial) Summaries (Set A) and Update Summaries (Set B), according to the part of the Guided Summarization Task they fall into[3]. Two baseline summarizers were included in the set of automatic summarizers. In 2010 a total of 9 participants submitted 27 different AESOP metrics, in addition to 3 baselines.

The AESOP task is "to create an automatic scoring metric for summaries, that would correlate highly with two manual methods of evaluating summaries, as applied in the TAC 2010 Guided Summarization task" (see TAC 2010 task description), namely the Pyramid method (modified pyramid score) [19] and Overall Responsiveness (see [4]). The scoring metrics (better "measures") are to evaluate summaries including both model (i.e., human generated) and peer (non-model) summaries, produced within the TAC 2010 Guided Summarization task. We note that there were some differences between the 2009 and 2010 datasets:

- In 2009 only ROUGE-SU4 and BE were used as baselines, while ROUGE-2 was added in 2010. In order to provide comparable results across datasets we omitted the ROUGE-2 metric when judging the performance of combined baselines. However, we did not remove it in the cases where all systems were combined ("All" case in the tables of the following section). Thus, we considered it another competing system for the purposes of the experiments.
- The responsiveness grade in 2009 was from 1 to 5, while in 2010 from 1 to 10.

In our experiments we used the TAC provided data for the AESOP task (files: "aesop_allpeers_[A—B]", "manual_allpeers_[A—B]")[4]. We combined the per summary data, by aligning manual grades to their corresponding automatic data lines. In our resulting data, each line contained the following fields:

- Pyramid and Responsiveness scores.
- AESOP SystemID and Topic.
- Evaluation results from each AESOP system.

Below we elaborate on the measures we used, in accordance to current literature, to determine the performance of the evaluation systems we propose.

---

[3] See http://www.nist.gov/tac (Last visit: Dec 20, 2012) for more info on the Guided Summarization Task of TAC 2010.
[4] The data are provided by NIST on request.

4.2   Measuring Correlation – Evaluation Method Performance

In the automatic evaluation of summarization systems we require automatic grades to correlate to human grades. The measurement of correlation between two variables provides an indication of whether two variables are independent or not. Highly correlated variables are dependent on each other, often through a linear relationship. There are various types of correlation measures, called *correlation coefficients*, depending on the context they can be applied. Three types of correlation will be briefly presented here, as they are related to the task at hand:

- The Pearson's product moment correlation coefficient reflects the degree of linear relationship between two variables[5]. The value of Pearson's product moment correlation coefficient ranges from -1 to 1, where 1 indicates perfect positive correlation and -1 perfect negative correlation. Perfect positive correlation indicates that there is a linear relationship between the two variables and that when one of the variables increases, so does the other in a proportional manner. In the case of negative correlation, when one of the two variables increases, the other decreases. A value of zero in Pearson's product moment correlation coefficient indicates that there is no *obvious* correlation between the values of two variables.
- The Spearman's rank correlation coefficient [38] performs a correlation measurement over the ranks of values that have been ranked before the measurement. In other words, it calculates the Pearson's product moment correlation of the ranking of the values of two variables. If two rankings are identical, then the Spearman's rank correlation coefficient will amount to 1. If they are reverse to each other, then the correlation coefficient will be -1. A value of zero in Spearman's rank correlation coefficient indicates that there is no obvious correlation between the rankings of values of two variables. It is important to note that this coefficient type does not assume linear relation between the values, as it uses rankings.
- The Kendall's tau correlation coefficient [39] relaxes one more limitation of the previous methods: it does not expect subsequent ranks to indicate equal distance between the corresponding values of the measured variable.

The above correlation coefficients have all been used as indicators of performance for summary systems evaluation (see, e.g., [1, 15]). To clarify how this happens, consider the case where an *automatic evaluation method* is applied on a set of summarization systems, providing a quantitative estimation of their performance by means of a grade. Let us say that we have assigned a number of *humans* to the task of grading the performance of the same systems as well. If the grades appointed by the method correlate strongly to the grades appointed by humans, then we consider the evaluation method good.

---

[5] The linear relationship of two correlated variables can be found using methods like linear regression.

4.3   Results — Correlation to Manual Measures

In this first experiment we try the three different sets of $\bar{x}$ features, to determine how well each individual set can perform. We stress that the evaluation we perform is *per summary*. We do this to go more in depth and see whether we can predict the quality of a single summary. If so, we will be able to use the resulting measure as an optimization factor when generating summaries (which is not possible when you have a per system evaluation).

To combine measures we used the WEKA software, as indicated in Section 3. We removed the fields of SystemID and Topic and performed 10-fold cross-validation, using as target variable the corresponding human assigned grade. We used the output file provided by the software as input to the R software [40] and applied correlation tests (cor.test command) between the estimated and true values of the grades.

In Table 1 we show the results of combination, and also provide the performance of the individual baselines (no combination) for reference. We judge performance by all measures of correlation to the human Responsiveness grading. We note that, in all the tables below, the statistical significance p-value of the correlation tests is much lower than 0.001. In the tables below the combination of n-gram graph methods is described as *NPowER*.

In Table 2 we judge performance by all measures of correlation to the human Pyramid grading.

The results of the experiments show the following:

- By combining measures one can significantly improve the estimation of a summary grade, regardless of the underlying measure (responsiveness or Pyramid in our case).
- It appears that using all the measures of AESOP as features, we get the best results in all the cases. However, it might prove impossible to combine all evaluations in a timely manner, since each evaluation represents a completely different system run.

Table 1. Per summary correlation of evaluation measures to Responsiveness

Year	Setting $\bar{x}$ features	Set A			Set B		
		Pearson	Spearman	Kendall	Pearson	Spearman	Kendall
	Baseline: ROUGE-SU4	0.33	0.30	0.22	0.39	0.40	0.29
	Baseline: BE	0.26	0.27	0.20	0.33	0.37	0.27
2009	Baseline comb.	0.34	0.29	0.21	0.39	0.39	0.29
	NPowER	0.60	0.42	0.32	0.61	0.50	0.38
	All	0.83	0.80	0.65	0.67	0.57	0.43
	Baseline: ROUGE-SU4	0.61	0.61	0.47	0.48	0.48	0.38
	Baseline: BE	0.48	0.50	0.38	0.45	0.47	0.36
2010	Baseline comb.	0.61	0.60	0.47	0.50	0.50	0.39
	NPowER	0.72	0.68	0.54	0.73	0.59	0.47
	All	0.75	0.72	0.58	0.74	0.62	0.50

Table 2. Per summary correlation of evaluation measures to Pyramid score

Year	Setting $\overline{x}$ features	Set A			Set B		
		Pearson	Spearman	Kendall	Pearson	Spearman	Kendall
	Baseline: ROUGE-SU4	0.64	0.65	0.47	0.62	0.60	0.43
	Baseline: BE	0.55	0.58	0.41	0.57	0.59	0.42
2009	Baseline comb.	0.64	0.65	0.47	0.62	0.62	0.45
	NPowER	0.80	0.73	0.55	0.77	0.69	0.51
	All	0.84	0.79	0.61	0.81	0.76	0.58
	Baseline: ROUGE-SU4	0.70	0.72	0.53	0.60	0.62	0.44
	Baseline: BE	0.61	0.64	0.46	0.55	0.58	0.42
2010	Baseline comb.	0.71	0.73	0.54	0.61	0.64	0.47
	NPowER	0.83	0.80	0.61	0.79	0.72	0.54
	All	0.85	0.83	0.64	0.81	0.75	0.56

- By only using baseline combination we do better than by using individual baselines; but not much better in most cases.
- By only using n-gram graph based methods combined (AutoSummENG and MeMoG in our case) we can significantly outperform the combination of baselines and even approach the performance of using all the systems (in most cases). In other words, 2 measures combined are performing close to more than 20 measures combined.

Using only the n-gram graph systems we performed another experiment to see how transferable the learnt regression models are across summary groups or different data (years):

- In the first experiment we train the regression model with all the 2009 data (from both sets) and test the model on the 2010 data (both sets). Then, we switch training and test sets and repeat the experiment. We describe this experiment as the "across years" experiment.
- In the second experiment we train the regression model with all the Set A data (from both 2009 and 2010) and test the model on all the Set B data (from both 2009 and 2010). Then, we switch training and test sets and repeat the experiment. We describe this experiment as the "across sets" experiment.

We illustrate the results of both experiments on Table 3. The experiment across years for Responsiveness offers good results, since the correlation scores remain high in both cases. We remind the reader that the correlation is judged on a per summary basis, which means that the per system performance is expected to be higher. The experiment across sets is equally interesting and promising. We see that the results are still high and the performance is almost identical (when rounding to the second deciman) in the two different settings. Of course, to be able to judge the robustness of the method with certainty, more experiments must be run (starting possibly from sampling the existing datasets). However, these first results are indications of acceptable stability on the examined tasks.

Table 3. Correlations between NPowER grades and Responsiveness (left), Pyramid (right) across years (top half) and sets (bottom half)

Setting		Target: Responsiveness			Target: Pyramid score		
Train Year	Test Year	Pearson	Spearman	Kendall	Pearson	Spearman	Kendall
2009	2010	0.72	0.65	0.52	0.72	0.74	0.55
2010	2009	0.61	0.47	0.35	0.78	0.72	0.74
Setting		Target: Responsiveness			Target: Pyramid score		
Train Set	Test Set	Pearson	Spearman	Kendall	Pearson	Spearman	Kendall
A	B	0.64	0.55	0.42	0.75	0.69	0.51
B	A	0.63	0.55	0.42	0.76	0.71	0.53

The results while optimizing for Pyramid scores were even better, as we illustrate on the right-hand side of Table 3. Overall, the method appears to be very effective across sets and years, forming a very interesting and useful estimator of summary quality.

On the other hand, deeper analysis shows that NPowER is not a perfect measure. This is clearly shown from the Kendall's tau value: an ideal measure that would indicate which summary of a pair is better, like a human, would have a value very close to 1.0. Thus, there is still much space for improvement. But, how can we improve? Can we determine which features are important and which are missing? To start addressing these questions we perform a feature study in the following section.

4.4   Results — Feature Selection

In this section, we examine which features are most informative for the estimation of a responsiveness grade of a system. In order to be able to apply information theory methods, such as Information Gain on the $\bar{x}$ features, we consider the grading problem as a *classification problem*. In the case of responsiveness we have 10 different possible classes, one per assignable grade (from 1 to 10).

The Information Gain (IG) measure is a measure of how "predictive" of a class a single feature is: $IG(\text{Class}, \text{Attribute}) = H(\text{Class}) - H(\text{Class}|\text{Attribute})$, where $H(x)$ is the entropy of the x values, and $H(x|y)$ is the entropy of x given y.

We use the 2009 and 2010 datasets Set A. We only focus on set A, because for set B (update task) information from set A should be used and we are trying to avoid inter-set dependecies at this point. The features with the top 10 information gain (IG) values are as follows:

The drawback of the IG measure is that it judges one feature at a time and does not offer combination information. Furthermore, by converting the regression problem to a classification problem we apply the same penalty to grades estimations that are not on-target, regardless of how different the grade was from the target value. However, it provides a hint at which features are more likely to help when determining the right grade: the n-gram graph features are

Table 4. Top 10 Information Gain features on TAC 2009 Responsiveness score classification problem

		Set A		
	2009		2010	
IG	System	IG	System
---	---	---	---
0.3852	MeMoG	0.5446	MeMoG
0.3814	AutoSummENG	0.5388	S7
0.3131	S17	0.4644	S9
0.3129	S22	0.453	S21
0.2989	S12	0.4497	AutoSummENG
0.2984	S20	0.4357	S17
0.2562	S19	0.4155	S18
0.2561	S14	0.4094	S10
0.2485	S16	0.4052	ROUGESU4
0.2425	S18	0.3472	S12

consistently highly graded. It is also noticeable that ROUGESU4 is within the table for the case of 2010, illustrating that baselines are important.

In order to see why combining methods offers additional information we studied the correlation between the baselines and the n-gram graph methods within NPowER. The results showed that the features were not too strongly correlated (0.70 Pearson correlation). We believe that the fact that each of the methods is correlated to the target features (e.g., responsiveness), but they are not highly correlated to each other makes their combination useful. It would make sense to determine, ideally orthogonal, evaluation measures to maximize the combination effect. Equivalently, it would make sense to *analyze human answers to orthogonal axes*, which would in turn be estimated by automatic measures.

5   Conclusion

In this paper we proposed a novel summary evaluation method, based on the linear combination of surface (n-gram graph) methods. The method is termed NPowER. We showed that combining several measures improves the estimation of summary quality. We then showed that NPowER is highly competitive when aiming to estimate two different, human evaluation measures (responsiveness and Pyramid score) on the summary level. We briefly studied the improtance of evaluation measures in term of information theory, by viewing the grading of a summary as a classification problem.

Our study showed that combining measures can prove effective, but there is significant space for improvement if we want to be able to confidently judge a summary automatically. Our future aims are to see whether combining state-of-the-art methods covering a variety of qualitative aspects (such as linguistic quality and coherence). We aim to examine whether these aspects are indeed uncorrelated enough to provide complementary information towards the best evaluation possible. We furthermore will try to examine, by viewing the evaluation process as a classification process, how one can improve the performance by viewing each grade as a different class.

References

[1] Lin, C.Y.: Rouge: A package for automatic evaluation of summaries. In: Proceedings of the Workshop on Text Summarization Branches Out (WAS 2004), pp. 25–26 (2004)

[2] Papineni, K., Roukos, S., Ward, T., Zhu, W.: Bleu: a method for automatic evaluation of machine translation. In: Proceedings of the 40th Annual Meeting on Association for Computational Linguistics, pp. 311–318. Association for Computational Linguistics (2002)

[3] Dang, H.T.: Overview of DUC 2005. In: Proceedings of the Document Understanding Conf. Wksp. (DUC 2005) at the Human Language Technology Conf./Conf. on Empirical Methods in Natural Language Processing, HLT/EMNLP 2005 (2005)

[4] Dang, H.T., Owczarzak, K.: Overview of the TAC 2008 update summarization task. In: TAC 2008 Workshop - Notebook Papers and Results, Maryland MD, USA, pp. 10–23 (2008)

[5] Conroy, J.M., Dang, H.T.: Mind the gap: Dangers of divorcing evaluations of summary content from linguistic quality. In: Proceedings of the 22nd International Conference on Computational Linguistics (Coling 2008), Manchester, UK, Coling 2008 Organizing Committee, pp. 145–152 (2008)

[6] Rankel, P., Conroy, J., Schlesinger, J.: Better metrics to automatically predict the quality of a text summary. Algorithms 5, 398–420 (2012)

[7] Giannakopoulos, G., El-Haj, M., Favre, B., Litvak, M., Steinberger, J., Varma, V.: TAC 2011 MultiLing pilot overview. In: TAC 2011 Workshop, Maryland MD, USA (2011)

[8] Owczarzak, K., Conroy, J., Dang, H., Nenkova, A.: An assessment of the accuracy of automatic evaluation in summarization. In: NAACL-HLT 2012, p. 1 (2012)

[9] Mani, I., Bloedorn, E.: Multi-document summarization by graph search and matching. In: Proceedings of AAAI 1997, pp. 622–628. AAAI (1997)

[10] Allan, J., Carbonell, J., Doddington, G., Yamron, J., Yang, Y.: Topic detection and tracking pilot study: Final report. In: Proceedings of the DARPA Broadcast News Transcription and Understanding Workshop, vol. 1998 (1998)

[11] Van Halteren, H., Teufel, S.: Examining the consensus between human summaries: Initial experiments with factoid analysis. In: Proceedings of the HLT-NAACL 2003 on Text Summarization Workshop, vol. 5, pp. 57–64. Association for Computational Linguistics, Morristown (2003)

[12] Lin, C.Y., Hovy, E.: Manual and automatic evaluation of summaries. In: Proceedings of the ACL 2002 Workshop on Automatic Summarization, vol. 4, pp. 45–51. Association for Computational Linguistics, Morristown (2002)

[13] Jones, K.S.: Automatic summarising: The state of the art. Information Processing & Management, Text Summarization 43, 1449–1481 (2007)

[14] Baldwin, B., Donaway, R., Hovy, E., Liddy, E., Mani, I., Marcu, D., McKeown, K., Mittal, V., Moens, M., Radev, D.: Others: An evaluation roadmap for summarization research. Technical report (2000)

[15] Nenkova, A.: Understanding the Process of Multi-Document Summarization: Content Selection, Rewriting and Evaluation. PhD thesis (2006)

[16] Radev, D.R., Jing, H., Budzikowska, M.: Centroid-based summarization of multiple documents: Sentence extraction, utility-based evaluation, and user studies. In: ANLP/NAACL Workshop on Summarization (2000)

[17] Marcu, D.: Theory and Practice of Discourse Parsing and Summarization, The. The MIT Press (2000)

[18] Saggion, H., Lapalme, G.: Generating indicative-informative summaries with sumum. Computational Linguistics 28, 497–526 (2002)

[19] Passonneau, R.J., McKeown, K., Sigelman, S., Goodkind, A.: Applying the pyramid method in the 2006 document understanding conference. In: Proceedings of Document Understanding Conference (DUC) Workshop 2006 (2006)

[20] Papineni, K., Roukos, S., Ward, T., Zhu, W.J.: Bleu: a method for automatic evaluation of machine translation. In: Proceedings of the 40th Annual Meeting on Association for Computational Linguistics, pp. 311–318 (2001)

[21] Hovy, E., Lin, C.Y., Zhou, L., Fukumoto, J.: Basic elements (2005)

[22] Hovy, E., Lin, C.Y., Zhou, L., Fukumoto, J.: Automated summarization evaluation with basic elements. In: Proceedings of the Fifth Conference on Language Resources and Evaluation, LREC (2006)

[23] Owczarzak, K.: Depeval (summ): dependency-based evaluation for automatic summaries. In: Proceedings of the Joint Conference of the 47th Annual Meeting of the ACL and the 4th International Joint Conference on Natural Language Processing of the AFNLP, vol. 1, pp. 190–198. Association for Computational Linguistics (2009)

[24] Giannakopoulos, G., Karkaletsis, V., Vouros, G., Stamatopoulos, P.: Summarization system evaluation revisited: N-gram graphs. ACM Trans. Speech Lang. Process. 5, 1–39 (2008)

[25] Giannakopoulos, G., Karkaletsis, V.: Summarization system evaluation variations based on n-gram graphs. In: TAC 2010 Workshop, Maryland MD, USA (2010)

[26] Schilder, F., Kondadadi, R.: A metric for automatically evaluating coherent summaries via context chains. In: IEEE International Conference on Semantic Computing, ICSC 2009, pp. 65–70 (2009)

[27] Conroy, J., Schlesinger, J., O'Leary, D.: Nouveau-rouge: A novelty metric for update summarization. Computational Linguistics 37, 1–8 (2011)

[28] Amigó, E., Gonzalo, J., Verdejo, F.: The heterogeneity principle in evaluation measures for automatic summarization. In: Proceedings of Workshop on Evaluation Metrics and System Comparison for Automatic Summarization, pp. 36–43. Association for Computational Linguistics, Stroudsburg (2012)

[29] Louis, A., Nenkova, A.: Automatically evaluating content selection in summarization without human models. In: Proceedings of the 2009 Conference on Empirical Methods in Natural Language Processing, vol. 1, pp. 306–314. Association for Computational Linguistics (2009)

[30] Saggion, H., Torres-Moreno, J., Cunha, I., SanJuan, E.: Multilingual summarization evaluation without human models. In: Proceedings of the 23rd International Conference on Computational Linguistics: Posters, pp. 1059–1067. Association for Computational Linguistics (2010)

[31] Vadlapudi, R., Katragadda, R.: Quantitative evaluation of grammaticality of summaries. In: Gelbukh, A. (ed.) CICLing 2010. LNCS, vol. 6008, pp. 736–747. Springer, Heidelberg (2010)

[32] Lloret, E., Palomar, M.: Text summarisation in progress: a literature review. Artificial Intelligence Review (2011)

[33] Pitler, E., Louis, A., Nenkova, A.: Automatic evaluation of linguistic quality in multi-document summarization. In: Proceedings of the 48th Annual Meeting of the Association for Computational Linguistics, pp. 544–554. Association for Computational Linguistics (2010)

[34] Menard, S.: Applied logistic regression analysis, vol. 106. Sage Publications, Incorporated (2001)

[35] Chang, C.C., Lin, C.J.: Libsvm: a library for support vector machines. Software 80, 604–611 (2001), http://www.Csie.Ntu.Edu.Tw/cjlin/libsvm

[36] Akaike, H.: Likelihood of a model and information criteria. Journal of Econometrics 16, 3–14 (1981)

[37] Witten, I.H., Frank, E., Trigg, L., Hall, M., Holmes, G., Cunningham, S.J.: Weka: Practical machine learning tools and techniques with java implementations. In: ICONIP/ANZIIS/ANNES, pp. 192–196 (1999)

[38] Spearman, C.: Footrule for measuring correlation. British Journal of Psychology 2, 89–108 (1906)

[39] Kendall, M.G.: Rank Correlation Methods. Hafner New York (1962)

[40] Team, R.C.: R: A Language and Environment for Statistical Computing. In: R Foundation for Statistical Computing, Vienna, Austria (2012) ISBN 3-900051-07-0

Explanation in Computational Stylometry

Walter Daelemans

CLiPS, University of Antwerp, Belgium
walter.daelemans@ua.ac.be

Abstract. Computational stylometry, as in authorship attribution or profiling, has a large potential for applications in diverse areas: literary science, forensics, language psychology, sociolinguistics, even medical diagnosis. Yet, many of the basic research questions of this field are not studied systematically or even at all. In this paper we will go into these problems, and suggest that a reinterpretation of current and historical methods in the framework and methodology of machine learning of natural language processing would be helpful. We also argue for more attention in research for explanation in computational stylometry as opposed to purely quantitative evaluation measures and propose a strategy for data collection and analysis for achieving progress in computational stylometry. We also introduce a fairly new application of computational stylometry in internet security.

1 Meta-knowledge Extraction from Text

The form of a text is determined by many factors. Content plays a role (the topic of a text determines in part its vocabulary), text type (genre, register) is important and will determine part of the writing style, but also psychological and sociological aspects of the author of the text will be sources of stylistic language variation. These psychological factors include personality, mental health, and being a native speaker or not; sociological factors include age, gender, education level, and region of language acquisition.

Writing style is a combination of consistent decisions in language production at different linguistic levels (lexical choice, syntactic structures, discourse coherence, ...) that is linked to specific authors or author groups such as male authors or teenage authors. It remains to be seen whether this link is consistent over time and whether there are style features that are unconscious and cannot be controlled, as some researchers have argued. The basic research question for computational stylometry seems then to describe and *explain* the causal relations between psychological and sociological properties of authors on the one hand, and their writing style on the other. These theories can be used to develop systems that generate text in a particular style, or perhaps more usefully, systems that detect the identity of authors (authorship attribution and verification) or some of their psychological or sociological properties (profiling) from text.

A limit hypothesis arising from this definition is that style is unique for an individual, like her fingerprint, earprint or genome. This has been called the *human stylome hypothesis*:

A. Gelbukh (Ed.): CICLing 2013, Part II, LNCS 7817, pp. 451–462, 2013.

'(...) authors can be distinguished by measuring specific properties of their writings, their stylome as it were.' [1]

Reliable authorship attribution and profiling is potentially useful in many areas: literary science, sociolinguistic research, language psychology, social psychology, forensics, medical diagnosis (detecting schizophrenia and Alzheimer's), and many others. In Sect. 3 we describe results in the context of an internet security case study as an example of useful computational stylometry. However, the current state of the art in computational stylometry seems not advanced enough to always guarantee the levels of reliability expected.

There are many excellent introductions to modern computational methods in stylometry [2–5] describing the methods and feature types used. Feature types include simple character n-grams, punctuation, token n-grams, semantic and syntactic class distributions and patterns, parse trees, complexity and vocabulary richness measures, and even discourse features.

Computational stylometry should be investigated in a Natural Language Processing (NLP) framework, more specifically as one of three levels of text understanding. The goal of text understanding is to extract knowledge from text and present it in a reusable format. NLP has seen significant progress in the last decade thanks to a switch to statistical and machine learning based methods in research and increased interest because of commercial applicability (Apple's SIRI and Google translate are only two examples of recent high impact commercial applications of NLP). The three types of knowledge we distinguish that can be extracted from text are: (i) objective knowledge (answering the who, what, where, when, ... questions), (ii) subjective knowledge (who has which opinion about what?), and (iii) metaknowledge (what can we extract about the text apart from its contents, mainly about its author?). Computational stylometry belongs in the latter category.

Core research in NLP addresses the extraction of objective knowledge from text: which concepts, attributes, and relations between concepts can be extracted from text, including specific relations such as causal, spatial and temporal ones. Research is starting also on the Machine Reading loop (how to use background knowledge in text analysis and conversely how to build up background knowledge from text). See work on Watson for state of the art research at this first level [6]. In addition to the extraction of objective knowledge, the large amount of text produced in social networks has motivated research to focus also on the extraction of subjective knowledge (sentiment and opinion). Never before have so many non-professional writers produced so much text, most of it subjective and opinionated (reviews, blogs, e-mail, chat, ...) [7]. Extraction of *metaknowledge* is conceptually a different type of knowledge extraction from text than the other two types. Where objective and subjective knowledge extraction try to make explicit and structure knowledge that is present in unstructured textual information, metaknowledge concerns knowledge about the author of the text (psychological and sociological properties, and ultimately identity), so outside the text. Recent advances in knowledge extraction from text at all these three levels have been made possible thanks to the development of robust and fairly accurate text analysis pipelines for at least some languages. These pipelines make possible the three types of knowledge extraction described earlier thanks to morphological analyzers, syntactic parsers, sentence semantics (including semantic roles and the analysis of negation and modality), and discourse processing (e.g. coreference resolution). Of course, the point is that by integrating in this process also analyses from objective and subjective knowledge extraction, more interesting theories about the extraction of metaknowledge become possible in principle.

For all types of knowledge extraction, supervised machine learning methods have been a powerful solution. Based on annotated corpora, various properties of text are encoded in feature vectors, associated with output classes, and machine learning methods are used to learn models that generalize to new data. It is surprising that much computational stylometry research is still explicitly linked to the idea of automatic text categorization [8] (as used in document filtering and routing applications) rather than to supervised machine learning of language in general (unsupervised and semi-supervised learning methods will not be discussed here). It makes sense to treat computational stylometry within the same methodological paradigm as other knowledge extraction from text tasks. For example, making a distinctions between similarity-based methods and machine learning methods as in [9] is unproductive as the former is a type of machine learning method as well (lazy learning as opposed to eager learning) [10]. All techniques proposed before in the long history of stylometry can be reinterpreted as machine learning methods to our advantage. A good example of this is Burrow's delta which through its reinterpretation as memory-based learning [11] leads to increased understanding of the method and to new useful variations. It would be equally productive if new methods like *unmasking* [12] and variants would be framed as instances of stacked classifiers and ensemble learning, which they are, thereby providing more clarity.

In a supervised machine learning approach to computational stylometry we have to consider the features to be used to describe our objects of interest (complete texts or text fragments), feature selection, weighting and construction methods, machine learning algorithm optimization, and the usefulness of techniques like ensemble methods, active learning, joint learning, structured learning, one-class learning etc. We can also rely on proven evaluation methods and methodological principles for comparing features and methods. Systematic studies in such a framework will go a long way in coming up to Rudman's [13, 14] criticism that after more than 40 years of research and almost a thousand papers (many more counting conference contributions), modern authorship studies "have not yet passed a 'shake-down' phase and entered one marked by solid, scientific, and steadily progressing studies."

2   Problems in Computational Stylometry

Computational stylometry is an exciting field with a promise of many useful applications, but initial successes have underplayed the importance of many remaining problems. So far, we already have encountered a number of unsolved basic research questions we will not go into in this paper, but that deserve more systematic study.

- Is style invariant or does it change with age and language experience? There is some work in this area (see [15] for an overview), but no large-scale systematic studies. If individual style changes over time, which seems to be the case, this is a confounding factor for attribution.
- Is style largely unconscious or can it be imitated? Again, there is some work on *adversarial stylometry*, but not enough for clear conclusions. Initial work [16] is not optimistic and shows that obfuscation reduces authorship identification methods to random behaviour.

Unless style markers can be found that are robust to aging and conscious manipulation, the human stylome hypothesis should be discarded. But there are other problems that need urgent attention as well.

2.1 Scalability and Character n-Grams

Another problem that has only relatively recently received attention is the issue of *scalability*. Authorship attribution and profiling work reasonably well when large amounts of text are available, and in the case of authorship attribution, few candidate authors for an unattributed text are present, one of which is the author (the closed case). This model fits literary disputed authorship cases with a small set of candidate authors, for example. In more realistic situations, we have short texts (for example letters or e-mails), and many potential authors. In [17], we showed, using a corpus of same-topic essays by 145 different authors, that with many potential authors or with short texts, attribution accuracy quickly decreases to levels that are still above baseline but nevertheless too low for practical applications. We also saw that simple character n-grams are more scalable than more complex (lexical and syntactic) feature sets. More work on scalability has been done (with better reported results) in [9]. The same scalability issues apply to profiling applications in computational stylometry as well.

The superiority of character n-grams is something which is often attested in stylometry: character n-grams often outperform more complex feature sets [18]. There is a good reason for this. They provide an excellent tradeoff between sparseness and information content. Because of their higher frequency compared to other feature types such as tokens, better probability estimates are possible for character n-grams, while at the same time they combine information about punctuation, morphology (character n-grams can represent morphemes as well as roots), lexicon (function words are often short), and even context (when extracting n-grams at sentence level rather than at token level). In addition they are tolerant to spelling variation and errors. On top of that, from a practical point of view, models based on character n-grams are very easy to construct and they are language-independent. There may also be a more negative explanation for their success in computational stylometry: it may be the case that the language processing tools that have to provide the more sophisticated linguistic analysis are not accurate enough and generate too much noise in the document representations.

The supervised machine learning context also helps us in understanding that scalable authorship attribution should not be framed as a multi-class learning problem, but as a binary or even one-class learning problem[19]. The real problem in authorship attribution is not to decide who from a limited number of authors, for all of whom we have training material, has written a particular text (the closed case), but to decide whether the new text was written by a particular author (for whom we have training material), or not, a task known as *authorship verification* (the open case). Very recently, this was defined in [20] as the fundamental problem of authorship attribution:

'Given two (possibly short) documents, determine if they were written by a single author or not.'

We will return to their solution in Sect. 2.2.

Successes with the closed case have lead to overoptimistic ideas about the possibilities of computational stylometry because of overfitting. When learning a model to distinguish between two or a few authors, there is no guarantee that the predictive features selected by the model will generalize to distinguishing from additional authors. Compare it to a fruit classification application: color will be a great feature to distinguish between apple and banana, but as soon as lemon and pear are added to the task, the model breaks down.

The human stylome hypothesis is trivially correct: given an unlimited supply of text from each person speaking a language, some combination of features can probably be found that uniquely discriminates anyone from all others. But we expect a stylome of an author to consist of a limited combination of features that are frequent enough to be found in all text written by that author so that generalization is possible.

2.2   Cross-Genre Stylometry

One of the most basic problems to be solved for computational stylometry is finding out how style, content, and genre interact in the generation of style. A straightforward strategy for avoiding topic detection rather than style detection is to exclude content words as features. However, topic words can be predictive as well (e.g. consistent selection of one word from a set of synonyms by authors or groups of authors). Although there is some work in this area (see for example chapter 4 in [21] and references therein), more systematic research is needed.

An even less researched aspect of computational stylometry is the effect of genre on attribution. To which extent do stylistic properties of individual authors or groups of authors transfer from one genre to the other? Can we expect that a model trained on essays written by someone will be able to identify his suicide note or blackmail letter? Again this is a well known problem in machine learning for the case where training and test data come from different distributions (the domain adaptation problem [22]). Domain adaptation problems exist both for genre and for topic (in the case where features based on content words are used).

In a recent study [23] we tackled both the problem of verification (rather than attribution, i.e. the open case) and the problem of cross-genre generalization. As machine learning method we tested the "unmasking" technique, recently proposed [12] and well-received [24]. Suppose we want to verify that a text X with unknown authorship was written by the author of a text A. We could split both texts in chunks, and train a classifier to distinguish between both. If the resulting classifier turns out to have low generalization accuracy, X and A were probably written by the same author, if it turns out to be easy to distinguish then not. The approach turned out not to work very well because a limited number of features can wrongfully maximize the differences in writing style between two texts written by the same author. As a solution, [12] proposed a stacked classifier approach, in which a new classifier is built on the basis of a previous classifier by removing those features that are most discriminative between the two texts. The degradation curves that can be attested by applying these subsequent classifiers to the task are indicative of whether the two texts were written by the same author. In the case of a few features being responsible for most differences (same author), the degradation curve would fall quickly. In the case of many features being responsible for the differences (different authors), the drop is less dramatic. It has been attested that the approach works well for longer texts and for related tasks such as intrinsic plagiarism detection, but not for shorter texts below 10,000 words in size [25]. We tested whether the approach works for the cross-genre authorship verification task in the expectation that the genre markers would be limited and superficial and would therefore be among the first to be discarded in the unmasking approach, leading to a clear degradation curve indicative of same authorship. We refer to the paper [23] for a detailed description of the operationalization of the unmasking approach to our cross-genre case. We applied the approach to theatre and prose texts of five authors. Whereas for the within-genre case the approach worked as expected, it didn't work very well for the cross-genre case. Although some of the most discriminative features discarded were indeed genre-related (names of principal characters, stage directions, colloquialisms, ...), the approach did not hold. Further research with optimization of the many parameters in the approach is still needed, but it seems clear that we will need new methods for coping with cross-genre cases.

In conclusion, we have argued that many of the basic problems in computational stylometry are not being investigated at all or not sufficiently systematically. Good features for authorship verification and profiling should be robust against genre variation, topic variation, individual style change over time, and conscious manipulation. Methods should also be scalable to short texts. Arguably, it is the feature selection (or feature construction) problem which is most important in this field rather than the choice of machine learning method, although the specific problem of authorship verification may call for ensemble methods such as unmasking. But overall, what is lacking is explanation.

2.3   Explanation

One aspect of current machine learning of NLP research that the field of computational stylometry should not adopt is its unidimensional focus on quantitative evaluation. The goal of research should be to increase understanding rather than maximizing performance (which is an engineering criterion). In profiling, the field started in an excellent way regarding explanation with the gender assignment studies of [26]. They provided a plausible explanation for their success in distinguishing male from female authors in written text by hypothesising that women use more relational language, and men more informative (descriptive) language. That men are prone to more descriptive language use is reflected in text by a more frequent use of nouns, determiners, prepositions etc. Figure 1 shows some similar frequent features (part of speech tags, Pennebaker LIWC classes, tokens) related to male and female language use in Dutch. A darker colour under male or female indicates more frequent use. The hypothesis "men use more descriptive language" then explains a number of (correlated) lower level text features, and provides *insight* into how male and female gender is realized into text.

Unfortunately, examples like this are rare. More frequently, a study will provide some new best result on a benchmark dataset using some clever feature engineering or classifier optimization, without attempting to provide an explanation for

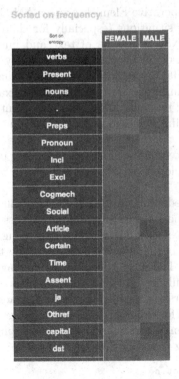

Fig. 1. Frequent Feature Types correlated with gender in Dutch

the results in a broader framework. At best there is some superficial error analysis. The current focus on challenges (also called shared tasks) using hastily compiled low quality "benchmark datasets" is an important culprit for this. There is seldom time for intelligent reflection on the construction of the datasets and the interpretation of the results, and there are no prizes for explanation, only for achieving the highest accuracy.

It could be argued that what is especially needed for improving understanding and explanation is (for each language) a real reference corpus which is carefully balanced according to genre, topic, age, and gender (and if possible also other psychological and sociological properties of the authors). Only then can real progress be made in solving the fundamental problems of computational stylometry. If we take the human stylome seriously as a hypothesis, we should start doing stylome-wide association studies (in analogy to genome-wide association studies) associating linguistic properties with author traits, and inferring explanatory concepts from the bottom-up interpretation of correlated sets of features. As in genetic studies, population stratification (i.e. balanced corpora) is a necessary precondition in such studies.

3   Detecting Harmful Content in Social Media

In a recently started cooperative Flemish project AMiCA[1], our goal is to identify possibly threatening situations (especially for children and adolescents) in social networks sites (SNS) by means of text and image analysis. The three critical situations targeted are cyberbullying, sexually transgressive behavior, and depression and suicidal behavior. For text-based analysis we see these tasks partly as instances of computational stylometry. For example, for the detection of transgressive behavior by pedophiles[2] it is important not only to be able to detect the typical grooming stages in pedophile behavior, but also to be able to detect age and gender of the text in order to check the information provided in the SNS profiles. For detecting suicidal emotions and insults in cyberbullying, similar computational stylometry tasks can be defined. Some early results of our team can be found in [27–29].

For the detection of pedophiles in SNS we have available some data from the Belgian SNS *Netlog* in the form of interaction with associated profile information (age, gender, and location). The data is challenging because the utterances are short and written in chat language which has properties completely different from standard language. The properties of chat language are based on the fact that the interactions should be quick and informal (spoken language like). This leads to omission of letters and words, abbreviations, acronyms, non-verbal and suprasegmental mimicry (for example character flooding, concatenation and even merger of words, emoticons), and many other strange phenomena. Investigating this, we found interesting reflections in our data of claims about sociolinguistic

---

[1] http://www.amicaproject.be

[2] A preparatory PhD project, DAPHNE, about this was started before AMiCA. See http://www.clips.ua.ac.be/projects/daphne

language variation in spoken language. For example, Fig. 2 from a submitted paper shows how much different chat language is from standard language for different age groups, genders and regions in the chat data. It is clear from this data that non-standard language use in chat is especially a property of adolescents, and that in their twenties, chatters revert to more standard language. Also, some attested facts about sociolinguistic variation in the Flemish Dutch region can be clearly shown in this data: for example that men use more non-standard language than women, that Western Flanders uses more non-standard language than other regions and so on.

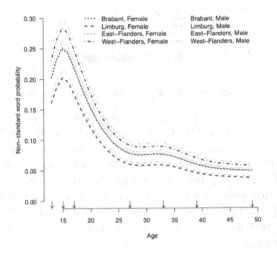

Fig. 2. Use of non-standard language in chat by Flemish sociological groups

More important for our purposes is that this data can be used to train accurate classifiers for assigning age and gender. Our strategy is to develop two classifiers, one based on age and gender to check for mismatches between profiles provided and information extracted from the text of the interactions, and a second one to detect grooming behaviour, which can be detected to some extent by typical types of language use, for example directive language, and specific topics, for example 'coast is clear' checks.

In Fig. 3, the architecture we are working on is given.

By optimizing the classifiers for legally relevant age groups (minus 16 and plus 21 for example), very high f-scores (in the nineties) can be reached. Incidentally, this data is one example of a task where n-grams don't do very well. Unfortunately, because of the non-standard characteristics of the text, standard language text analysis tools cannot be used, so that we had to restrict ourselves to word tokens in these experiments. Current work on automatic normalization of chat language should make additional levels of analysis available soon.

Increasingly, the field has become interested in these more peripheral applications of computational stylometry. For example, in the context of CLEF, a

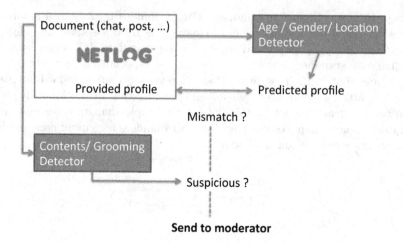

Fig. 3. Architecture for a pedophile detection system

shared task was organized in 2012 on pedophile detection[3]. There were many participating systems and some good very good detection results. However, the event illustrates many of the problems with data collection in shared tasks alluded to earlier. By collecting negative and positive data from different sources (the perverted justice website for the positive data and unrelated sources for the negative data), the task turns out to be artificially easy and generalization to other datasets very low. Also in this case, more work should be done on population stratification.

4   Conclusion

With our case study in guarding security of children and adolescents in SNS, we hope to have shown that computational stylometry has large application possibilities and is, thanks to advances in Natural Language Processing and Machine Learning, in a state where useful applications are already possible. But many fundamental problems of computational stylometry remain unsolved or even largely ignored. We are looking not just for a system that reaches a certain target accuracy in a task, but for explanations, and for systems that are scalable, and that generalize over different genres and topics in their author identification and profiling results. It seems clear that a systematic study of the components and concepts of style will only be possible by collecting a large balanced dataset for each language of a type that doesn't yet exist in current benchmark efforts.

Acknowledgements. I gratefully acknowledge the support of various research funds, most notably FWO (Research Foundation Flanders), and the EWI

---

[3] http://www.uni-weimar.de/medien/webis/research/events/
pan-12/pan12-web/authorship.html

ministry for supporting research in the CLiPS Computational Linguistics Group on computational stylometry. The research described in Sect 3 is sponsored by IWT (Flemish Agency for Innovation by Science and Technology) and the University of Antwerp research fund. Some of the research described explicitly in this paper was done in cooperation with (former) CLiPS colleagues Kim Luyckx, Mike Kestemont, Vincent Van Asch, and Claudia Peersman. The possible errors in interpretation of the results are my own. I am grateful to all CLiPS members for providing an intellectually stimulating research environment.

References

1. van Halteren, H., Baayen, H., Tweedie, F., Haverkort, M., Neijt, A.: New machine learning methods demonstrate the existence of a human stylome. Journal of Quantitative Linguistics 12(1), 65–77 (2005)
2. Stamatatos, E.: A survey of modern authorship attribution methods. JASIST 60(3), 538–556 (2009)
3. Koppel, M., Schler, J., Argamon, S.: Computational methods in authorship attribution. JASIST 60(1), 9–26 (2008)
4. Juola, P.: Author attribution. Foundations and Trends in Information Retrieval 1(3), 233–334 (2008)
5. Pennebaker, J.: The Secret Life of Pronouns. Bloomsbury Press, New York (2011)
6. Fan, J., Kalyanpur, A., Gondek, D., Ferrucci, D.: Automatic knowledge extraction from documents. IBM Journal of Research and Development 56(3/4), 1–10 (2012)
7. Liu, B.: Sentiment Analysis and Opinion Mining, 180 pages. Morgan & Claypool Publishers(2012)
8. Sebastiani, F.: Machine Learning in Automated Text Categorization. ACM Computing Surveys 34(1), 1–47 (2002)
9. Koppel, M., Schler, J., Argamon, S.: Authorship attribution in the wild. Language Resources and Evaluation 45, 83–94 (2011)
10. Daelemans, W., Van den Bosch, A.: Memory-based language processing. Cambridge University Press, Cambridge (2005)
11. Argamon, S.: Interpreting Burrow's Delta: Geometric and Probabilistic Foundations. Literary and Linguistic Computing 23(3), 131–147 (2008)
12. Koppel, M., Schler, J., Bonchel-Dokov, E.: Measuring differentiability: unmasking pseudonymous authors. Journal of Machine Learning Research 8, 1261–1276 (2007)
13. Rudman, J.: The state of authorship attribution studies: some problems and solutions. Computers and the Humanities 31(4), 351–365 (1997)
14. Rudman, J.: The satet of non-traditional authorship studies 2010: some problems and solutions. In: Proceedings of the Digital Humanities, pp. 217–219 (2010)
15. Stamou, C.: Stylochronometry: stylistic development, sequence of composition, and relative dating. Literary and Linguistic Computing 23(2), 181–199 (2008)
16. Brennan, M., Afroz, S., Greenstadt, R.: Adversarial Stylometry: circumventing authorship recognition to preserve privacy and anonymity. ACM Transactions on Information and System Security 15(3), 12:1–22 (2012)
17. Luyckx, K., Daelemans, W.: The effect of author set size and data size in authorship attribution. Literary and Linguistic Computing 26(1), 35–55 (2011)
18. Grieve, J.: Quantitative authorship attribution: an evaluation of techniques. Literary and Linguistic Computing 22(3), 251–270 (2007)

19. Koppel, M., Schler, J.: Authorship verification as a one-class classification problem. In: Proceedings 21st International Conference on Machine Learning, pp. 489–495 (2004)
20. Koppel, M., Schler, J., Argamon, S., Winter, Y.: The Fundamental Problem of Authorship Attribution. English Studies 93(3), 284–291 (2012)
21. Luyckx, K.: Scalability Issues in Authorship Attribution. UPA, Antwerp (2010)
22. Daumé III, H.: Marcu. D.: Domain Adaptation for Statistical Classifiers. Journal of Artificial Intelligence Research 26, 101–126 (2006)
23. Kestemont, M., Luyckx, K., Daelemans, W., Crombez, T.: Cross-genre authorship verification using unmasking. English Studies 93(3), 340–356 (2012)
24. Stein, B., Lipka, N., Prettenhofer, P.: Intrinsic plagiarism analysis. Language Resources and Evaluation 45(1), 63–82 (2011)
25. Sanderson, C., Guenter, S.: Short text authorship attribution via sequence kernels, markov chains and author unmasking: an investigation. In: Proceedings of the 2006 EMNLP, pp. 482–491 (2006)
26. Koppel, M., Argamon, S., Shimoni, S.: Automatically categorizing written texts by author gender. Literary and Linguistic Computing 17(4), 401–412 (2003)
27. Peersman, C., Daelemans, W., Van Vaerenbergh, L.: Predicting Age and Gender in Online Social Networks. In: 3rd International Workshop on Search and Mining User-generated Contents (SMUC 2011), pp. 37–44 (2012)
28. Peersman, C., Vaassen, F., Van Asch, V., Daelemans, W.: Conversation Level Constraints on Pedophile Detection in Chat Rooms. In: CLEF 2012 Conference and Labs of the Evaluation Forum, pp. 1–13 (2012)
29. Luyckx, K., Vaassen, F., Peersman, C., Daelemans, W.: Fine-Grained Emotion Detection in Suicide Notes: A Thresholding Approach to Multi-Label Classification. Biomedical Informatics Insights 5(suppl. 1), 61–69 (2012)

The Use of Orthogonal Similarity Relations in the Prediction of Authorship

Upendra Sapkota[1], Thamar Solorio[1], Manuel Montes-y-Gómez[2], and Paolo Rosso[3]

[1] University of Alabama at Birmingham, Birmingham, AL 35294, USA
{upendra,solorio}@cis.uab.edu
[2] Instituto Nacional de Astrofísica, Optica y Electrónica , Puebla, Mexico
mmontesg@ccc.inaoep.mx
[3] NLE Lab - ELiRF, Universitat Politècnica de València, Valencia, Spain
prosso@dsic.upv.es

Abstract. Recent work on Authorship Attribution (AA) proposes the use of meta characteristics to train author models. The meta characteristics are orthogonal sets of similarity relations between the features from the different candidate authors. In that approach, the features are grouped and processed separately according to the type of information they encode, the so called linguistic modalities. For instance, the syntactic, stylistic and semantic features are each considered different modalities as they represent different aspects of the texts. The assumption is that the independent extraction of meta characteristics results in more informative feature vectors, that in turn result in higher accuracies. In this paper we set out to the task of studying the empirical value of this modality specific process. We experimented with different ways of generating the meta characteristics on different data sets with different numbers of authors and genres. Our results show that by extracting the meta characteristics from splitting features by their linguistic dimension we achieve consistent improvement of prediction accuracy.

1 Introduction and Background

Authorship Attribution (AA) is the task of identifying the author of a given anonymous text, or a text whose authorship is in doubt. Although the authorship attribution task is often solved as a multi-class, single-label text categorization task, the purpose of AA is to model each author's writing style rather than modeling thematic content of the available documents, as in the case of the typical text classification task. There are many relevant applications of AA in Forensic Linguistics. For instance, AA can help fight spam filtering [26], cyber bullying, and other forms of cyber crime (e.g., identifying authors of malicious code, or potential pedophiles). Other applications include plagiarism detection [22], author recognition of a given program [7], and web information management.

As described in the Stamatatos survey [23], there are two main frameworks that have been successfully used in the relevant literature: the profile-based approach, and the standard machine learning one. Both of them assume the availability of some number of documents with known authorship that can be used to build the models. In a profile based approach, all documents from the same author in the training set are concatenated. Then profiles are created for each author by extracting several features from

A. Gelbukh (Ed.): CICLing 2013, Part II, LNCS 7817, pp. 463–475, 2013.
© Springer-Verlag Berlin Heidelberg 2013

these merged files. These approaches rely mostly on low level features, such as character n-grams. To predict authorship of a new document, a similarity score between the new document and each author profile needs to be computed. The document will then be assigned to the author whose profile yields the highest similarity score. Because the similarity between the test document and the profiles is computed independently for each author, this approach allows to use profile-specific features. Typically the features are selected based on their frequency of appearance in the profile. Examples of a profile based approach include [10,20,11].

In contrast, machine learning approaches to AA use a feature vector representation where each single document from the training set is represented individually by the same set of features. The feature vectors are then used to train a machine learning algorithm. These feature vectors are usually a varied combination of lexical, character, and syntactic features such as average word length, average sentence length, content words, function words, word n-grams, character n-grams, and parts-of-speech (POS) n-grams. Recent approaches have reported good prediction performance for this task using Support Vector Machines [6], memory based learners [13,14], and Probabilistic Context Free Grammars [17].

In a recent work, Solorio et al. [19] proposed an AA approach that explicitly exploits the differences in the nature of the features representing the documents to generate informative meta features. The key assumption in their work is that by breaking down the document representation into a set of orthogonal dimensions[1], meaningful similarity patterns among authors could emerge. Then these similarity patterns can be exploited by the machine learning algorithms to boost authorship prediction accuracy. This approach is loosely related to well known machine learning approaches, such as the co-training algorithm by Blum and Mitchell [3] where two classifiers are trained on different views of the data. However, the goal of having different views of the data in Solorio et al.'s work is to extract disjoint similarity relations among the instances from different classes and not to train classifiers on disjoint subsets of features.

In this paper we set out to investigate the value of extracting the meta features following the framework proposed by [19]. The main contribution of this paper is the empirical evidence gathered that shows we can model the writeprint of authors by combining standard lexical and stylistic features with modality specific similarity relations among the writing preferences of different authors. Although the idea of these meta features was proposed by previous work, the empirical evaluation was done on a single corpus and with a single train/test partition of the data. Moreover, the authors in that paper left an important question unanswered: *Is the notion of linguistic modalities really needed?* In other words, similarity relations from disjoint sets of features seem to help boost prediction accuracy. Do we need to partition the feature set by their linguistic nature, or is it sufficient to just partition this set randomly? Because the implications of these questions are relevant to explore a more general application of this approach, we consider necessary to search for answers and report our findings. This study presents the first statistically significant results supporting the need for linguistic modalities in several datasets using a cross validation setting. We also report on results of

---

[1] We use the term *orthogonal* loosely in this paper to refer to sets of features that are coming from different linguistic dimensions and that are disjoint from one another.

experiments that allow for a direct comparison with state-of-the-art AA approaches. New in this paper is also a study of the individual modalities that are being used. In sum, we aim to provide a better understanding of the value of adding the meta characteristics to the representation of documents in the AA task.

2 Document Representation

Following the formulation in [19], we exploit the notion of linguistic modalities, where each linguistic modality refers to a set of features representing different aspects of the text. For instance, features related to syntax are considered a different modality from features related to semantics. Therefore, rather than representing each instance directly by a feature vector x, we represent it by a set of M smaller feature vectors $\{x_1, x_2..., x_m\}$, where $m = |M|$, the number of modalities, and x_i is the feature vector in modality i. The combination (union) of these smaller vectors (sub-vectors) forms the single feature vector representation of the instance in the standard scenario. We call this set of vectors first level features (FLF) following the same convention as in [19]. After the extraction of FLF we proceed with the generation of modality specific meta features (MSMF) as follows:

1. The first step in this meta feature extraction is the unsupervised clustering of all the feature vectors in the training set belonging to the same modality. We do this for each modality in the training set, which results in k clusters per modality, i.e., $m \times k$ total clusters. Because the training instances are clustered by modality, each modality will have its own clustering solution.
2. For each ith clustering solution, we compute the centroid of each cluster $c_{i,j}$ by averaging the feature vectors belonging to that cluster.
3. For each document, from the training and testing sets, we compute its similarity to the centroids of each cluster using the cosine similarity function. These similarity scores are the meta features.

As meta features are calculated on a modality basis, each modality gives us as many meta features as the number of clusters in that modality. Each sub-vector of FLF, x_i, has a corresponding meta feature vector x_i' with the length equal to the number of clusters k in the given modality. We use same k for each modality, therefore, $m \times k$ meta features are extracted from m modalities. All FLF and meta features are joined (concatenated) into a single feature vector that is used to train a machine learning algorithm. In Figure 1 we show a graphical representation of the computation of MSMF. In that figure, $\{c_{sty,1}..c_{sty,k}\}$ are k clusters formed from the stylistic feature vectors, and likewise, $\{c_{syn,1}..c_{syn,k}\}$ are k clusters formed from the syntactic feature vectors. The stylistic meta feature vectors $x_{sty}' = \{x_{sty,1}'..x_{sty,k}'\}$ and syntactic meta feature vectors $x_{syn}' = \{x_{syn,1}'..x_{syn,k}'\}$ are formed after computing the cosine similarity of the document instance to the cluster centroid on their own modality. $x_{p(sty)}$ and $x_{p(syn)}$ are vector representations of each instance in the *Stylistic* and *Syntactic* modality, respectively.

In this work we consider four different types of feature groups –stylistic, lexical, perplexity values from character level n-gram language models, and syntactic features, for a total of four modalities ($m = 4$). It is worth noting that the notion of linguistic

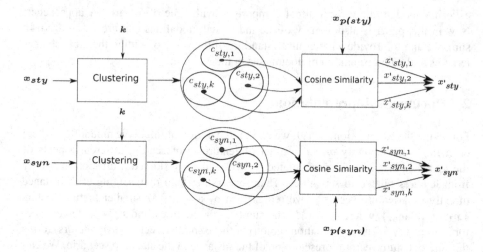

Fig. 1. Diagram showing the computation of modality specific meta features from two modalities: *Stylistic (sty)* and *Syntactic (syn)*

modalities as used in this and previous work has a connection with the notion of linguistic dimensions defined by Biber's work on genre analysis [2]. The contrast is at the level of abstraction. Biber's dimensions define a set of features common at a discourse level, while in this work linguistic modalities refer to different lower levels of analysis.

Note that since no class information is used during the clustering process, the MSMF approach is clearly different from other well studied methods for reducing data dimensionality [1,18,4]. The goal of the clustering step, in our AA framework, is to generate new meaningful features. The clustering allows us to generate similarity patterns from the posts of different authors on individual modalities. Some authors use the emoticons in a similar way, while some share the use of punctuations marks. We believe that the encoding of these similarities in the meta features complements the information provided by the first level features to the machine learning algorithm.

2.1   Features

The FLF features used in our work are refinement (addition and modification) of [19]. The final list of features is shown in Table 1. The first column shows how these features are categorized by the type of information they are extracting from the document. The first modality (*Stylistic*) tries to capture writing choices that reflect authors preferences and thus contains features related to the use of punctuation marks, length of sentences, and use of contractions, among others. The *Syntactic* modality focuses on the grammatical patterns of the authors. It includes n-grams from POS tags and bag of syntactic relations. In the *Semantic* modality the goal is to capture the topic/author correlation, as well as the information related to word choices for each author. This modality then uses the standard bag of words representation used in text classification tasks where stop words are removed from the documents before generating the feature vectors.

The last modality (*Perplexity*) contains perplexity values from language models. We train one language model per author. We expect that perplexity scores will be lower for the documents belonging to the corresponding author's model, similar to the intuition in [17] of using probabilistic context free grammars. We trained character-level 4-gram language models for each author in the training set. Then we compute the perplexity values for each document in a leave-one-out setting.

There are some differences between the final feature set used in our work and that used by Solorio *et al.* (2011). We added new features in the *Stylistic* modality: total number of sentences, percentage of words without a vowel, number of balanced parenthesis, and number of tokens containing at least one capital letter. We also modified the feature for the use of quotations in the same *Stylistic* modality. Instead of having a binary feature we use here the total number of quotations. Because several datasets are coming from social media, we thought vowel-less words would be a common feature and might improve the performance. The features such as number of sentences have been successfully used in previous research. The goal is to distinguish authors that produce long and wordy documents from those that tend to be more succinct. For the *Perplexity* modality, we used higher order language models, 4-grams, instead of 3-grams. Character 4-grams have been successfully used for AA tasks and in our case, we believe that 4-grams allow us to better capture not only patterns from the endings of the words but also the lemmas of the words as well as patterns about the use of functional words. Features that are different in this paper are highlighted in Table 1.

Table 1. First level features used in the representation of documents for the AA task. The '+' after a feature indicates new features not present in previous work. The '*' indicates a modified feature.

Modality	First Level Features (FLF)
Stylistic	Total number of sentences[+]
	Average number of tokens per sentence
	Percentage of words without vowel[+]
	Average number of punctuations per sentence
	Percentage of contractions
	Total number of balanced parenthesis[+]
	Percentage of two consecutive punctuation marks
	Percentage of three consecutive punctuation marks
	Total number of alphabetic characters
	Average number of tokens with at least a capitalized letter per sentence[+]
	Toal number of sentence initial words with first letter capitalized
	Total number of quotations*
Syntactic	Top 1000 POS tag unigrams
	Top 1000 POS tag bigrams
	Top 1000 POS tag trigrams
	Top 1000 Grammatical relations from the dependency parses
Semantic	Top 1000 bag-of-words
Perplexity	All the perplexity values from character 4-grams*

3   Data Sets

We tried to consider a collection of test sets with varied challenging characteristics to provide a more comprehensive data that will also allow us to benchmark our results. We selected three collections used by state-of-the-art approaches to AA. Table 2 shows several statistics of the different data sets. The first collection is from Solorio *et al.* (2011) [19]. This collection consists of five datasets with a different number of authors taken from forums of The Chronicle of Higher Education (CHE). As shown in Table 2, this data set contains very short documents (~6 sentences per post), which imposes an interesting challenge for the AA task. Another important characteristic of this data set is the imbalanced distribution of documents per author. In the data sets with 20, 50 and 100, there are some authors that are heavily represented and some for which only a few documents are available. This setting is closer to what one would expect to see in real world scenarios, since we cannot control how much each user interacts on the forum. However the nice characteristic about this collection is that all posts are coming from the same topic. We expect this will reduce the chances of having a strong topic/author correlation that will be reflected in the value of the *Semantic* modality.

Another collection is from Raghavan *et al.* (2010) [17]. They collected five datasets from material downloaded from the Internet. Four of them contain news articles on topics related to Business, Travel, Football, and Cricket. The fifth data set contains poems from the Project Gutenberg website[2]. We chose Raghavan *et al.*'s collection because it contains data from different topics and different genres, and because we can do a direct comparison with their results. The datasets in this collection have a varied number of authors ranging from 3 to 6.

The last collection is the CCAT topic class, a subset of the Reuters Corpus Volume 1 [12]. This collection was not gathered for the goal of doing authorship analysis studies. But the common use of this data set in previous studies provides a unique opportunity to benchmark our results. Previous work has reported results for AA with 10 and 50 authors [21,16,6] and we follow this lead to experiment as well with 10 and 50 authors.

We do not expect to have a single best approach that outperforms all other results in such a diverse collection of benchmark data. The goal is to study whether the benefit of using linguistic modality framework generalizes to different datasets, and to try to tease apart how different modalities have varied performance accross collections.

4   Experiments

In this paper we report results that are the overall average accuracy from 5-fold cross-validation, along with the statistical significance of our results. But to provide a one to one reference for comparison, we also performed experiments with the fixed train/test partitions used in state-of-the-art systems whenever this information was available.

We used support vector machines (SVMs) implemented in Weka [27] with default parameters as the underlying classifier. For the *Syntactic* modality, the POS tags were generated by the Stanford tagger [25]. We used the Stanford parser to generate the

---

[2] http://www.gutenberg.org/wiki/Main_Page

Table 2. Some statistics, including distribution of the documents across authors, from the collections we used in our AA experiments. Figures shown in Columns 3 and 4 are averages over the entire collection. **min** shows the minimum number of documents for any single author and **max** shows the maximum number of documents belonging to a single author.

dataset	#auth	#words/doc	# sent/doc	min #docs	max #docs	#docs
CHE	5	75.88	6.26	434	693	2,889
CHE	10	78.24	6.82	321	914	5,579
CHE	20	84.60	7.20	173	1,369	9,779
CHE	50	79.27	6.86	33	2,369	15,543
CHE	100	79.89	6.89	6	2,369	16,171
Football	3	877.00	44.00	31	34	97
Business	6	827.00	40.00	25	30	175
Travel	4	908.00	40.00	37	45	172
Cricket	4	978.00	50.00	30	48	158
Poetry	6	271.00	13.00	19	56	200
CCAT	10	507.24	21.07	100	100	1,000
CCAT	50	505.65	21.54	100	100	5,000

dependency parsers [15]. The SRILM toolkit [24] was used to train the character 4-gram language models. The clustering of the FLF on modality basis was done using CLUTO [9]'s $vcluster$ clustering program with parameter $clmethod = rbr$.

We performed a set of experiments designed to answer the question posed in Section 1: *Is the notion of linguistic modalities really needed?* To answer this question we run experiments where instead of fragmenting the FLF by linguistic modality, we randomly generate m subsets of FLF, simulating "random modality meta features" (RMMF). Then we compare results of this approach against generating meta features by linguistic modality, as described in Section 2. For the sake of completeness we also show figures for using only the FLF to train the machine learning algorithm.

The results for all the datasets are presented in Table 3. In 10 out of the 12 datasets the approach using the modality specific meta features in combination with first level features (MSMF+FLF) yields the best results. This was expected since previous work showed this combination to be the best setting in the CHE dataset. The results on additional datasets show the consistency of the approach.

The table also highlights the results that are statistically significant with a 95% confidence level using a two-tailed t-test. All the results, with the exception of the Cricket dataset, show the differences between randomly splitting the feature set and using the notion of linguistic modality to be statistically significant. From these results we can conclude that the boost in accuracy results from the discriminative power of the orthogonal similarities extracted, and not because of a simple decomposition of the feature vector into disjoint subvectors.

There is also a notable difference between the margin of increase in accuracy among the different collections. It seems the CHE collection benefits the most out of the MSMF+FLF setting. There are two possible reasons for this, document length and number of authors. On average, the CHE posts have a length of 80 words and $\approx$ 7 sentences, while all other datasets have as much as 8 times more tokens and sentences (see Table 2)

Table 3. Accuracies on 12 datasets from 3 different collections used in previous work for AA. Each column shows accuracies for different feature sets: FLF, Randomized Modality Meta Features combined with FLF (RMMF+FLF), and MSMF combined with FLF (MSMF+FLF). Statistical significance between RMMF+FLF and MSMF+FLF, using a two-tailed t-test, is marked with *. Similarly, differences between FLF and MSMF+FLF that are statistically significant are noted with [b]. Gain is given by 100(Col5 - Col3)/Col3.

dataset	#auth	FLF	RMMF+FLF	MSMF+FLF	Gain(%)
CHE	5	72.24	71.86	**79.00**^{*b}	+9.36
CHE	10	71.04	70.56	**76.07**^{*b}	+7.07
CHE	20	65.94	64.35	**71.79**^{*b}	+8.88
CHE	50	57.92	56.31	**65.09**^{*b}	+12.39
CHE	100	55.53	52.10	**63.50**^{*b}	+14.35
Football	3	89.30	88.35	**92.75***	+3.41
Business	6	**86.66**	83.05	86.29*	-0.66
Travel	4	83.84	81.46	**86.70***	+3.41
Cricket	4	**96.20**	93.23	95.59	-0.63
Poetry	6	64.27	63.05	**78.29**^{*b}	+21.81
CCAT	10	83.50	80.5	**84.20***	+0.83
CCAT	50	74.42	69.06	**76.12**^{*b}	+2.28

per document. It is possible that for documents like the ones on the CHE collection, the FLF do not carry enough information for accurate classification due to the short length of these posts, and therefore there is much more to gain from the meta features. This is also supported by the larger increase in accuracy with the MSMF+FLF approach for the Poetry dataset, which is the one with shorter documents from Raghavan *et al.*'s collection.

But another possible reason why the MSMF+FLF yields higher gains in accuracy could be the number of authors. In a small pool of authors the potential relationships that can emerge from comparing writing styles is limited. In a sufficiently large pool of authors it is clear that there are many more possible combinations, and it is more likely that some of the authors will share writing styles in specific dimensions with different authors. The meta features in such a setting will then carry new and more discriminative value than in the setting with a small number of authors.

To allow a comparison with state-of-the-art approaches we run additional experiments using the same train/test partitions as those reported on recent work. It should be noted that the figures we report for each data set and existing approaches are the highest accuracies we found on those papers. These results are shown in Table 4. It is clear that the MSMF+FLF approach is competitive across the different collections, reaching very similar results to those reported earlier, and in some cases outperforming previous approaches.

5   Analysis of Results

The previous section presented interesting results on different AA tasks where the notion of linguistic modalities and a framework designed to exploit similarity scores in

Table 4. Benchmark comparison with recent AA approaches using the same collections and on the same train/test partitions. The numbers in parenthesis show our results from the 5-fold cross-validation setting. For each dataset, bold figure represents the best performance.

dataset	#auth	MSMF+FLF (5fcv)	Benchmark Comparison
CHE	5	74.30 (79.00)	**75.47** [19]
CHE	10	**77.96** (76.07)	77.38 [19]
CHE	20	**72.48** (71.79)	71.42 [19]
CHE	50	**67.00** (65.09)	63.79 [19]
CHE	100	**63.61** (63.50)	62.10 [19]
Football	3	91.11 (92.75)	**93.34** CNG-WPI [5] 91.11 PCFG-E [17]
Business	6	86.66 (86.29)	**91.11** PCFG-E [17] 80.00 CNG-WPI [5]
Travel	4	90.00 (86.70)	**91.67** PCFG-E [17] 73.33 CNG-WPI [5]
Cricket	4	91.11 (95.59)	**95.00** PCFG-E [17] 90.00 CNG-WPI [5]
Poetry	6	63.63 (78.29)	**87.27** PCFG-E [17] 85.45 CNG-WPI [5]
CCAT	10	78.80 (84.20)	**86.4** BOLH Diffusion Kernel [6] 79.40 Char n-grams SVM [21] 78.00 STM-Asymmetric cross [16] 73.60 CNG-WPI [5]
CCAT	50	69.48 (76.12)	**74.04** Char n-grams SVM [8]

a modality specific way yields higher prediction rates than simply using the first level features. We run and analysed an additional set of experiments to explore how much these linguistic modalities are contributing to the models.

Table 5 shows the results on training a SVM using a single modality at a time. In this set of experiments we used the single train/test partition as that used in previous work, and in the results reported in Table 4. These results clearly show that the different characteristics of the data sets have a notable effect on the usefulness of the different linguistic modalities. The *Stylistic* modality has a considerable contribution to the final classification for all the CHE datasets, and is the one with one of the lowest accuracies for all other data sets. This was somewhat expected as several of the stylistic features in that set were crafted in [19] with web forum data as the focus. If we go back to the description of the features (see Table 1), we can easily identify some of the stylistic features that are most likely to not carry any discriminative value for the other data sets, such as the ones related to punctuation marks and capitalization information, as these other data sets have a very uniform pattern for them.

For the *Semantic* modality we have a different result. In the CHE collection, this modality was the second best one in accuracy, but in the CCAT collection this modality reached the highest accuracy. We believe this is a good indication that in this data set there is a stronger correlation between the topic of the document and the authors. Similarly, there seems to be a strong author/topic effect in the Raghavan et al.'s collection,

Table 5. Comparison on accuracies obtained by individual modalities on various datasets. For each dataset, the bold figure indicates the accuracy of the best modality obtained by the best feature set (one of the three feature sets: FLF, MSMF, and MSMF+FLF).

Dataset	#Author	Feature Set	Modality			
			Semantic	Perplexity	Syntactic	Stylistic
CHE	5	MSMF	38.02	23.95	34.89	54.86
		FLF	45.86	16.36	36.42	59.44
		MSMF+FLF	46.40	41.09	36.87	**65.11**
CHE	10	MSMF	30.75	40.73	23.02	60.70
		FLF	45.86	16.36	36.42	60.70
		MSMF+FLF	46.40	40.64	36.87	**65.10**
CHE	20	MSMF	31.46	14.73	22.17	56.05
		FLF	39.01	14.06	30.13	52.92
		MSMF+FLF	39.83	14.88	29.67	**60.42**
CHE	50	MSMF	30.50	15.57	19.68	51.45
		FLF	35.36	15.34	25.38	45.88
		MSMF+FLF	37.07	15.54	25.35	**54.04**
CHE	100	MSMF	31.21	14.84	20.72	50.93
		FLF	32.46	14.84	23.70	45.27
		MSMF+FLF	32.87	15.02	24.20	**52.09**
Football	3	MSMF	80.00	75.55	60.00	44.44
		FLF	86.66	**91.11**	82.22	64.44
		MSMF+FLF	86.66	77.77	82.22	73.33
Business	6	MSMF	73.33	77.77	40.00	32.22
		FLF	80.00	63.33	73.33	57.77
		MSMF+FLF	80.00	**83.33**	73.33	53.33
Travel	4	MSMF	80.00	**86.66**	36.66	35.00
		FLF	76.66	76.66	81.66	43.33
		MSMF+FLF	76.66	85.00	81.66	46.66
Cricket	4	MSMF	73.33	91.66	58.33	63.33
		FLF	80.00	61.66	90.00	66.66
		MSMF+FLF	80.00	**91.66**	91.66	80.00
Poetry	6	MSMF	40.00	**80.00**	27.27	18.18
		FLF	34.54	52.72	40.00	18.18
		MSMF+FLF	34.54	78.18	43.63	20.00
CCAT	10	MSMF	68.40	68.60	28.80	24.60
		FLF	74.80	51.00	74.00	33.20
		MSMF+FLF	**76.00**	69.60	73.60	31.80
CCAT	50	MSMF	57.92	56.92	15.96	13.28
		FLF	62.76	34.00	55.20	10.96
		MSMF+FLF	**66.08**	57.56	55.36	14.60

as the results from this modality are among the highest ones. This could also explain why their PCFG-E approach gave the best results in those data sets as the use of lexical features that carry the semantic content could help boost accuracy of their system.

The results on the *Syntactic* modality seem to indicate a correlation with document length. This modality yields some of the lowest results for those data sets with shorter documents. Overall, these features seem to have a limited contribution to identify authors for the CHE collection, the one with the shortest documents. In the Raghavan et al.'s collection, the data sets with longer documents have higher accuracies when the SVM is trained on only these features. The datasets Football, Business, Travel and Cricket yield accuracies higher than 70% when using the *Syntactic* modality. But for the Poetry dataset this same modality reached an accuracy of 40% in the best case. This latter dataset has an average of 250 words per document, while the former datasets have between 800 and a little over 950 words (see Table 2). Another plausible explanation for these results can be the genre of the datasets. In the CHE datasets there could be a lot of noise in the parser output because of the spontaneity and casual writing style. Although since it is a forum tied to academe, the level of noise from typos, emoticons, abbreviations and slang is not as high as in a typical web forum. In the Poetry dataset it is possible that the format from the prose in there can cause the syntactic analysers to break and output noisy tags and parses.

The same document length effect can be observed in the *Perplexity* modality. Overall, higher accuracies can be seen for datasets with longer documents (CCAT collection and Raghavan et al.'s Football, Business, Cricket, and Travel data sets). Since we are using character 4-grams, there will be some semantic content included here. It is likely too that this is also playing a role in reaching better results for collections that were not deliberately controlled for topic.

In summary, the differences in accuracy reached by the individual modalities indicate that the genre of the documents should guide the selection of features for building the models. For the MSMF approach studied here this conclusion motivates the need for a more sophisticated way to combine the features from the individual modalities. It is possible that higher overall results could be attained if we allow the more discriminative modalities to have a higher weight than other less meaningful modalities in the final author models. A framework like this must be adaptable to the peculiarities of the target datasets and this could be reached with the help of a validation set where such parameters could be fine tuned.

6 Conclusions and Future Work

In this paper we set out to the task of investigating the empirical value of extracting orthogonal similarity patterns in authors writing style to improve AA accuracy. Most approaches rely on finding distinguishable markers in each author's writing style to perform the task, whereas this approach explicitly exploits the notion that authors share writing patterns across specific linguistic dimensions. This idea has been explored by previous work, but without a comprehensive evaluation across different datasets, a one to one comparison with state-of-the-art approaches, and without a necessary comparison with a random generation of linguistic dimensions. The sets of experiments reported here resolve those remaining questions. Our findings show this is a competitive AA framework that seems to be especially useful for datasets with larger number of authors and shorter documents.

The findings from this work also underscore the need for a better modelling of the genre for the AA task. Our results show that significant differences can be attained by the linguistic modalities depending on the nature of the documents. Therefore, a promising line for future work concerns the investigation of an adaptable model where the meta features from different linguistic dimensions will have different weights on the final decisions to reflect their discriminative value.

One of the trends identified in our experimental results refers to observing higher gains in accuracy when adding the modality specific meta features when the number of candidate authors increases. This trend was consistent for all but one of the CHE datasets and both of the datasets from the CCAT collection. It indicates the possibility that the differences in the writeprint of authors reach the ceiling of their discriminative power as the number of candidate authors increases. At the same time, the richer set of authors allows to extract more powerful similarity coefficients when following the modality specific framework. Further experiments are needed to support this claim and we plan to focus on this in the coming months.

It is possible that the notion of orthogonal similarity patterns could be useful in other classification tasks beyond authorship analysis. One potential task is genre classification. Clearly there must be several similarities between different genres across different dimensions. Some genres share stylistic features, consider data from social media, while some others share a different set of modalities, *Semantic* or *Syntactic*. It is then possible that a MSMF approach could yield competitive results.

Acknowledgements. This research was partially supported by ONR grant N00014-12-1-0217 and by NSF award 1254108. It was also supported in part by the CONACYT grant 134186 and by the European Commission as part of the WIQ-EI project (project no. 269180) within the FP7 People Programme.

References

1. Baker, L.D., McCallum, A.: Distributional clustering of words for text classification. In: SIGIR 1998: Proceedings of the 21st Annual International ACM SIGIR, pp. 96–103. ACM, Melbourne (1998)
2. Biber, D.: The multi-dimensional approach to linguistic analyses of genre variation: An overview of methodology and findings. Computers and the Humanities 26, 331–345 (1993)
3. Blum, A., Mitchell, T.: Combining labeled and unlabeled data with co-training. In: Proceedings of the 1998 Conference on Computational Learning Theory (1998)
4. Dhillon, I.S., Mallela, S., Kumar, R.: A divisive information-theoretic feature clsutering algorithm for text classification. Journal of Machine Learning Research 3, 1265–1287 (2003)
5. Escalante, H.J., Montes-y-Gómez, M., Solorio, T.: A weighted profile intersection measure for profile-based authorship attribution. In: Batyrshin, I., Sidorov, G. (eds.) MICAI 2011, Part I. LNCS, vol. 7094, pp. 232–243. Springer, Heidelberg (2011)
6. Escalante, H.J., Solorio, T., Montes-y-Gomez, M.: Local histograms of character n-grams for authorship attribution. In: Proceedings of the 49th Annual Meeting of the Association for Computational Linguistics: Human Language Technologies, pp. 288–298. Association for Computational Linguistics, Portland (2011)
7. Hayes, J.H.: Authorship attribution: A principal component and linear discriminant analysis of the consistent programmer hypothesis. I. J. Comput. Appl., 79–99 (2008)

8. Houvardas, J., Stamatatos, E.: N-gram feature selection for authorship identification. In: Euzenat, J., Domingue, J. (eds.) AIMSA 2006. LNCS (LNAI), vol. 4183, pp. 77–86. Springer, Heidelberg (2006)

9. Karypis, G.: CLUTO - a clustering toolkit. Tech. Rep. #02-017 (November 2003)

10. Keselj, V., Peng, F., Cercone, N., Thomas, C.: N-gram based author profiles for authorship attribution. In: Proceedings of the Pacific Association for Computational Linguistics, pp. 255–264 (2003)

11. Koppel, M., Schler, J., Argamon, S.: Authorship attribution in the wild. Language Resources and Evaluation 45, 83–94 (2011)

12. Lewis, D.D., Yang, Y., Rose, T.G., Li, F.: Rcv1: A new benchmark collection for text categorization research. Journal of Machine Learning Research 5, 361–397 (2004)

13. Luyckx, K., Daelemans, W.: Authorship attribution and verification with many authors and limited data. In: Proceedings of the 22nd International Conference on Computational Linguistics (Coling 2008), Manchester, UK, pp. 513–520 (August 2008)

14. Luyckx, K., Daelemans, W.: The effect of author set size and data size in authorship attribution. In: Literary and Linguistic Computing, pp. 1–21 (August 2010)

15. Marneffe, M.D., MacCartney, B., Manning, C.D.: Generating typed dependency parses from phrase structure parses. In: LREC 2006 (2006)

16. Plakias, S., Stamatatos, E.: Tensor space models for authorship identification. In: Darzentas, J., Vouros, G.A., Vosinakis, S., Arnellos, A. (eds.) SETN 2008. LNCS (LNAI), vol. 5138, pp. 239–249. Springer, Heidelberg (2008)

17. Raghavan, S., Kovashka, A., Mooney, R.: Authorship attribution using probabilistic context-free grammars. In: Proceedings of the ACL 2010 Conference Short Papers, pp. 38–42. Association for Computational Linguistics, Uppsala (2010)

18. Slonim, N., Tishby, N.: The power of word clusters for text classification. In: 23rd European Colloquium on Information Retrieval Research, ECIR (2001)

19. Solorio, T., Pillay, S., Raghavan, S., Montes-y-Gómez: Generating metafeatures for authorship attribution on web forum posts. In: Proceedings of the 5th International Joint Conference on Natural Language Processing, IJCNLP 2011, pp. 156–164. AFNLP, Chiang Mai (2011)

20. Stamatatos, E.: Author identification using imbalanced and limited training texts. In: 18th International Workshop on Database and Expert Systems Applications, DEXA 2007, pp. 237–241 (September 2007)

21. Stamatatos, E.: Author identification: Using text sampling to handle the class imbalance problem. Information Processing and Managemement 44, 790–799 (2008)

22. Stamatatos, E.: Plagiarism detection using stopword n-grams. Journal of the American Society for Information Science and Technology 62(12), 2512–2527 (2011)

23. Stamatatos, E.: A survey on modern authorship attribution methods. Journal of the American Society for Information Science and Technology 60(3), 538–556 (2009)

24. Stolcke, A.: SRILM - an extensible language modeling toolkit, pp. 901–904 (2002)

25. Toutanova, K., Klein, D., Manning, C.D., Singer, Y.: Feature-rich part-of-speech tagging with a cyclic dependency network. In: Proceedings of the 2003 Conference of the North American Chapter of the Association for Computational Linguistics on Human Language Technology, NAACL 2003, vol. 1, pp. 173–180 (2003)

26. de Vel, O., Anderson, A., Corney, M., Mohay, G.: Multi-topic e-mail authorship attribution forensics. In: Proceedings of the Workshop on Data Mining for Security Applications, 8th ACM Conference on Computer Security (2001)

27. Witten, I.H., Frank, E.: Data Mining: Practical Machine Learning Tools and Techniques, 2nd edn. Morgan Kaufmann (2005)

ERNESTA: A Sentence Simplification Tool for Children's Stories in Italian

Gianni Barlacchi[1] and Sara Tonelli[2]

[1] Information Engineering Department, Università di Siena, Italy
barlacchi@student.unisi.it
[2] Fondazione Bruno Kessler, Via Sommarive 18, Povo (Trento), Italy
satonelli@fbk.eu

Abstract. We present ERNESTA (Enhanced Readability through a Novel Event-based Simplification Tool), the first sentence simplification system for Italian, specifically developed to improve the comprehension of factual events in stories for children with low reading skills. The system performs two basic actions: First, it analyzes a text by resolving anaphoras (including null pronouns), so as to make all implicit information explicit. Then, it simplifies the story sentence by sentence at syntactic level, by producing simple statements in the present tense on the factual events described in the story. Our simplification strategy is driven by psycholinguistic principles and targets children aged 7 - 11 with text comprehension difficulties. The evaluation shows that our approach achieves promising results. Furthermore, ERNESTA could be exploited in different tasks, for instance in the generation of educational games and reading comprehension tests.

Keywords: Sentence simplification, Anaphora Resolution, Children Stories, Italian Language.

1 Introduction

Developing children's capabilities to comprehend written texts is key to their development as young adults. Nowadays, around 5–10% of comprehenders aged between 7 and 11 demonstrate text comprehension difficulties, despite proficiency in word decoding and well developed vocabulary knowledge [19]. The consequences of poor reading comprehension extend beyond literacy skills, and can have a negative impact, for instance, on motivation to reading, performances in school curricula and a child's self-esteem [6].

While methodologies for enhancing reading comprehension of English texts have been widely investigated, and research on English simplification has continuously progressed thanks to advances in psycholinguistic research and in natural language processing (see e.g. [3], [24]), only limited efforts have been made to extend current approaches to other languages, such as French [5], Brazilian Portuguese [1] and German [14].

In this work, we present ERNESTA (Enhanced Readability through a Novel Event-based Simplification Tool), the first sentence simplification for Italian,

A. Gelbukh (Ed.): CICLing 2013, Part II, LNCS 7817, pp. 476–487, 2013.

specifically developed to enhance the comprehension of factual events in children's stories. The system performs two basic actions: (i) it resolves pronominal and zero anaphoras, so as to make all implicit information explicit, and (ii) it simplifies the story sentence by sentence at syntactic level, by producing simple statements in the present tense on the factual events described in the story.

The system integrates different NLP modules, such as a PoS-tagger, a dependency parser and an anaphora resolution module. The latter is based on a maximum entropy classifier and has been specifically developed for ERNESTA.

Our simplification approach is targeted on children aged between 7 and 11 with poor reading capabilities. Our methodology leans on psycholinguistic studies focusing on the relationship between the different linguistic dimensions of a text and the cognitive skills of poor comprehenders. Since psycho-linguistic studies suggest that poor comprehenders are impaired in their ability to *select the main point and the main events of a story* [20] the primary goal of our simplification approach is to focus only on factual events, in order to enhance children's ability to understand the story structure. Poor comprehenders have also problems in *resolving anaphors* [12,29,13] therefore our system includes a module for the resolution of personal pronouns and zero anaphora. Besides, many of the skills involved in reading comprehension depend on working memory, i.e. the ability to maintain and process information simultaneously. If sentences are particularly complex and long, their readability is affected because the children's working memory is overloaded [8,7,22]. Therefore, the output of our system is a list of simple sentences describing single events in the present tense and containing only the mandatory arguments of factual verbs.

2   Related Work

Numerous approaches to text compression and simplification have been proposed in the past (see [9] and [11] for reviews of various techniques). Nevertheless, the majority of such approaches has been tested on English, and also the data sets available for evaluation and comparison are mainly composed of English texts (see for instance Simple Wikipedia). Few works related to simplification have focused on Italian, mainly dealing with readability assessment [10,18,27]. In this work, instead, we describe a system for text simplification developed for Italian which, to our knowledge, is the first simplification system for this language. Besides, we also develop and make available specific data sets for evaluation. To this purpose, we take advantage of progress on text simplification presented in similar works [3,4], while coping with language-specific issues.

As for syntactic simplification, our approach is closely related to the method discussed in [16], which extracts simple sentences from each verb in the syntactic dependency trees of complex sentences. However, the authors do not provide code or evaluate their system on realistic texts, so it is unclear how robust and effective their approach is (in a small evaluation with only one text, 55% of the simplifications extracted by their system were acceptable). Also [24] and [25] proved that using typed dependencies for text simplification is a well-founded approach, and that they are preferable to constituency-based information. Nevertheless, the authors evaluate their approach only in a post-hoc setting, by asking judges to consider a simplified sentence acceptable or not. ERNESTA, instead, will be evaluated at different stages in a fully automatic way.

Our approach is tailored to the needs of children aged between 7 and 11, similar to [3], which explicitly addresses text simplification for children. However, we focus here on syntactic simplification, while in [3] the authors combine syntactic and lexical simplification through integer linear programming. Another difference is that the authors' simplification approach reduces the complexity of a sentence by eliminating appositions and replacing complex constructions with a list of single clauses, while we also prune the syntactic trees to eliminate adjuncts and retain only the mandatory arguments of factual verbs.

3   System Architecture

ERNESTA is composed by several processors, displayed in Fig. 1.

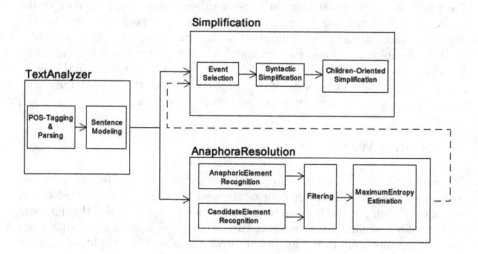

Fig. 1. Sentence simplification process

The Text Analyzer component performs the linguistic pre-processing needed for the two following steps. Then, the Anaphora Resolution component identifies the antecedent of pronouns / subjectless verbs. Finally, the Simplification component performs sentence simplification. It can take in input the text with resolved anaphoras, as produced by the previous module, or operate directly on the pre-processed text. Details on the different modules composing the overall architecture are reported in Section 4.

4 From Syntactic Analysis to Sentence Simplification

ERNESTA sentence simplification has been inspired by some linguistic and cognitive principles (see e.g. [8,7,22,12,29,13], as reported in Section 1), in order to adapt it to the reading difficulties of children:

- **Extract only factual events:** The goal of simplification is to obtain a list of simple statements containing the events described in a story. This should support children in understanding the sequence of events that actually took place in the story. For this reason, we focused on *factual* events, i.e. those that really happened in the narrative. This led us, for instance, to discard as candidate events verbs in the conditional mood and in the future tense.
- **Make information explicit:** Each simplified statement should be understandable, self-contained, with no implicit information. This is particularly crucial in Italian, because it is a pro-drop language in the case of subject pronouns, and the sentences of a story in isolation may miss the information on who performed the action. Therefore, we introduced an anaphora resolution module aimed at replacing personal pronouns with their overt antecedent.
- **Retain all and only necessary information:** Each simplified statement should describe an event and its participants in a well-formed sentence. Each piece of information reported should be necessary to make the event understandable, avoiding all superfluous elements. At syntactic level, this corresponds to eliminating all adjuncts, which however is not a trivial task for Italian, since existing parsers cannot distinguish between arguments and adjuncts with good precision. We devised a simplification strategy to cope also with possible parsing errors (for details, see Section 4.3).
- **Report events in the present tense:** The output of the simplification process should be a list of self-contained statements, which can be used to generate questions on the events, or serve as captions for the graphic illustration of the story. In order to use them in isolation in any of such contexts, the event has been reported in the present tense.

The tools and the algorithms employed to implement these simplification principles are detailed in the following subsections.

4.1 Pre-processing with the Text Analyzer Component

Given an input story, the first step towards its simplification consists in applying a set of NLP tools to extract basic lexical, morphological, syntactic and semantic information. We use TextPro [21], a freely available NLP suite for Italian, to tokenize and lemmatize the text, as well as to provide morphological information on each token. Besides, we run also the in-built named entity recognizer, whose information will be used in the anaphora resolution step.

Next, the Italian version of the Malt parser [17] is used to obtain a syntactic analysis of each sentence, with the tokens being pairwise connected by labeled dependencies[1]. While past works have mainly relied on constituency information to perform syntactic simplification of sentences (see e.g. [3,4]), we implement a simplification strategy based on argument/adjunct information, and therefore we integrate a dependency parser in the processing pipeline. Our choice is corroborated by recent findings on dependency-based simplification [24,25].

4.2   Anaphora Resolution

Anaphora resolution is crucial in ERNESTA because each simplified sentence must be understandable in isolation, with no relevant information left implicit. This is particularly relevant for Italian texts, in which zero-anaphora is very frequent. Therefore, ERNESTA includes an anaphora resolution module aimed at identifying the antecedent of (i) personal pronouns and (ii) null subjects (zero anaphora). Other anaphoric elements may be present in a text, for instance relative pronouns or possessives. However, we focus primarily on these two types of anaphora resolution, which are particularly relevant and frequent in Italian, while we leave the extension and refinement of this module to future work.

Since no anaphora resolution system is freely available for Italian, we developed it by combining a rule-based selection of anaphoric elements and possible antecedents with supervised classification.

Anaphora resolution is performed in four steps, as displayed in Fig. 1: first, anaphoric elements, which need to be resolved by finding an antecedent, are recognized. We basically extract all personal pronouns and verbs without subject. We further exclude from the verbs to resolve some typical verbs which do not have a subject in Italian, for instance meteorological verbs such as 'to rain', 'to snow', etc. Then, the candidate antecedents are collected by considering all nominal phrases that are directly depending on a verbal argument and that appear in a window up to three sentences before the anaphoric element to resolve. Then, we filter out candidate antecedents whose person, noun (and possibly gender) do not match with those of the anaphoric element. Finally, we classify each candidate pair including a possible antecedent and an anaphoric element as co-referring or not.

The features considered take into account lexical and morphological information, the distance between the anaphoric element and the antecedent, as well as the dependency label of candidate antecedents in the dependency tree. The complete list is reported in Table 1. All linguistic information needed to extract the features relies on the output of the pre-processing phase.

We train the maximum entropy classifier implemented in OpenNLP[2] with the data made available at Evalita 2011 evaluation campaign[3]. A first evaluation of the anaphora resolution module in ERNESTA is reported in Section 5.2.

---

[1] The parser achieved 86.5% accuracy on relation labeling and 90.96% accuracy on head finding in the parsing task at Evalita 2009 evaluation campaign, see http://www.evalita.it/2009/tasks/parsing

[2] http://opennlp.apache.org/

[3] http://www.evalita.it/2011

Table 1. Feature list for anaphora resolution

Feature class	Features
Grammatical role of candidate antecedent	Dependency label (e.g. SUBJ, OBJ, etc.)
Grammatical role of anaphoric element (if pronoun)	Dependency label (e.g. SUBJ, OBJ, etc.)
Position	Candidate antecedent is at sentence begin
Position	Anaphoric element is at sentence begin
Position	Anaphoric element and antecedent in the same sentence
Position	Token-based distance between element and antecedent
Dependency	Anaphoric element and antecedent depend on same verb
PoS of anaphoric element	Verb or pronoun
Entity	Entity type of antecedent (PER, LOC, GEO, OTHER)

Note that the anaphora resolution module is independent from ERNESTA and can be easily integrated in other NLP applications.

4.3   Simplified Statement of Factual Events

In the last step of the workflow, the proper simplification process is carried out exploiting the linguistic information extracted in the previous steps. In particular, three subtasks are performed:

- **Identification and selection of factual events:** Factual events correspond to actions that took place in the story. In order to identify them, we select verbs based on their tense and mood information. Specifically, we discard all verbs in conditional mood, because they typically describe potential actions that may not happen. We also discard events expressed in the future tense, because the reader cannot assess if they will actually take place.
- For each factual event, **identification and selection of the mandatory arguments:** for this step, a set of rules has been implemented according to the verb valence. For transitive verbs (specified in a separate list), we discard arguments that are neither a subject nor a direct object. For other verb types, we identify the subject and the first modifier as mandatory, in case the latter is present. This rule has been introduced in order to alleviate an issue with the parser in distinguishing between arguments and adjuncts: this is a very difficult task in Italian, and the parser performance on this specific distinction is quite poor. Therefore, we implement a general rule that does not distinguish between the two.
- **Event reformulation:** after the sentence has been simplified, the verb needs to be expressed in the present tense indicative mood. This task may seem trivial, nevertheless some issues need to be taken into account, for instance the distinction between a passive verb at present tense (e.g. 'sono visto', transl. 'I am seen') and an intransitive verb at the past tense (e.g. 'sono andato', transl. 'I have gone'), because in Italian they are both composed

by the auxiliary 'essere' ('to be') followed by a past participle. Again, we implement some rules to alleviate TextPro errors in distinguishing between active and passive verbs, and then we conjugate the verb accordingly. The conjugation is performed by FSTAN, a generator of morphological forms included in TextPro.

As an example, we show the simplification process of the following sentence:

(1) Ernesta stava mangiando la torta con i suoi amici.
   Ernesta was eating the cake with her friends.

The parser output is displayed in Fig. 2: after the main event 'mangiando' ('eating') has been extracted, ERNESTA looks for its arguments labeled by the dependency parser. In this example, the verb is the head of a SUBJ, a DOBJ and a RMOD. Since the simplification rule implemented for transitive verbs retains only the subject and the direct object, the RMOD subtree is first removed. Then, 'stava mangiando' ('was eating') is replaced by the present indicative form 'mangia' ('eats') . The simplified sentence is displayed in Fig. 3.

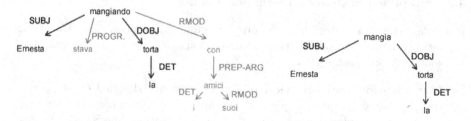

Fig. 2. Sentence before simplification: '*Ernesta was eating a cake with her friends*'

Fig. 3. Simplified sentence: '*Ernesta eats the cake*'

5   Experimental Setup and Evaluation

We perform evaluation on three different tasks, namely (i) event extraction, (ii) anaphora resolution and (iii) sentence simplification. The details are reported in the following subsections.

5.1   Evaluation of Event Extraction

The first subtask we evaluate is the performance of ERNESTA on event extraction: since we want to extract only factual events, describing what really happened in a story, we first assess whether the correct verbs are selected. We create a gold standard composed of 6 children stories[4], asking an annotator to

---

[4] Both the original stories and the gold standard are available at the link
   http://terence.fbk.eu/services/api/Ernesta

select only the sentences whose main verb describes a factual event. Specifically, she was asked to select an event only if it would be possible to put it in a timeline and connect it temporally to the other events in the story (an intuitive way to select only factual events). The total number of events in the gold data set is 320. Then, we compare the events recognized as factual by ERNESTA with those in the gold data set and we compute standard precision, recall and F1. The system achieves **P 0.88, R 0.79** and **F1 0.83** (macro average).

Although the system performance is generally good, there are few sources of errors which negatively affect event extraction. For instance, the PoS-tagger may assign a wrong label (mainly Noun) to the verb denoting the event, which causes the extraction algorithm to discard it. Another possible source of error is the morphological analysis, because imperatives in Italian have the same form as verbs in the present indicative, therefore some misclassifications may arise: imperatives should be discarded from the event list, while verbs in the present indicative are generally included. Imperatives are quite frequent in our stories, because the characters tend to have short, direct interactions between them. In general terms, however, the implemented strategy seems to be effective in extracting factual events from children's stories.

5.2 Evaluation of Anaphora Resolution

The only Italian data set available with manually annotated anaphora is, to our knowledge, the one released for the coreference resolution task at 2011 Evalita evaluation campaign [28]. Since other existing data sets (for instance [23]) are not freely available, we train our classifier on Evalita data. We rely on the same training and test split used in the task, although we focus only on resolution of pronouns and zero anaphora, while the data set includes all coreferring entities. The maximum entropy classifier included in the OpenNLP suite was trained with 250 iterations.

The number of anaphoric elements in the test set is 515. The anaphora resolution module of ERNESTA identifies 236 (46%) of them as anaphoric. The main source of error is the fact that our selection strategy is sometimes too greedy, discarding possible antecedents if they are more than 3 sentences distant from the anaphoric element. Also, agreement of person, noun and gender between the pronoun and the antecedent is a strong constraint, which is probably not robust enough to cope with errors in the morphological analysis.

Considering only the subset of these 236 anaphoric elements, the resolution algorithm achieves **P 0.43, R. 0.16** and **F1 0.23**. Most of the classification mistakes depend on the complexity of the documents in the text set. In fact, they often show convoluted syntactic structures, full of appositions and subordinating / coordinating constructions. Given the lower complexity of children's stories, we expect that anaphora resolution on this kind of texts will be less problematic. However, this requires the development of a domain-specific test set with stories being manually annotated with anaphoras. This is currently under development.

5.3   Evaluation of Sentence Simplification

Evaluating sentence simplification is not a trivial task, because a sentence may be simplified in several ways. Therefore, past approaches to the task have been usually evaluated by asking human raters to judge the correctness / plausibility of simplified sentences. In [3], for instance, the authors use Amazon's Mechanical Turk, asking three judges to indicate if the resulting simplified sentences are still correct in English.

In our case, the simplification process is constrained by the fact that *all and only* mandatory arguments of the verbs should be retained, which makes it possible to create a gold standard where all required participants in the event are present and no much space is left to human interpretation. This allows us to use an evaluation approach which is stricter though more objective than post-hoc human judgment. We apply the Translation Error Rate (TER) measure [26], usually employed to evaluate automatic machine translation, which is computed as the number of edits needed to transform the output of a machine translation system into a reference translation, normalised by the number of words in the reference translation. The possible edit operations are the insertion, deletion and substitution of single words as well as changes in the position (shifts) of word sequences. We applied this measure to pairs made up of an automatically simplified statement (*hypothesis*) and a manually simplified statement containing the same event as the hypothesis (*reference*), using the TER-Plus evaluation tool.[5] Again, we do not perform a post-hoc evaluation, but we create the gold standard independently from the system output. For each event in the gold data created for the previous evaluation (Section 5.1) which was correctly extracted by ERNESTA, an annotator was asked to rewrite it in the present tense and remove all verb arguments which are not mandatory. For instance, in case of transitive verbs, only the subject and the object should be retained. In case of doubts, the annotator was allowed to look up the verb valence in the Sabatini – Colletti dictionary of Italian.[6] This manually simplified version of the sentences in the story are considered our *reference* in computing TER. Note that, while in machine translation evaluation several possible references are admissible, in our case only one reference is considered correct.

Table 2. Evaluation of simplification process (the lower, the better)

	Translation Error Rate
Baseline	0.73
ERNESTA: All Weights = 1	0.51
ERNESTA: Insertion = 0.5	0.41

The evaluation results are reported in Table 2. Note that TER ranges from 0 to 1, with 0 being a perfect simplification (i.e. the simplified version and the

---

[5] http://www.cs.umd.edu/~snover/tercom/

[6] http://dizionari.corriere.it/dizionario_italiano/

reference are identical, no edit operations are needed) and 1 being completely wrong. The baseline is obtained by removing all adjuncts (labeled as RMOD) to simplify the sentences. This means that we delete all arguments identified by the parser as optional (RMOD), without additional processing. We further compute TER with two different weighting schemes: in the first case, the weight of each edit operation is set to 1. In the second case, we set insertions to 0.5, meaning that if some argument is present in the system output and not in the reference, it is less penalized than the opposite case. In fact, since the reference contains all and only necessary information to describe an event, removing some words would probably produce an ungrammatical sentence. The opposite may lead to having a sentence only partially simplified but still grammatical.

In general, we notice that our simplification strategy combining lexical, syntactic and morphological information is more effective than a strategy purely based on syntactic information as produced by the parser. However, the current evaluation involves the simplification model in isolation. We leave the joint evaluation of the simplified sentences *with* resolved anaphora to future work.

6   Future Research Directions and Conclusions

In this paper, we presented ERNESTA, the first sentence simplification system for Italian. We showed that a strategy based on rich linguistic information is effective in producing simplified statements targeting children in elementary schools with poor reading skills. The system, however, strongly relies on the parser output, therefore it would be interesting to test it using other existing dependency parsers for Italian, e.g. [2]. Another open issue is the problem of anaphora: current results of the resolution module are clearly insufficient to be integrated in a tool that is supposed to be used in schools. Alternative approaches using the web as external knowledge source for antecedent selection are currently being explored.

The output of ERNESTA can be used in a number of tasks. For instance, it can be given in input to a question generation system that automatically creates reading comprehension tests. In order to generate questions on the subject of a simple declarative clause, the NP starting the sentence and preceding the verb can be easily identified as the subject. Then its corresponding entity type has to be found (for instance by using a NER and WordNet), and the corresponding question word ('who' or 'what') can be joined to the following verb. The process is displayed in Example (2), with the declarative sentence being the simplified clause generated by our system. This extension would be in line with [15], who transform sentences into questions by first expanding the source text into a set of derived declarative clauses.

(2) <u>Ernesta</u> mangia la torta. → <u>Chi</u> mangia la torta?
    <u>Ernesta</u> eats the cake → <u>Who</u> eats the cake?

Another application of our system is the creation of captions for figures in children's storybooks and the generation of games based on these stories. This direction is already being explored in the Terence European project[7] and has shown

---

[7] http://www.terenceproject.eu/

promising results. Although the generated sentences cannot be used as is, they currently undergo a manual revision to avoid inconsistencies and are then fed to the game generation system. In the near future, we plan to exploit this revised version of the sentences to improve our current output by using machine learning techniques.

Acknowledgments. The work described in this paper has been partially funded by the European Commission under the contract number FP7-ICT-2009-5, Terence project. We thank Laura Carniel for preparing the gold data, Marco Guerini for useful comments on an early version of the paper and Davide Alocci for technical support. We will always be grateful to Emanuele Pianta for inspiring this work and for being a brilliant, patient and friendly advisor. He will be missed greatly.

References

1. Aluisio, S., Specia, L., Gasperin, C., Scarton, C.: Readability assessment for text simplification. In: Proceedings of the 5th Workshop on Innovative Use of NLP for Building Educational Applications, Stroudsburg, PA, USA, pp. 1–9 (2010)
2. Attardi, G., Dell'Orletta, F., Simi, M., Chanev, A., Ciaramita, M.: Multilingual Dependency Parsing and Domain Adaptation using DeSR. In: Proceedings of the CoNLL Shared Task Session of of EMNLP-CoNLL (2007)
3. Belder, J.D., Moens, M.F.: Text simplification for children. In: Proceedings of the SIGIR 2010 Workshop on Accessible Search Systems (2010)
4. Bernhard, D., de Viron, L., Moriceau, V., Tannier, X.: Question Generation for French: Collating Parsers and Paraphrasing Questions. Dialogue and Discourse 3(2), 43–74 (2012)
5. Brouwers, L., Bernhard, D., Ligozat, A.L., François, T.: Simplification syntaxique de phrases pour le français. In: Actes de la Conférence Conjointe JEP-TALN-RECITAL, Montpellier, France, pp. 211–224 (2012)
6. Cain, K.: Making sense of text: skills that support text comprehension and its development. Perspectives on Language and Literacy (2009)
7. Cain, K., Oakhill, J.: Profiles of children with specific reading comprehension difficulties. British Journal of Educational Psychology 76(683-696) (2006)
8. Carretti, B., Borella, E., Cornoldi, C., De Beni, R.: Role of working memory in explaining the performance of individuals with specific reading comprehension difficulties: A meta-analysis. Learning and Individual Differences 19(2), 246–251 (2009)
9. Clarke, J.: Global Inference for Sentence Compression: An Integer Linear Programming Approach. Ph.D. thesis, University of Edinburgh (2008)
10. Dell'Orletta, F., Montemagni, S., Venturi, G.: READ-IT: Assessing Readability of Italian Texts with a View to Text Simplification. In: Proceedings of the Second Workshop on Speech and Language Processing for Assistive Technologies, Edinburgh, Scotland, UK, pp. 73–83 (July 2011)
11. Dorr, B., Zajic, D.: Hedge Trimmer: A parse-and-trim approach to headline generation. In: Proceedings of the Workshop on Automatic Summarization (2003)
12. Ehrlich, M.F., Rémond, M.: Skilled and less skilled comprehenders: French children's processing of anaphoric devices in written texts. British Journal of Developmental Psychology 15, 291–309 (1997)

13. Ehrlich, M.F., Remond, M., Tardieu, H.: Processing of anaphoric devices in young skilled and less skilled comprehenders: Differences in metacognitive monitoring. Reading and Writing 11, 29–63 (1999)
14. Glöckner, I., Hartrumpf, S., Helbig, H., Leveling, J., Osswald, R.: An architecture for rating and controlling text readability. In: Proceedings of KONVENS 2006, Konstanz, Germany, pp. 32–35 (2006)
15. Heilman, M., Smith, N.A.: Extracting Simplified Statements for Factual Question Generation. In: Proceedings of QG 2010: The Third Workshop on Question Generation, Pittsburgh, Pennsylvania, USA (2010)
16. Beigman Klebanov, B., Knight, K., Marcu, D.: Text simplification for information-seeking applications. In: Meersman, R. (ed.) OTM 2004. LNCS, vol. 3290, pp. 735–747. Springer, Heidelberg (2004)
17. Lavelli, A., Hall, J., Nilsson, J., Nivre, J.: MaltParser at the EVALITA 2009 Dependency Parsing Task. In: Proceedings of EVALITA Evaluation Campaign (2009)
18. Lucisano, P., Piemontese, M.E.: Gulpease. Una formula per la predizione della difficoltà dei testi in lingua italiana. Scuola e Città 3, 57–68 (1988)
19. Lyon, G.R., Fletcher, J.M., Barnes, M.C.: Learning Disabilities. In: Mash, E., Barkley, R.A. (eds.) Child Psychopathology, The Guilford Press, NY (2003)
20. Nation, K., Clarke, P., Marshall, C.M., Durand, M.: Hidden language impairments in children: parallels between poor reading comprehension and specific language impairment? Journal of Speech, Language and Hearing Research 47, 199–211 (2004)
21. Pianta, E., Girardi, C., Zanoli, R.: The TextPro tool suite. In: Proc. of the 6th Language Resources and Evaluation Conference (LREC), Marrakech, Morocco (2008)
22. Pimperton, H., Nation, K.: Suppressing irrelevant information from working memory. Evidence for domain-specific deficits in poor comprehenders. Journal of Memory and Language 62(380–391) (2010)
23. Poesio, M., Delmonte, R., Bristot, A., Chiran, L., Tonelli, S.: The VENEX corpus of anaphora and deixis in spoken and written Italian, University of Essex (2004) (manuscript)
24. Siddharthan, A.: Complex lexico-syntactic reformulation of sentences using typed dependency representations. In: Proceedings of the 6th International Natural Language Generation Conference (INLG 2010), Dublin, Ireland (2010)
25. Siddharthan, A.: Text Simplification using Typed Dependencies: A Comparision of the Robustness of Different Generation Strategies. In: Proceedings of the 13th European Workshop on Natural Language Generation (2011)
26. Snover, M., Dorr, B., Schwartz, R., Micciulla, L., Makhoul, J.: A study of translation edit rate with targeted human annotation. In: Proceedings of Association for Machine Translation in the Americas (2006)
27. Tonelli, S., Tran Manh, K., Pianta, E.: Making readability indices readable. In: Proceedings of the First Workshop on Predicting and Improving Text Readability for Target Reader Populations, Montréal, Canada, pp. 40–48 (June 2012)
28. Uryupina, O., Poesio, M.: Evalita 2011: Anaphora Resolution Task. In: Proceedings of EVALITA Evaluation Campaign (2011)
29. Yuill, N.M., Oakill, J.V.: Effects of inference awareness training on poor reading comprehension. Applied Cognitive Psychology 2, 33–45 (1988)

Automatic Text Simplification in Spanish: A Comparative Evaluation of Complementing Modules

Biljana Drndarević[1], Sanja Štajner[2], Stefan Bott[1],
Susana Bautista[3], and Horacio Saggion[1]

[1] Universitat Pompeu Fabra, Barcelona, Spain
{biljana.drndarevic,stefan.bott,horacio.saggion}@upf.edu
[2] University of Wolverhampton, Wolverhampton, UK
sanjastajner@wlv.ac.uk
[3] Universidad Complutense de Madrid, Madrid, Spain
subautis@fdi.ucm.es

Abstract. In this paper we present two components of an automatic text simplification system for Spanish, aimed at making news articles more accessible to readers with cognitive disabilities. Our system in its current state consists of a rule-based lexical transformation component and a module for syntactic simplification. We evaluate the two components separately and as a whole, with a view to determining the level of simplification and the preservation of meaning and grammaticality. In order to test the readability level pre- and post-simplification, we apply seven readability measures for Spanish to three sets of randomly chosen news articles: the original texts, the output obtained after lexical transformations, the syntactic simplification output, and the output of both system components. To test whether the simplification output is grammatically correct and semantically adequate, we ask human annotators to grade pairs of original and simplified sentences according to these two criteria. Our results suggest that both components of our system produce simpler output when compared to the original, and that grammaticality and meaning preservation are positively rated by the annotators.

1 Introduction

Automatic text simplification as an NLP task arose from the necessity to make electronic textual content equally accessible to everyone. Organisations such as Inclusion Europe[1] point out to the essential right for every person to take active part in the life of their society through access to information. Nevertheless, numerous people experience difficulties reading government reports, laws, news articles and other written material that enables their inclusion in the community. Some Internet portals have created simplified variants of their content, as is the case with Simple English Wikipedia[2]. However, simplifying text manually

---

[1] http://inclusion-europe.org/en
[2] http://simple.wikipedia.org/wiki/Main_Page

A. Gelbukh (Ed.): CICLing 2013, Part II, LNCS 7817, pp. 488–500, 2013.

is time-consuming and not cost-effective, especially in the case of news articles, which are constantly being generated and updated. That is why attempts have been made to automate the laborious process of text simplification. So far, systems have been developed for English [1], Portuguese [2] and Japanese [3], with recent attempts at Basque [4] and Swedish [5] text simplification.

Automatic text simplification is a complex task which encompasses a number of operations applied at different linguistic levels. The aim is to turn a complex text into its simplified variant, taking into consideration the specific needs of a particular target user. Our Simplext project is one such aspiration [6]. We have been developing a system for automatic text simplification in Spanish, aimed at producing more readable news articles for people with cognitive disabilities. We conduct simplification at the syntactic and the lexical levels of the input text. Easy-to-read guidelines indicate that a single idea should be expressed per sentence [7], so we divide a complex sentence into as many simple sentences as possible, as part of our syntactic simplification strategy. The guidelines also suggest that common and simple words should be used to express the desired idea, and that the use of technical and complex vocabulary should be avoided [8]. This entails treatment of the lexical items of the input text. We here describe one component of our lexical simplification module, which applies rule-based transformations to phrases and expressions that cannot be simplified through a more traditional synonym substitution approach. However, we do employ the latter approach for the second component of our lexical module [9], currently under development and not presented on this occasion. Our main goal here is to concentrate on the evaluation of the two existing components of our system, and test their performance in terms of the grade of simplification, the grammaticality of the output, and the preservation of original meaning.

The remainder of the article is organised as follows: in Section 2 we present an overview of the most relevant work in the field of automatic text simplification; in Section 3 we outline our approach to the task at hand, describe in some detail the different components of our system, and present the experimental setting, while section 4 discusses the results of our evaluation experiments; we conclude the article with a summary and plans for future work in Section 5.

2 Related Work

Automatic text simplification has traditionally had a double purpose. It can be used as a preprocessing tool for other NLP applications [10], where it serves the purpose of improving their performance. On the other hand, it has been widely used to offer simpler reading material for target users, such as foreign language learners [11], readers with aphasia [12], low literacy individuals [13], etc. The first attempts are rule-based syntactic simplification systems [14]. Carroll et al. [15] contribute with an additional lexical simplification module, and introduce the paradigm, often repeated thereafter, of simplification based on synonym substitution. They use WordNet to obtain a set of potential synonyms of content words in the input text, and determine the simplest out of the set by looking up Kucera-Francis frequencies in the Oxford Psycholinguistic Database [16]. Word frequency is, therefore, seen as a measure of lexical complexity, and this approach has been adopted in a number of works that follow. Bautista et al. [17] use a similar approach but introduce word length as an additional indicator of word difficulty. De Belder et al. [18] were the first to introduce a word-sense-disambiguation element to their lexical simplification system in order to account for numerous cases of polysemy, especially common among the more frequent words.

The availability of large parallel corpora, such as the "original" and the Simple English Wikipedia, has made recent approaches to automatic text simplification more data-driven. Biran et al. [19] apply an unsupervised method for learning pairs of complex and simple synonyms from a corpus of texts from the original Wikipedia and Simple English Wikipedia. Their approach is called context-aware because they calculate cosine similarity between the given context of a lexical item and the context vector of that item from a trained model. Recently, text simplification has been likened to machine translation, and techniques traditionally used in the latter have been exploited for the purpose of developing novel automatic text simplification systems [20], [21].

3   Methodology

Our methodology consists of: (1) an analysis of a parallel corpus of original and manually simplified news articles, aimed at extracting types of simplification operations to be automated; (2) building our system accordingly; and (3) evaluating the output of the automatic simplification, with regards to the grade of simplification, the grammaticality of the output, and the preservation of meaning in the simplification process.

3.1   Corpus Analysis

We have compiled a corpus of 200 original and manually simplified news articles in Spanish, provided by the Spanish news agency Servimedia[3]. Simplifications have been applied by trained human editors, familiar with the particular needs of our target user (a person with cognitive disabilities) and following a series of easy-to-read guidelines suggested by Anula [22]. We examine the said corpus in order to target different types of simplification operations and, subsequently, prepare their possible computational implementation.

The simplification changes observed in the corpus can largely be grouped as follows:

1. **Syntactic operations**: changes applied at the sentence level, such as sentence splitting or quotation inversion.
2. **Lexical operations**: infrequent, long or technical terms are substituted with their simpler synonyms, and certain expressions are paraphrased or otherwise modified.

---

[3] http://www.servimedia.es/

3. **Content reduction**: a significant portion of original content is eliminated through summarisation and paraphrases, in accordance with the guidelines that indicate that only the most essential piece of information should be preserved.
4. **Clarification**: certain complex terms and concepts, for which no synonym can be found, are explained by means of a definition.

Even though we have explored the possibility of automating all four strategies employed by human editors, we have so far implemented the first two: (1) a syntactic simplification module, which conducts a series of transformations at the sentence level, based on operations observed in the parallel corpus; (2) lexical rule-based transformations, which transform certain phrases and expressions that cannot be simplified through the traditional synonym substitution, such as numerical expressions or ethnic adjectives. These operations are also a subset of operations applied by human editors when building the parallel corpus. We are also working on developing a second component of the lexical module, one based on synonym substitution, in which we employ a word vector model to find possible substitutes for difficult original words, and we compute the difficulty (or simplicity) of a word based on its frequency and length. Even though this is intended as a significant component of our system, this module is currently under development and will not be discussed in further detail in this paper.

3.2   Syntactic Simplification Module

We developed a rule-based system for syntactic simplification [23] which is dedicated to several types of sentence splitting operations. These operations turn subordinate and coordinate structures, such as relative clauses, gerundive constructions and VP coordinations into separate sentences, producing shorter and syntactically less complex outputs. The module operates on syntactic dependency trees and tree manipulation is modelled as graph transduction. The following pair of (1) original and (2) simplified sentences are an example of the simplification of a participle construction.[4]

1. The participants (...) will be presented with a book, edited by the town council (...)
2. The participants (...) will be presented with a book. This book is edited by the town council (...)

The grammar comprises five groups of rules, which are dedicated to different syntactic target phenomena. The grammar was previously evaluated, looking at correct rule applications, but so far it has not been evaluated for its contribution to the simplicity of its output. The evaluation in Section 4 includes this second aspect of evaluation.

---

[4] All examples in the paper are translated into English so as to make it more legible.

3.3   Lexical Rule-Based Transformations

Corpus analysis has revealed that human editors pay special attention to certain types of expressions and that they consistently apply simlification operations to them. Although the operations are applied at the lexical level of the text, synonym substitution is not sufficient in these cases, since the simplification strategies we have observed are somewhat more complex. After carefully examining all such cases, we eventually prepared the computational treatment of the following expressions:

1. **Numerical Expressions (NumExp).** We here define numerical expressions as consisting of a number, expressed either in figures or in letters (*in [2009]*) and additional elements, such as modifiers (*[around] 370,000 children*), measurements (*120,000 [square kilometres]*), quantified objects (*almost 700 [crimes]*), etc. We treat nine different types of numerical expressions. Simplification operations include rounding, insertion of modifiers to account for the loss of precision, eliminating optional elements in dates, etc. Detailed description of all simplification operations is beyond the scope of this paper (see [24] for details), though some have been illustrated in Table 1, together with the nine types of Numerical Expressions we treat.

Table 1. Types of NumExp and examples of original and simplified expressions

Type of NumExp	Example Orig.	Example Simpl.
General quantities	*451 attacks*	*almost 500 attacks*
Decimal numbers and fractions	*1.5 million Pakistanis*	*almost 2 million Pakistanis*
Monetary expressions	*1,400 euros*	*more than 1,000 euros*
Percentages	*13.4% of the doctors*	*more than 13% of the doctors*
Dates	*from the 1st of February of 2011*	*from 2011*
Years	*2010*	*the year 2010*
Numbers in letters	*nine million*	*9 million*
Decades	*two decades*	*20 years*
Centuries	*four centuries*	*400 years*

2. **Parenthetical Expressions.** Any information contained in parentheses is eliminated, as it is seen as additional content not essential for the core message of the text. The following sentence is an illustration from our corpus, where the eliminated content is in boldface:

    'Ana María Matute had previously won the National Award for Children's Literature (**"Just a bare foot", 1987**) and the Spanish Literature National Award **(2007)**.'

3. **Ethnic Adjectives.** We have observed that ethnic adjectives, such as *Tunisian*, have been substituted with the construction [from/of + <ORIGIN>] rather consistently in our corpus. So, for example, *the Tunisian authorities* has been transformed into *the authorities of Tunisia*. The same is true of nominalisations of these adjectives, where the combination of the definite article and the adjective, e.g. *the Pakistanis*, is substituted with the construction [person from/of + <ORIGIN>], e.g. *the people from/of Pakistan*.

4. **Reporting Verbs.** The various reporting verbs found in the original texts of our corpus have been repeatedly substituted with *decir* (*say*), which is perceived as the simplest option. Such decision is in accordance with the WCAG guidelines that indicate that one and the same term should be consistently used to express the same concept [8]. In order to apply this transformation, we have compiled a list of 32 reporting verbs, based on our corpus and using the Web as a resource. Although this list is by no means an exhaustive list of reporting verbs in Spanish, it serves the purpose better than looking up synonyms in a dictionary. In the Spanish OpenThesaurus dictionary[5], which we use for the development of the module based on synonym substitution, only a third of the verbs from our list is synonymous with *decir*.

These transformations have been implemented with JAPE rules [25], but given space constraints, we here cannot provide a full account of the implementation procedure.

3.4   Experimental Setting

We evaluate both the different components of our system and the system as a whole. The aim of the evaluation is to test (1) the degree of the simplification of the system and its components; and (2) the grammaticality of the output and the preservation of meaning with respect to the original. To achieve the former, we use a set of Spanish readability formulae, and we simultaneously carry out evaluation with human annotators, who rate the degree of grammaticality and meaning preservation in a Likert-scale type of questionnaire[6].

The first evaluation step consists of applying a series of readability formulae for Spanish [22], [26] to the original and simplified texts. The readability formulae intend to capture complexity at the syntactic and lexical levels and are presented in Table 2 (where N = number, w = words, s = sentences, cs = complex sentences[7], dcw = different content words, lfw = low-frequency words, cw = content words, dw = different words, rw = rare words, NumExp = numerical expressions, punct = punctuation marks, and char = characters). It is important to point out that, following Anula [22], we consider as *low frequency words* those words whose frequency rank in the Referential Corpus of Contemporary Spanish[8] is lower than 1,000. Similarly, *rare words* are, according to Spaulding [26], the words that do not appear on the list of 1,500 most commonly used Spanish words. Both lists were lemmatised using Connexor's parser in order to retrieve the frequency of the lemma and not a word form (action carried out manually in the two cited works).

The decision about the grammaticality of the output of our system and the meaning preservation in the process of simplification was entrusted to a group of

---

[5] http://openthes-es.berlios.de/

[6] http://nil.fdi.ucm.es/surveysimp

[7] We here consider a complex sentence one that contains multiple finite predicates according to the output of Connexor's parser.

[8] http://corpus.rae.es/lfrecuencias.html

Table 2. Readability formulae applied to the data sets

Formula	Calculation
Average Sentence Length (ASL)	N(w)/N(s)
Index of Complex Sentences (ICS)	N(cs)/N(s)
Sentence Complexity Index (SCI)	(ASL+ICS)/2
Lexical Density Index (LDI)	N(dcw)/N(s)
Index of Low-Frequency Words (ILFW)	(N(lfw)/N(cw))*100
Lexical Complexity (LC)	(LDI+ILFW)/2
Spaulding Density (SD)	N(w)/N(rw)
Spaulding Spanish Readability (SSR)	1.609*ASL+331.8*SD+22.0
Average Word Length (AWL)	N(char)/N(w)
Number of NumExp (NUM)	N(NumExp)
Number of punctuation marks (PUNC)	N(punct)

25 human annotators. They were presented with a questionnaire consisting of 38 pairs of original (O) sentences taken from the corpus of 100 texts used to test the formulae (see Section 4.1), and their simplified (S) equivalents obtained by our system. Every O-S pair contained at least one syntactic and one lexical change. The order of O and S sentences in the 38 pairs was alternated randomly. For every pair of sentences, three questions had to be answered choosing the degree of agreement on a scale from 1 (completely disagree) to 5 (completely agree): (1) *Paragraph A is grammatical*; (2) *Paragraph B is grammatical*; (3) *Paragraphs A and B have the same meaning*[9]. All annotators were native speakers of Spanish and did not include the authors of this paper. Inter-annotator agreement was not calculated, given the elevated number of annotators, and a wide range of options to choose from when grading (five-point scale).

4   Results and Discussion

The results of the two evaluation experiments are discussed separately in the sections that follow, upon which a joint conclusion is presented in Section 5.

4.1   Evaluating the Degree of Simplification

In the first instance, we applied the formulae to the pairs of original and manually simplified texts in our corpus (see Section 3.1), in order to test whether the formulae are a good indicator of the degree of simplification. The results of this experiment are presented in Table 3 (where higher values indicate higher complexity, and the individual formulae that combine into a single complexity index are left out). Differences between all features were reported to be statistically significant at a 0.001 level of significance (paired t-test implemented in SPSS).

After we confirmed the validity of all formulae as indicators of text complexity, the formulae were applied to 100 randomly chosen news articles from the

---

[9] We used the word "paragraph" since some original sentences were transformed into two simplified ones, and these could not be called a sentence.

Table 3. Formulae applied to the original and manually simplified texts

	LC	SSR	AWL	ASL	SCI	PUNC	NUM
Original	9.71	184.20	4.97	33.42	17.09	12.90	5.20
Simplified	5.28	123.82	4.75	13.69	7.14	1.61	1.80
Rel. diff.	-46.25%	-32.60%	-4.27%	-57.15%	-57.88%	-46.95%	-87.97%

categories of national news, international news, culture, and society, which had been simplified in three stages:

– applying only lexical rule-based transformations (*Lexical*);
– applying only syntactic simplification (*Syntactic*);
– applying both components of our system (*Both*).

We thus obtain three different data sets to be evaluated in comparison with the original texts (Table 4). Differences between the outputs of automatic simplification systems and the original texts which are statistically significant at a 0.001 level of significance (paired t-test implemented in SPSS) are shown in bold. Those marked with an '*' are statistically significant at a 0.002 level of significance, which is still a reasonable result. One important observation is that both original sets (Table 3 and Table 4) achieve practically identical scores on all formulae, meaning that the 100 texts used to test the system are structurally close to the 200 texts simplified manually and used to test the formulae. We can, therefore, expect the selected formulae to be a reliable indicator of complexity of the output produced by our system.

Table 4. Comparison of the original texts and the three simplified text sets

Corpus	LC	SSR	AWL	ASL	SCI	PUNC	NUM
Original	10.10	182.21	4.93	33.48	17.14	13.92	6.41
Lexical	10.08	**174.85**	**4.81**	33.65	17.22	**10.18**	5.73*
Syntactic	**9.92**	**174.40**	4.94	**28.15**	**14.43**	**13.50**	6.41
Both	9.90	**167.21**	**4.82**	**28.36**	**14.54**	**10.64**	5.73*

Averaged relative differences between the corresponding text pairs are given in Table 5. Two general conclusions can be made: (1) both the syntactic simplification and the lexical transformations generally produce simpler output with respect to the original; (2) the combination of the two simplfication processes generally produces a simpler output than either one individually. We have to acknowledge the considerable distance between the relative differences of automatically simplified texts and the ones simplified manually. This, however, is to a large extent due to the fact that manual simplification employs summarisation and paraphrases as most common simplification operations (44%), which results in the loss of a large number of structural elements of the original, among them the punctuation marks and numerical expressions taken into account by the formulae. Our system in its current state does not perform comparable content reduction. We have previously investigated the possibility of using some summarisation techniques for the purpose of text simplification [27], and intend to accordingly expand the system in the future. However, it is important to point out that manual transformations applied to the original texts in our corpus are often highly idiosyncratic and dependent on world knowledge, and, as a result, it would be difficult to expect to achieve the same grade of simplification automatically.

Table 5. Averaged relative differences between the corresponding text pairs

Comparing	LC	SSR	AWL	ASL	SCI	PUNC	NUM
Original vs. Lexical	+1.31%	-3.97%	-2.55%	+0.65%	+0.66%	-25.22%	-6.66%
Original vs. Syntactic	-1.94%	-4.25%	+0.16%	-14.97%	-15.08%	-2.54%	0
Original vs. Both	-0.36%	-8.13%	-2.27%	-14.22%	-14.34%	-19.86%	-6.66%

4.2   Evaluating Grammaticality and Meaning Preservation

The obtained results were grouped in such a way so as to measure: (1) the annotators' attitude towards the grammaticality of original sentences; (2) the annotators' attitude towards the grammaticality of simplified sentences; and (3) the annotators' attitude towards the differences in meaning between O and S sentences. For each of the sets we measured the average, mean and median value, as indicators of central tendency, and frequency distribution, as an indicator of variability [28]. Table 6 contains the said data. We combined the two lower scores (1-2) into one, to indicate a generally negative attitude towards the grammaticality/meaning, the higher two scores (4-5) into the one indicating a generally positive attitude towards grammaticality/meaning, while the central score (3) represents a neutral attitude.

Table 6. Grammaticality and meaning preservation – central tendency and variability

Measure	Gramm. of O	Gramm. of S	Meaning
Average	4.60	3.58	3.83
Mode	5	4	4
Median	5	5	5
1	2.00%	10.53%	7.47%
2	2.63%	15.26%	10.74%
3	5.47%	16.53%	14.11%
4	13.26%	21.37%	26.53%
5	76.63%	36.32%	41.16%
Negative	4.63%	25.79%	18.21%
Neutral	5.47%	16.53%	14.11%
Positive	89.89%	57.69%	67.69%

Even though the grammaticality of original texts was, expectedly, rated more positively than that of their simplified equivalents (though not at the expected rate of 100%), the latter were also rated rather positively on the whole (average score for the entire set of simplified sentences being in the neutral category). The sentences that received individual average score lower than 3 (i.e. the grammatiacality of these sentences was generally negatively rated), contained 18 grammatical errors, ten resulting from poor syntactic simplification and the remaining eight from bad application of lexical transformation rules. The most recurrent grammatical error was incorrect treatment of different types of coordinate structures, such as coordination of relative clauses. The following pair of sentences is an illustration, with the coordination in question highlighted:

1. The Defence Minister announced that the museum (...) is going to achieve wider presence *in Spain and outside our borders* establishing itself as (...)
2. The Defence Minister said that the museum (...) is going to achieve wider presence in Spain. Outside our borders establishing itself as (...)

We found that one third of the errors were traceable to previous parsing errors. Correcting this bad input is beyond the scope of our system. Another third of the errors were attributed to slight errors of the grammar which can be reliably remedied with minor changes in the rules. The remaining errors were related to more complicated syntactic phenomena, which could, in principle, be treated by syntactic rules, but which would require more extensive grammar engineering.

As for the lexical errors, all but one resulted from poor inclusion of the output structure into the existing context. When rounding numbers and using modifiers to account for the loss of precision, we sometimes obtain an ungrammatical combination consisting of a determiner and an adverb, as in *another almost 30 houses*. Given that the majority of numerical expressions from the 100 text set (see Section 3.4) are accompanied by some kind of determiner, restricting the application of the rule to cases other than these would result in considerable drop in recall. What could be done is round the number without the use of modifier, since the loss of precision in meaning is seen as less problematic for our target user than is the actual complexity of the content (see Section 3.1). What is significant is that these two types of errors account for 80% of the S sentences with poor grammaticality, and addressing these two issues in the future should considerably improve the performance of the system.

Meaning preservation was quite positively rated, with the annotators stating that the meaning of the two sentences in the O-S pair was the same in almost 70% of the cases. Only three pairs of sentences were rated negatively (1 or 2). In all three cases, the distortion of meaning is due to syntactic simplification errors, similar to the one previously discussed. The said syntactic errors in combination with the previously mentioned lexical error, account for 60% of the pairs rated neutrally. Therefore, meaning preservation is seen as directly dependent on grammaticality, and the latter is perceived as more important than the loss of precision, even for users without cognitive disabilities (i.e. the participants in the questionnaire). With that in mind, future fine-tuning of certain elements of our system, such as the aforementioned rounding of numerical expressions, seems like a feasible task, and one to favourably affect overall system performance.

5   Conclusions and Future Work

In this paper we presented two components of an automatic text simplification system for Spanish, and we evaluated them from two perspectives: (1) employing seven readability measures developed for Spanish, we tested the degree of the simplification of our system and its components; (2) in a Likert-scale type of questionnaire, we asked 25 human annotators to rate the grammaticality of the automatic simplification output and the grade to which meaning was preserved in the process.

Our results indicate that both components of our system (syntactic simplification module and rule-based lexical transformations) produce simpler output compared to the original, and that the combination of the two achieves a higher degree of simplification than either of the elements individually. Our system does not reach the simplification degree of manual transformations, but this is largely due to the fact that summarisation and paraphrases are two most commonly applied techniques in the process of manual simplification (they account for as much as 44% of all manual transformations), and as a result, a significant portion of the original content is eliminated. Given that easy-to-read guidelines for people with cognitive disabilities indicate that complexity reduction has preference over the preservation of informational precision, we intend to incorporate a summarisation component into future versions of our system, with the aim of increasing the degree of simplification.

As for linguistic accuracy of the output, our system was rather positively rated by the annotators, 60% of whom considered the simplified sentences to be grammatical, while around 70% of them agreed on the fact that the meaning was preserved reasonably well in the process of simplification. The qualitative analysis of the results revealed that most common errors that result in poor grammaticality of the output were bad treatment of coordinate structures in the syntactic simplification stage, and infelicitous treatment of context when applying lexical transformations. Meaning was seen as directly dependent on the grammaticality of the output, so addressing the two previously mentioned aspects of our system components in the future, should positively influence its overall performance. Nevertheless, the problems resulting from parsing errors remain out of our control for the time being.

Acknowledgments. We present this work as part of a project entitled Simplext: An automatic system for text simplification, with file number TSI-020302-2010-84[10] and partially supported by the Spanish Ministry of Economy and Competitiveness (Project Number TIN2012-38584-C06-03). We are also grateful to the fellowship RYC-2009-04291 from *Programa Ramón y Cajal 2009, Ministerio de Economía y Competitividad, Secretaría de Estado de Investigación,*

---

[10] http://www.simplext.es

Desarrollo e Innovación, Spain. Part of this research is funded by the Spanish Ministry of Education and Science (TIN2009-14659-C03-01 Project), *Universidad Complutense de Madrid*, Banco Santander Central Hispano (GR58/08 Research Group Grant), and the FPI grant program.

References

1. Medero, J., Ostendorf, M.: Identifying Targets for Syntactic Simplification. In: Proceedings of Speech and Language Technology in Education Workshop (2011)
2. Aluísio, S.M., Specia, L., Pardo, T.A.S., Maziero, E., De Mattor Fortes, R.P.: Towards Brazilian Portuguese Automatic Text Simplification systems. In: Proceedings of the ACM Symposium on Document Engineering (2008)
3. Inui, K., Fujita, A., Takahashi, T., Iida, R., Iwakura, T.: Text Simplification for Reading Assistance: A Project Note. In: Proceedings of the 2nd International Workshop on Paraphrasing: Paraphrase Acquisition and Applications, IWP (2003)
4. Aranzabe, M.J., Díaz De Ilarraza, A., González, I.: First Approach to Automatic Text Simplification in Basque. In: Proceedings of the First Natural Language Processing for Improving Textual Accessibility Workshop, NLP4ITA (2012)
5. Rybing, J., Smithr, C., Silvervarg, A.: Towards a Rule Based System for Automatic Simplification of Texts. In: The Third Swedish Language Technology Conference (2010)
6. Saggion, H., Gómez Martínez, E., Etayo, E., Anula, A., Bourg, L.: Text Simplification in Simplext: Making Text More Accessible. In: Revista de la Sociedad Española Para el Procesamiento del Lenguaje Natural (2011)
7. Freyhoff, G., Hess, G., Kerr, L., Menzel, E., Tronbacke, B., Van Der Veken, K.: Make it Simple, European Guidelines for the Production of Easy-to-Read Information for People with Learning Disability; for authors, editors, information providers, translators and other interested persons (1998)
8. Cooper, M., Reid, L., Vanderheiden, G., Caldwell, B.: Understanding WCAG 2.0. A guide to understanding and implementing Web Content Accessibility Guidelines 2.0. In: World Wide Web Consortium, W3C (2010)
9. Bott, S., Rello, L., Drndarević, B., Saggion, H.: Can Spanish Be Simpler? LexSiS: Lexical Simplification for Spanish. In: Proceedings of Coling 2012: The 24th International Conference on Computational Linguistics (2012)
10. Chandrasekar, R., Doran, D., Srinivas, B.: Motivations and Methods for Text Simplification. In: COLING, pp. 1041–1044 (1996)
11. Burstein, J., Shore, J., Sabatini, J., Lee, Y.W., Ventura, M.: The Automated Text Adaptation Tool. In: HLT-NAACL (Demonstrations), pp. 3–4 (2007)
12. Devlin, S., Unthank, G.: Helping aphasic people process online information. In: Proceedings of the 8th International ACM SIGACCESS Conference on Computers and Accessibility (2006)
13. Specia, L.: Translating from complex to simplified sentences. In: Proceedings of the 9th International Conference on Computational Processing of the Portuguese Language, Berlin, Heidelberg, pp. 30–39 (2010)
14. Siddharthan, A.: An Architecture for a Text Simplification System. In: Proceedings of the Language Engineering Conference (LEC 2002), 64–71 (2002)
15. Carroll, J., Minnen, G., Canning, Y., Devlin, S., Tait, J.: Practical Simplification of English Newspaper Text to Assist Aphasic Readers. In: Proc. of AAAI 1998 Workshop on Integrating Artificial Intelligence and Assistive Technology (1998)

16. Quinlan, P.: The Oxford Psycholinguistic Database. Oxford University Press (1992)
17. Bautista, S., Gervás, P., Madrid, R.: Feasibility Analysis for SemiAutomatic Conversion of Text to Improve Readability. In: Proceedings of the Second International Conference on Information and Communication Technologies and Accessibility (2009)
18. De Belder, J., Deschacht, K., Moens, M.F.: Lexical simplification. In: Proceedings of the 1st International Conference on Interdisciplinary Research on Technology, Education and Communication (2010)
19. Biran, O., Brody, S., Elhadad, N.: Putting it Simply: a Context-Aware Approach to Lexical Simplificaion. In: Proceedings of the ACL (2011)
20. Zhu, A., Bernhard, D., Gurevych, I.: A Monolingual Tree-based Translation Model for Sentence Simplification. In: Proceedings of The 23rd International Conference on Computational Linguistics, Beijing, China, pp. 1353–1361 (2010)
21. Coster, W., Kauchak, D.: Simple English Wikipedia: a New Text Simplification Task. In: Proceedings of the 49th Annual Meeting of the Association for Computational Linguistics: Human Language Technologies (2011)
22. Anula, A.: Tipos de Textos, Complejidad Lingüística y Facilicitación Lectora. In: Actas del Sexto Congreso de Hispanistas de Asia, pp. 45–61 (2007)
23. Bott, S., Saggion, H.: Automatic Simplification of Spanish Text for e-Accessibility. In: Proceedings of the 13th International Conference on Computers Helping People with Special Needs, pp. 54–56 (2012)
24. Bautista, S., Drndarević, B., Hervás, R., Saggion, H., Gervás, P.: Análisis de la Simplificación de Expresiones Numéricas en Español mediante un Estudio Empírico. Linguamática 4 (2012)
25. Maynard, D., Tablan, V., Cunningham, H., Ursu, C., Saggion, H., Bontcheva, K., Wilks, Y.: Architectural Elements of Language Engineering Robustness. Journal of Natural Language Engineering – Special Issue on Robust Methods in Analysis of Natural Language Data 8, 257–274 (2002)
26. Spaulding, S.: A Spanish Readability Formula. Modern Language Journal (1956)
27. Drndarević, B., Saggion, H.: Reducing Text Complexity through Automatic Lexical Simplification: an Empirical Study for Spanish. SEPLN Journal, 13–20 (2012)
28. Boone Jr., H., Boone, D.: Analizing Likert Data. Journal of Extension (2012)

The Impact of Lexical Simplification by Verbal Paraphrases for People with and without Dyslexia

Luz Rello[1,2], Ricardo Baeza-Yates[1,3], and Horacio Saggion[2]

[1] Web Research Group, Dept. of Information and Communication Technologies,
Universitat Pompeu Fabra, Barcelona, Spain
[2] Natural Language Processing Research Group, Dept. of Information and
Communication Technologies, Universitat Pompeu Fabra, Barcelona, Spain
[3] Yahoo! Research Barcelona, Spain
{luzrello,rbaeza}@acm.org, horacio.saggion@upf.edu

Abstract. Text simplification is the process of transforming a text into an equivalent which is easier to read and to understand, preserving its meaning for a target population. One such population who could benefit from text simplification are people with dyslexia. One of the alternatives for text simplification is the use of verbal paraphrases. One of the more common verbal paraphrase pairs are the one composed by a lexical verb (*to hug*) and by a support verb plus a noun collocation (*to give a hug*). This paper explores how Spanish verbal paraphrases impact the readability and the comprehension of people with and without dyslexia dyslexia. For the selection of pairs of verbal paraphrases we have used the *Badele.3000* database, a linguistic resource composed of more than 3,600 verbal paraphrases. To measure the impact in reading performance and understandability, we performed an eye-tracking study including comprehension questionnaires. The study is based on a group of 46 participants, 23 with confirmed dyslexia and 23 control group. We did not find significant effects, thus tools that can perform this kind of paraphrases automatically might not have a large effect on people with dyslexia. Therefore, other kinds of text simplification might be needed to benefit readability and understandability of people with dyslexia.

Keywords: Lexical simplification, verbal paraphrases, readability, understandability, eye-tracking, dyslexia.

1 Introduction

The goal of this paper is to present the impact of lexical simplification through verbal paraphrases in readability and understandability for people with and without dyslexia.

Dyslexia has been defined as a specific reading disability [39] and as a learning disability [20]. It is characterized by difficulties with accurate and/or fluent word recognition and by poor spelling and decoding abilities. These difficulties

A. Gelbukh (Ed.): CICLing 2013, Part II, LNCS 7817, pp. 501–512, 2013.
© Springer-Verlag Berlin Heidelberg 2013

typically result from a deficit in the phonological component of language that is often unrelated to other cognitive disabilities. Secondary consequences may include problems in reading comprehension and reduced reading experience that can impede growth of vocabulary and background knowledge [20]. According to cognitive neuroscience studies, people with dyslexia find difficulties with functional [26] and short words [37]. Functional and short words are present in the verbal paraphrases (support verb plus a noun collocation, *dar un paseo, 'to go for a walk'* to be simplified by a lexical verb *pasear, 'to walk'*).

In this study, we distinguish between **readability** and **understandability**. Readability refers to the legibility of a text, that is, the ease with which text can be read (that is, the person can reproduce it even though does not understand it) while understandability refers to comprehensibility, the ease with which text can be understood. Since readability strongly affects text comprehension [5], sometimes both terms have been used interchangeably [21]. However, previous research with people with dyslexia has shown that both concepts need to be taken into consideration separately. For instance, in [31] the inclusion of graphical schemes in the text improved their readability in terms of reading speed, but had a negative effect on the comprehension for people with dyslexia. Moreover, for people with dyslexia, comprehension has been found to be independent of the lexical quality of the text. While errors in text affect negatively readability and understandability of people without dyslexia, they do not affect that much in people with dyslexia [29].

This research is motivated by (1) its novelty and (2) by the social relevance of its results. First, lexical complexity such as word frequency, verb complexity and lexical ambiguity has an effect on the readability and understandability for people with dyslexia [18] and without this condition [28]. In this study, we try to enrich previous findings exploring how practical examples of verbal paraphrases impact readability and understandability to find out whether lexical simplification systems targeted for people with dyslexia shall include verbal paraphrases. To measure readability we analyze eye movements of readers with and without dyslexia using eye tracking and for addressing reading comprehension, we used questionnaires with inferential questions. To the best of our knowledge, this is the first time that the effect of verbal paraphrases is measured in terms of readability and understandability for people with and without dyslexia using this methodology.

Second, since dyslexia is universal and frequent, people with dyslexia are a relatively large group of users. The Interagency Commission on Learning Disabilities [19] states that 10 to 17.5% of the population in the U.S.A. have dyslexia and between 7.5 to 11.8 % of the Spanish speaking population has dyslexia [30]. Also, dyslexic-related difficulties are shared by other groups with special needs such as low vision [16] and symptoms of dyslexia are common to varying degrees among most people [14]. Thus, the results of this research may be applicable to general usability problems and other target groups.

This paper is organized as follows. Next section covers the related work while Section 3 covers lexical simplification by using verbal paraphrasing. In Section 4, we present our experimental methodology while in Section 5 we show the results of it. We end in Section 6 with some concluding remarks and our future work.

2   Related Work

Related work to our study belong to different fields: (a) natural language processing (NLP) literature about paraphrases and their use in lexical simplification, and (b) experimental psychology studies which takes into account the impact of language complexity in reading comprehension and performance of people with dyslexia.

In **NLP** a paraphrase is an alternative surface form in the same language expressing the same semantic content as the original form [24]. The use of automatic methods for generating paraphrases has been successfully applied for text simplification among other NLP tasks. For instance, in [21] paraphrasing is used to remove difficult syntactic structures for deaf learners of written English and Japanese. Paraphrasing methods were applied to simplify newspaper texts for people with aphasia [10,11] and Down syndrome [33] as well as to simplify online information for people with aphasia [13].

Text complexity and dyslexia also has been studied in **experimental psychology**. Word frequency, verb complexity and lexical ambiguity are related to the processing time of words [28,34]. Hyönä and Olson measure the effect of word length and word frequency in relation with eye fixation patterns and show that low frequency and long words present longer gaze durations and more reinspections in both, readers with and without dyslexia [18]. In that work, the analysis is focused on target words [18] while we measure the whole text and the integration of target words in the overall text. The rationale behind this is that readability and understandability pertain to longer segments of texts [17]. Comprehension in people with dyslexia was studied in correlation with syntax complexity including long sentences with complex structures [35], the sentence context [25], or the word fluency [12], among others.

However, there are no studies for Spanish which approach readability and comprehension of people with dyslexia taking into consideration one common verbal paraphrasing pair [2] used for lexical simplification. That is, the pair composed of a lexical verb (*abrazar*, 'to hug') and by a support verb plus a noun collocation (*dar un abrazo*, 'to give a hug').

3   Lexical Simplification by Verbal Paraphrases

Under 18% of manual simplification operations made by experts in newspaper articles are lexical changes [6]. One of the most common simplification solutions done manually in Spanish is the substitution of the combination of the support verb and a deverbal noun by the corresponding verb alone [15]. That is, *dar un paseo*, 'to go for a walk' by *pasear*, 'to walk' or *dar un abrazo*, 'to give a hug' by *abrazar*, 'to hug'. Although these kind of lexical simplifications are frequent in manual simplifications, their automatic computational process is still challenging

[15]. Thus, there are specific linguistic resources developed for such tasks, such as the *Badele.3000* database [3].

Badele.3000 is a database that contains more than 3,600 high frequency Spanish nouns and 2,800 high frequency Spanish verbs, including 23,000 collocations made from the combinations of both kinds of words. The paraphrase pairs consisting of a verb and a verb-noun collocation were manually extracted [4]. As *Badele.3000* was created manually by an expert, the linguistic validity of the paraphrases pairs used in our study is guaranteed.

The selected pairs of synonymic paraphrases are composed of a support verb plus a noun collocation and a lexical verb. According to the manual simplifications [15], the lexical verb alone is considered to be simpler; for instance:

> [−simple] Sus lectores *tenían confianza* en ella.
>
> Her readers *had trust* in her.
>
> [+simple] Sus lectores *confiaban* en ella.
>
> Her readers *trusted* her.

According to cognitive neuroscience studies, it would also be expected that people with dyslexia might find more difficult to read the [−simple] option since they have more frequent errors with functional [26] and short words [37]. However, from a linguistic point of view it is not clear which option is simpler.

Linguists agree in differencing lexical words and functional words [23]. Lexical words have a lexical meaning which is less ambiguous than the grammatical meanings expressed by functional words. Functional words are prepositions, pronouns, auxiliary verbs, conjunctions, among others. Support verbs have been considered as functional words because they are semantically empty, for instance verb *dar*, *'to give'* is a support verb in *dar un abrazo*, *'to give a hug'*.[1]

Since functional words do not have a lexical representation their processing is different than lexical words [8]. There are still many open questions about the difference levels of word processing by the human brain. However, in the case of dyslexia a special emphasis have been made for errors in functional words [26]. To the best of our knowledge, there is no formal explanation behind errors in functional words. They could be due to their nature (i.e. lack of lexical content) or could be simply due to the fact that higher errors rates are observed for shorter words [37].

On the other hand, word processing depends on the complexity of the morphological components of the word [9]. For instance, *paseo*, *'walk'* is simpler than *pasear*, *'to walk'* because it is composed by one lexeme while *pasear* is made by one lexeme plus one derivative morpheme *pasear = paseo + ar*. Since it is not trivial to access the complexity of the paraphrase pairs from a linguistic point of view, we take as our criteria the empirical analysis observed in manual simplifications performed by experts [15].

---

[1] However, Barrios [2] analyzed extensively the meaning of support verbs concluding that some of them are not fully empty.

4   Experimental Methodology

We designed one experiment which combines reading tests, comprehension tests and semi-structured interviews. Twenty three participants with dyslexia and a comparable control group undertook the experiment.

4.1   Participants

Twenty-three native Spanish speakers with a confirmed diagnosis of dyslexia took part in the study, twelve of whom were female and eleven male. Their ages ranged from 13 to 37, with a mean age of 20.74. Three of the participants were also diagnosed with attention deficit disorder. All participants were frequent readers; eleven read less than four hours per day, nine read between four and eight hours per day, and three participants read more than eight hours daily. Ten people were studying or already finished university degrees, eleven were attending school or high school and two had no higher education. All the participants were asked to bring their diagnoses to the experiment. Therefore, we can guarantee that the participant was diagnosed in an authorized center or hospital. A control group of 23 participants without dyslexia with the same age range and similar age average (20.91) also took part the experiment.

4.2   Design

The experiment is composed of: (1) a questionnaire designed to collect demographic information, (2) two reading tests with their corresponding target words, (3) two tests designed to control the comprehension, and (4) a semi-structured interview about their impression and opinions about the readability of the texts. The experiment followed a within-subjects design, so every participant contributed to each of the conditions, [+simple] and [−simple], in both experiments. The order of the conditions was counter-balanced to cancel out sequence effects, guaranteeing that the person never reads the same text twice (see Figure 1).[2]

With the reading tests we collect the quantitative data to measure readability, while with the comprehension tests we measure understandability. At the end, with the semi-structured interviews we gather information about the participant preferences.

We selected two very similar newspaper texts from the Spanish Simplex corpus [7]. To meet the comparability requirements among the texts, we slightly adapted the texts maintaining as much as possible the original text. To determine these comparability requirements we took into account the parameters that different complexity measures take into consideration [15]. Next, we present the characteristics shared by the texts of the experiment:

---

[2] We do not need to consider the two texts in different order as they have similar text complexity.

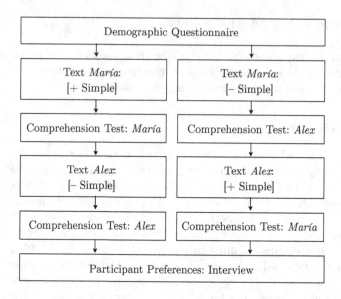

Fig. 1. Structure of the experiment

(a) They are about similar topics: a literature award (Text *María*) and a cinema award (Text *Alex*). See the Appendix for the texts used in the experiment.
(b) They have the same number of target lexical substitutions: nine verbal paraphrase pairs [±simple]. See the Appendix for all the paraphrase pairs used in the experiment.
(c) They share the same genre: culture news.
(d) They have the same number of sentences per text, five sentences.
(e) They have the same number of words per text, 100 words.
(e) All the texts have a similar word length average ranging from 4.87 to 5.19 letters per word.
(f) They contain the same number of named entities mentioned for the first time.
(i) The texts do not contain numerical expressions, foreign words or acronyms.

Since the presentation of the text has an effect on reading speed of people with dyslexia [32], we used the same layout for all the texts. We chose a recommended font type, sans serif arial [1], unjustified text [27], a big size of 20 points, 62 characters per column, and recommended color and brightness contrast using a black font with creme as background[3] [32].

To control the comprehension, after each text we designed a test including inferential items related to the main idea. We did not include items referred to details because they involve memory more than comprehension [36]. Each of the items has three choices where one is correct, one is partially correct (normally

---

[3] The CYMK are creme (FAFAC8) and black (000000). Color difference: 700, Brightness difference: 244.

containing details), and one is incorrect. We gave 100, 50 and 0 points for each type of answer, respectively, to compute a comprehension score.

The test finishes with one semi-structured interview to learn the participant preferences. The participant was asked which text seemed to be more readable. After this, we asked face-to-face the reasons leading to the selected answer, which difficulties they encountered when reading the texts, and which options would they like to find to achieve a better understanding.

4.3 Equipment

The eye tracker used was the Tobii T50 [38, 17-inch TFT monitor] using a resolution of 1024x768 pixels. The eye tracker was calibrated for each participant and the light focus was always in the same position. The distance between the participant and the eye tracker was constant (approximately 60 cm. or 24 in.) and controlled by using a fixed chair.

4.4 Procedure

The sessions were conducted at Pompeu Fabra University and they took from 20 to 30 minutes each, depending on the amount of information given by the participant. In each session, the participant was alone with the interviewer (first author) in the quiet room prepared for the study, and had to do the following three steps: (1) answer a questionnaire designed to collect demographic information; (2) perform the eye-tracking experiment; and (3) answer the semi-open interview. For (2) the participant was asked to read the texts in silence and to complete the comprehension tests. We obtained 40 test samples out of the 44 possible that were successfully recorded.

4.5 Data Analysis

The software used for analyzing the eye tracking data was Tobii Studio 3.0 and the R 2.14.1 statistical software. The dependent variables used for the comparison of the text passages were the means of the fixation duration and the total duration of reading. Differences between groups and dependent variables were analyzed by means of matched-pairs, and two-way Student t-tests.

5 Results

To measure the impact of verbal simplification in readability we analyzed two variables derived from eye-tracking data: the average fixation duration and the total visit duration of the text passages. In general, shorter fixations are preferred to longer ones because according to previous studies [22,28,34], readers make longer fixations at points where processing loads are greater. Also, shorter reading durations are preferred to longer ones since faster reading is related to more readable texts [40]. We compare readability with understandability through

Table 1. Experimental results of the eye-tracking and comprehension user study for the texts using paraphrases (none of the differences are statistically significant)

Measure (ave. ± std.dev.)	[+simple]	[−simple]
	Group with Dyslexia	
Fixation Duration	0.229 ± 0.063	0.226 ± 0.054
Visit Duration	44.403 ± 17.225	47.425 ± 14.610
Correct Answers	67.5%	67.5%
	Group without Dyslexia	
Fixation Duration	0.180 ± 0.040	0.178 ± 0.039
Visit Duration	25.172 ± 5.482	27.825 ± 6.993
Correct Answers	75%	77.5%

the inferential items of the comprehension test which are assessed by the percentage of correct answers.

All our results are given in Table 1. As expected, comprehension for people with dyslexia is slightly lower than those for people without dyslexia.

First, we studied the differences between participants with dyslexia and control group. The average fixation duration of people with dyslexia (0.228 ± 0.058) was significantly higher than for people without dyslexia (0.179 ± 0.039), with $t(80) = 4.4583$, and $p < 0.001$ (see Figure 2).

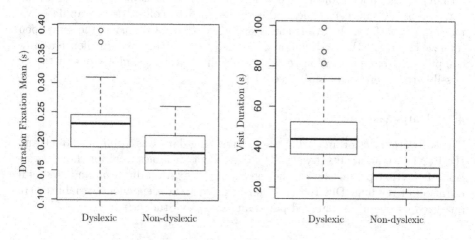

Fig. 2. Box plots for the average fixation and the total duration for the two groups

The results for fixation duration do not show statistical significance, because we obtain $t(40) = 0.1613$, $p < 0.873$. The same happens with visit duration with $t(40) = 0.1753$, $p < 0.862$.

To estimate the likelihood that we missed revealing an existing effect of verbal paraphrases on the mean of fixation durations, we calculated the achieved statistical power. Given a p-value of 0.873, an effect size of 0.052 (Cohen's d), and

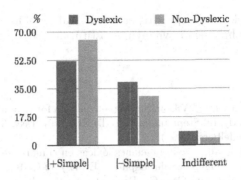

Fig. 3. Preferences of the two groups

a sample size of 40, the achieved power is 0.880. Hence, the probability of not committing a Type II Error is 88%, that is, the likelihood that an unrevealed effect exists is only 12%.

Second, we studied the visit duration time. The statistical results were similar to the ones for the average duration time and hence corroborated the negative finding. The larger range of values for visit time in the group with dyslexia compared with the control group probably indicates the individual variations in reading methods which would make correlations between individuals difficult to observe. On the other hand, the results of the semi-open interview did not matched the analysis of the quantitative variables as shown in Figure 3. That is, the perception of the people is that the simpler text was indeed simpler, although they did not read faster.

6 Conclusions

In this paper we have studied the impact of verbal paraphrases in lexical simplification for people with and without dyslexia.

We chose to study these kind of paraphrases because of two reasons. First, there are already linguistic resources for NLP including these type of Spanish paraphrases [3], which can serve as a starting point. Second, according to cognitive neuroscience studies, this kind of verbal simplification might be especially suitable for people with dyslexia because they find difficulties with functional [26] and short words [37].

The effect of the verbal paraphrases is concluded to be insignificant. Our results are negative in the sense that verbal paraphrases neither improved readability nor understandability in our experiment. However, we can argue a few reasons that may explain this result, implying that further research is needed. The main reason might be that the impact of paraphrasing may depend on the complexity of the text (e.g. in more complex texts verbal paraphrases might be beneficial). Another reason is that the impact is small and hence we need larger texts and a larger number of people to detect it.

As people with dyslexia do have problems with reading most texts, including simple texts, as shown by our results, a more promising line of future research is studying more complex techniques to perform lexical simplification. For example, other types of paraphrasing or synonym substitution.

References

1. Al-Wabil, A., Zaphiris, P., Wilson, S.: Web navigation for individuals with dyslexia: an exploratory study. In: Stephanidis, C. (ed.) HCI 2007. LNCS, vol. 4554, pp. 593–602. Springer, Heidelberg (2007)
2. Barrios, M.A.: El dominio de la funciones léxicas en el marco de la Teoría Sentido-Texto. Estudios de Lingüística del Español (ELiEs) 30 (2010)
3. Barrios, M.A., Aguado de Cea, G., Ramos, J.A.: Enriching a lexicographic tool with domain definition problems and solutions. In: Sierra, G., Pozzi, M., Torres, J. (eds.) 1st International Workshop on Definition Extraction, RANLP 2009, Bulgaria, INCOMA Ltd, Shoumen (2009)
4. Barrios Rodríguez, M., Rello, L.: False paraphrase pairs in Spanish for verbs and verb + noun collocations. Procesamiento del Lenguaje Natural 46, 107–112 (2011)
5. Barzilay, R., Elhadad, N., McKeown, K.R.: Inferring strategies for sentence ordering in multidocument news summarization. Journal of Artificial Intelligence Research 17, 35–55 (2002)
6. Bott, S., Rello, L., Drndarevic, B., Saggion, H.: Can Spanish be simpler? LexSiS: Lexical simplification for Spanish. In: Proceedings of the 24th International Conference on Computational Linguistics (Coling 2012), Mumbai, India (December 2012)
7. Bott, S., Saggion, H.: Text simplification tools for Spanish. In: Proceedings of the Eighth International Conference on Language Resources and Evaluation (LREC 2012), ELRA, Istanbul (2012)
8. Caramazza, A.: How many levels of processing are there in lexical access? Cognitive Neuropsychology 14(1), 177–208 (1997)
9. Caramazza, A., Laudanna, A., Romani, C.: Lexical access and inflectional morphology. Cognition 28(3), 297–332 (1988)
10. Carroll, J., Minnen, G., Canning, Y., Devlin, S., Tait, J.: Practical simplification of English newspaper text to assist aphasic readers. In: Proceedings of the AAAI 1998 Workshop on Integrating Artificial Intelligence and Assistive Technology, pp. 7–10. Citeseer (1998)
11. Carroll, J., Minnen, G., Pearce, D., Canning, Y., Devlin, S., Tait, J.: Simplifying text for language-impaired readers. In: Proceedings of EACL, pp. 269–270 (1999)
12. Cutting, L., Materek, A., Cole, C., Levine, T., Mahone, E.: Effects of fluency, oral language, and executive function on reading comprehension performance. Annals of Dyslexia 59(1), 34–54 (2009)
13. Devlin, S., Unthank, G.: Helping aphasic people process online information. In: Proceedings of the 8th International ACM SIGACCESS Conference on Computers and Accessibility, pp. 225–226. ACM (2006)
14. Dixon, M.: Comparative study of disabled vs. non-disabled evaluators in user-testing: dyslexia and first year students learning computer programming. In: Stephanidis, C. (ed.) HCI 2007. LNCS, vol. 4554, pp. 647–656. Springer, Heidelberg (2007)
15. Drndarevic, B., Saggion, H.: Towards automatic lexical simplification in Spanish: an empirical study. In: Proceedings of the NAACL HLT 2012 Workshop Predicting and Improving Text Readability for Target Reader Populations, PITR 2012 (2012)

16. Evett, L., Brown, D.: Text formats and web design for visually impaired and dyslexic readers-clear text for all. Interacting with Computers 17, 453–472 (2005)

17. Huenerfauth, M., Feng, L., Elhadad, N.: Comparing evaluation techniques for text readability software for adults with intellectual disabilities. In: Proceedings of the 11th International ACM SIGACCESS Conference on Computers and Accessibility, pp. 3–10. ACM (2009)

18. Hyönä, J., Olson, R.: Eye fixation patterns among dyslexic and normal readers: Effects of word length and word frequency. Journal of Experimental Psychology: Learning, Memory, and Cognition 21(6), 1430 (1995)

19. Interagency Commission on Learning Disabilities: Learning Disabilities: A Report to the U.S. Congress. Government Printing Office, Washington DC, U.S (1987)

20. International Dyslexia Association: Definition of dyslexia, http://interdys.org/DyslexiaDefinition.htm (2011); Based in the initial definition of the Research Committee of the Orton Dyslexia Society, former name of the IDA, done in 1994

21. Inui, K., Fujita, A., Takahashi, T., Iida, R., Iwakura, T.: Text simplification for reading assistance: A project note. In: Proceedings of the Second International Workshop on Paraphrasing, vol. 16, pp. 9–16. Association for Computational Linguistics (2003)

22. Just, M., Carpenter, P.: A theory of reading: From eye fixations to comprehension. Psychological Review 87, 329–354 (1980)

23. Lyons, J.: Semantics, vol. 2. Cambridge Univ. Pr. (1977)

24. Madnani, N., Dorr, B.J.: Generating phrasal and sentential paraphrases: A survey of data-driven methods. Computational Linguistics 36(3), 341–387 (2010)

25. Nation, K., Snowling, M.: Individual differences in contextual facilitation: Evidence from dyslexia and poor reading comprehension. Child Development 69(4), 996–1011 (1998)

26. Patterson, K., Marshall, J., Coltheart, M.: Surface dyslexia: Neuropsychological and cognitive studies of phonological reading. Lawrence Erlbaum Associates, London (1985)

27. Pedley, M.: Designing for dyslexics: Part 3 of 3 (2006), http://accessites.org/site/2006/11/designing-for-dyslexics-part-3-of-3

28. Rayner, K., Duffy, S.: Lexical complexity and fixation times in reading: Effects of word frequency, verb complexity, and lexical ambiguity. Memory & Cognition 14(3), 191–201 (1986)

29. Rello, L., Baeza-Yates, R.: Lexical quality as a proxy for web text understandability. In: The 21st International World Wide Web Conference (WWW 2012), Lyon, France (April 2012)

30. Rello, L., Baeza-Yates, R.: The presence of English and Spanish dyslexia in the Web. New Review of Hypermedia and Multimedia 8, 131–158 (2012)

31. Rello, L., Baeza-Yates, R., Saggion, H., Graells, E.: Graphical schemes may improve readability but not understandability for people with dyslexia. In: Proceedings of the NAACL HLT 2012 Workshop Predicting and Improving text Readability for Target Reader Populations, PITR 2012 (2012)

32. Rello, L., Kanvinde, G., Baeza-Yates, R.: Layout guidelines for web text and a web service to improve accessibility for dyslexics. In: International Cross Disciplinary Conference on Web Accessibility (W4A 2014). ACM Press, Lyon (2012)

33. Saggion, H., Martínez, E., Etayo, E., Anula, A., Bourg, L.: Text simplification in Simplext. Making text more accessible. Procesamiento de Lenguaje Natural 47, 341–342 (2011)

34. Sereno, S., Rayner, K.: Measuring word recognition in reading: eye movements and event-related potentials. Trends in Cognitive Sciences 7(11), 489–493 (2003)
35. Simmons, F., Singleton, C.: The reading comprehension abilities of dyslexic students in higher education. Dyslexia 6(3), 178–192 (2000)
36. Sinatra, R., Stahl-Gemake, J., Berg, D.: Improving reading comprehension of disabled readers through semantic mapping. The Reading Teacher 38(1), 22–29 (1984)
37. Sterling, C., Farmer, M., Riddick, B., Morgan, S., Matthews, C.: Adult dyslexic writing. Dyslexia 4(1), 1–15 (1998)
38. Tobii Technology: Product description Tobii 50 Series (2005)
39. Vellutino, F., Fletcher, J., Snowling, M., Scanlon, D.: Specific reading disability (dyslexia): What have we learned in the past four decades? Journal of Child Psychology and Psychiatry 45(1), 2–40 (2004)
40. Williams, S., Reiter, E., Osman, L.: Experiments with discourse-level choices and readability. In: Proceedings of the 9th European Workshop on Natural Language Generation (ENLG 2003), Budapest, Hungary (2003)

Appendix

The text and the corresponding paraphrases pairs used are shown below.

Text María: Se [premia/otorga un premio] premia a Ana María Matute

Sus lectores [confiaban/tenían la confianza] en ella a pesar de que la humildad de Ana María no [ambicionara/tuviera ambición de] más premios. Tras [aparecer/hacer aparición] en las quinielas como la principal aspirante, finalmente, el Ministerio de Cultura [ha galardoneado/otorgó el galardón] con el Premio Cervantes a la escritora. Ana María Matute [ha contribuido/ha hecho una contribución] a la literatura española con novelas y relatos aunque también [ha atendido/ha prestado atención] al público más joven con cuentos para niños. Ana María tenía diez años cuando [comenzó/ dio comienzo] la Guerra Civil Española. Luciérnagas fue su primera obra premiada, pero la [censuraron/impusieron censura] censuraron y no fue publicada hasta años más tarde.

Text Alex: Se [premia/otorga un premio] a Álex de la Iglesia

El Ministerio de Cultura [concedió el/hizo la concesión del] Premio Nacional de Cinematografía al director Álex de la Iglesia. Este premio del Instituto Nacional de la Cinematografía y de las Artes Audiovisuales [contribuye/hace una contribución] a [recompensar/dar una recompensa] a la aportación más sobresaliente en el ámbito cinematográfico español [manifestado/puesta en manifiesto] , a través de una obra durante el año. En casos excepcionales como éste también se [reconoce/ofrece un reconocimiento] a una trayectoria profesional. El jurado [valoró/dió valor] a la trayectoria profesional de álex de la Iglesia, que [enriquecido/ha aportado riqueza] al lenguaje de nuestro cine, así como su gran labor por poner [acercar/poner más cerca] el cine español a la sociedad.

Detecting Apposition
for Text Simplification in Basque

Itziar Gonzalez-Dios*, María Jesús Aranzabe,
Arantza Díaz de Ilarraza, and Ander Soraluze**

IXA NLP Group, University of the Basque Country (UPV/EHU),
Manuel Lardizabal 1 48014 Donostia
{itziar.gonzalezd,maxux.aranzabe,a.diazdeilarraza,ander.soraluze}@ehu.es
http://ixa.si.ehu.es/Ixa

Abstract. In this paper we have performed a study on Apposition in
Basque and we have developed a tool to identify and to detect automatically these structures. In fact, it is necessary to detect and to code this
structures for advanced NLP applications. In our case, we plan to use the
Apposition Detector in our Automatic Text Simplification system. This
Detector applies a grammar that has been created using the Constraint
Grammar formalism. The grammar is based, among others, on morphological features and linguistic information obtained by a named entity
recogniser. We present the evaluation of that grammar and moreover,
based on a study on errors, we propose a method to improve the results.
We also use a Mention Detection System and we combine our results with
those obtained by the Mention Detector to improve the performance.

Keywords: Apposition Detector, Basque, Automatic Text Simplification,
Mention Detection.

1 Introduction

Automatic Text Simplification (TS) is a Natural Language Processing (NLP) task
whose aim is to simplify texts automatically, keeping the meaning of original text,
or at least avoiding information loss. TS is a necessary research line in NLP since
the texts which are simplified are easier to process both for people and advanced
NLP applications.

TS systems have already been proposed for people with disabilities [1], illiter-
ate [2] or people who learn foreign languages [3] [4] among others. There are TS
systems for advanced applications such us machine translation [5], Q&A systems
[6], information extraction systems [7], and so on.

Our main motivation for TS is that long sentences cause problems in advanced
applications like machine translation [8]. Apposition is a phenomenon that increases the length of the sentences and it has been reported in the context of TS as

---

* Itziar Gonzalez-Dios's work is funded by a PhD grant from the Basque Government.
** Ander Soraluze's work is funded by PhD grant from Euskara Errektoreordetza.

A. Gelbukh (Ed.): CICLing 2013, Part II, LNCS 7817, pp. 513–524, 2013.
© Springer-Verlag Berlin Heidelberg 2013

a complex phenomenon and rules to simplify these structures have been studied e.g. in [9] and [10] and for Basque in [11]. The information that an appositional phrase contains is not syntactically necessary and therefore it can be taken out of the sentence. This will mean the loss of some information, unless we create a new sentence out of the apposition. So if we remove apposition out of the sentence and create shorter sentences, for example, the task of machine translation will be more affordable.

In NLP, apposition detection has been mainly studied in the context of its integration in other general tools. However, there are tools that identify apposition explicitly [12] by means of machine learning techniques. Other techniques that have been used to detect apposition are heuristics [13] or full parse information [14]. In [15] appositive detection is applied as preprocess of a mention detection system and they use patterns to identify these structures. In [16] they use sequence mining to detect linguistic patterns in French like appositive qualifying phrases.

There are two tools in Basque that can be useful to detect Apposition. The first is a named entity recogniser and classifier, *Eihera* [17] and the second is the combination of the rule based (*IXAti* [18]) and the statistical-based (*ML-IXAti* [19]) shallow syntactic parsers for Basque. These tools consider apposition inside a noun phrase (restrictive) as a chunk, and apposition, that is expressed by noun phrase as appositive (non-restrictive), as more than an independent chunk. Since there is no explicit way to mark the apposition, we need a special tool to detect them.

So, in this paper we present a rule based Apposition Detector, based on linguistic knowledge, that is able to identify these structures and classify them according to their type. The output of this tool is human friendly, but it can be easily coded for machines as well. Although the first use of this Detector is TS, the Apposition Detector can be useful for other NLP advanced applications like mention detection, coreference resolution, parsing, textual entailment, text summarisation, Q&S systems, information extraction, event extraction, opinion mining etc. In the evaluation, we obtain 0.80 in F-measure. However, we analyse the errors and to improve the results, we use a Mention Detection System [20].

This paper is structured as follows: in section 2 we present the apposition types in Basque language. In section 3 the framework and the formalism of the Apposition Detector is explained. In section 4 we show the evaluation results. To improve this result we show in section 5 the experiments we carried out using the Mention Detector. In section 6 we describe how this tool will be used for Automatic Text Simplification and finally, in section 7 we expose the conclusion and the future work.

2   Apposition in Basque

Basque is Pre-Indo-European language and differs considerably in grammar from the languages spoken in surrounding regions. It is, indeed, an agglutinative head-final pro-drop isolated language whose case system is ergative-absolutive. Basque displays a rich inflectional morphology. Basque is still undergoing the normalisation process, and in charge of that, among others, there is *Euskaltzaindia* (Royal Academy of the Basque Language).

Apposition detection grammar has been built according to *Euskaltzaindia* [21]. As regulated, there are two types of apposition in Basque:

- **First type (restrictive):** Apposition that occurs inside a noun phrase. There are two ways to realise this type: a) example (1), the named entity *Luis Uranga* precedes the common name *presidenteak* (henceforth, type 1A):

(1) *Luis Uranga presidenteak   (...)*
    Luis Uranga president_the

    'The president Luis Uranga (...)'

  or b) example (2), the common name *presidente* precedes the named entity *Luis Uranga* (henceforth, type 1B):

(2) *Errealeko   presidente Luis Uranga (...)*
    Real_Sociedad_of president   Luis Uranga

    'The president of Real Sociedad Luis Uranga (...)'

- **Second type (non-restrictive):** A noun phrase as appositive like (3)[1]:

(3) *Jakinduria hori, guretzat harrapezina   dena,   (...)*
    Wisdom   that, us_for   unattainable is_which_the,

    'That wisdom, that is unattainable for us, (...) '

It is possible as well to combine both types (4):

(4) *Simon   Peres laborista,   Israelgo lehen ministro izana,*
    Shimon Peres Labour_the, Israel_of Prime Minister have_been_the

    'Labour Shimon Peres, the former Prime Minister of Israel, (...)'

and to merge the both structures (1A and 1B), example (5):

(5) *Vatikanoko Estatuekiko Harremanetarako idazkari Jean Louis Tauran*
    Vatican_of states_with relations_for   secretary Jean Louis Tauran
    artzapezpikuak (...)
    archbishop_the

    'The archbishop Jean Louis Tauran, Secretary for Relations with States of The Vatican, (...)'

Parenthetical structures are not considered as apposition by *Euskaltzaindia*, since there is no agreement. However, some kind of parenthetical structures follow the same pattern as apposition in the simplification rules [11], so we have included rules to treat them in this grammar. For non simplification uses, these rules can be omitted. In (6) we see an example of a parenthetical structure the grammar covers.

---

[1] Notice that the equivalent translation is a relative clause.

(6) *Durangon   (Bizkaia)*
     Durango_in (Biscay)

     'in Durango (Biscay)'

These are the target structures for our Apposition Detector. Each structure is given a tag, so they are classified.

If we applied only our shallow syntactic parser *IXAti* [18], type one apposition (both 1A and 1B) will be considered as a chunk, which is correct and valid for shallow parsing. But for some tasks like Automatic Text Simplification they should be distinguished. Apposition type two is considered by *IXAti* as more than one chunk. In both cases there is no explicit tag to express the appositional relation. This way Apposition Detector accomplishes this tagging task before the chunker *IXAti* is applied.

3   Architecture of the Apposition Detector

In this section we explain how our Apposition Detector works. Having as input a text, we perform the following analysis before we apply the Apposition Detector:

– **Morpho-syntactic analysis:** *Morpheus* [22] makes word segmentation and part of speech tagging. Syntactic function identification is made by *Constraint Grammar* formalism [23].
– **Lemmatisation and syntactic function identification:** *Eustagger* [24] resolves the ambiguity caused at the previous phase.
– **Multi-words items identification:** The aim is to determine which items of two or more words are always next to each other [25] [26].
– **Named entity recognition:** *Eihera* [17] identifies and classifies named-entities in the text (person, organisation, location).

To detect the apposition we have written a grammar following Constraint Grammar formalism [23]. The linguistic features we have used to write the rules in grammar are category, subcategory, and named entity tags.

Our detection system works in two phases: first, a grammar tags the named entities that are candidates to be a part of an apposition and secondly, based on the previous tags another grammar tags the second part of the apposition, if it fulfils the conditions of being a real apposition. The phrase with both tags is an apposition. There are 37 rules for the first phase, and 21 rules for the second phase. The rules are classified according to the entity type as well.

Each structure presented in section 2 has a tag (Table 1). This is the way apposition classification is made. This classification is valid, for example to know what kind of structures are used frequently or which rule should be applied for Text Simplification.

Once the apposition has been tagged we apply the rule based chunker *IXAti* [18] and *ML-IXAti* [19], which identifies chunks and clauses by combining rule-based grammars and machine learning techniques, exactly the version implemented in

Table 1. Tags applied by the grammar

Type	1 appositional phrase	2 appositional phrases
1A	]APOS1	[APOS2
1B	]APOS1_KONTRA	[APOS2_KONTRA
2	]APOS1SINT	[APOS2SINT
Parenthetical structures	]APOS1_EGON	[APOS2_EGON

[20] to get the both appositional phrases. The algorithm is the following: the first appositional phrase begins where the chunker has tagged the phrase begin and it finishes with the word that has the first tag by our grammar. The second appositional phrase is formed by the word(s) between the first tag and second tag.

Let see this process with example (1), *Luis Uranga presidenteak*. The first rule (Figure 1) tags *]APOS1* and targets the end boundary of a named entity classified as person *Luis Uranga*, that is composed only by two words[2] and that is in the context of an apposition, in this example *Uranga*.

```
MAP (]APOS1) TARGET (ENTI_BUK_PER) IF (-1 ENTI_HAS_PER) (0 NEXT_KM)
                                      (1 IZE + ARR) (NOT 1 NEXT_KM);
```

Fig. 1. CG rule to tag a candidate appositional phrase

The second rule (Figure 2) tags *[APOS2* and targets a common name, if previously an apposition candidate has been tagged (i.d. there is previously *]APOS1* tag), that is not followed by a adjective, in this example *presidenteak*.

```
MAP ([APOS2) TARGET (IZE) IF (0C ARR) (NOT 0 PUNT_KARDI OR LEKU )
                             (NOT 0 HM OR DM) (-1 APO) (NOT 1 ADJ) ;
```

Fig. 2. CG rule to tag second appositional phrase and confirm the apposition

Taking into account the information of *IXAti* and *ML-IXAti* and the previously mentioned tags, the whole appositional phrases are *Luis Uranga* and *presidenteak*. In figure 3 we see the output of example (1) in text version (human-friendly).

4   Evaluation and Error Analysis

The corpus that has been used to develop and to evaluate the grammar has been EPEC (*Euskararen Prozesamendurako Erreferentzia Corpusa*-Reference Corpus for the Processing of Basque) [27]. EPEC Corpus is interesting for this task since

---

[2] *ENTI_HAS_PER* and *ENTI_HAS_PER* tag the beginning and the ending of a named entity, and the other tags express morphological features.

> [Luis Uranga]APOS1 [presidenteak]APOS2

Fig. 3. Output of Apposition Detector in Text Version

it compiles text from newspapers, where apposition is a normal feature. In the first column of table 2 we see the quantities of the apposition found in the evaluation part of the corpus, in general and classified according to their type. To evaluate this grammar we have created a gold standard, where the apposition has been manually tagged.

In table 2 we also show the results[3] obtained by Apposition Detection and the quantities that are in the corpus. We show the results according to the apposition type as well.

Table 2. Evaluation results of the Apposition Detection

	Quantities	Precision	Recall	F measure
All types	336	0.87	0.74	0,80
1A type	286	0.90	0.62	0.73
1B type	30	0.85	0.73	0.79
2 type	9	1	0.44	0.62
Parenthetical structures	11	1	0.64	0.78

Except for a case, appositions were classified correctly. It was the case of a parenthetical structure that was considered as 1A type.

These results have been analysed qualitatively and we found out following errors and missing structures:

- Due to errors in named entity detection, rules were not applied or misapplied
- Apposition was detected, but a tag was not in the correct place. For example, the tag was in the substantive, when it should be in the adjective
- Complex appositional phrases that were already dismissed in development phase because they made a lot of errors for a correct one, like coordination in appositional phrases.

5   Improving Apposition Detection Using a Mention Detector

By analysing the results (section 4) we noticed that in some cases Apposition Detector has tagged the candidate (first tag) but due to the complexity of the

---

[3] Precision = correctly detected apposition/detected apposition; Recall = correctly detected apposition/all apposition; F-measure = 2 * precision * recall / (precision + recall).

appositional phrases, the tag for the second appositional phrase has been omitted (rule failed or dismissed rule) and in other cases nothing was retrieved. Those were considered as errors. This is the case of example (7).

(7) *Manuel Contreras Inteligentzia Nazionaleko Zuzendaritzako (DINA)*
     Manuel Contreras Intelligence national_of direction_of   (DINA)

     buruzagi ohiak
     head   former

     'Manuel Contreras, former head of the National Intelligence Directorate (DINA), '

In order to get this complex structures (e.g, (7)), we have carried out an experiment with the Mention Detector [20]. This system identifies mentions that are potential candidates to be part of coreference chains in Basque written texts. The aim of this experiment is to see if the Mention Detector can help to improve the results, without making changes in the system. In other words, we want to combine the output of the grammar and the output of the Mention Detector to see if we can get the discarded instances. This process is illustrated in figure 4.

We have formed two hypotheses that we explain next and developed a technique for each one. To test these hypotheses we made a subcorpus with the errors the grammar made, that is, we used the phrases which the first candidate was tagged,

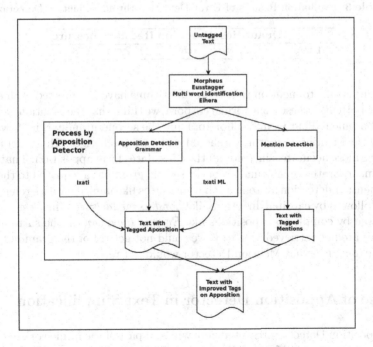

Fig. 4. Architecture of Apposition Detector and Improvement Process through Mention Detection

but the second one was omitted. There are 25 instances in this subcorpus. We only used 1A type, because other type quantities were insignificant.

Taking into account that this subcorpus was formed by the structures the grammar failed, we form **the first hypothesis**: If inside a mention is an appositional phrase candidate according to the grammar, it may be an apposition. So, the algorithm we implemented is next: if a mention has first tag inside (candidate), the rest of the mention is given second tag, and therefore considered as an apposition. Out of 25 instances 5 were were retrieved correctly.

To continue improving the results and taking into account the results of the first hypothesis, we formed **the second hypothesis**: if a mention is an appositional phrase candidate, the following mention in text should be its appositive. The technique we use to track is the mention identification number. If the candidate mention has identification number 1 in text, mention with identification number 2 should be its appositive. Applying this method, the 13 instances of the 20 left were correctly retrieved. Three instances more were retrieved, but as the whole appositional phrase was not correct, they were consider as errors.

So, we concluded that Mention Detection, without having been tuned, can improve the detection of apposition, retrieving 18 instances out of 25 and obtaining following results in the error subcorpus (Table 3). This approach using the Mention Detector is above all helpful to retrieve the cases which grammarians had discarded the rule due to error increasing.

Table 3. Evaluation Results of Error Detection through Mention Detection

	Quantities	Precision	Recall	F measure
A1 type	25	0.86	0.72	0.78

It is important to mention that these algorithms have been tested with errors. To prove both hypotheses in a normal corpora, we think that the Mention Detector should be tuned. That is, instead of applying the second grammar, if we want to use only the Mention Detector, we should make severe changes. These algorithms should be more accurate, since not all the candidates form apposition. That is, we should incorporate the information of the second grammar adequated to the rules of the mention detection system, so that instances like named entities referring to a place followed by cardinal directions like *Londres mendebalean* (in West London) or followed by complex postpositions like *Erroma inguruan* (in the surroundings of Rome) are not retrieved. Anyway, we could not get rid of the grammar, since there are instances that Mention Detector would not retrieve.

6   Use of Apposition Detector in Text Simplification

The Apposition Detector presented here will be a part of the framework in our TS system, together with *Mugak* [28], the clause identifier. Based on its output apposition follows the simplification process [29], that will be explained by means of example (8): *Jasser Arafat buru palestinarra Egiptoko presidente Hosni Mubarak-ekin bildu zen atzo Kairon* (Palestinian Chairman Jasser Arafat met President of Egypt Hosni Mubarak yesterday in Cairo).

1. **Splitting:** First apposition is detected: there are two in sentence (9): [*Jasser Arafat buru palestinarra*] (Palestinian Chairman Jasser Arafat)and [*Egiptoko presidente Hosni Mubarak-ekin*] (President of Egypt Hosni Mubarak). Secondly, a chunk is created for each appositional phrase in each apposition (figure 5). This is the task that the Apposition Detector presented in section 3 carries out.

[[Jasser Arafat] [buru palestinarra]] [[Egiptoko presidente][Hosni Mubarak-ekin]]

Fig. 5. Appositional phrases in sentence (8)

2. **Reconstruction:**
    (a) Removing: The chunks with the second tag (second appositional phrase) will be removed from the original sentence, obtaining following output: [*Jasser Arafat Hosni Mubarak-ekin bildu zen atzo Kairon*] (Jasser Arafat met Hosni Mubarak yesterday in Cairo). If a chunk has a suffix like *-ekin* (with) in *Hosni Mubarak-ekin* it should be removed.
    (b) Adding: Chunks with both tags will be added together with the copula, in these examples *da* (is), to form simple sentences: The absolutive suffix *-a* should be added in the phrase *Egiptoko presidentea* (the President of Egypt).
    This is the output of this operation: [*Jasser Arafat buru palestinarra da*] (Jasser Arafat is a Palestinian chairman) and [*Egiptoko presidentea Hosni Mubarak da*] (The president of Egypt is Hosni Mubarak). In this operation sentences have been created, but the simplification process is not yet fulfilled.

3. **Reordering:**
    (a) Internal word reordering in sentence: First the internal order will be checked: the order of former original sentence is kept untouched, the new sentences follow this rule pattern: Chunk with first tag (SUBJ), chunk with second tag (PRED), copula in present tense, 3 person, singular or plural depending on the subject. The first apposition follows the pattern of the rule, so it is left untouched but the second should be reordered to follow that pattern[4]: [*Hosni Mubarak Egiptoko presidentea da*] (Hosni Mubarak is the president of Egypt).
    (b) Sentence reordering in text: First, the former original sentence; then, new simple sentences following the order they appear in the original sentence.

---

[4] Before reordering this sentence was already grammatically correct, since Basque is a free word order language. But according to the simplification rule, the order should change.

4. **Correction:** There is no grammatical error to correct but sentences should be punctuated. This will be the final output: [*Jasser Arafat Hosni Mubarak-ekin bildu zen atzo Kairon. Jasser Arafat buru palestinarra da. Hosni Mubarak Egiptoko presidentea da.*] (Jasser Arafat met Hosni Mubarak yesterday in Cairo. Jasser Arafat is a Palestinian chairman. Hosni Mubarak is the President of Egypt.).

Following this process we have got shorter sentences which are useful for advanced applications like machine translation. Anyway, as simplification rules can be tuned according to the target audience, another option is to make a coordinate sentence with *eta* (and) to unify the new simple sentences. This will be the final output: *Jasser Arafat Hosni Mubarak-ekin bildu zen atzo Kairon. Jasser Arafat buru palestinarra da eta Hosni Mubarak Egiptoko presidentea da.* (Jasser Arafat met Hosni Mubarak yesterday in Cairo. Jasser Arafat is a Palestinian chairman and Hosni Mubarak is the President of Egypt.).

7 Conclusion and Future Work

In this paper we have presented an Apposition Detector based on linguistic knowledge. Moreover, it is able to classify the apposition in corpora according to their type and structure, which is helpful for linguistic analysis and research on apposition.

We have evaluated this tool and looking at the results (F-measure 0.80), we realised that they could be improved. So we have made an experiment on errors with another tool, the Mention Detector. We have formed two hypotheses and created to techniques to combine the output of the grammar and the output of the Mention Detector. This way, the instances that were not covered by the grammar were retrieved (F-Measure 0.78), without having changed the Mention Detection system.

We have explained as well how we are going to use the output of the Mention Detector in Automatic Text Simplification by means of an example. Performing the syntactic simplification process, we get shorter sentences that are easier to process for NLP advanced applications such us machine translation.

Although the first use of the Apposition Detector is Automatic Text Simplification, it can be used for other tasks like coreference resolution, information extraction, lexicon elaboration or text summarisation. Indeed, we plan to implement this Detector to improve the mention detection system and in the coreference resolution system.

Acknowledgments. Itziar Gonzalez-Dios's work is funded by a PhD grant from the Basque Government and Ander Soraluze's work is funded by PhD grant from Euskara Errektoreordetza, the University of the Basque Country (UPV/EHU). This research was also supported by the the the Basque Government (IT344-10).

References

1. Carroll, J., Minnen, G., Pearce, D., Canning, Y., Devlin, S., Tait, J.: Simplifying Text for Language-Impaired Readers. In: 9th Conference of the European Chapter of the Association for Computational Linguistics (1999)
2. Candido Jr, A., Maziero, E., Gasperin, C., Pardo, T.A.S., Specia, L., Aluisio, S.M.: Supporting the adaptation of texts for poor literacy readers: a text simplification editor for Brazilian Portuguese. In: Proceedings of the Fourth Workshop on Innovative Use of NLP for Building Educational Applications. EdAppsNLP 2009, pp. 34–42. Association for Computational Linguistics, Stroudsburg (2009)
3. Petersen, S.E., Ostendorf, M.: Text Simplification for Language Learners: A Corpus Analysis. In: Electrical Engineering (SLaTE), pp. 69–72 (2007)
4. Burstein, J.: Opportunities for Natural Language Processing Research in Education. In: Gelbukh, A. (ed.) CICLing 2009. LNCS, vol. 5449, pp. 6–27. Springer, Heidelberg (2009)
5. Poornima, C., Dhanalakshmi, V., Anand, K., Soman, K.: Rule based Sentence Simplification for English to Tamil Machine Translation System. International Journal of Computer Applications 25(8), 38–42 (2011)
6. Bernhard, D., De Viron, L., Moriceau, V., Tannier, X.: Question Generation for French: Collating Parsers and Paraphrasing Questions. Dialogue and Discourse 3(2), 43–74 (2012)
7. Jonnalagadda, S., Gonzalez, G.: Sentence simplification aids protein-protein interaction extraction. Arxiv preprint arXiv:1001.4273 (2010)
8. Labaka, G.: EUSMT: Incorporating Linguistic Information into SMT for a Morphologically Rich Language. Its use in SMT-RBMT-EBMT hybridation. PhD thesis, UPV-EHU (2010)
9. Siddharthan, A.: Syntactic simplification and text cohesion. Research on Language & Computation 4(1), 77–109 (2006)
10. Specia, L., Aluisio, S.M., Pardo, T.A.: Manual de Simplificação Sintática para o Português. Technical Report NILC-TR-08-06, So Carlos-SP (2008)
11. Gonzalez-Dios, I.: Euskarazko egitura sintaktikoen azterketa testuen sinplifikazio automatikorako: Aposizioak, erlatibozko perpausak eta denborazko perpausak. Master's thesis, University of the Basque Country (September 2011)
12. Freitas, M.C., Duarte, J.C., Santos, C.N., Milidiú, R.L., Rentería, R.P., Quental, V.: A machine learning approach to the identification of appositives. In: Sichman, J.S., Coelho, H., Rezende, S.O. (eds.) IBERAMIA 2006 and SBIA 2006. LNCS (LNAI), vol. 4140, pp. 309–318. Springer, Heidelberg (2006)
13. Phillips, W., Riloff, E.: Exploiting strong syntactic heuristics and co-training to learn semantic lexicons. In: Proceedings of the ACL-02 Conference on Empirical Methods in Natural Language Processing, vol. 10, pp. 125–132. Association for Computational Linguistics (2002)
14. Roth, D., Sammons, M.: Semantic and logical inference model for textual entailment. In: Proceedings of the ACL-PASCAL Workshop on Textual Entailment and Paraphrasing, pp. 107–112. Association for Computational Linguistics (2007)
15. Kummerfeld, J.K., Bansal, M., Burkett, D., Klein, D.: Mention detection: heuristics for the OntoNotes annotations. In: Proceedings of the Fifteenth Conference on Computational Natural Language Learning: Shared Task. CONLL Shared Task 2011, pp. 102–106. ACL, Stroudsburg (2011)
16. Béchet, N., Cellier, P., Charnois, T., Crémilleux, B.: Discovering linguistic patterns using sequence mining. In: Gelbukh, A. (ed.) CICLing 2012, Part I. LNCS, vol. 7181, pp. 154–165. Springer, Heidelberg (2012)

17. Fernandez Gonzalez, I.: Euskarazko Entitate-Izenak: identifikazioa, sailkapena, itzulpena eta desanbiguazioa. PhD thesis, UPV-EHU (2012)
18. Aduriz, I., Aranzabe, M.J., Arriola, J.M., de Ilarraza, A.D., Gojenola, K., Oronoz, M., Uria, L.: A cascaded syntactic analyser for basque. In: Gelbukh, A. (ed.) CICLing 2004. LNCS, vol. 2945, pp. 124–134. Springer, Heidelberg (2004)
19. Arrieta, B.: Azaleko sintaxiaren tratamendua ikasketa automatikoko tekniken bidez: euskarako kateen eta perpausen identifikazioa eta bere erabilera koma-zuzentzaile batean. PhD thesis, UPV-EHU (2010)
20. Soraluze, A., Arregi, O., Arregi, X., Ceberio, K., Díaz de Ilarraza, A.: Mention Detection: First Steps in the Development of a Basque Coreference Resolution System. In: Proceedings of KONVENS 2012, pp. 128–163 (2012)
21. Euskaltzaindia: Euskal gramatika laburra: perpaus bakuna. Euskaltzaindia (2002)
22. Alegria, I., Aranzabe, M.J., Ezeiza, A., Ezeiza, N., Urizar, R.: Robustness and customisation in an analyser/lemmatiser for Basque. In: LREC-2002 Customizing Knowledge in NLP Applications Workshop, pp. 1–6 (2002)
23. Karlsson, F., Voutilainen, A., Heikkila, J., Anttila, A.: Constraint Grammar, A Language-independent System for Parsing Unrestricted Text. Mouton de Gruyter (1995)
24. Aduriz, I., Aldezabal, I., Naki Alegria, I., Arriola, J.M., de Ilarraza, A.D., Ezeiza, N., Gojenola, K.: Finite State Applications for Basque. In: EACL 2003 Workshop on Finite-State Methods in Natural Language Processing, pp. 3–11 (2003)
25. Ezeiza, N.: Corpusak ustiatzeko tresna linguistikoak. Euskararen etiketatzaile morfosintaktiko sendo eta malgua. PhD thesis, UPV-EHU (2002)
26. Urizar, R.: Euskal lokuzioen tratamendu konputazionala. PhD thesis, UPV-EHU (2012)
27. Aduriz, I., Aranzabe, M.J., Arriola, J.M., Atutxa, A., Díaz de Ilarraza, A., Ezeiza, N., Gojenola, K., Oronoz, M., Soroa, A., Urizar, R.: A corpus of written Basque tagged at morphological and syntactic levels for automatic processing. In: Methodology and Steps Towards the Construction of EPEC, vol. 56, pp. 1–15. Rodopi (2006)
28. Aranzabe, M.J., Díaz de Ilarraza, A., Gonzalez-Dios, I.: Transforming Complex Sentences using Dependency Trees for Automatic Text Simplification in Basque (manuscript)
29. Aranzabe, M.J., Díaz de Ilarraza, A., Gonzalez-Dios, I.: First Approach to Automatic Text Simplification in Basque. In: Rello, L., Saggion, H. (eds.) Proceedings of the Natural Language Processing for Improving Textual Accessibility (NLP4ITA) Workshop (LREC 2012), Istanbul, Turkey, pp. 1–8 (2012)

Automation of Linguistic Creativ*itas* for Ads*logia*

Gözde Özbal and Carlo Strapparava

FBK-irst, Trento, Italy
gozbalde@gmail.com, strappa@fbk.eu

Abstract. In this paper, we propose a computational approach to automate the generation of neologisms by adding Latin suffixes to English words or homophonic puns. This approach takes into account both semantic appropriateness and sound pleasantness of words. Our analysis of the generated neologisms provides interesting clues for understanding which technologies can successfully be exploited for the task, and the results of the evaluation show that the system that we developed can be a useful tool for supporting the generation of creative names.

1   Introduction

The interest in the use of foreign languages in advertising has recently grown due to various reasons. The global presence of English is exploited to target English speaking and young people [1], or a product is named in a language of a country which is well-known and successful for a specific domain (e.g. using French for a non-French beauty product). The assumption here is that foreign languages are used in this context for their symbolic and visual values. For instance, a foreign language like Latin can help to underline various features of a product (e.g. exoticness or antiquity).

An interesting trend in creative naming which is most probably built on top of these reasons is mixing languages in one name. As Özbal et al. [2] described after an analysis of creativity devices used in creative naming, a non-Latin word can be *Latinized* by concatenating a Latin suffix to it (e.g. *Machinarium*), or more generally, roots and suffixes from different languages can be combined (e.g. *Vueling*, the name of the Spanish airline company where an English suffix is embedded into the Spanish lexeme *vuel* meaning 'to fly').

Two studies in the literature propose computational approaches for creative naming with the goal of reducing time and labor requirements of this process. While Stock and Strapparava [3] introduce an ironic acronym re-analyzer and generator, Özbal and Strapparava [4] propose a system that generates homophonic puns by taking semantic, phonetic, lexical and morphological knowledge into consideration.

In this paper, we take a further step and propose a system to automatically *Latinize* non-Latin words. After analyzing the most common Latin suffixes, we select the ones which are suitable for automation. The *Latinization* process is

A. Gelbukh (Ed.): CICLing 2013, Part II, LNCS 7817, pp. 525–536, 2013.

applied to an English word or a homophonic pun that is produced by following the approach proposed by Özbal and Strapparava [4]. Then, an appropriate Latin suffix is added by conforming to the semantic, phonetic and morphological harmony between the root and the suffix.

The rest of the paper is structured as follows. In Section 2, we review the state-of-the-art relevant to creative naming. In Section 3, we describe the system that we have developed for automatically generating *Latinized* words. In Section 4, we analyze the performance of this system with an extensive manual annotation, provide concrete output examples and discuss the main virtues and limitations of our approach. Finally in Section 5, we draw conclusions and outline ideas for possible future work.

2   Related Work

With the motivation of automating the naming process in a systematic way, Özbal et al. [2] conduct an annotation task on a dataset of brand and company names collected from various resources to determine both the creativity devices used in each name and the effects that these names provoke. *Latinizing* a non-Latin word by concatenating a Latin suffix to it, which is our exact focus in this paper, is one of the latent devices that their analysis found to be effective for creative naming.

Özbal and Strapparava [4] give a detailed summary of state-of-the-art linguistic and computational approaches, and commercial systems related to creative naming. The authors state that the random generation approach used in existing online systems often result in names with bad quality, and eventually obtaining an appropriate name requires a long time. Among the ones listed, the closest generator to ours (www.naming.net) can combine an input word with Greek and Latin affixes and roots. However, this combination is done without any linguistic analysis and consideration of semantic, phonetic or morphological appropriateness. Concerning naming agencies and branding firms providing professional service for creative naming, they are found to require high financial resources and processing time.

To the best of our knowledge, only two computational studies in the literature make an attempt of automating the name generation process to deal with the shortcomings previously mentioned:

Stock and Strapparava [3] propose an ironic acronym re-analyzer and generator called HAHAcronym. This system both makes fun of existing acronyms, and produces funny acronyms that are constrained to be words of the given language by starting from concepts provided by users. HAHAcronym is mainly based on lexical substitution via semantic field opposition, rhyme, rhythm and semantic relations such as antonyms retrieved from WordNet [5].

Özbal and Strapparava [4] propose a system which combines several linguistic resources and natural language processing (NLP) techniques to generate neologisms based on homophonic puns and metaphors. In this system, a user is required to determine the category of the service to be advertised and the properties to be emphasized. Afterwards, common sense knowledge about the category is obtained by using a set of assertions coming from ConceptNet [6]. *Direct hypernym* and *synonym* relations in WordNet are also used to retrieve semantically related words to the category, new words coming from the ConceptNet assertions and input properties. Later on, metaphors for the input properties are generated by using both Google Suggest and ConceptNet. The authors refer to all these words as the *ingredient*s of the pun generation. Throughout the rest of this paper, we will adopt the same terminology. All possible ingredient pairs are analyzed to generate neologisms with homophonic puns based on the phonetic similarity of a short ingredient to a substring of a longer ingredient. Finally, the likelihood and well-formedness of the puns are checked using a language model.

Table 1. The list of the suffixes used for *Latinization*

Suffix	Meaning	Example	POS
-end(a/um/us)	worthy of, required to	*pagare* (pay) → *paganda* (payable)	verb
-issimus, -issima	added to adjectives to form superlatives	*bonus* (good) → *bonissimus* (very good)	adjective
-philia	forms words meaning "an abnormal liking for or tendency towards a given thing	*libro* (book) → *librophilia* (love for books)	noun
-phila	someone who loves something abnormally	*homo* (man) → *homophila* (someone passionate about men/people)	noun
-abilis	-able; able or worthy to be (the recipient of an action)	*amabilis* (lovable) from *amo* (I love), *durabilis* (durable, lasting) from *duro* (I make hard)	verb
-cida	one who kills	*arboricida* (tree killer), *fraticida* (brother killer)	noun
-cidio	the action of killing	*fraticidio* (the act of killing someone's brother)	noun
-(el/il/o/u)lus	used to form a diminutive of a noun, indicating small size or youth	*porculus* (little pig) from *porcus* (pig)	noun
-ismus	-ism in English	*Atheismus* (Atheism)	noun
-ista	-ist in English; one who practices or believes	*batterista* (batterist), *Marxista* (Marxist)	noun
-itas, -tas	-ity, -ness, -ship; used to form nouns indicating a state of being	*amaritas* (bitterness) from *amarus* (bitter, pungent), *difficultas* (difficulty, trouble) from *difficilis* (difficult, troublesome)	adjective noun
-itia	-ness, -ity; used to form nouns describing the condition of being something	*duritia* (hardness) from *durus* (hard)	adjective noun
-tudo	-itude, -ness; used to form abstract nouns indicating a state or condition	*magnitudo* (greatness) from *magnus* (great)	adjective
-ium	1) forms the names of metal elements, 2) appended to common words to create scientific-sounding or humorous-sounding fictional substance names, 3) indicates the setting where a given activity is carried out	*uranium*, *auditor* (hearer) → *auditorium*	-
-logia	-logy, the study of	*ecologia* (ecology), *cronologia* (chronology)	-

3   Latinization

In this section, we will explain the steps that we follow to automatically add Latin suffixes to English words or to homophonic puns. We generate the homophonic puns with the same approach as in [4]. Our system reuses the input (the *category* of the service to be advertised and the desired *properties*), semantically related words and metaphors generated by theirs. On top of them, we add other *ingredients* based on the semantic connotation of each Latin suffix as we will describe in detail below.

For the *Latinization* process, we analyzed all the Latin suffixes and selected the ones that seemed more interesting for naming, as listed in Table 1. The meanings of the suffixes are stated in the second column (main source: http://en.wiktionary.org/wiki/Category:Latin_suffixes), while examples depicting their usage in Latin are given in the third column. The required part-of-speech of the roots that can receive the suffixes are shown in the last column.

3.1   Root Selection

In this section, we describe the methods that the Latinization system uses to automatically decide when and how to add each Latin suffix in our list to the English words and homophonic puns.

1. **-end(a/um/us):** For these suffixes, the relation *ReceivesAction* from ConceptNet (described as "What can you do to it?" in the documentation of the resource) is exploited. Among the argument pairs based on this relation, we only take into account the ones where the category word determined by the user is the first argument and the part-of-speech of the newly discovered word is a verb. (e.g. *ReceivesAction (tea,drink)* or *ReceivesAction (tea,brew)* for the category word *tea*). The new words (e.g. *drink* and *brew* according to the previous assertions) obtained from these assertions are added to the list of potential words that *-enda*, *-endum* and *-endus* can be added to.

2. **-issimus, -issima:** They are added to i) the properties determined by the user ii) the newly discovered properties of the category according to the ConceptNet relation *HasProperty*. As an example, for an energizing and healthy British tea, these suffixes can be added to the words *British*, *energizing* and *healthy* based on the requested properties; and to *hot* and *addictive* based on the assertions *HasProperty(coffee,hot)* and *HasProperty(coffee,addictive)*.

3. **-phil(ia/a):** They are added to i) the category word (e.g. *shampoo*), ii) nouns satisfying *IsA($\langle category \rangle$,*)* (e.g. *soap*), iii) nouns satisfying either *ConceptuallyRelatedTo($\langle category \rangle$,*)* or *ConceptuallyRelatedTo(*,$\langle category \rangle$)* (e.g. *conditioner*), iv) metaphors coming from properties as explained for the pun generation module (e.g. *silk*), v) the noun which is most frequently related to the target word according to ConceptNet when the category is used as a query word (e.g. *hair*).

4. **-abilis:** It can be used as a replacement for the English suffix *-able* in all ingredients or puns ending with *-able* (e.g. *reliable* or *delendable* as a combination of *dependable* and *lend* for a *bank*). Alternatively it can be added to

i) verbs retrieved through *CapableOf(⟨category⟩,*)* (e.g. *transport* for a *car*)
iii) puns in which the longer ingredient satisfies *CapableOf(⟨category⟩,*)* (e.g. *transpearth* as a combination of *transport* and *earth* for a car with huge capacity).

5. **-cid(a/io):** They are added to the metaphors for the antonyms of the words that are obtained for the suffixes *-issimus* and *-issima* (e.g. for a *fast* car, *turtle* as a metaphor for *slow*).

6. **-(el/il/o/u)lus:** For these suffixes, we compiled a list of properties conveying a meaning of smallness, loveliness or cuteness. This list includes words such as *tiny, slender, adorable* and *fragile*. These suffixes are used if the user explicitly requests one of the listed properties or if any element of the list is a property of the category word according to the *HasProperty* assertions in ConceptNet. They can be added to i) the category, ii) other noun ingredients including metaphors, iii) homophonic puns where the longer ingredient is a noun.

7. **-ismus, -ista:** They are added to all noun ingredients. Besides, *-ismus* is used as a replacement for the English suffix *-ism* in all ingredients and puns, while *-ista* is used as a replacement for *-ist*.

8. **-itas, -tas, -itia, -tudo:** While *-tudo* is added to only adjective ingredients, the rest is added to both noun and adjective ingredients. Besides, they are used as a replacement for *-ity, -ness, -ship* in all ingredients and puns ending with these English suffixes.

9. **-ium:** It is added to all ingredients and puns independently from the part-of-speech.

10. **-logia:** It is added to all noun ingredients and the most frequent noun occurring in all ConceptNet assertions.

The compatibility of ConceptNet concepts with a given part of speech is established by querying WordNet. When a compatible word is found, we also use WordNet to retrieve its synonyms and apply the Latin suffix to them as well.

3.2   Building the Latin Phonetic Model

While generating a *Latinized* word, we need a mechanism to assess its phonetic pleasantness. Accordingly, we build a Latin language model to measure the likelihood of a sequence of phonemes to precede any given suffix.

To build the language model, we use the avaiable dumps of the Latin Wikipedia[1]. From the text, we remove all the words containing non-Latin characters, the ones that appear in the English lexicon (we check whether the word exists in WordNet) and the ones that do not end with one of the suffixes that we are interested in. Since we need a unique standard for making comparisons between English and Latin phonemes, we first map all Latin letters to the corresponding international phonetic alphabet (IPA) phonemes. Afterwards, for each Latin word, we create a phonetic representation of the whole word and the suffix

---

[1] http://dumps.wikimedia.org/lawikisource/20120719/

so that we can replace the part coming from the suffix with the suffix string itself (i.e. we treat the whole suffix as a single unit). With these data, we train a four-gram phonetic model with unigram smoothing by using Kylm (The Kyoto Language Modeling Toolkit) [2].

3.3 *Latinizing* Roots

After selecting the appropriate English words or homophonic puns for each Latin suffix, we start the *Latinization* procedure. In this phase, we try to find ways of mixing the root and the suffix in the most pleasant sounding way. To achieve that, we use the Latin phonetic model and also take very simple heuristics into account.

If the root ends with a consonant and the suffix begins with a vowel or vice versa, a straightforward concatenation takes place. As an example, the pun root *comfurt* and the suffix *-ium* concatenate to form the *Latinized* neologism *comfurtium*.

If the root ends and the suffix starts with a vowel, it is checked whether the vowel coming from the root contributes to the pronunciation of the root. To achieve that, we use the same alignment between letters and phonemes created for the pun generation as [4]. If the vowel is aligned to a phoneme, it is not removed and the root is concatenated to the Latin suffix without any modification. Otherwise, the vowel is deleted from the word before the concatenation. If the root being considered is a pun, the alignment of the ingredient at the end of the pun is used for the last vowel check. To illustrate, for the pun *peatza* and the suffix *-ista*, the alignment of *pizza* is referred to and since the vowel 'a' is important for the pronunciation of the word, it cannot be removed and the neologism *peatzaista* is generated. However, during the generation of a neologism from the root *robe* and suffix *-ista*, since 'e' does not map to a phoneme in the phonetic representation of *robe*, it is removed and the new word *robista* is generated.

If the root ends and the suffix begins with a consonant, one of the five vowels ('a', 'e', 'i', 'o', 'u') or just the empty string is inserted in the middle. Then, the phonetic quality of each resulting neologism is checked in the Latin phonetic model. English phonemes that do not exist in Latin (such as *"ER"*, *"DH"* and *"OW"* in the CMU phonetic dictionary) are mapped onto the closest Latin phonemes. The surface form resulting in the highest language model score is selected as the new *Latinized* neologism. As an example, for the root *cotton* and the suffix *-phila*, the insertion of the letter 'o' gives the highest score among all possibilities and *cottonophila* is generated.

4 Evaluation

To evaluate the performance of the *Latinization* module, we conducted a manual annotation in which 5 annotators judged a set of neologisms along 5 dimensions:

---

[2] http://www.phontron.com/kylm/

1. **Appropriateness of the English words:** A binary decision concerning the suitableness of the English word(s) (1 or 2 according to whether the root of the neologism in question is a pun or not) for the given category and properties
2. **Appropritateness of the suffix:** A binary decision concerning the suitableness of the Latin suffix for the English words occurring in the root and given properties in terms of semantics and part-of-speech
3. **Pleasantness:** A binary decision concerning the conformance of the *Latinized* neologism to the sound patterns of English
4. **Humor/wittiness:** A binary decision concerning the wittiness of the *Latinized* neologism
5. **Success:** An assessment of the fitness of the *Latinized* neologism as a name for the target category/properties (unsuccessful, neutral, successful)

The annotators in this task were selected among people speaking at least one Latin based or influenced language. They were also provided with the list of suffixes occurring in the dataset together with their meanings, examples for their usage in Latin and part-of-speech of the words that they can be added to.

For the dataset, we used the same 50 category and property lists as in [4]. We restricted the number of *Latinized* neologisms generated for each input to reduce the effort of the annotators. We built 13 suffix groups based on their proximity, namely: 1) -end(a/um/us); 2) -issim(us/a); 3) -phil(ia/a); 4) -abilis; 5) -cid(a/io); 6) -(el/il/o/u)lus; 7) -ismus; 8) -ista; 9) -itas -tas; 10) -itia; 11) -tudo; 12) -ium; 13) -logia.

For each input, we first ran the pun generation system [4] and used the ranking mechanism with a hybrid scoring method by giving equal weights to the language model and the phonetic similarity between the pun and the longer ingredient. Among the sorted list, 10 generated puns were processed by our *Latinization* system to build the dataset. Among the *Latinized* neologisms that could be generated for the 50 input data, we sorted the ones ending with the same suffix group according to their scores from the Latin phonetic model. We picked the highest ranked *Latinized* word based on an English root and, if available, the highest ranked *Latinization* of a pun. In this manner, we obtained a dataset of 878 *Latinized* neologisms.

For the inter-annotator agreement, we calculated the majority class for each dimension. Since 5 annotators are included in this task, a majority class greater than or equal to 3 means that the absolute majority of the annotators agreed on the same decision. Table 2 shows the distribution of majority classes along the five dimensions of the annotation. Next to each group of columns, we also show the chance of random agreement among the same number of annotators for the binary (*Rnd-2*) and ternary (*Rnd-3*) decisions, respectively. For appropriateness of the English words (APP-E), appropriateness of the suffix (APP-S), pleasantness (PLE) and humor (HUM), we always have an absolute majority (i.e 3/5) decision due to binary nature of the decision (i.e., given only two options, at least three annotators must take the same decision). As for the success (SUX) dimension, in ~24% of the cases it is not possible to take a majority decision.

Table 2. Inter-annotator agreement (in terms of majority class, MC) on the five annotation dimensions

MC	APP-E	APP-S	PLE	HUM	Rnd-2	SUX	Rnd-3
2	-	-	-	-	-	23.92	37.04
3	46.58	41.91	52.16	55.24	62.50	44.65	49.38
4	39.07	46.58	36.79	36.90	31.25	24.60	12.35
5	14.35	11.51	11.05	7.86	6.25	6.83	1.23

However, in ∼76% of the cases the absolute majority of the annotators agreed on the annotation.

Table 3 shows the micro and macro-average of the percentage of cases in which at least 3 annotators have labeled the English words as appropriate (APP-E), the suffix as appropriate (APP-S) and the *Latinized* neologisms as pleasant (PLE), humorous (HUM) or successful (SUX). While the system selects appropriate English words for the roots in approximately 68% of the cases, the suffix is found to be appropriate by at least 3 annotators in ∼96% of the cases. The *Latinized* neologisms sound pleasant in ∼77% of the cases. Almost 31% of the names are found witty or humorous, ∼36% are labeled as successful by the majority, which is a big improvement in comparison to the success of the pun generation (∼24%) as reported in [4]. In addition, the system manages to generate at least one successful and one witty name for all 50 input categories according to the majority of the raters.

Table 3. Accuracy comparison along the five dimensions

	APP-E	APP-S	PLE	HUM	SUX
micro	67.77	95.67	76.77	30.64	36.22
macro	68.18	95.52	76.50	30.75	36.23

In Table 4, we compare the accuracy of the neologisms with English roots versus pun roots for each dimension. While there is no noticeable difference in the appropriateness of suffixes, the appropriateness of puns is generally higher than the one of English roots. For the annotators, it might not be so easy to realize the semantic connection between a single word and the input category and properties. Adding one more word and combining them into a pun might help the raters to establish such connection. In all other dimensions, English words perform overall better than puns as roots. The difference in PLE (93.09 vs. 54.36) can be ascribed to the difficulty of combining an invented word (the neologism) with a Latin suffix.

Table 5 lists the suffixes showing significantly different results for English and pun roots. When -*issimus* and -*issima* are combined with English roots, the ingredients in the root are found to be more appropriate. This kind of combination

Table 4. Accuracy comparison, English vs. pun roots

	APP-E	APP-S	PLE	HUM	SUX
En	63.32	96.03	93.09	35.25	44.64
Pun	71.09	96	54.36	24	23.88

also results in more pleasant and humorous neologisms, and it has a significantly higher success rate. *-(el/il/o/u)lus* is the only suffix group where using puns instead of English words for the root gives better results for humor and success. This seems to be related to the fact that these suffixes generate especially humorous and possibly ironic results when combined with a pun. Regarding *-tudo*, adding this suffix to English roots instead of puns improves the appropriateness and leads to higher success due to higher pleasantness and humor. Lastly, concerning *-abilis*, even though the English ingredients are generally found to be more appropriate when this suffix is added to puns, the resulting neologisms are generally less pleasant and humorous, and hence less successful.

Table 5. Suffixes showing significantly different results for English vs. pun roots

	APP-E		APP-S		PLE		HUM		SUX	
	En	Pun	En	Pun	En	Pun	En	Pun	En	Pun
-issim(us/a)	**94**	80	100	100	**100**	54	**52**	28	**84**	30
-(el/il/o/u)lus	33.33	**50**	100	100	**100**	33.33	16.67	**33.33**	16.67	**33.33**
-tudo		88	76.19	94	90.48	88	61.90	**30**	16.67	**62**	11.90
-abilis		87.09	**100**	100	100	**100**	85.71	**64.52**	28.57	**87.10**	42.86

Table 6 presents an accuracy comparison of all suffix groups along the five dimensions. For each dimension, the highest accuracy is highlighted in bold while the lowest is both highlighted in bold and shown in italic. As can be observed from the table, for the suffix group *-end(a/um/us)*, no output can be obtained for any category since no assertion using the relation *ReceivesAction* could be found in ConceptNet for these categories. Therefore, we need to determine another method to find semantically appropriate words for these suffixes. The English words combined with *-abilis* were found to be the most appropriate, whereas the majority could not find a semantic connection in most of the cases for *-cid(a/io)*. Suffixes were generally found to be appropriate for most of the suffix groups. Especially for *-issim(us/a)*, *-(el/il/o/u)lus*, *-phil(a/ia)* and *-abilis*, 100% accuracy was acquired, while the lowest accuracy was obtained with *-cid(a/io)*. *-phil(a/ia)* resulted in most pleasant neologisms, whereas the least pleasant were obtained with *-(el/il/o/u)lus*. The most humorous names (%59.18 accuracy) were products of the suffix group *-phil(a/ia)*. However, only %12.94 accuracy could be obtained with the suffix *-itia*. Finally, the highest accuracy in terms of success was achieved by the suffix group -*abilis*. The lowest accuracy (%8.11) obtained with -*cid(a/io)* shows that the semantic reasoning behind it needs to be improved.

Table 6. Accuracy comparison among suffix groups

Suffix Group	APP-E	APP-S	PLE	HUM	SUX
-issim(us/a)	87	**100**	77	40	57
-ismus	57.29	96.88	73.96	19.79	26.04
-ium	67	99	76	29	31
-cid(a/io)	*18.92*	*78.38*	89.19	21.62	*8.11*
-(el/il/o/u)lus	41.67	**100**	*66.67*	25	25
-ista	50.56	98.88	73.03	30.34	24.72
-phil(a/ia)	75.51	**100**	**97.96**	**59.18**	57.14
-tudo	88	94	88	30	62
-abilis	**89.47**	100	97.36	57.89	**78.95**
-logia	56.25	97.5	77.5	43.75	26.25
-itia	84.71	94.12	67.06	*12.94*	35.29
-itas, -tas	65	89	70	24	32
-end(a/um/us)	-	-	-	-	-

In Table 7, we show a selection of successful and unsuccessful *Latinized* neologisms according to the majority of annotators (i.e. 3 or more). These neologisms were generated for the category and properties listed under the block of columns labeled as *Input*.

To analyze the shortcomings of the current methods that we use and decide whether they can be improved, we can focus on some of the unsuccessful results. A common class of unsuccessful outputs includes neologisms generated from puns that are already too long or that do not sound pleasant (e.g. *copracticaletudo* from *comfortable*, *practical* and *tudo* as a name for a *brassiere* or *frankfleatherismus* from *frankfurter*, *leather* and *ismus* as a name for a *glove*). This finding implies that the sorting mechanism of pun generation should definitely be improved for *Latinization* for example by promoting shorter or more pleasant sounding outputs.

Another finding from the analysis of the unsuccessful outputs is that ConceptNet can result in ingredients that are not appropriate for the category. This problem especially occurs due to the lack of part-of-speech tags and word sense disambiguation in this resource. As an example, for an economical, sexy, comfortable, informal and chic dress, we obtain a word like *bareylliumista*. This neologism is a combination of: *beryllium*, the synonym of the first sense of the noun *be* which we obtain from the ConceptNet relation *UsedFor(dress,be)*; *bare*; and the suffix -*ista*, where the first ingredient is clearly a bad choice. To improve in this respect, we need to accurately determine the part-of-speech and sense of words appearing in the ConceptNet assertions.

As another example for an unsuccessful output, for a sparkling, hot, energetic and tasty beverage, our Latinization system outputs *generalium*. This word is a

Table 7. A selection of successful and unsuccessful *Latinized* neologisms

Input		Successful output		Unsuccessful output	
Category	Properties	Word	Ingredients	Word	Ingredients
brassiere	sexy feminine attractive colourful elegant practical comfortable	bracticalium attractivitas	bra, practical, ium attractive, itas	copracticaletudo	comfortable, practical, tudo
beverage	sparkling hot energetic tasty	hotissimus	hot, issimus	generalium	general, ium
dress	elegant economical sexy comfortable informal chic	missylogia womanophila	missy, logia woman, phila	bareylliumista	beryllium, bare, ista
perfume	attractive strong intoxicating unforgettable feminine mystic sexy audacious provocative	mysteelissimus occultitia inscrutabilis	mysterious, steel, issimus occult, itia inscrutable, abilis	H2Oismus	H2O, ismus
hat	colourful comfy trendy elegant warm outstanding	shadowilogia headophila hatressista	shadow, logia head, phila headdress, hat, ista	begorationismus	decoration, beg, ismus
robe	soft comfortable elegant colorful cozy relaxing warm	furmentium cottonophila featherista	garment, fur, ium cotton, phila feather, ista	furvetista	velvet, fur, ista
glove	warm fashionable resistant eclectic casual comfortable	pawlogia resishandium icecida	paw, logia resistand, hand, ium ice, cida	frankfleatherismus	frankfurter, leather, ismus

combination of the suffix *-ium* and the word *general* coming from the Concept-Net assertion *ConceptuallyRelatedTo(general,beverage)*. The relation *ConceptuallyRelatedTo* generally causes noise and we need to apply more sophisticated filtering techniques to avoid injecting ingredients which are semantically too far from the target category.

Lastly, as a name for a *hat* we obtain the neologism *begorationismus* as a combination of *decoration* and *beg*, which comes from the ConceptNet assertion *UsedFor(hat,beg)*. This name is not considered as a successful choice for the use-case most probably due to the negative connotation of the word *beg*. Therefore, it might be a good idea to filter out the ingredients with negative connotation. On the other hand, learning how to exploit such words can be relevant to inject humor and irony in neologisms.

5   Conclusion

In this paper, we focused on the task of automating the naming process and generating original, creative and witty names for brands, companies and products. We presented in detail the set of techniques that we used to automate Latin suffixation. We carried out an annotation task on the output of this system to validate the quality and potential utility of the generated neologisms. It was seen after the annotation that the user response to the output was generally positive. In fact, in all use-cases the system managed to output at least one successful and one humorous name according to the majority of the annotators.

As future work, we plan to improve the output quality by considering word sense disambiguation techniques to reduce the effect of inappropriate ingredients.

We also need to design a filtering mechanism based on semantic relatedness metrics to remove the noise coming from ConceptNet. Thereby, we will be able to better exploit the concept associations in relations such as *ConceptuallyRelatedTo*, which are effected by considerable noise but which have the potential of establishing interesting and original connections. As another improvement, we plan to implement other modules to automate other classes of creative devices such as rhyming and oxymorons.

Finally, we plan to make the system that we have developed publicly available so that we can test its performance in a more systematic way and use the feedback of users for future improvement. Differently from our current design, we want to build an interactive system that collaboratively works with users on the generation, instead of only receiving input from them and outputting generated neologisms. In this respect, the system would mostly work as an aid to establish semantic associations and to explore the applicability of creative devices based on lexical, phonetic and semantic analysis to support the creativity process of even expert copywriters. We believe that this kind of semi-automated design opens interesting venues in human computer interaction for creative linguistic tasks.

References

1. Garca Vizcano, M.J.: Code-breaking/code-making: A new language approach in advertising. Journal of Pragmatics 43, 2095–2109 (2011)
2. Ozbal, G., Strapparava, C., Guerini, M.: Brand pitt: A corpus to explore the art of naming. In: Proceedings of LREC 2012, Istanbul, Turkey, pp. 1822–1828 (2012)
3. Stock, O., Strapparava, C.: Laughing with hahacronym, a computational humor system. In: Proceedings of the 21st National Conference on Artificial Intelligence, vol. 2, pp. 1675–1678. AAAI Press (2006)
4. Özbal, G., Strapparava, C.: A computational approach to the automation of creative naming. In: Proceedings of ACL 2012: Long Papers, Stroudsburg, PA, USA, vol. 1, pp. 703–711 (2012)
5. Stark, M.M., Riesenfeld, R.F.: Wordnet: An electronic lexical database. In: Proceedings of 11th Eurographics Workshop on Rendering. MIT Press (1998)
6. Liu, H., Singh, P.: Conceptnet — a practical commonsense reasoning tool-kit. BT Technology Journal 22, 211–226 (2004)

Allongos: Longitudinal Alignment for the Genetic Study of Writers' Drafts

Adrien Lardilleux, Serge Fleury, and Georgeta Cislaru

SYLED, Université Sorbonne Nouvelle—Paris 3, France
`firstanme.lastname@univ-paris3.fr`

Abstract. We present *Allongos*, a procedure capable of aligning multiple drafts for genetic text analysis purposes. To our knowledge, this is the first time a complete alignment is attempted on the longitudinal axis in addition to the textual axis, i.e. all drafts that lead to the production of a text are consistently aligned together, taking word shifts into account. We propose a practical interface where differences between successive drafts are highlighted, giving the user control over the drafts to be displayed and automatically adapting the display to the current selection. Our experiments show that our approach is both fast and accurate.

1 Introduction

Textual genetics is a subfield of linguistic studies that aims at analyzing the genesis of a text through the observation of writers' drafts [1, 2]. By examining elementary operations (insertions, deletions, substitutions, and shifts, as have long been identified by philologists, e.g. [3]), it is possible to display linguistic rewriting operations that account for the transition between drafts produced at different stages of the writing process: adding, deleting, or moving an adverb or an adjective, replacing a substantive or a radical by a hyponym or a hypernym, correcting a typo, etc. [4]

Traditionally, the comparison of successive drafts of a text, from manuscripts to typescripts corrected by hand, was done manually by linguists, who had to go back and forth multiple times between drafts. This is all the more difficult as some segments may differ by a single character only, demanding a lot of attention from the reader. In addition, finding text segment shifts, that typically account for the ordering of ideas or stylistic optimizations, constitutes by itself a very difficult challenge, as the number of possible moves increases exponentially with the size of the window they are searched in.

In order to alleviate for those problems, propositions to use the computer to automatically align successive versions of a text have been put forward in the past years (see a survey in [5]). The advantages are twofolds: it saves linguists from the burden to manually detect edit operations, through an appropriate interface where differences are highlighted; and an exhaustive listing makes it possible to collect various statistics about the writing process as a whole.

A. Gelbukh (Ed.): CICLing 2013, Part II, LNCS 7817, pp. 537–548, 2013.

One of the most significant works in the field is MEDITE [6–8], which is also one of the most recent. Building on a qualitative study of existing alignment software (Word, Beyond Compare, WinMerge, TUSTEP, among others), the authors have proposed an algorithm inspired from bioinformatics and implemented it in an application that copes with drawbacks found in those software: high alignment error rate when the texts are too different (especially at the character level), inability to detect substitutions and/or shifts, poor user interface, impossibility to gather statistics, and long processing time on large texts [7].

The present work is part of a project which aims at studying professional writing under discursive, genetic, and textometric angles. Our goal here is not to propose alignments of *higher quality* than those produced by MEDITE, but rather to extend the alignment process to a *complete longitudinal* one, i.e. *all* drafts of a text will be consistently aligned together, instead of limiting the process to a pair of drafts. This will allow for a global representation of the writing process, by observing the evolution of every text segment along all stages of the text production. To our knowledge, this is the first time a global alignment has been attempted on the longitudinal axis, in addition to the textual axis.

This paper focuses on the design of the longitudinal alignment procedure (named *Allongos*) from a computer point of view; no linguistic or textometric aspect is discussed as it will constitute the continuation of this work. The paper is organized as follows: Sect. 2 gives an overview of the corpus of our project, Sect. 3 describes the alignment procedure in detail, Sect. 4 presents the interface we specifically designed for the visualization of longitudinal alignments, Sect. 5 evaluates Allongos according to two criteria, and Sect. 6 concludes this work.

2   Corpus Description

A number of materials are available as part of this project. Examples and experiments presented in this paper use the "core" corpus, which consists of drafts of educational reports written by social workers from a French organization specialized in the protection of endangered children.

Social workers write reports with a standard word processing software (Word), regularly saving intermediary versions in separate files, hereafter referred to as states. The date at which each draft is written is saved along with the file so that chronological order is preserved in all subsequent operations. Currently, a total of 25 reports have been delivered by the organization, through two distinct deliveries. Amongst those, two reports have been specifically produced with a logging tool, Inputlog[1] [9]. An overview of the reports is displayed in Table 1, along with various figures. The last two characters of each report's name indicate the origin of the report: first delivery (v1), second delivery (v2), or Inputlog (IL). The "length" column indicates the number of word tokens in the longest state of the report, which happens to be the last one most of the time since reports tend to get naturally longer as the writing process goes on. Figures in the last

---

[1] http://www.inputlog.net

Table 1. The 25 reports collected within the framework of this project

Report name	States	Length in thousand of tokens	Total proc. time in seconds	Proc. time for TER in seconds	Anchors / seg.
Adele-Ravet-v2	6	1.5	3.6	2.8 (78%)	0/129
Annie-Pauty-Bilan-v2	4	2.5	2.6	2.0 (77%)	2/141
Annie-Pauty-Synthese-v2	41	1.8	102.0	27.8 (27%)	11/311
Annie-Pauty-v1	12	1.4	9.5	6.9 (73%)	2/ 64
Anthony-Viti-Bilan-v2	6	2.5	5.3	4.2 (79%)	2/189
Anthony-Viti-Synthese-v2	30	1.6	22.5	15.3 (68%)	16/220
Anthony-Viti-v1	24	1.4	12.9	9.1 (71%)	10/160
Charlene-Baillo-v1	3	1.5	2.9	2.4 (83%)	0/114
Clara-Serpereau-v2	2	1.4	1.9	1.5 (79%)	0/101
Damien-Desmoulins-v1	3	2.3	5.1	4.3 (84%)	4/226
Didier-Lescot-v2	7	2.1	6.6	5.4 (82%)	1/199
Elise-Seyvet-v1	4	1.0	2.4	1.9 (79%)	0/115
Emmanuel-Perrot-v2	5	2.9	10.0	8.5 (85%)	9/274
Fabrice-Lemesle-v2	4	3.2	5.6	4.9 (88%)	10/158
Germain-Correia-Dethiere-v1	6	1.8	9.8	8.7 (89%)	4/111
Houria-Ucar-v1	5	1.2	4.1	3.3 (80%)	0/154
Jacqueline-Chapel-v2	4	1.2	3.7	3.0 (81%)	0/ 85
Jade-Malpartida-IL	12	3.4	26.0	19.5 (75%)	10/369
Lea-Mebarek-v1	5	1.1	2.3	1.8 (78%)	0/ 22
Marine-Dumont-v2	10	2.1	9.3	7.6 (82%)	4/185
Marjorie-Cabarry-v1	2	1.3	1.6	1.2 (75%)	1/ 95
Nathalie-Pourtois-v1	5	1.0	2.5	2.0 (80%)	2/102
Olivier-Viti-IL	14	3.2	48.4	35.4 (73%)	15/601
Robert-Afara-v2	4	1.1	2.2	1.7 (77%)	0/ 65
Thierry-Roux-v1	16	5.6	125.3	105.0 (84%)	10/656

three columns (processing times and number of anchors and segments) will be discussed in Sect. 5.

We converted the reports into plain text, removing all formatting information except paragraph segmentation, using the Antiword[2] software. Then we normalized various characters, including spaces, and manually changed named entities in all reports in order to maintain anonymity, allowing for a safe publication of those sensitive data. In the end, each report consists in a unique text file containing all states of the report, in chronological order, distinguished using a light markup (<state=01>, etc.) as used by textometric software Lexico 3[3] [10] and Le Trameur[4] [11]. These text files constitute the starting point of our alignment procedure.

---

[2] http://www.winfield.demon.nl
[3] http://www.tal.univ-paris3.fr/lexico
[4] http://www.tal.univ-paris3.fr/trameur

3   Procedure Description

In this section, we detail our longitudinal alignment procedure. Our aim is to automatically segment and align all states of a report, taking word shifts into account. The complete longitudinal alignment will be obtained by performing a standard alignment between all pairs of successive states of the report, then further processing the set of alignment pairs obtained. A state is assimilated to a unique string of characters (hereafter "state-sentence").

Our approach differs from that used in MEDITE on two main points. First, the base unit is larger than the typographical word, while MEDITE's primary algorithm is character-based—see a discussion in next subsection. This is done for alignment purposes only, and a second, finer, pass will be performed before the alignment is presented to the user in the visualization interface.

Second, shifts are dealt in a particular way: we do not rely on the natural segmentation that results from the alignment to highlight shifted segments, leaving them on the spot and "linking" them to their successor in the next state through means such as visual arrows or identifiers, because this tends to quickly become hard to read in the visualization interface. Instead, we adopt an approach whereby a shift is circumscribed within a single segment. In other words, a word must remain in the same segment in all states it appears in, i.e. words should not "jump" between segments from state to state. As a result, the resulting segment size will be variable, ranging from a single word to several paragraphs in some rare cases. Shifts within a segment will nevertheless be highlighted in the visualization interface through the above-mentioned second pass.

3.1   First Step: Pseudo-word Based Alignment

Our first step consists in aligning all states of a report on a pairwise basis. Among the various existing comparison algorithms, we use one that is able to handle shifts. In the present work, we rely on TER (Translation Error Rate) [12], which was roughly proposed at the same time as MEDITE, but for a different purpose. It was designed as a metric to evaluate the quality of translations produced by a machine translation system. Given an output sentence produced by the system (hereafter "hypothesis") and a reference sentence, TER calculates the number of edit operations required to turn the former into the latter: the lower, the better the translation. String comparisons are word based and case sensitive.

The main reason we chose this algorithm is because a reference implementation is freely available,[5] in a form that allows for a quick integration in Allongos. It is also fast and scales up easily, as the processing time required by the underlying algorithm is linear in the size of the two texts to be compared. It has shown to perform well in monolingual alignment tasks [13]. We perform a similar alignment between each pair of successive state-sentences, each of them being potentially very long, after introducing a separation word "§" at paragraph boundaries in order to keep track of the initial segmentation in paragraphs.

---

[5] http://www.cs.umd.edu/~snover/tercom

Fig. 1. An excerpt of the pseudo-word based alignment of the first three states from report "Anthony-Viti-v1"

In order to lower processing time, as well as to improve alignment quality, we pre-process the text so that function words are agglutinated to the following word, using a predefined list of French grammatical words. Agglutinated forms will be referred to as *pseudo-words* in the rest of this paper. For instance, the following sentence from first state of report "Anthony-Viti-v1:"

> Observation depuis son arrivée sur le groupe *(7 words)*
> *'Observation since his arrival in the group'*

becomes:

> Observation depuis_son_arrivée sur_le_groupe *(3 pseudo-words)*

This reduces the number of tokens in state-sentences, allowing for faster processing, and compensates for a problem inherent to comparison algorithms, which is the detection of false positives during the search for shifts. This frequently happens with function words, as they may occur several times within a single sentence, resulting in a large number of erroneous alignment links. This is especially true in our case since our "sentences" are actually long state-sentences (up to 5.6 thousand words, see Table 1). Similarly, we do *not* pre-process punctuations (no "tokenization"), which means that a character string such as "_(demande_parentale,_" yields two tokens: "(demande" and "parentale,".

An example of pseudo-word based alignment is visible in Fig. 1, where the original state-sentences have been simplified in order to maintain readability. We only show the first three states, but the actual graph extends downwards over 30 lines, as each line corresponds to a state. Pseudo-words that are edited from one state to another are marked with crosses: below a cell for a deletion, above for an insertion, and on the alignment link for a substitution.

3.2   Second Step: Merging Compatible Adjacent Cells

The next step consists in creating a super-segmentation based on the pseudo-word alignments obtained in previous step, in order to ensure that the final segmentation will not contain superfluous runs of pseudo-words with identical edit operations. For instance, pseudo-words of a paragraph that remains constant in all states of the report can safely be merged into a single segment, as a finer segmentation would not bring any further information as long as this segment is marked as being constant. This will allow the user to quickly differentiate between constant and modified segments in the final visualization interface.

To this end, we group together adjacent pseudo-words with compatible associated edit operations (operations from previous to current state, and from current to next state, hereafter referred to as pre- and post-operations). The resulting cell's associated edit operations depend on the operations of the two merged cells: we compare the two pre-operations on the one hand, and the two post-operations on the other hand.

We define the following set of compatible operation pairs:

1. similarity + similarity = similarity
2. substitution + substitution = substitution
3. substitution + insertion = substitution
4. substitution + deletion = substitution
5. insertion + insertion = insertion
6. deletion + deletion = deletion

as well as the following set of *non*-compatible operation pairs:

1. similarity + substitution
2. similarity + insertion
3. similarity + deletion

If the two pre-operations are not compatible or the two post-operations are not compatible, then no merging is performed. In addition, a cell made up of the paragraph separator (§) is never merged.

If the two pre-operations are compatible *and* the two post-operations are compatible, then the two cells will *possibly* be merged. To be merged, they have to fulfill an additional constraint: they must remain adjacent, in the same order, in all states they appear in. For instance, inserting a word between two other words in any state, or swapping them, will prevent them from merging in all preceding and following states. We thus perform compatibility tests recursively on all states, progressively following alignment links from state to state.

Figure 2 gives the result of the merging step from the states shown in Fig. 1, from which adjacent cells with compatible pre- and post-operations are merged together, providing their positions are constant in all states they appear in. Note how the insertion cells in second state of Fig. 1 have been merged with the two last substitution cells, resulting in a unique substitution cell; or how the two cells "du_placement" and "Connaissance" have *not* been merged in the first state, because of their change of position in the next state.

3.3   Third Step: Circumscribing Shifts

The last step consists also in a merging phase: one that will guarantee that shifts are circumscribed within a single cell from state to state. To this end, we merge adjacent cells involved in a shift, i.e. those which alignment links intersect. In order to simplify this step, we use an intermediary representation where each state contains the same number of cells: empty cells are introduced to fill in the gaps resulting from insertions and deletions (Fig. 3).

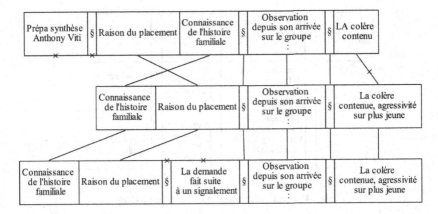

Fig. 2. Result of the merging step on the first three simplified steps of report "Anthony-Viti-v1"

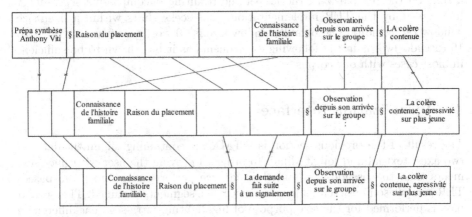

Fig. 3. Intermediary representation whereby empty cells are introduced so that all states have the same number of cells

Adjacent cells involved in a shift, including empty ones, are merged, and the merging propagates onto all states, as was the case in second step. In Fig. 3 for instance, cells 3 to 6 (counting from the left) are merged into a unique cell in the three states: since each state has now the same number of cells, cells 3 to 6 in state n correspond to cells 3 to 6 in state $n+1$, possibly in a different order. Upon completion of this process, alignment links between cells are no more required, and a tabular view such as Fig. 4 suffices. Paragraph separators will eventually be removed.

In practice, we store the structure of this rectangular piece of data within a single TMX file (Translation Memory eXchange, a XML format designed to store translations of textual segments in multiple languages simultaneously [14]), assimilating each state of a report to a different language. This allows for easy processing and sharing of our alignments.

Prépa synthèse Anthony Viti	§	Raison du placement Connaissance de l'histoire familiale	§	Observation depuis son arrivée sur le groupe :	§	LA colère contenue
		Connaissance de l'histoire familiale Raison du placement	§	Observation depuis son arrivée sur le groupe :	§	La colère contenue. agressivité sur plus jeune
		Connaissance de l'histoire familiale Raison du placement § La demande fait suite à un signalement	§	Observation depuis son arrivée sur le groupe :	§	La colère contenue. agressivité sur plus jeune

Fig. 4. Final alignment and segmentation

The final segmentation unit is not fixed: the larger the distance covered by a shift on the textual axis, the longer the resulting encompassing segment. A limit is set by the TER implementation, that seeks shifts within a fixed-size window (-d option). The default window size is 50 words, however we set it to 10 (pseudo-)words in the following experiments as it has shown to be sufficient in most cases with our corpus.

4   Visualization Interface

The result of the previous section is a TMX file containing alignment data on two axes: textual and longitudinal. In order to provide the user with finer segments than the pseudo-words used so far, we perform a second, *word*-based, TER comparison between all pairs of successive segments obtained. This second pass is performed for the sole purpose of highlighting intra-segment differences on the longitudinal axis in the final interface.

The original TMX segments are converted into a HTML table, viewable in any Web browser. The textual axis is represented vertically and the longitudinal axis horizontally. The background color of the text is used to stress the differences with the preceding segment (pre-operations), while the color of the text itself stresses the differences with the following segment (post-operations). This distinction allows for a practical display of *all* intra-segment modifications, even when pre- and post-operations overlap, and to match corresponding intra-segments from one state to another at a glance.

A screenshot is shown in Fig. 5. The user can select the states to be displayed (here, states 7 to 9 of report "Marine-Dumont-v2"). The differences are highlighted by colors: green: insertion; blue: substitution; red: deletion; yellow: shift (none in this example). Rows and columns made empty because of the current state selection (here, only line 63) are automatically hidden, replacing segments by small cells whose color reflects the most frequent edit operation within them. This becomes a full "map" of the report when no state is selected, allowing the

49	Marine	Marine	Marine	1 insertion
50	arrive	réussit	réussit [seg. 50 état 9]	1 insertion / 1 substitution
51	à quitter une place de jeune fille sérieuse assumant une place d'aînée	à quitter une place de jeune fille sérieuse assumant une place d'aînée	à quitter une place de jeune fille sérieuse assumant une place d'aînée	1 insertion
52	responsable	responsable,	responsable,	2 insertions
53	dans laquelle elle a	dans laquelle elle a	dans laquelle elle a	1 insertion
54	souvent été mise pour être une adolescente	souvent été mise, pour être une adolescente	souvent été mise, pour être une adolescente	2 insertions
55	comme les autres qui vit sa vie en essayant de ne pas avoir à toujours supporter le poids d'une lourde histoire familiale. Elle sait profiter de ses relations	comme les autres qui vit sa vie en essayant de ne pas avoir à toujours supporter le poids d'une lourde histoire familiale. Elle sait profiter de ses relations	comme les autres qui vit sa vie en essayant de ne pas avoir à toujours supporter le poids d'une lourde histoire familiale. Elle sait profiter de ses relations	1 insertion
56	d'école et amicales et s'autorise des débordements	amicales et s'autorise des débordements, sans gravité ni conséquence néfastes pour elle,	amicales et s'autorise des débordements, sans gravité ni conséquence néfastes pour elle,	2 insertions / 1 suppression
57	qui lui permettent de sortir de cette place de jeune fille trop sérieuse qu'elle semblait vouloir endosser	qui lui permettent de sortir de cette place de jeune fille trop sérieuse qu'elle semblait vouloir endosser	qui lui permettent de sortir de cette place de jeune fille trop sérieuse qu'elle semblait vouloir endosser	1 insertion
58	un moment.	auparavant.	auparavant.	1 insertion / 1 substitution
59	OBJET CULTUREL	OBJET CULTUREL	OBJETS CULTURELS	1 insertion / 2 substitutions
60	Marine pratique depuis cette année la danse « Modern Jazz », activité dans laquelle elle s'épanouit totalement.	Marine pratique depuis cette année la danse « Modern Jazz », activité dans laquelle elle s'épanouit totalement.	Marine pratique depuis cette année la danse « Modern Jazz », activité dans laquelle elle s'épanouit totalement.	1 insertion
61	VISITE	VISITE	VISITE	1 insertion / 1 substitution
62	FAMILLE ET RELATION FAMILIALE	FAMILLE ET RELATION FAMILIALE	FAMILLE ET RELATION FAMILIALE	1 insertion
64	Avec Madame DUMONT	Avec Madame DUMONT	Avec Madame DUMONT	1 insertion

Fig. 5. A screenshot of the dynamic HTML visualization interface

user to quickly find out not only *where* (textual axis, vertical), but also *when* (longitudinal axis, horizontal) edits take place in the writing process. Note that the two axes are now transposed compared to all previous figures. The total number of each edit operation performed within a state (resp. segment) is displayed in the last row (resp. column). The original files are available on the project's website for a better visualization: http://www.univ-paris3.fr/anr-ecritures.

5 Evaluation

All experiments reported in the following were run on a single processor (1.83 GHz). Allongos was implemented in Python, while the TER implementation is written in Java. The TER beam width (-b option) was set to 500, allowing for large searches on the textual axis.

5.1 Evaluation in Speed

The processing time required by Allongos for each report of our corpus is shown in Table 1. On average, only 17 seconds are required to run the complete procedure, which is quite fast despite our modest machine processor. 77% of this time is used up by TER. Note however that our texts here are all relatively short. As one would expect, the total processing time and the length of the texts are moderately correlated (Pearson's coefficient = 0.64).

In order to accurately assess the impact of the number of states on the processing time, we focus on report "Annie-Pauthy-Synthese-v2," which contains the highest number of states (41). It also appears to be the one for which TER takes the least time relatively to global processing time (only 27%). We perform the following experiment: we create numerous sub-reports by randomly selecting states from the original report. 10 sub-reports of each "width" (from 2 to

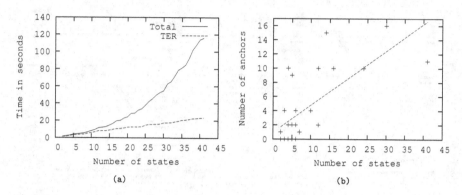

Fig. 6. Impact of the number of states on the processing time (a) and on the number of anchors required to get a "perfect" alignment (b)

41 states) are constituted, and we measure the time required to align them with Allongos. We then report the average processing time for each sub-report width.

Results are presented in Fig. 6.a. As expected, the higher the number of states, the longer the processing time. The time required by TER grows linearly with the number of states, as it only depends on the number of times it is called. This is faster than the rest of the procedure for sub-reports with less than roughly 20 states, while we observe the opposite trend for sub-reports with more states, which is in accordance with the figures in Table 1. This suggests that a really large number of states might hinder the speed of our approach. Further experiments with more states would be required to confirm this, but our current results are sufficient for the time being as we do not have any such corpus at our disposal.

5.2   Evaluation in Quality

As stated by [8], it is difficult to evaluate the quality of monolingual alignment for genetic studies, because no reference corpus exists for this task. This is all the more true in our case because we aligned numerous drafts simultaneously.

That being said, we can afford a manual evaluation here because the reports that make up our corpus are relatively short for a human reader: 194 segments on average (between 22 and 656—see last column of Table 1). Although this number is to be multiplied by the number of states in the worst case, our visualization interface allows us to check for the correctness of alignments of about ten states simultaneously, depending on the user's screen width and display configuration; it thus remains a fast process because the majority of the text is constant from one state to another, so we only need to focus on colored cells.

In fact, since our corpus has to be fully aligned with a minimum of errors to ensure a safe study by the various participants of our project (majority of linguists), this manual verification happened to be a necessity. Therefore, we manually checked and corrected all 25 reports. Only the primary alignment was checked, i.e. the one reflected by the HTML table rows, because the intra-segment highlighting is a product of TER and does not result from Allongos itself. In addition, the small size of the segments TER is run on makes the resulting intra-segment alignment quite reliable.

Whenever an alignment error was encountered, which mostly happens in case of alignment ambiguities, we inserted an anchor in the input text file and restarted the complete alignment procedure. Anchors enforce the correct alignment of a particular point in all states simultaneously. In practice, they are made up of runs of specific characters that are automatically deleted after the pseudo-word based alignment step.

In some reports, the writer made long distance shifts that fall out of the TER search window. When such a shift occurs, Allongos reports a deletion followed by an insertion instead of a single shift, like basic string comparison algorithms do. This happens in four of our 25 reports. This could be easily corrected by increasing the width of the search window, but it might make the resulting segment very coarse, thus defeating the purpose of alignment (in the worst case, one single segment could encompass a whole state because a string has been moved from the beginning to the end of the text). We decided not to count those as errors in our manual evaluation as it follows from a conceptual choice, and the phenomenon remains quite rare. We keep improvements on this matter for further researches.

The number of anchors inserted in each report is shown in last column of Table 1, along with the number of segments. About a third are null, meaning the alignment does not contain any error; and about a half are strictly lower than 3, with a maximum of 16. This is low in regard to the number of segments: between 0% and 7.3% error rate, with an average of 2%, thus validating the efficiency of our approach. The number of corrections is moderately correlated to the length of the report (Pearson's coefficient = 0.56), and highly correlated to the number of input states (Pearson's coefficient = 0.88; see Fig. 6.b for the corresponding scatter plot and linear regression). As one would expect, the longer a text and the more drafts available, the higher the number of manual corrections needed to get a "perfect" alignment; however this number remains very small with Allongos.

6   Conclusion

We have described Allongos, a procedure capable of aligning multiple drafts for genetic text analysis purposes. It is able to perform a complete alignment on the longitudinal axis in addition to the textual axis. We have proposed a dynamic interface specifically designed for the comparison of several drafts simultaneously, allowing the user to quickly find out where and when edits take place within the writing process. In our experiments, a handful of seconds have shown to be sufficient to run the complete procedure on a typical social worker's report, which constitute the core of our project's corpus. It is also very accurate, revealing only 2% of errors on average in our manual evaluation.

All materials mentioned in this paper are freely available at the following address: http://www.univ-paris3.fr/anr-ecritures. These alignments will constitute the starting point of further textometric analyses, in the line of the primary study proposed by [15]. Further improvements might include the integration of different algorithms for the completion of the first stage, e.g. similar to that used by MEDITE, which would allow for a more robust detection of long distance shifts on longer texts such as books.

Acknowledgments. This work was conducted as part of the Écritures project, funded by the French National Research Agency.

References

1. Bellemin-Noël, J.: Le texte et l'avant-texte. Larousse, Paris, 143 pages (1972)
2. Deppman, J., Ferrer, D., Groden, M. (eds.): Genetic Criticism: Texts and Avant-textes. Material Texts, 272 pages. University of Pennsylvania Press (2004)
3. Grésillon, A., Lebrave, J.L.: Manuscrits Écriture. Production linguistique. Number 69 in Langages. Larousse, Paris, 125 pages (1983)
4. Boucheron-Pétillon, S., Fenoglio, I.: Processus d'écriture et marques linguistiques. Number 147 in Langages. Larousse, Paris, 128 pages (2002)
5. Ganascia, J.G., Lebrave, J.L.: Trente ans de traitement informatique des manuscrits de genèse. In: Anokhina, O., Pétillon, S. (eds.) Critique génétique: concepts, méthodes, outils, Paris, IMEC, pp. 68–82 (2009)
6. Ganascia, J.G., Fenoglio, I., Lebrave, J.L.: EDITE MEDITE: un logiciel de comparaison de versions. In: Actes de JADT 2004, Leuven, pp. 468–478 (2004)
7. Bourdaillet, J., Ganascia, J.-G.: MEDITE: A unilingual textual aligner. In: Salakoski, T., Ginter, F., Pyysalo, S., Pahikkala, T. (eds.) FinTAL 2006. LNCS (LNAI), vol. 4139, pp. 458–469. Springer, Heidelberg (2006)
8. Bourdaillet, J.: Alignement monolingue avec recherche de déplacements pour la critique génétique. Traitement Automatique des Langues 50, 61–85 (2009)
9. Leijten, M., Waes, L.V.: Inputlog: New Perspectives on the Logging of On-Line Writing. In: Computer Key-Stroke Logging and Writing: Methods and Applications. Studies in Writing, vol. 18, pp. 73–94. Elsevier (2006)
10. Lebart, L., Salem, A.: Statistique Textuelle. Dunod, Paris, 342 pages (1994)
11. Söze-Duval, K.: Pour une textométrie opérationnelle, 10 pages (2008), http://issuu.com/sfleury/docs/pour_une_textometrie_operationnelle
12. Snover, M., Dorr, B., Schwartz, R., Micciulla, L., Makhoul, J.: A Study of Translation Edit Rate with Targeted Human Annotation. In: Proc. of AMTA 2006, Cambridge, pp. 223–231 (2006)
13. Bouamor, H., Max, A., Vilnat, A.: Monolingual Alignment by Edit Rate Computation on Sentential Paraphrase Pairs. In: Proc. of ACL-HLT 2011, Portland, pp. 395–400 (2011)
14. The Localisation Industry Standards Association: TMX 1.4b Specification (2005), http://www.gala-global.org/oscarStandards/tmx
15. Née, E., MacMurray, E., Fleury, S.: Textometric Explorations of Writing Processes: A Discursive and Genetic Approach to the Study of Drafts. In: Actes de JADT, Liège, 12 pages (2012)

A Combined Method
Based on Stochastic and Linguistic Paradigm
for the Understanding of Arabic Spontaneous Utterances

Chahira Lhioui, Anis Zouaghi, and Mounir Zrigui

ISIM of Medenine, Gabes University, Road Djerba, 4100 Medenine Tunisia,
ISSAT of Sousse, Sousse University, Taffala city (Ibn Khaldoun), 4003 Sousse,
FSM of Monastir, Monastir University, Avenue of the environnement 5019 Monastir,
LATICE Laboratory, ESSTT Tunis, Tunisia
Chahira_m1983@yahoo.fr, Anis.Zouaghi@gmail.com,
Mounir.Zrigui@fsm.rnu.tn

Abstract. ASTI is an Arabic-speaking spoken language understanding (SLU)
system which carries out two kinds of analysis which are relatively opposed. It
is designed for touristic field to tell trippers about something that interests them.
Based on a dual approach, the system adapts the idea of stochastic approach to
the probabilistic context free grammar (PCFG) (approach based on rules). This
paper provides a detailed description of ASTI system as well as well as results
compared with several international ones. The observed error rates suggest that
our combined approach can stand a comparison with concept spotters on larger
application domains.

Keywords: Hidden Markov Model (HMM), Probabilistic Grammar free
Context PCFG, corpus, Wizard of Oz.

1   Introduction

This work is part of Arabic automatic spoken understanding language (SLU) and in
the context of highly spontaneous speech and human-machine (HM) communication
relatively opened (travel's field information). In fact, the automatic SLU is an
essential step in Oral Dialogue Systems. It consists of extracting the meaning of
utterances that are in the most of time ambiguous and uncertain. Hence a certain level
of robustness in the analysis of utterances is needed to overcome spontaneous
difficulties of oral communication.

Most SLU systems follow one of the two main approaches (not mutually
exclusive): a rule-based approach [6] or probabilistic approach [8].

The rule-based parsing requires a grateful work to analyze corpus by experts in
order to extract concept spotting and their predicates. This method is limited to
specific fields using restrictive language. Thus, it leads to many difficulties like
portability and extension. However, this formal approach is rapid due to ATN and
RTN [2] encouraging precision when it is a limited language and where the words
used in the utterances are known as in the case of systems implementing the guided

A. Gelbukh (Ed.): CICLing 2013, Part II, LNCS 7817, pp. 549–558, 2013.

dialogue strategy. Semantic voice interaction analyzer systems, in case of ATIS, MASK, and RAILTEL ARISE [6] developed at LIMSI-CNRS and implement the grammar uses cases as rules, are a good example of rule-based parsers.

Stochastic analysis has many encouraging points among them such as: the decrease and the acceleration of experts work due to the training techniques. This reduces development time. However, this approach suffers from noticeable limitations. Indeed, because of the number of parameters, this method has difficulties in estimating small probabilities accurately from limited amounts of training data. In addition, it doesn't support infrequently phenomena unlike the rule-based approach. Thus, the choice of technology depends much more on the considered application.

It is common in the automatic processing of natural language (NLP), to oppose two approaches. From an historical perspective, the first is based on formal rules, grammars constructed by linguistic experts. The second (currently the most used in the speech recognition) is based on n-grams models. The gap between these two schools is diminishing. Several trends tried to combine the best of both approaches. This is the case for example of [4], [5] or [9]. It is in this context that this hybrid language model, based on the integration of linguistic rules (local grammars) in a statistical model, is proposed.

As an application, the case of an Interactive Voice Server services for travel information and hotel reservation was chosen. The aim of this server is to enable tourists to communicate with the machine via the standard Arabic spontaneous word, to get information about a city staying (restaurants, hotels, location houses etc.), a route, a touristic event, or a price constraint or date, etc. Note that there are no voice server is able to communicate with tourists in Arabic in the field of tourism.

2   Problems of Parsing Spontaneous Speech

Currently, formal languages are a good mean of communication between Human and machine. However, they have significant differences compared to natural human language. As a result, many researchers are working to reduce these differences. The oral NLP is a field of multidisciplinary researches involving electronics, computer sciences, artificial intelligence, linguistics, cognitive sciences, etc. However, techniques have been developed for understanding written language but do not adapt well to oral problems. This is due to:

- Intrinsic characteristics of spontaneous speech: ellipses, anaphora, hesitations, repetitions, and repairs. Here is an example of a tourist who hesitates, does apologize and repairs his utterance:

آه أحب الذهاب إذا كان ممكنا بعد عفوا قبل الساعة 17 بين الساعة
16 و17 إلى تونس

(Euh, I would like to go if it's possible after sorry before 17 O'clock between 16 and 17 O'clock to Tunis)

- Errors related to the non language fluency

أريد ترسيم (متعد فعل)

(I want to register (transitive verb))

Oral speech is characterized by ungrammatical appearances. Therefore, face to the problem of ungrammaticality and oral language, it would be absurd to reject a false syntactically utterances because the goal is not to check the conformity of user's utterances to syntactic rules but to extract rather the semantic content. In fact, to master a language (natural language, a programming language, etc.), both syntax and semantic should have to be controlled in order to don't cause understanding problems. Hence, semantic level is important to control meaning of utterances.

Indeed, a syntactic grammar leads to some rigidity at the analysis of a sentence. Thus, it will reject such a word or phrase does not belong to the language (i.e. you cannot produce the grammar defining this language).

3   Formal Grammars vs. Stochastic Language Models

The two main approaches to automatic NLP, which are rule-based approach and the stochastic approach, have different qualities and limitations. In this section, similarities and differences between both of the mentioned approaches is presented.

3.1   Coverage

Grammars, as complete as they are, do not describe a natural language in its entirety. This aspect is even more pronounced for spoken language processing as many grammatically incorrect phrasing can be used in an oral conversation. Stochastic models do not have this coverage problem: they accept all the sentences of a language. Even incorrect sentences are accepted. Stochastic models are more permissive than formal grammars, which is useful for processing spontaneous speech despite the acceptance of erroneous recognition hypotheses.

3.2   Construction

In terms of construction, both of the mentioned approaches are very different: the formal approach is based on language expertise that is to say on the linguists' skills. The stochastic approach is, in principle, completely automated. However, it should be noted that the amount of necessary corpus to train a robust stochastic model language is not always available.

The qualities and weaknesses of these approaches seem to be complementary: this observation is the starting point of this work which aims at combining the formal and stochastic approach.

4   The Used Methodology

To understand the problem of understanding Arab oral utterances recognized by the Automatic Recognition System (ASR), a hybrid method combining the syntactic and the stochastic is proposed. The decision taken for a combination of these two approaches is guided by combining the fruits of both approaches to improve further the performance of systems for automatic NLP.

4.1   Architecture of the Hybrid Model

The principle of this work is to design a system based on a stochastic method for determining the meaning of user's queries in a syntactic context. This new approach was evaluated in the tourism domains relatively opened. The problem of extracting meaning is solved in steps. Like any stochastic method, semantic analysis is performed in two basic steps which are training and decoding. Both techniques are applied in most semantic parsers and are quite similar [7]. Before training and decoding steps, two other interesting ones are required. The two steps are preprocessing and syntactic parsing. It is in these modules that the difference between analyzers exists.

4.2   Principle of the Used Method

Fig. 1 below illustrates the architecture of the system proposed of an automatic understanding of the Arabic spontaneous utterances.

Learning or Training

The estimation of parameters is to establish a Hidden Markov Model (HMM) (Fig. 2) if a pretreated and transcribed sequence of words (this words are obviously the output of recognition module) and their annotated corresponding sequences was taken. These sequences were generated during the annotation process data.

Decoding

The decoding step provides the most likely sequence when a test query is taken.

Parsing

Noting that a detailed parsing becomes essential for the proper treatment of utterances, including certain phenomena, such as ellipses. It relies on the use of a rule base: context free grammar augmented with probabilities associated to the rules (see at section 4.3). These grammars are a refinement of formal grammars.

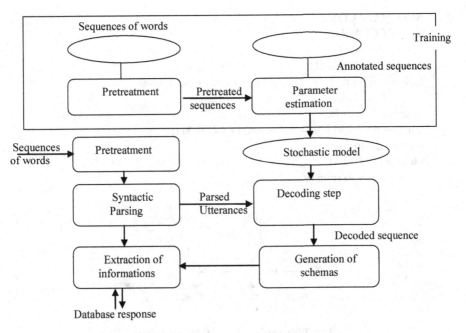

Fig. 1. Overall architecture of the hybrid model

Each rule for producing a probabilistic grammar is associated to a probability. This additional information aims at reducing the syntactic ambiguities that may arise during parsing sentence. The advantage of this statistical information increases with the number of production rules which constitute the whole grammar. The probability of branch (that is to say the application of a sequence of production rules r_i) can be written as follows:

$$P(S \xrightarrow{r_1, r_2, ..., r_n} x) = P(r_1)P(r_2)...P(r_n)$$

Probabilistic grammars are an extension of formal grammars. Their construction is done in two phases. Firstly, a set of production rules had to be retained, as in a formal grammar. From a corpus containing sentences already parsed, the simplest approach to calculate probabilities of occurrence of rewrite rules is to count the number of times of each used rule. The probability of applying a grammar's rule type $A \to \alpha$ may be denoted by $P(A \to \alpha \mid G)$ or $P(r \mid G)$. The following example provides a context-free grammar for the following sentence using successive derivations of production rules.

أريد حجز تذكرة إلى مدينة قربص

I want to reserve a ticket to Korbos city

The grammar generated by this sentence is as follows:

G: S→ GN GV COMP
    S→ GV COMP
    GN→ pronoun | ε
    COMP → GNominal
    GNominal → prep GNominal | noun GNominal | noun
    GV → vloc verb
    pronoun → I (أ) | ε
    noun → مدينة (city) | قربص (korbos) | تذكرة (ticket)
    prep → إلى(to)
    vloc → أريد (want)
    verb → حجز(to reserve)

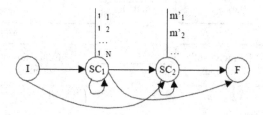

Fig. 2. View of a 1-level HMM modeling

Syntactic and Semantic Annotation

The purpose of this step consists on associating each word in a sentence to a grammatical label (or tag), depending on the context, as ADJ (adjective) NOMP (Proper Name) NOMC (Common Name), DET (determinant).... For example:

أريد الذهاب إلى سوسة

I want to go to Sousse

This will be easier if:

- First of all, an automatic reduction step to canonical form of words can be used (الذهاب (edhaha:ba) to ذهب (dhahaba));
- Second, information is available in the dictionary

A semantic information which is useful for decoding later like (DC: destination_city, TD: departure_time...) is added. This step can be automated through Brill's tagger based rules [13].

Pretreatment of Transcribed Utterances

An oral statement is inherently rigid and difficult to control. This is mainly due to the spontaneous nature of the statement that contains various types of dysfluency (i.e. repetitions, hesitations, self-corrections, etc), which are frequently phenomena of spontaneous speech. Here is an example of hesitation and self correction statement:

هل يوجد مطعم خاص بالكباب هنا آه بالبيتزا عفوا

Is there a restaurant special kabab here, ah pizza sorry?

These phenomena lead to ambiguities that can produce analysis errors. The pretreatment step is required to facilitate the processing of utterances transcribed by the step below. This step removes duplication and unnecessary information, to convert numbers written in all letters, and to determine the canonical forms [2] of words. To achieve this, the statement undergoes standardization [10], a mo pho-lexical parsing and repetition processing [1].

4.3   PCFG and Probabilistic Grammar

A grammar rich enough to accommodate natural language, including rare and sometimes even 'ungrammatical' constructions, fails to distinguish natural from unnatural interpretations. But a grammar sufficiently restricted so as to exclude what is unnatural fails to accommodate the scope of real language. These observations led to a growing interest in probabilistic approaches to natural language.

Obviously natural language is rich and diverse, broad coverage is desirable, and not easily held to a small set of rules. But it is hard to achieve broad coverage without massive ambiguity (a sentence may have tens of thousands of parses), and this of course complicates applications like language interpretation, language translation, speech recognition and speech understanding. This is the dilemma of coverage that we referred at section 3.1, and it sets up a compelling role for probabilistic and statistical methods.

A probabilistic context-free grammar (PCFG; also called stochastic context-free grammar, SCFG) is a Context-Free Grammar. The key idea in the PCFG is to extend a context free grammar (CFG) definition to give a probability distribution over possible derivations. That is, we will find a way to define a distribution over parse derivations. For example:

أنا لا أريد أن أسا فر

1.0 S → PV PN
0.5 PN → أنا
0.3 vloc → لا أريد أن
0.6 verb → أسا فر

نحن نريد أن نحجز

1.0 PV → vloc verb
0.5 PN → نحن
0.7 vloc → نريد أن
0.4 verb → نحجز

The probabilistic context-free grammar is formally defined as follows:

1. A context-free grammar G (N, $\sum$, S, R) having rules of the form, A $\rightarrow$ α , α (NU$\sum$)$^+$

2. A parameter q(α→β), for each rule α→β$\in$ R. The parameter q(α→β) can be interpreted as the conditional probability of choosing rule α→β in a *right-most derivation*, given that the non terminal being expanded is α. For any X$\in$N , we have the constraint

$$\sum_{\alpha \to \beta \in \mathcal{R}:\alpha=X} q(\alpha \to \beta).$$

Having defined PCFGs, we derive a PCFG from a *corpus*. We will assume a set of training data, which is simply a set of parse derivations.

The maximum-likelihood [12] parameter estimates are:

$$q(\alpha \to \beta) = \frac{count(\alpha \to \beta)}{count(\alpha)}.$$

Where count(α→β) (resp. count(α)) is the number of times that the rule α→β (resp. the non terminal α) is seen in corpus training derivations.

The EM algorithm can also estimate PCFGs from a corpus of utterances.

5   Corpus Establishments

This used corpus is dedicated to the study of touristic applications accessing to databases. It is composed of an Arabic spontaneous dialogues stemming from the simulation of tourist information server and hotel reservations. Dialogues aimed at booking one or more rooms in one or more hotels are performed in the context of organizing a weekend, holiday or business trips. Thus, the dialogue may be about different themes: choice of living city, finding a route or a tourist event, a satisfaction of a price or date constraint. The system had to provide information on transportation as well as hotels, restaurants, shops and cinemas around hotels, museums and monuments, the services enjoying the tourist, cities of staying, tourist events and staying days. Indeed, a tourist can learn about the following details:

-   Hotels (price, address, services, classes, path),
-   Restaurants (price, address, benefits, types, path),
-   Monuments (address, opening hours, description, path),
-   Museums (address, hours, prices, description, path),
-   Stores (address, hours, prices, description, path),
-   Cinemas (address, hours, prices, description, path),
-   Services (information, coffe_wifi, pharmacie, gym, ...),

- Stay_city (location, transportation, reservation, route),
- Touristic_event (tour, festival ...),
- Period (weekend, holiday, business trip).

5.1 Collection of Corpus

This corpus was collected by asking ten different people to make written utterances relating to tourist information, using the method of the Wizard of Oz. The following table provides information about the complexity of this task.

Table 1. Statistics from touristic corpus.

Complexity indices	Value
Number of utterance	140
Number of speakers	10
Queries types	14

6 Tests and Results

Some languages such as English, French, and German have platforms for evaluation understanding modules of dialogue systems. These platforms give to the community a large set of corpus of real annotated dialogues. However, this is not the case for the Arabic language where these resources are absents, with the exception of a few corpus distributed by ELDA/ELRA [1]. Thus, a proper evaluation corpus using the same technique of Wizard of Oz used to build test corpus have to be built. The evaluation of corpus involves 100 queries of different types (negation, affirmation, interrogation and acceptance), uttered spontaneously and manually transcribed. These requests correspond to scenarios dealing with information on the tourism fields. These scenarios are inspired from corpus MEDIA [3] and try to cover the input space The evaluation of the understanding module, with this evaluation corpus showed that this system generates 20 errors (average one error by 5 items). Measures of recall, precision and F-measure are respectively 70.00%, 71.00% and 73.79% and the average time to execute an utterance of 12 words is 0.279 seconds. Comparing these results with results obtained by other understanding modules [6], ASTI system has provided fewer errors than many official sites such as UNISYS and MITRE.

Table 2. Comparison of ASTI system results with official sites

	AT&T	CMU	BBN	UNISYS	MYTRE	ASTI
%ERROR	3.8	3.8	9.4	23.6	30.6	20

In fact, as it is shown in Table 2, the error response rate, obtained by the 445 transcript requests, was reached 20% in the case of ASTI system which is less than CMU-PHOENIX system.

7   Conclusion and Perspectives

When the ASTI system was implemented, one of the supervised objectives was to achieve robust parsing of spontaneous spoken Arabic language while making the application domain much wider than is currently done. Syntactic formalisms are not usually viewed as efficient tools for pragmatic applications. That's the two interesting approaches (syntactic and stochastic) are combined. Another objective was to have a rather generic system, despite the use of a domain-based syntactic knowledge. This constraint is fulfilled through the definition of generic rules as well as their probabilities training the HMM model which makes it possible to estimate efficiently its parameters. The performances of ASTI show that a combination of the two divergent approaches can bear comparison with international system.

References

1. Bahou, Y., Belguith, H.L., Ben Hamadou, A.: Towards a Human-Machine Spoken Dialogue in Arabic. In: 6th Language Resources and Evaluation Conference (LREC 2008),Workshop HLT Within the Arabic World. Arabic Language and Local Languages Processing Status Updates and Prospects, Marrakech, Morocco (2008b)
2. Hadrich Belguith, L., Bahou, Y., Ben Hamadou, A.: Une méthode guidée par la sémantique pour la compréhension automatique des énoncés oraux arabes. International Journal of Information Sciences for Decision Making (ISDM) 35 (Septembre 2009)
3. Bonneau Maynard, H., Rosset, S., Ayache, C., Kuhn, A., Mostefa, D.: Semantic Annotation of the French Media Dialog Corpus. In: 9th European Conference on Speech Communication and Technology (2005)
4. Chelba, Jelinek, F.: Structured language modeling. Computer, Speech and Language 14(4), 283–332 (2000)
5. El-Bèze, Sprit, T.: Stratégie mixte d'étiquetage syntaxique : Statistiques et connaissances Revue TAL 36(1-2), 47–66 (1995)
6. Minker, W.: Compréhension Automatique de la Parole Spontanée. L'Harmattan (1999)
7. Minker, W., Bennacef, S.: Speech And Human-Machine Dialog. Kluwer Academic Publishers Group, Pays-Bas (2004)
8. Riccardi, Gorin, A.L.: Stochastic language adaptation over time and state in natural spoken dialogue systems. IEEE Transactions on Speech and Audio Processing 8(1), 3–10 (2000)
9. Sallomaa: Probabilistic and weighted grammars. Information and Control, 15, 529–544 (1969)
10. Zouaghi, A., Zrigui, M., Antoniadis, G.: Compréhension Automatique de la Parole Arabe Spontanée. Une Modélisation Numérique, Traitement Automatique des Langues (TAL 2008) 49(1), 141–166 (2008)
11. Zouaghi, A., Zrigui, M., Ben Ahmed, M.: Un Étiqueteur Sémantique des Énoncés en Langue Arabe. In: Actes de la 12ème Conférence sur le Traitement Automatique des Langues Naturelles (TALNRECITAL 2005), Dourdan, France (2005)
12. Collins, M.: Probabilistic Context-Free Grammars (PCFGs),
    http://www.cs.columbia.edu/
13. Brill, E.: Transformation-Based Error-Driven Learning and Natural Language Processing: A Case Study in Part of speech Tagging. Computational Linguistics 21, 4 (1995b)

Evidence in Automatic Error Correction Improves Learners' English Skill

Jiro Umezawa[1], Junta Mizuno[1], Naoaki Okazaki[1,2], and Kentaro Inui[1]

[1] Tohoku University, 6-3-09 Aramaki Aza Aoba, Aobaku, Sendai 980-8579, Japan
[2] Japan Science and Technology Agency (JST)
{umezawa,junta-m,okazaki,inui}@ecei.tohoku.ac.jp

Abstract. Mastering proper article usage, especially in the English language, has been known to pose an extreme challenge to non-native speakers whose L1 languages have no concept of articles. Although the development of correction methods for article usage has posed a challenge for researchers, current methods do not perfectly correct the articles. In addition, proper article usage is not taught by these methods. Therefore, they are not useful for those wishing to learn a language with article usage. In this paper, we discuss the necessity of presenting evidence for corrections of English article usage. We demonstrate the effectiveness of this approach to improve the writing skills of English learners.

Keywords: English Article, Automatic Correction, Grammatical Error.

1 Introduction

In recent years, there is a growing need for assisting with the improvement of English composition for non-native speakers, who have an increasing number of opportunities for writing English [1]. Particularly, non-native speakers often make mistakes in article and preposition usage. For instance, the two major errors in the Cambridge Learner Corpus [2], which comprises thousands of exam scripts written by students around the world, are related to prepositions (13.4%) and determiners (11.7%), excluding spelling errors. As a result, researchers have proposed various approaches for automatic correction of these mistakes (See Section 2 for details).

The methods proposed by researchers, however, have not achieved perfect correction. Dahlmeier *et al.* [3], which gained the highest performance in Helping Our Own (HOO) 2012 shared task [4], reported 62.93% precision and 31.88% recall for determiner error correction in the real learners' dataset. In addition, these methods do not teach the usage of articles and prepositions. Therefore, learners might have to choose suggested corrections without knowing the reason why they made a mistake. It is important for automatic correction methods to present the evidence for correction, i.e., *why suggested corrections are better than the original ones*. Based on correction evidence, learners can make final decisions for choosing articles and prepositions in their compositions and improve their understanding of proper usage.

A. Gelbukh (Ed.): CICLing 2013, Part II, LNCS 7817, pp. 559–571, 2013.

As described in this paper, we propose an approach for presenting evidence for correcting articles. Then we investigate its effect on improving learners' skills. Following the approaches of previous studies, we formalize the task of correction as a classification problem: The classifier chooses an article (either an indefinite, definite, or zero article) for a given noun phrase. Additionally, we propose several types of evidence that might be useful for learners to make a final decision for choosing an article. The correctness of users is measured for choosing appropriate articles with different types of evidence presented. The experimentally obtained results confirm that users obtained higher scores in choosing articles when they were presented with correction evidence.

2   Related Work

A number of researchers have presented various approaches to automatic error correction in learners' English. The great attention attracted by this research area is demonstrated by Helping Our Own (HOO) [5], the shared task for grammatical error detection and correction. Several end-user services of error correction have been built, e.g., ESL Assistant [6] and Criterion Online Writing Evaluation Service [7].

Previous studies specifically examined article and preposition errors because articles and prepositions have a closed set of vocabulary and because they account for the major errors in learners' English. Most methods incorporate a supervised learning approach such as Memory-Based Learning [8], decision tree [9, 10], decision list [11], Maximum Entropy Modeling [6, 12–17], Conditional Random Fields [18], Alternating Structure Optimization (ASO) [19], and Statistical Machine Translation [20, 21]. Some methods are characterized by language models [9, 22], corpora built from text in L2-language users [12, 17], and domain adaptation between training datasets from both L1 and L2 language users [19, 23]. However, these studies do not present evidence for a correction but only a correction suggestion.

A few previous works made attempts to provide evidence for corrections. Liu et al. [24] proposed a machine-aided English writing system that incorporates an information retrieval engine for providing suggestive example sentences. Gamon et al. [9] also described a system that shows real-world examples as additional information to suggestions. However, the experiments in these studies were not designed to verify the effect of presenting evidence to improve learners' skill.

3   Proposed Method

3.1   Task Definition

As described in this paper, we consider presenting evidence for an error correction, such as knowledge about article usage and example sentences. One can consider the following sentence.

A remainder of this section describes ...

In this example, an automatic correction method is expected to detect the error in the article usage in the underlined noun phrase and to suggest the definite article *the* instead of the indefinite article *a*.

Some evidence for choosing the definite article is that the noun *remainder* usually describes a relative position to a known entity or concept. In addition, the phrase *of this section* explicitly states the target to which the noun *remainder* refers. Expressing this evidence as 'features', an automatic correction method might predict the definite article easily and confidently. Furthermore, the features contributing to the prediction might provide a good hint for explaining the usage.

We can also infer from the sentences below that articles of the word *remainder* tend to be definite.

> – *The remainder of this paper is organized ...*
> – *Formulas for the remainder term of Taylor ...*

These example sentences might also improve the confidence and understanding of the learners.

As described in this paper, we build a support system for English writing. As the initial attempt for demonstrating evidence for correction, we specifically examine correcting article usage, which is relatively easier for computers to correct than other kinds of mistakes.

In addition to the error correction of article usage, the system presents evidence for correcting article usage. More specifically, we propose evidence of three kinds: *confidence scores of predictions*, *features contributing to predictions*, and *example sentences that are relevant to the target text*. The subsequent subsection presents a detailed description of the design of the automatic article corrector with subsequent presentation of evidence of three types.

3.2   Correction Model of Article Usage

In this study, we use Maximum Entropy Modeling to build a correction model. Let x be the feature vector extracted from the context of the target noun phrase x. Let $y \in \{a, the, \phi\}$ be the article corresponding to the noun phrase (ϕ indicates zero article). The Maximum Entropy Modeling defines the conditional probability of an article y for a given context of a noun phrase x,

$$P(y|\boldsymbol{x}) = \frac{\exp\left(\boldsymbol{w_y} \cdot \boldsymbol{x}\right)}{\displaystyle\sum_{y' \in \{a, the, \phi\}} \exp\left(\boldsymbol{w_{y'}} \cdot \boldsymbol{x}\right)} \tag{1}$$

where $\boldsymbol{w_y}$ is the weight vector for the feature vector $\boldsymbol{x}$ for predicting the article y. The most probable article $\hat{y}$ for a given article x is given as

$$\hat{y} = \underset{y}{\operatorname{argmax}} P(y|\boldsymbol{x}). \tag{2}$$

We train the weight vectors $\boldsymbol{w_y}$ using Classias[1], a machine learning toolkit.

---

[1] http://www.chokkan.org/software/classias/

As described in greater detail in Section 3.4, we want to keep the features human-readable without losing prediction accuracy because we will present contributing features to users as evidence for correction. For this reason, we carefully design features that are sufficiently simple for humans and sufficiently discriminative for the correction model.

Table 1 shows a list of features used in our study. These features are inspired by Dahlmeier *et al.* [3] which achieved the top performance at Helping Our Own 2012 shared task [4]. In addition to commonly used features such as the word feature and POS feature, we use countability features extracted from the SPECIALIST Lexicon[2]. To reduce the running time and overfitting, we remove features that do not appear more than once in the training data.

It is preferred to learn the error correction model directly from a learner corpus including both article errors and their corrections, e.g., Cambridge Learner Corpus (CLC)[3], NICE[4] and The NICT JLE Corpus[5]. However, these corpora are too small for the machine learning approach. Therefore, instead of using a small learner corpus, we build a correction model only from large and correct English corpus.

We split the text in the corpus into multiple sentences using the Natural Language Toolkit (NLTK) [25]. We use the GENIA tagger [26] to perform part-of-speech (POS) tagging and chunking. For each noun phrase in the text, we create a training instance by 'hiding' the article in the noun phrase[6]. Each training instance is assigned a reference label $y \in \{a, the, \phi\}$ based on the article in the original noun phrase.

3.3 Conditional Probability as 'Confidence' of Predictions

In addition to the predicted label $\hat{y}$, we present the conditional probability $P(y|\boldsymbol{x})$ as correction evidence. This evidence provides users with the confidence of the correction model. A user might trust an article prediction more if the model confidence is high. Alternatively, users might respect their own decisions if the model confidence is low.

---

[2] The SPECIALIST Lexicon contains lexical information of both biomedical vocabulary and common English words. The lexicon entry for each lexical item records syntactic, morphological, and orthographic information. Each entry for noun includes information about countability. http://lexsrv3.nlm.nih.gov/LexSys Group/Projects/lexicon/2011/web/index.html

[3] http://www.cambridge.org/gb/elt/catalogue/subject/custom/item3646603/ Cambridge-International-Corpus-Cambridge-Learner-Corpus/

[4] http://sugiura5.gsid.nagoya-u.ac.jp/~sakaue/nice/

[5] http://alaginrc.nict.go.jp/nict_jle/

[6] We exclude a noun phrase containing either of PRP, PRP$, WDT, WP or WP$ POS tags from the training instances because it always has a zero article ϕ. We also remove instances with determiners (e.g., *this* and *that*) other than the article.

Table 1. List of features used for this study. Word features are lowercased lemmas. The 'Example' field shows feature values for the underlined noun phrase in the sentence, *Each test sentence is parsed by a bottom-up chart parser using initially the indexed subtrees only.*

Feature	Example
First word in NP	bottom-up
Second word in NP	chart
Third word in NP	parser
Word 1 before NP	by
Word 2 before NP	parse
Word+POS 1 before NP	by+IN
Word+POS 2 before NP	parse+VBN
Word+POS 3 before NP	be+VBZ
Word after NP	use
Word+POS 1 after NP	use+VBG
Word+POS 2 after NP	initially+RB
Bag of words in NP	{bottom-up, chart, parser}
N-grams around article position(N=2,3,4,5)	{by_X, X_bottom-up, parse_by_X, ...}
Word before + NP	by+bottom-up_chart_parser
NP + N-gram after NP(N=1,2,3)	{bottom-up_chart_parser+use, ...}
Noun compound	chart_parser
Adjective + Noun compound	bottom-up+chart_parser
Adjective POS + Noun compound	JJ+chart_parser
NP POS + noun compound	JJ_NN_NN+chart_parser
First POS in NP	JJ
Second POS in NP	NN
Third POS in NP	NN
POS 1 before NP	IN
POS 2 before NP	VBN
POS 3 before NP	VBZ
POS after NP	VBG
Bag of POS in NP	{JJ, NN, NN}
POS N-grams around article position(N=2,3,4)	{IN_X, X_JJ, VBN_IN_X, ...}
Head of NP	parser
Countability of Head	countable
NP POS + Head	JJ_NN_NN+parser
Word before + Head	by+parser
Head + N-gram after NP(N=1,2,3)	{parser+use, parser+use_initially, ...}
Adjective + Head	bottom-up+parser
Adjective POS + Head	JJ+parser
Word before + Adjective + Head	by+bottom-up+parser
Word before + Adjective POS + Head	by+JJ+parser
Word before + NP POS + Head	by+JJ_NN_NN+parser

3.4 Presenting the Reason Why the Correction Model Chooses $\hat{y}$

We would like to be informed of the reason why the correction model chooses $\hat{y}$ for a given noun phrase x. To do this, we show the features that contribute to the classification the most. The error correction model of the articles includes the variables $w_{a,i}$, $w_{the,i}$ and $w_{\phi,i}$ as the weight of the feature x_i of each article with the noun phrase, respectively. We define the strength $d(i, y)$, where the feature x_i recommends the article y, as

$$d(i, y) = \frac{\exp(w_{y,i})}{\exp(w_{a,i}) + \exp(w_{the,i}) + \exp(w_{\phi,i})}. \tag{3}$$

For each article y, we choose three features which acquired a high value $d(i, y)$ as evidence for the article correction. When presenting the evidence to English learners, the feature representation is converted to an easy-to-understand text explanation by a template translation rule for each feature type. In the model, a feature is composed of a feature type (e.g., HEAD-OF-NP) and argument(s) (e.g., *method*). We apply the template rule for a feature type to obtain a human-readable explanation of a feature, filling the template slot(s) with the argument(s). For example, a feature type HEAD-OF-NP has a translation rule "HEAD-OF-NP=*arg*" → "The head of the NP is *arg*", where *arg* is an argument. Then the feature "HEAD-OF-NP=*method*", which is an instantiation of the feature type HEAD-OF-NP is translated to "The head of the NP is *method*", filling the argument with *method*.

3.5   Presenting Example Sentences

We also present example sentences that have similar context to the target noun phrase. We retrieve example sentences in the following procedure:

This procedure is performed for each article with the target noun phrase.

1. The system chooses three key features that acquired high value $d(i, y)$ in the features generated for the target noun phrase in the same way as Section 3.4. The system extracts noun phrases in the training data as candidates having the same head and containing the key features.
2. The system calculates the cosine similarity between the feature vector of the target noun phrase and that of each candidate. The system chooses three sentences that yielded high similarity values.
3. If the system cannot find three example sentences, then the system repeats the same process by reducing the number of the key feature by one until the system obtains three example sentences.

4   Evaluation

To evaluate our method, we conducted two experiments. The first experiment verifies the classification accuracy of the article correction model. The second experiment evaluates the performance of English learners in choosing articles with suggestions and evidence presented by the proposed method.

4.1   Dataset

The data used in our experiment consisted of technical passages on Natural Language Processing (NLP). Although ACL Anthology Reference Corpus [27] exists as an NLP literature corpus, it contains a number of errors because of OCR. Therefore, in both experiments, we use a dataset that was prepared by crawling papers published in 2000–2012 on the ACL Anthology website[7]. We extracted plain text from the downloaded PDF files. Then we removed unwanted regions such as author names and figure captions. In both experiments, we use the papers of the ACL 2012 to build a test dataset and the other for training. The training data includes 3,442,940 sentences and 9,468,343 training instances for article correction. The test data consist of 73,430 sentences and 168,900 instances of noun phrases.

4.2   Evaluation of Classifier Performance

We measured the performance of the article correction model. The accuracy of the baseline, which always classifies any noun phrase as zero article ϕ, was 65.02%. The article correction model gained accuracy of 86.35%, which was significantly higher than the baseline.

4.3   Evaluation of Correction Performance by Users

To demonstrate the effectiveness of the proposed approach, we measure the performance of real users in choosing articles with suggestions and evidence presented by the proposed method. In this experiment, we asked a user to select the most suitable article for a given noun phrase in a sentence. We also presented evidence of different kinds to the users, and recorded the performance of the user. We chose four Japanese and one Chinese student for this experiment, all of whom learn English as a second language.

The following procedures were used in this experiment:

1. A sentence containing a noun phrase without an article is presented to the user. The user then selects an article that they believe is correct for the noun phrase.
2. The confidence score of the correction model for the noun phrase is presented by the system as additional information. At this point, the user can change the article selected in (1) if the user likes the suggestion.
3. Additionally, the evidence of the reason and example sentences is presented to the user, who then has another chance to change the article selected in (2).
4. Repeat 1–3 until the user completes answers to all test examples. for a selected number of examples.

---

[7] http://aclweb.org/anthology-new/

In addition to the article selected in 1–3, the user must also provide their self-confidence for their selection. During the experiment, users were not able to view the correct answer.

Compared with the case of viewing only the sentence from (1) above, the user's accuracy and self-confidence in predicting the article was investigated when the system's confidence score and knowledge of article usage was presented to the user. In the experiment, we prepared the phase of looking over only the confidence score as additional information because this evidence can be regarded as contributing most to the accuracy of the users' prediction. Our approach is proven to help English learners' writing if the degree of improvement of accuracy is significantly higher than the case of viewing the target sentence only, or if the self-confidence improved properly.

We used 248 instances of the test data described in Section 4.1 for the evaluation experiment. These instances were extracted randomly from the test data. Unnecessary instances were removed manually when target noun phrases fell under one of the following conditions:

- surrounded by symbols
- included either a number, a proper noun, a numerical formula, or the word "there"
- extracted incorrectly
- target noun phrase or surrounding context included errors such as spelling errors
- included in a short sentence

The purpose of choosing these rules was to eliminate instances that might interfere with both the user's and system's prediction.

The accuracy of the classifier was 85.48% for the data used in the evaluation experiment. Fig. 1 portrays the moving average accuracy for the averaged confidence score of every 30 instances. It can be confirmed that the confidence score does not disturb a user's prediction because the classifier's accuracy is proportional to the confidence score.

Fig. 2 presents a screenshot of the system. The first column on the far left includes the classifier's confidence percentage. The immediate right field displays its corresponding article. The third column shows the target sentence displaying the noun phrase for selecting a correct article. Finally, the remaining rows below the sample sentence inform the other evidence (reason and example sentences). For each problem, the target sentence and articles only are displayed. After a user selects an article, they can press the "Confidence" button to see the classifier's confidence. Then, the user can press the "All" button to see the other evidence in the third column.

Table 2 shows the accuracy of each user's prediction. The accuracy of every user was roughly comparable to the classifier's accuracy after looking at the confidence score. Some users achieved accuracy exceeding that of the classifier. This result implies that it is valuable for learners' writing to present the confidence score of the classifier rather than merely presenting the classifier's prediction

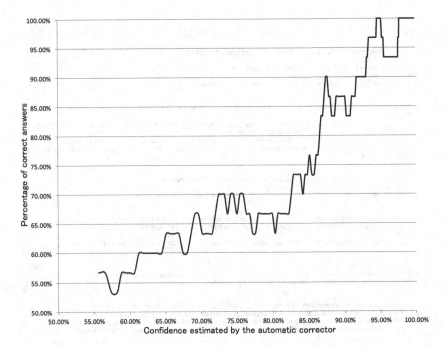

Fig. 1. Relation between confidence score and accuracy

Table 2. Accuracy of users

	Sentence only	+Confidence score	+All evidence
user 1	59.27%	85.89%	85.48%
user 2	64.92%	85.89%	81.05%
user 3	37.10%	79.03%	77.42%
user 4	62.10%	79.84%	83.87%
user 5	64.52%	85.08%	84.27%

because it is difficult to imagine that learners blindly follow the classifier's prediction which is incorrect sometimes.

4.4   Discussion

The previous section described that the confidence score of the classifier can raise the accuracy of a user's decision for a correct article. However, the accuracy of a user's prediction did not change much when presenting the evidence about the articles usage and example sentences after displaying the classifier's confidence level (1 individual's accuracy slightly increased, whereas the remaining individuals' accuracy decreased a little). It is most likely that the user was led to predict the correct answer when the classifier's confidence level was presented. Therefore, it can be assumed that the accuracy of a user's prediction did not change much because all of the proposed evidence is based on the feature

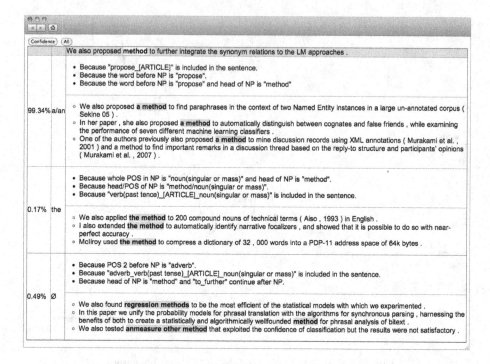

Fig. 2. Example of the system output

weights of the model. For detailed evaluation of our proposed evidence, for each case transitioning from the phase of presenting confidence score to the phase of presenting all reason information, we classified the test instances of each user's result of the experiment in accordance with the following definition:

- Positive Evidence Instance
  - From the incorrect to the correct choice.
  - From the correct to the correct choice accompanied by gain of self-confidence.
  - From the incorrect to the incorrect choice accompanied by loss of self-confidence.
- Negative Evidence Instance
  - From the correct to the incorrect choice.
  - From the incorrect to the incorrect choice accompanied by gain of self-confidence.
  - From the correct to the correct choice accompanied by loss of self-confidence.
- Neutral Evidence Instance
  - From the correct to the correct choice accompanied by no change of self-confidence.
  - From the incorrect to the incorrect choice accompanied by no change of self-confidence.

Table 3. Validity of evidence

	Positive	Negative	Neutral
user 1	79.44%	13.71%	6.85%
user 2	48.79%	21.77%	29.44%
user 3	18.15%	9.27%	72.58%
user 4	83.47%	16.13%	0.40%
user 5	81.05%	14.11%	4.84%

Table 3 presents results of these counts for each user. Although the tendency of the self-confidence rate of change is uneven by each user, because self-confidence is based on the user's subjective evaluation, the number of positive evidence instances was significantly higher than the number of negative evidence instances. In this way, we were able to confirm that presenting all evidence contributes to selection of the correct article with understanding than presenting only the confidence score as evidence. In addition, the reason and example sentences related to article usage is expected to give a high learning effect to learners such as the prevention of the similar error. We infer that the evidence is more helpful to improve both writing and learning for English learners.

5 Conclusion

As described in this paper, we presented a method to display evidence of correction (confidence of each article, reason about the article usages, example sentences relevant to the target context) as a support for learners' decisions. This approach improves the effectiveness of writing skills for English learners not only by presenting the correction article, but by also presenting the evidence. Although our main emphasis was article usage, the approach in this paper is applicable to grammatical errors of other kinds such as preposition usage error. The future directions include verifying the effectiveness of the proposed method for prepositions. Furthermore, we are planning to improve the feature design to provide more useful evidence and to conduct long-term evaluations for presenting evidence for error correction.

Acknowledgments. This research was partly supported by JST, PRESTO. This research was partly supported by JSPS KAKENHI Grant Numbers 23240018 and 23700159.

References

1. Leacock, C., Chodorow, M., Gamon, M., Tetreault, J.R.: Automated Grammatical Error Detection for Language Learners. Synthesis Lectures on Human Language Technologies. Morgan & Claypool Publishers (2010)
2. Nicholls, D.: The cambridge learner corpus - error coding and analysis for lexicography and elt. In: Corpus Linguistics 2003, pp. 572–581 (2003)

3. Dahlmeier, D., Ng, H.T., Ng, E.J.F.: NUS at the HOO 2012 Shared Task. In: Proceedings of the Seventh Workshop on Building Educational Applications Using NLP, pp. 216–224 (2012)

4. Dale, R., Anisimoff, I., Narroway, G.: HOO 2012: A report on the preposition and determiner error correction shared task. In: Proceedings of the Seventh Workshop on Building Educational Applications Using NLP, pp. 54–62 (June 2012)

5. Dale, R., Kilgarriff, A.: Helping our own: Text massaging for computational linguistics as a new shared task. In: Proceedings of the 6th International Natural Language Generation Conference, pp. 261–265 (2010)

6. Chodorow, M., Gamon, M., Tetreault, J.: The utility of article and preposition error correction systems for English language learners: Feedback and assessment. Language Testing 27(3), 419–436 (2010)

7. Han, N.R., Chodorow, M., Leacock, C.: Detecting errors in English article usage by non-native speakers. Natural Language Engineering 12, 115–129 (2006)

8. Minnen, G., Bond, F., Copestake, A.: Memory-based learning for article generation. In: Proceedings of the 2nd Workshop on Learning Language in Logic and the 4th Conference on Computational Natural Language Learning, CoNLL 2000, pp. 43–48 (2000)

9. Gamon, M., Gao, J., Brockett, C., Klementiev, A., Dolan, W.B., Belenko, D., Vanderwende, L.: Using contextual speller techniques and language modeling for ESL error correction. In: Proceedings of the Third International Joint Conference on Natural Language Processing (IJCNLP 2008), pp. 449–456 (2008)

10. Gamon, M.: Using mostly native data to correct errors in learners' writing: a meta-classifier approach. In: Human Language Technologies: The 2010 Annual Conference of the North American Chapter of the Association for Computational Linguistics (HLT 2010), pp. 163–171 (2010)

11. Nagata, R., Nakatani, K.: Evaluating performance of grammatical error detection to maximize learning effect. In: Proceedings of the 23rd International Conference on Computational Linguistics (COLING 2010): Posters, pp. 894–900 (2010)

12. Izumi, E., Uchimoto, K., Saiga, T., Supnithi, T., Isahara, H.: Automatic error detection in the japanese learners' English spoken data. In: Proceedings of the 41st Annual Meeting on Association for Computational Linguistics, vol. 2, pp. 145–148 (2003)

13. Lee, J.: Automatic article restoration. In: HLT-NAACL 2004: Student Research Workshop, pp. 31–36 (2004)

14. Chodorow, M., Tetreault, J., Han, N.R.: Detection of grammatical errors involving prepositions. In: Proceedings of the Fourth ACL-SIGSEM Workshop on Prepositions, Prague, Czech Republic. Association for Computational Linguistics, pp. 25–30 (June 2007)

15. De Felice, R., Pulman, S.G.: A classifier-based approach to preposition and determiner error correction in L2 English. In: Proceedings of the 22nd International Conference on Computational Linguistics (Coling 2008), pp. 169–176 (2008)

16. Tetreault, J.R., Chodorow, M.: The ups and downs of preposition error detection in ESL writing. In: Proceedings of the 22nd International Conference on Computational Linguistics (Coling 2008), pp. 865–872 (2008)

17. Han, N.R., Tetreault, J.R., Lee, S.H., Ha, J.Y.: Using an error-annotated learner corpus to develop an ESL/EFL error correction system. In: Proceedings of the Seventh International Conference on Language Resources and Evaluation (LREC 2010), pp. 763–770 (2010)

18. Gamon, M.: High-order sequence modeling for language learner error detection. In: Proceedings of the Sixth Workshop on Innovative Use of NLP for Building Educational Applications, pp. 180–189 (2011)

19. Dahlmeier, D., Ng, H.T.: Grammatical error correction with alternating structure optimization. In: Proceedings of the 49th Annual Meeting of the Association for Computational Linguistics: Human Language Technologies, pp. 915–923 (2011)

20. Brockett, C., Dolan, W.B., Gamon, M.: Correcting ESL errors using phrasal SMT techniques. In: Proceedings of the 21st International Conference on Computational Linguistics and 44th Annual Meeting of the Association for Computational Linguistics, pp. 249–256 (2006)

21. Hermet, M., Désilets, A.: Using first and second language models to correct preposition errors in second language authoring. In: Proceedings of the Fourth Workshop on Innovative Use of NLP for Building Educational Applications, pp. 64–72 (2009)

22. Yi, X., Gao, J., Dolan, W.B.: A web-based English proofing system for English as a second language users. In: Proceedings of the Third International Joint Conference on Natural Language Processing (IJCNLP 2008), pp. 619–624 (2008)

23. Rozovskaya, A., Roth, D.: Algorithm selection and model adaptation for ESL correction tasks. In: Proceedings of the 49th Annual Meeting of the Association for Computational Linguistics: Human Language Technologies, pp. 924–933 (2011)

24. Liu, T., Zhou, M., Gao, J., Xun, E., Huang, C.: PENS: a machine-aided English writing system for Chinese users. In: Proceedings of the 38th Annual Meeting on Association for Computational Linguistics, pp. 529–536 (2000)

25. Loper, E., Bird, S.: NLTK: the Natural Language Toolkit. In: Proceedings of the ACL 2002 Workshop on Effective Tools and Methodologies for Teaching Natural Language Processing and Computational Linguistics, vol. 1, pp. 63–70 (2002)

26. Tsuruoka, Y., Tateishi, Y., Kim, J.-D., Ohta, T., McNaught, J., Ananiadou, S., Tsujii, J.: Developing a robust Part-of-Speech tagger for biomedical text (chapter 36). In: Bozanis, P., Houstis, E.N. (eds.) PCI 2005. LNCS, vol. 3746, pp. 382–392. Springer, Heidelberg (2005)

27. Bird, S., Dale, R., Dorr, B., Gibson, B., Joseph, M., Kan, M.Y., Lee, D., Powley, B., Radev, D., Tan, Y.F.: The ACL Anthology Reference Corpus: A reference dataset for bibliographic research in computational linguistics. In: Proceedings of the Sixth International Conference on Language Resources and Evaluation, LREC 2008 (2008)

Author Index